Bionanotechnology

Biological Self-assembly and its Applications

Edited by

Bernd H.A. Rehm

Institute of Molecular Biosciences
Massey University
Palmerston North
New Zealand

Caister Academic Press

Caister Academic Press
Norfolk, UK

www.caister.com

British Library Cataloguing-in-Publication Data
A catalogue record for this book is available from the British Library

ISBN: 978-1-908230-16-4

Cover design adapted from Figure 11.2.

Printed and bound in Malta by Gutenberg Press Ltd.

Contents

Contributors

John O. Baker
National Renewable Energy Laboratory
Biosciences Center
Golden, CO
USA

john.baker@nrel.gov

François Baneyx
Department of Chemical Engineering
University of Washington
Seattle, WA
USA

baneyx@uw.edu

Edward A. Bayer
Department of Biological Chemistry
The Weizmann Institute of Science
Rehovot
Israel

ed.bayer@weizmann.ac.il

Mathieu Bennet
Department of Biomaterials
Max Planck Institute of Colloids and Interfaces
Potsdam
Germany

mathieu.bennet@mpikg.mpg.de

Satish Chander Bhatla
Department of Botany
University of Delhi
Delhi
India

bhatlasc@gmail.com

Robert R. Birge
Department of Molecular and Cell Biology;
Department of Chemistry
University of Connecticut
Storrs-Mansfield, CT
USA

rbirge@uconn.edu

Yannick J. Bomble
National Renewable Energy Laboratory
Biosciences Center
Golden, CO
USA

yannick.bomble@nrel.gov

Geoffrey D. Bothun
Department of Chemical Engineering
University of Rhode Island
Kingston, RI
USA

bothun@egr.uri.edu

James F. Conway
Department of Structural Biology
University of Pittsburgh
Pittsburgh, PA
USA

jxc100@pitt.edu

Brandon L. Coyle
Department of Chemical Engineering
University of Washington
Seattle, WA
USA

blcoyle@uw.edu

Anisha David
Department of Botany
University of Delhi
Delhi
India

anisha.icy@gmail.com

Shi-You Ding
National Renewable Energy Laboratory
Biosciences Center
Golden, CO
USA

shi.you.ding@nrel.gov

Jenny Draper
Institute of Molecular Biosciences
Massey University
Palmerston North
New Zealand

learithe@gmail.com

Jinping Du
Institute of Molecular Biosciences
Massey University
Palmerston North
New Zealand

j.du@massey.ac.nz

Damien Faivre
Department of Biomaterials
Max Planck Institute of Colloids and Interfaces
Potsdam
Germany

damien.faivre@mpikg.mpg.de

Daniel B. Fried
Department of Biological Chemistry
The Weizmann Institute of Science
Rehovot
Israel

daniel.fried@weizmann.ac.il

Juliet A. Gerrard
Industrial Research Limited
Lower Hutt;
Biomolecular Interaction Centre;
MacDiarmid Institute
University of Canterbury
Christchurch
New Zealand

j.gerrard@irl.cri.nz
juliet.gerrard@canterbury.ac.nz

Jordan A. Greco
Department of Chemistry
University of Connecticut
Storrs-Mansfield, CT
USA

jordangreco9@gmail.com

Anju Gupta
Department of Biology and Chemistry
Texas A&M International University
Laredo, TX
USA

anju.gupta@tamiu.edu

Michael E. Himmel
National Renewable Energy Laboratory
Biosciences Center
Golden, CO
USA

mike.himmel@nrel.gov

David O. Hooks
Institute of Molecular Biosciences
Massey University
Palmerston North
New Zealand

d.o.hooks@massey.ac.nz

Jason W. Lee
Institute of Molecular Biosciences
Massey University
Palmerston North
New Zealand

j.lee2@massey.ac.nz

Dong Li
Department of Chemistry and Biochemistry
Stephenson Life Sciences Research Center
University of Oklahoma
Norman, OK
USA

larry@ou.edu

Chuanbin Mao
Department of Chemistry and Biochemistry
Stephenson Life Sciences Research Center
University of Oklahoma
Norman, OK
USA

cbmao@ou.edu

Sarah Moraïs
Department of Biological Chemistry
The Weizmann Institute of Science
Rehovot
Israel

sarahv@weizmann.ac.il

Rob Noad
Centre for Emerging, Endemic and Exotic Diseases
Department of Pathology and Infectious Diseases
Royal Veterinary College
Hatfield
UK

rnoad@rvc.ac.uk

Natalie Parlane
AgResearch
Hopkirk Research Institute;
Institute of Molecular Biosciences
Massey University
Palmerston North
New Zealand

natalie.parlane@agresearch.co.nz

Teresa Perez-Gonzalez
Department of Biomaterials
Max Planck Institute of Colloids and Interfaces
Potsdam
Germany

teresa.perezgonzalez@mpikg.mpg.de

Matthew R. Preiss
Department of Chemical Engineering
University of Rhode Island
Kingston, RI
USA

matthewpreiss@my.uri.edu

Jasna Rakonjac
Institute of Molecular Biosciences
Massey University
Palmerston North
New Zealand

j.rakonjac@massey.ac.nz

Jared K. Raynes
Industrial Research Limited
Lower Hutt;
Biomolecular Interaction Centre
University of Canterbury
Christchurch
New Zealand

j.raynes@irl.cri.nz

Bernd H.A. Rehm
Institute of Molecular Biosciences
Massey University
Palmerston North
New Zealand

b.rehm@massey.ac.nz

Polly Roy
Department of Pathogen Molecular Biology
London School of Hygiene and Tropical Medicine
London
UK

polly.roy@lshtm.ac.uk

Nicole L. Wagner
Department of Molecular and Cell Biology
University of Connecticut
Storrs-Mansfield, CT
USA

nicole.wagner508@gmail.com

Dean Wood
School of Chemistry
The University of Edinburgh
Edinburgh
UK

dean.s.wood@gmail.com

Qi Xu
National Renewable Energy Laboratory
Biosciences Center
Golden, CO
USA

qi.xu@nrel.gov

Sunita Yadav
Department of Botany
University of Delhi
Delhi
India

response2sunita@gmail.com

Weibin Zhou
Department of Chemical Engineering
University of Washington
Seattle, WA
USA

zhouw@uw.edu

Preface

In living organisms a vast diversity of self-assembling building blocks and processes for the assembly of nano-scaled structures can be discovered. Harnessing this enormous natural design space towards the manufacture of highly functional nanomaterials is the subject of *Bionanotechnology*. Bionanotechnology is a particularly interdisciplinary field, which combines biological principles with physical and chemical procedures to generate nano-sized building blocks with specific functions and new properties. It involves the development of biologically-based procedures, the use of biological components and systems, the design of biocompatible objects and systems and the use of nanotechnology to support biotechnological processes.

In contrast to nanotechnology, which uses the 'top-down' approach, i.e. making devices smaller, the 'bottom-up' strategy is utilized in bionanotechnology, which harnesses the inherent properties of biological molecules and living organisms to form functional nanostructures from small building blocks. Biological nanostructures (rods, particles, beads, etc.) build from biological moleculaes such as e.g. biopolymers, proteins, DNA or lipids can be either directly functionalized by designing the self-assembly process and/or its building blocks or the preformed natural nanostructures are subjected to chemical modification or are used as templates for the formation of inorganic nanostructures called biomimetics. Microorganisms are often used to manufacture functionalized nanostructures often implying metabolic engineering, genetic engineering and protein design. *In vitro* the design space of self-assembled biomolecule complexes is strongly enhanced applying additional chemical modifications, which leads to almost unlimited functionalities considering medical and technological applications.

This book intends to provide a survey on the most striking and successful approaches to produce biogenic nanodevices considering not only living organisms as manufacturer but also in vitro processes applying the self-assembly of isolated biomolecules. Two chapters describe the microbial production of tailor-made self-assembled nanostructures such as bacterial biopolyester and magnetosomes, which can be processed into functional nanoparticles. Other chapters summarize recent developments in the use of protein-based assemblies for nanodevice/nanomaterials production. The book aims to provide an impression of the vast field of bionanotechnology by describing various biological nanostructures, the implied design space and the enormous potential for applications in medicine and technology.

Bernd H. A. Rehm

Polyhydroxyalkanoate Inclusions: Polymer Synthesis, Self-assembly and Display Technology

1

Jenny Draper, Jinping Du, David O. Hooks, Jason W. Lee, Natalie Parlane and Bernd H.A. Rehm

Abstract

Biopolyesters are a class of carbon storage polymers synthesized by a wide variety of bacteria in response to nutrient stress. Production of these polyhydroxyalkanoates (PHAs = polyesters) is catalysed by PHA synthases, which polymerize (*R*)-3-hydroxyacyl-CoA thioesters into polyester. There are several different classes of PHA synthases which preferentially utilize different CoA thioester precursors, generating PHAs with varying material properties such as elasticity and melting point. Genetic engineering and growth on varied carbon sources can be used to modify the type of polyester produced. The general biopolyester properties of biocompatibility, biodegradability, and production from renewable carbon sources have led to considerable interest in PHAs as biomaterials for medical applications as well as alternatives to petrochemical plastics.

Biopolyesters are generated in the cell as water-insoluble granules coated with structural, regulatory, and synthase proteins. Recently, the natural structure of the granules has been exploited to generate functionalized nanoparticles for use in a wide variety of applications, including bioseparation, drug delivery, protein purification, enzyme immobilization, diagnostics, and vaccine delivery.

Introduction

Polyhydroxyalkanoates (PHAs) are biopolyesters produced naturally as intracellular inclusions by a wide range of bacteria and archaea when a carbon source is available in excess and other nutrients are growth–limiting. These PHA granules serve as a carbon and energy reserve which can be accessed by depolymerizing enzymes during periods of carbon starvation.

The key enzyme for PHA production is the PHA synthase, which catalyses the enantiomer-selective conversion of (*R*)-3-hydroxyacyl-CoA thioesters into polyester, while releasing CoA (Fig. 1.1). The thioester precursors are generated from intermediates of primary metabolism. This is exemplified by the production of (*R*)-3-polyhydroxybutyrate (PHB) from acetyl–CoA in *Ralstonia eutropha*: first, the PhaA β-ketothiolase condenses two acetyl-CoA monomers into acetoacetyl-CoA; these are reduced into (*R*)-3-hydroxybutyryl-CoA by the acetoacetyl-CoA reductase PhaB, and finally the PhaC synthase uses the

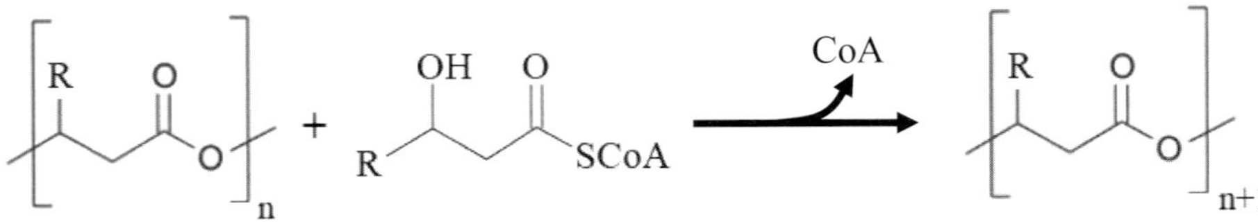

Figure 1.1 The PHA production reaction catalysed by PHA synthases. (Reproduced from Rehm, 2007.)

(*R*)-3-hydroxybutyryl-CoA monomers to synthesize PHB. Together, the *phaABC* genes are sufficient for production of PHB in the presence of acetyl-CoA.

However, the intermediates can also be diverted from the β-oxidation cycle or fatty acid *de novo* biosynthesis pathways when other carbon sources are used. Incorporation of (*R*)-3-hydroxy fatty acids with different monomer chain lengths generates PHAs with varied properties such as melting point and crystallinity. The four major classes of PHA synthases preferentially utilize different precursors, thus favouring formation of different PHA types. However, most of the PHA synthases studied are able to utilize a broad range of precursors. Genetic engineering to modify the precursor-generating pathways, such as *fadAB* knock-out mutants disrupting β-oxidation, can influence the type and amount of PHA produced. Mutagenesis of the PHA synthases or growth on varied carbon sources also affects PHA production.

Much research has been done on PHA production; PHAs are highly biocompatible, biodegradable, and produced naturally from renewable carbon resources, making them an attractive alternative to petrochemical-based plastics. However, due to the cost of fermentation and purification, PHAs currently remain more expensive to produce than conventional plastics. Thus, they are primarily manufactured for higher value niche applications, especially in the biomedical field where they are currently used as suture material, tissue scaffolds, or potentially for drug delivery.

PHAs are deposited in the cell cytoplasm as granules with a hydrophobic biopolyester core surrounded by attached or embedded surface proteins (Figs. 1.2 and 1.3). The surface-attached proteins include the PHA synthase, which remains covalently attached to the polyester chain it synthesized, as well as structural proteins (phasins), depolymerases, and regulatory proteins bound to the polyester core by hydrophobic interactions. However, aside from the PHA synthase these surface proteins are not necessary for granule formation. Granules can form *in vitro*, merely by providing purified PHA synthase with precursor (*R*)-3-hydroxyacyl-CoA thioesters. Additionally, PHA granules are produced efficiently in recombinant bacteria lacking the structural and regulatory proteins. In naturally PHA-producing cells the surface proteins are involved in regulation of granule size, number, and distribution during cell division.

The natural production of PHAs as protein-coated granules has recently been exploited to produce functionalized biopolyester 'nanobeads'. This new technology uses genetic fusions of naturally granule-attached proteins to functional proteins of interest; expression of these proteins in PHA-producing bacteria results in one-step production of functionalized granules. PHA nanobeads suitable for bioseparation, diagnostics and imaging, enzyme immobilization, protein purification, and delivery of drugs and vaccines have already been

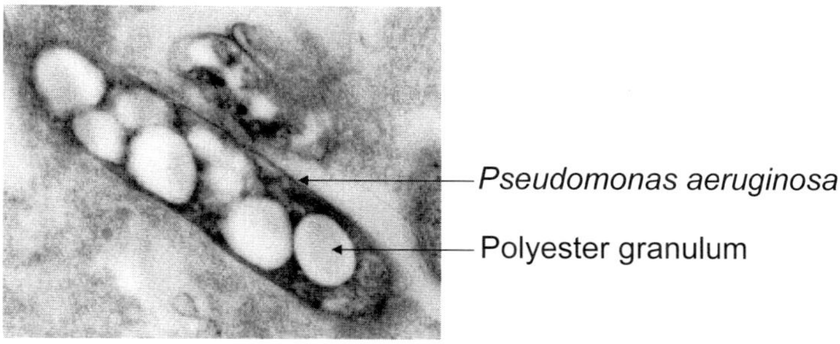

Figure 1.2 Electron microscopy image of bacteria containing PHA granules. (Reproduced from Rehm, 2007.)

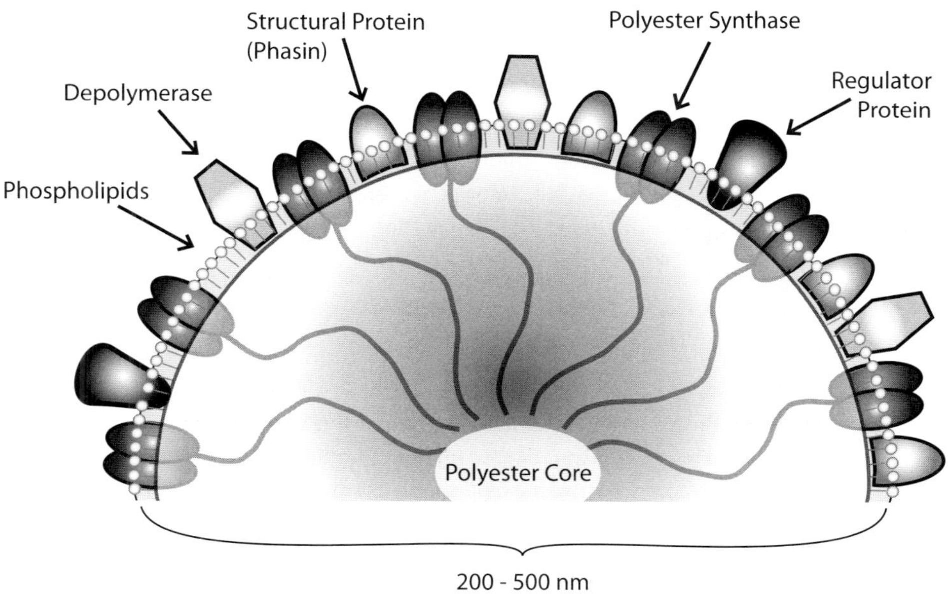

Figure 1.3 Schematic representation of a PHA granule and its associated proteins.

developed. The potential applications of this technology are only limited by the ability to express functional protein in a host cell. Most bacteria are capable of producing PHA either naturally or recombinantly; currently *Escherichia coli*, *Pseudomonas* spp., *Ralstonia eutropha* and *Lactococcus lactis* are the primary producers of recombinant functionalized polyester granules.

Polyester diversity

PHAs are high molecular weight (5×10^5 to 5×10^6) linear polyesters composed of (*R*)-3-hydroxy fatty acids with different monomer chain lengths (Fig. 1.4) (Anderson *et al.*, 1990; Rehm, 2010). Short-chain-length PHAs (PHA_{SCL}) comprise 3–5 carbon atoms

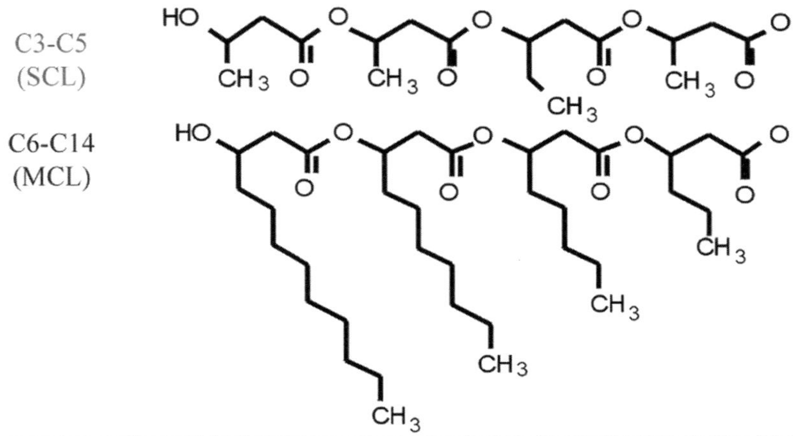

Material Properties	PHA_{SCL}	PHA_{MCL}	PP
T_m (°C)	177	61	176
T_g (°C)	2	-36	-10
Crystallinity (%)	70	30	60
Elongation at break (%)	5	300	400

PP, polypropylene

Figure 1.4 Chemical structure and material properties of the two major classes of bacterial polyesters, compared with polypropylene. (Reproduced from Rehm, 2007.)

and are produced by a wide-range of bacteria and archaea. These PHAs have a high melting point, crystallinity, and brittleness. Medium-chain-length PHAs (PHA_{MCL}) with 6–14 carbon atoms are produced primarily by pseudomonads; these PHAs are more elastomeric and have a lower melting point and crystallinity. Long chain-length PHAs (PHA_{LCL}) have more than 14 carbon atoms. Intracellular PHB was first discovered in *Bacillus megaterium* in 1925 (Lemoigne, 1925). Over 150 different biologically produced PHAs have since been described (Rehm, 2003; Steinbüchel and Valentin, 1995) which may occur as homopolymers or as co-polymers (Kessler *et al.*, 2001). Examples of other PHAs include poly 4-hydroxybutyrate (P4HB), copolymers of 3-hydroxybutyrate and 3-hydroxyvalerate (PHBV), copolymers of 3-hydroxybutyrate and 3-hydroxyhexanoate (PHBHHx), and poly 3-hydroxyoctanoate (PHO).

In contrast to the high-molecular weight carbon-storage PHA produced by prokaryotes, many prokaryotic and eukaryotic organisms produce non-storage, low molecular weight PHA (Reusch *et al.*, 1987, 1992; Reusch, 1989; Reusch and Sadoff, 1988). These 130–170 monomer unit molecules known as cPHB are complexed with other macromolecules such as polyphosphates and are found in lipoproteins, cell membranes, and cytoplasm. cPHBs can dissolve salts and facilitate their transfer across hydrophobic barriers. For example, cPHBs can form channels in the cytoplasmic membrane, allowing import of calcium ions (Seebach and Fritz, 1999). This plays a part in acquisition of competence by *E. coli*, most likely by replacing the inorganic polyphosphate in the PHB channel with organic DNA polyphosphate, resulting in DNA being drawn into the cell (Reusch, 1992). Reusch has also shown an involvement of cPHB in atherogenic plaques and diabetes and suggested plasma PHB levels may serve as a disease marker (Reusch *et al.*, 2003). No synthesis genes or enzymes for cPHB production have yet been identified.

Some eukaryotic organisms such as slime-moulds and fungi are able to produce the polyester poly(β,L-malic acid) (PMLA) (Rathberger *et al.*, 1999). PMLAs probably function as carrier molecules for proteins involved in synchronization of cell nucleus division (Holler *et al.*, 1992). Although PMLA can be produced chemically, an enantiomerically pure product is obtained by biological fermentation, often using myxomycetes such as *Physarum polycephalum*. PMLA is highly water soluble, biodegradable, biocompatible, and can be chemically modified. It is often produced as a co-polymer microspheres (Portilla-Arias *et al.*, 2007) for use as slow-release drug delivery agents combined with active drugs for cancer treatment (Huang *et al.*, 2012; Portilla-Arias *et al.*, 2010), bronchodilators (Lambov *et al.*, 1997), or to treat pulmonary hypertension (Yoncheva *et al.*, 2001).

In contrast to traditional petrochemical-based plastics, PHA is produced from renewable carbon sources such as glucose, cane or beet molasses (Liu *et al.*, 1998; Ramsay *et al.*, 1995), waste oil (Füchtenbusch *et al.*, 2000), or xylose and transgenic plants (John and Keller, 1996; Nakashita *et al.*, 1999; Nawrath *et al.*, 1994). Furthermore, the biodegradability of PHAs is desirable in many environments and ecosystems. Numerous microorganisms secrete extracellular polymerases that hydrolyse PHA into water-soluble oligomers and monomers and subsequently utilize these as cell nutrients.

Some PHAs have properties similar to the main commodity plastics (e.g. polypropylene and polystyrene), and they can be heat-processed using current plastic industry techniques (Sudesh and Iwata, 2008). PHAs have already been used in the packaging, pharmaceutical, and medical industries for a wide range of products. Additionally, the chiral hydroxy acids that compose PHA can be used as building blocks for the synthesis of enantiomerically pure fine chemicals such as antibiotics or vitamins (Ruth *et al.*, 2006). However, it is presently more expensive to produce plastic from renewable biological sources than non-renewable petrochemical sources. Therefore, the use of PHA is currently targeted at niche medical applications which benefit from its biocompatibility and biodegradability, such as drug delivery, suture material and bone scaffolds.

Polyester synthases: genetics

Polyester synthases are the key enzymes for PHA biosynthesis. The four major classes of PHA synthase are distinguished primarily by subunit composition and sequence similarity and to a lesser extent by substrate specificity (Fig. 1.5).

Class I

Class I synthases are composed of a single PhaC subunit with a relatively large molecular weight of 60–73 kDa. The class I PHA synthases from *R. eutropha* H16 preferentially utilize SCL CoA thioesters of (*R*)-3-hydroxy fatty acids of 3–5 carbon atoms. However, medium chain length monomers can also be incorporated by the class I PhaC from *R. eutropha* B5786 which shows a 99% sequence similarity to the PhaC from H16 (Kozhevnikov *et al.*, 2010). The class I synthases from *Aeromonas punctata* FA440 also produce polymers from SCL and MCL CoA thioesters of (*R*)-3-hydroxy fatty acids (Fukui and Doi, 1997).

Recently, a highly active PHA synthase from *Chromobacterium* sp. USM2 has been

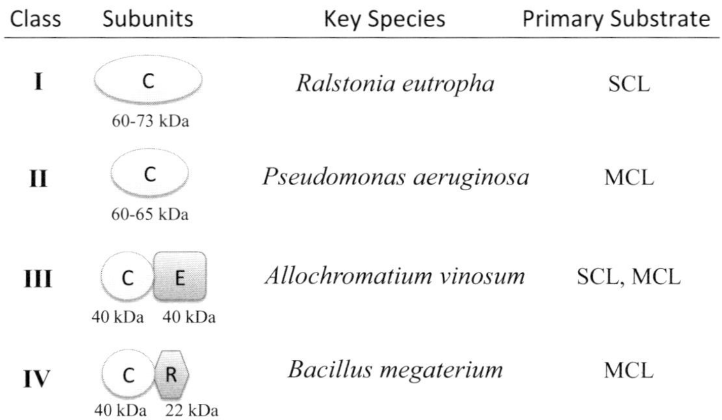

Figure 1.5 The four classes of polyester synthases.

isolated and characterized (Bhubalan *et al.*, 2011). This enzyme can utilize a broad substrate range (3HB, 3HV, and 3HHx). Compared with the activity of PhaC from *R. eutropha* (307 ± 24 U/g), the new synthase has an eightfold higher activity (2462 ± 80 U/g) with respect to the polymerization of 3-HB-CoA.

Class II

This class of synthases is found in *Pseudomonas* spp.; it is also composed of a single PhaC subunit with a slightly smaller molecular weight of 60–65 kDa. Class II synthases usually utilize MCL CoA thioesters of (*R*)-3-hydroxy fatty acids (e.g. *P. putida, P. aeruginosa, P. oleovorans*) (Ren *et al.*, 2000). However the PHA synthase from *Pseudomonas* sp. 61-3 produces polymers from both SCL and MCL CoA thioesters of (*R*)-3-hydroxy fatty acids (Matsusaki *et al.*, 1998). It should be noted that pseudomonads can also produce class I synthases (Solaiman and Ashby, 2005).

Class III

In contrast to the previous two classes, class III synthases are composed of two subunits, PhaC and PhaE. The ~40 kDa PhaC subunit from *Allochromatium vinosum* has only 24.7% similarity to PhaC from *R. eutropha* (Liebergesell *et al.*, 1992; Liebergesell and Steinbüchel, 1992). The ~40 kDa PhaE subunit shows no homology to PhaC. Class III synthases from *Allochromatium vinosum* prefer SCL CoA thioesters of (*R*)-3-hydroxy fatty acids of 3–5 carbon atoms. Phylogenetic trees of PhaC or PhaE sequences suggest the synthases found in a number of haloarchaeons belong to a subgroup of class III and possibly result from horizontal gene transfer (Lu *et al.*, 2008). The *Haloferax mediterranei* synthase from this subgroup synthesizes poly(3-hydroxybutyrate-co-3-hydroxyvalerate) (PHBV) (Lu *et al.*, 2008).

Class IV

Class IV synthases such as those from *Bacillus megaterium* are similar to class III synthases, except the PhaE subunit is replaced with a smaller 20 kDa PhaR subunit; they also prefer SCL CoA thioesters (McCool and Cannon, 1999). The class IV synthase from *Bacillus cereus* YB-4 can incorporate 3HHx when grown on longer-chain fatty acids (e.g. palm or soybean oil) whereas the synthase from *B. megaterium* could not (Hyakutake *et al.*, 2011).

The genetics of PHA production

PHA synthesis and regulatory genes are often clustered, although such clustering is not strictly conserved. Class I synthases such as *phaC1* from *R. eutropha* typically occur in a *phaCAB* operon with the *phaA* (β-ketothiolase) and *phaB* (acetoacetyl-CoA reductase) genes, which together generate the (*R*)-3-hydroxybutyryl-CoA monomers required for synthesis of PHB; additionally, a *phaR* regulator gene is often located just downstream of the *phaCAB* operon. Class II synthases such as PhaC1 and PhaC2 from *P. aeruginosa* are typically organized in a *phaC1–phaZ–phaC2–phaD* operon; PhaZ is a depolymerase, while PhaD is a structural protein. This operon is followed by another transcribed in the opposite direction which encodes the structural and regulatory proteins PhaI and PhaF. Class III synthases are typically encoded as a *phaC–phaE* synthase operon, adjacent to an operon (usually transcribed in the opposite direction) encoding *phaA, phaB,* and a phasin *phaP.* Class IV synthases are less consistently organized, but the two subunits (PhaRC) are typically encoded by a *phaR–phaB–phaC* operon; in *Bacillus* species, this is adjacent to genes encoding the PhaP phasin and PhaQ regulator genes.

Polyester synthases: structure and function

Structural features

Currently, no structural data for polyester synthases are available, although they do contain a conserved functional α/β hydrolase domain that can be threaded onto solved structures (Fig. 1.6) (Amara and Rehm, 2003; Jia *et al.*, 2000). The class I PHA synthase from *R. eutropha* has been studied in the most detail, and the majority of mutagenesis approaches have used the enzyme from this model PHA producer (Rehm, 2003). Mutational analysis of the non-conserved N-terminal region, including deleting the first 100 amino acid residues, indicated it was not essential for synthase activity. However, replacement of residues involved in a putative α-helix (D70-E88) with proline-enhanced PHA accumulation (Zheng

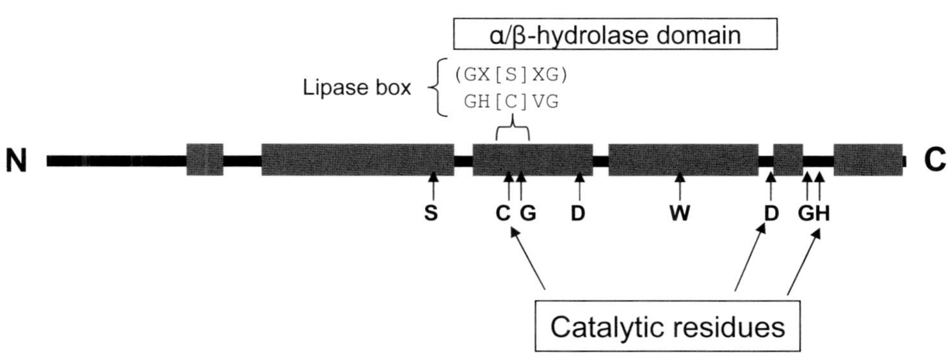

Figure 1.6 Primary structure of the *R. eutropha* PHA synthase. (Reproduced from Rehm, 2007.)

et al., 2006), suggesting a role for the N-terminus in regulatory protein–protein interactions. The high hydrophobicity of the conserved C-terminal region suggests it may associate with the hydrophobic PHA granule core. Deletions of five and twelve amino acids from the C-terminal region caused inactivation of the synthase, showing this region is essential for activity (Rehm *et al.*, 2002). Fusion proteins of class I and class II synthases with the fusion point inside the α/β hydrolase fold are not active, confirming the presence of an essential functional domain (Rehm *et al.*, 2002).

Polyester synthases exist in both monomeric and dimeric forms; however, dimerization is strongly induced in the presence of substrate or analogues such as $(3\text{-hydroxybutyryl})_3$-CoA (Wodzinska *et al.*, 1996). Dimerization of the PHA synthase from *R. eutropha* reduces enzyme lag phase and increases specific activity (Wodzinska *et al.*, 1996). Polyester synthases have been localized to the surface of PHA granules from *R. eutropha* by immunoelectron microscopy and gold-labelled anti-PHA synthase antibodies (Gerngross *et al.*, 1993). The class III enzyme (PhaEC) from *A. vinosum* has also been localized to the granule surface, where it exists as a complex with a native molecular mass of 400 kDa (representing about ten subunits) (Liebergesell *et al.*, 1994).

Catalytic reaction mechanism

Griebel and colleagues first identified the PHA synthase as a sulfhydryl enzyme, as it was inhibited upon addition of *N*-ethylmaleimide or *p*-mercuribenzoate (Griebel *et al.*, 1968). They proposed a mechanistic model for the PhaC enzyme that shared features with fatty acid synthesis (Griebel and Merrick, 1971). Their model for PHA chain elongation suggested two thiol groups from the PHA synthase would be involved in the polymerization reaction. The growing PHA chain would cycle between the two thiol groups as the other was loaded with the next HB monomer.

Initially, the conserved residues cysteine-319 and cysteine-459 from the class I PhaC of *R. eutropha* were thought to provide the two sulfhydryl groups necessary for PHB chain extension. However, site-directed mutagenesis revealed only cysteine-319 was involved in covalent catalysis, whereas cysteine-459 was clearly not required for enzyme activity (Gerngross *et al.*, 1994).

In fatty acid synthesis the second thiol is provided by post-translational modification. Therefore, a second thiol group was postulated to become available from the covalent modification of the conserved serine-260 from *R. eutropha* by 4-phosphopantetheine. However, expression of the PHA synthase gene in a β-alanine mutant of *R. eutropha* followed by detection of 4-phosphopantetheinylated protein did not reveal this post-translational modification of PhaC. Nonetheless, site-directed mutagenesis of serine-260 abolishes enzyme activity indicating an important role for this residue (Hoppensack *et al.*, 1999).

Currently, the active form of PHA synthase is considered to be a homodimer (class I and II) or a multimeric heterodimer (class III and IV). The dimerization of PhaC suggests that each monomer could provide one of the two necessary thiol groups, allowing the catalytic mechanism to proceed once the dimer is formed (Fig. 1.7). Tryptophan-398 in *P. aeruginosa* is hypothesized to generate a hydrophobic surface which allows for PhaC dimerization, due to its surface-exposure in threading models (Amara and Rehm, 2003). Replacing the highly conserved tryptophan-425 in *R. eutropha* and tryptophan-398 in *P. aeruginosa* with alanine caused inactivation of the respective synthases, suggesting that dimerization is indeed necessary for enzyme function (Amara and Rehm, 2003).

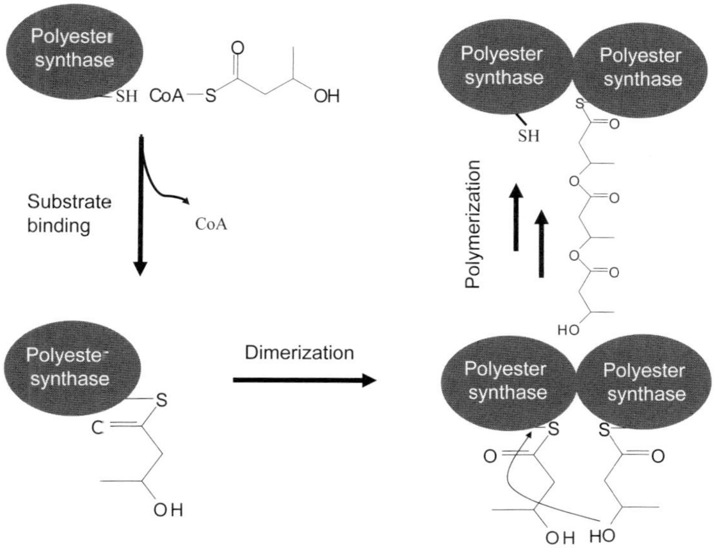

Figure 1.7 Model of the polyester synthase catalytic mechanism. (Reproduced from Rehm, 2007.)

All PHA synthases posses a lipase box-like sequence (G-X-[S/C]-X-G), and structural modelling comparing PhaC to lipases identified conserved residues with a potential role in covalent catalysis. The first enzyme compared to the lipase model was a class III PHA synthase from *Allochromatium vinosum*. Site-directed mutations in each of the conserved residues cysteine-149, histidine-331, and aspartate-302 almost abolished enzyme activity (Jia *et al.*, 2000). Thus, these three residues were hypothesized to form the catalytic mechanism, with cysteine-149 as the nucleophile for covalent catalysis and histidine-331 as its general base catalyst. Aspartate-302 appeared to function as a general base catalyst for the substrate, activating the 3-hydroxyl of 3-hydroxybutyryl CoA thioester for nucleophilic attack and allowing formation of the acylated thiolester intermediate. Further support for the role of aspartate-302 in elongation rather than activating cysteine-139 for acylation was provided by the detection of polymeric HB covalently bound only to peptides containing cysteine-139 after a tryptic digest of the mutant synthase (D302A-PhaCPhaE) from *A. vinosum*. Additional *in vivo* experiments using the same mutant synthase resulted in production of very small (<0.05 µm) diameter PHA granules in a *R. eutropha* host compared with the ~0.4 µm granules in the wild-type strain, possibly due to shorter HB polymer lengths (Tian *et al.*, 2005a).

The homologous amino acid residues cysteine-319, histidine-508, and aspartate-480 from the class I *R. eutropha* PhaC were also subjected to site-directed mutagenesis resulting in greatly diminished enzyme activity, suggesting a similar catalytic mechanism to the class III enzyme (Jia *et al.*, 2001). However, the use of a saturated trimer of 3-HB-CoA as an artificial primer of synthesis demonstrated the conserved aspartate-480 was not required for the acylation step in HB polymerization, while C319-H508 formed the catalytic dyad. Additionally, the class II synthase from *P. aeruginosa* has been examined using epoxide hydrolase from mice as a model (Amara and Rehm, 2003). The putative catalytic residues cysteine-296, aspartate-452, histidine-453, and histidine-480 were replaced by site-directed mutagenesis. Mutations to cysteine-296 and aspartate-452 abolished enzyme activity. Interestingly, mutation to histidine-480, which aligns with the conserved base catalyst in α/β hydrolases, did not impact *in vivo* enzyme activity and only reduced *in vitro* activity to 41% of the wild type. In contrast H453Q strongly reduced both *in vivo* and *in vitro* activity to 24% and 7% of the wild type respectively, indicating histidine-453 may functionally replace histidine-480 in class II synthases (Amara and Rehm, 2003). This study also replaced the active site cysteine-296 with a serine residue and found the synthase remained active, although at a reduced level (Amara and Rehm, 2003).

Synthesis of PHB has also been suggested to proceed in a manner analogous to some type III polyketide synthases. Rather than needing to bring two thiol groups from different synthase units into a dimer, this model requires only one active site cysteine. The PHB chain is extended by repeated acylation of the same cysteine with a non-covalent binding step necessary for extension of the polymer. To test this model, a class III C149S-PhaEC mutant exhibiting a slowed reaction rate was used to catch non-covalently bound (HB)nCoA ($n=2$ or 3) in the process of chain elongation. The model of PHB chain elongation by repeated acylation of cysteine-149 was supported by the detection of (HB)nCoA after rapid initiation and quenching of the reaction (Li *et al.*, 2009).

In all structural models the proposed catalytic triad is located adjacent to the enzyme core and the putative active site cysteine is always found within the strand–nucleophile–helix motif common to α/ hydrolases. Overall, the catalytic mechanism of α/β hydrolases, in particular lipases, has provided a useful model for examining the catalytic residues of PHA synthases. The mutational studies of class I-III PHA synthases from *R. eutropha*, *P. aeruginosa*, and *A. vinosum* respectively have provided strong support in favour of this model. The interfacial activation of lipases at lipid–water interfaces could be analogous to the activation of polyester synthase catalysis at the polyester–water interface of the growing PHA granule. This view is supported by the increased activity of attached polyester synthase from *R. eutropha* compared with the soluble enzyme (Gerngross and Martin, 1995). Upon binding of the substrate and extension of the PHA polymer, the soluble polyester synthase converts into an amphipathic molecule (Rehm, 2003). This change leads to self-assembly of the PHA granules with the hydrophobic PHA in the core surrounded by active polyester synthase on the surface.

Random mutagenesis

Polyester synthases have been subjected to directed evolution experiments to enhance enzyme activity, alter substrate specificity, and change the molecular weight of the PHA product (Amara *et al.*, 2002; Kichise *et al.*, 2002; Matsumoto *et al.*, 2006; Normi *et al.*, 2005; Tsuge *et al.*, 2007; Zheng *et al.*, 2006). Altering the properties of polyester synthases allows for more efficient production of specific tailor-made biopolyesters.

In vitro and *in vivo* evolution of PHA synthases from *Aeromonas punctata* by random mutagenesis has been performed to identify point mutations which enhance PHA synthase activity (Amara *et al.*, 2002; Kichise *et al.*, 2002). The first approach used PCR-mediated random mutagenesis targeting a limited region of the *A. punctata* FA440 *phaC* gene. This synthase produces a random co-polyester of 3-hydroxybutyrate and 3-hydroxyhexanoate [P(3HB-*co*-3HHx)], which has favourable toughness and flexibility compared to PHB homopolyester. Two mutations, N149S and D171G, were found to exhibit a 56% and 21% increase in activity towards 3HB-CoA, respectively. *In vivo* this increase in specific activity corresponded to an enhanced accumulation of the random co-polyester by 6.5-fold and 3-fold, respectively. Interestingly, both mutants showed an increase in the 3HHx fraction of the polymer – up to 18% and 16% respectively compared with the wild type (10%) (Kichise *et al.*, 2002). Neither of these mutations is found in conserved regions of the *phaC* gene. The synergistic effect of both mutations was examined by expressing a double mutant (N149S, D171G) in *R. eutropha*. The mutant showed enhanced incorporation (18.5%) of longer 3-hydroxyalkanoate monomers when octanoate was used as a carbon source and a very high molecular weight ($M_w = 3.68 \times 10^4$) of P(3HB) homopolymer when grown on a fructose carbon source (Tsuge *et al.*, 2007).

A second approach used the *E. coli* XL1-Red mutator strain to generate *in vivo* random mutants of the PHA synthase gene from *A. punctata*. Four mutants which displayed enhanced *in vivo* and *in vitro* activity were selected from the 200,000 mutants generated. The single mutation F518I showed a five-fold increase in PHA synthase activity but led to only a 20% increase in PHA accumulation compared to the wild type. A second mutation, V214G, displayed a two-fold increase in specific activity but only a 7% increase in PHA accumulation (Amara *et al.*, 2002). All mutants synthesized PHA with increased molecular weight averages, but the molar fractions of PHA monomers were not significantly affected.

Improved class II synthases have also been achieved by random mutagenesis. *In vitro* evolution of *Pseudomonas* sp. 61-3 PHA synthase I identified E130D as a positive mutant with 10-fold higher accumulation of PHB in *E. coli* harbouring the mutant PHA synthase compared to 0.1 wt% for the wild-type synthase. Additionally, *in vitro* activity tests revealed the mutant PhaC had a higher activity towards various 3-hydroxyacyl-CoAs of up to 10 carbons in length (Matsumoto *et al.*, 2005). Another mutation (S477R) led to increased average molecular weights of PHB homopolymers – up to sixfold over the wild-type enzyme – and a substrate preference shift towards smaller monomer units (Matsumoto *et al.*, 2006). Combinations of beneficial mutations such as E130D with S325T, S325C, and Q481L resulted in a synergistic effect on *in vivo* PHA production, *in vitro* enzyme activity, and substrate specificity (Matsumoto *et al.*, 2005, 2006).

Saturation mutagenesis of the PHA synthase from *R. eutropha* was carried out on the glycine-4 residue. The N-terminal mutation G4D increased PhaC levels *in vivo*, thereby enhancing the PHB content of cells by 23 percentage points (Normi *et al.*, 2005). Further mutagenesis to a putative amphipathic α-helix region (D70-E88) enhanced activity of the synthase *in vitro*. The mutations Y75P and A81P produced an activity of 137% and 105% of the wild type respectively. The double mutant Y75P/A81P had a synergetic effect, enhancing *in vitro* synthase activity to 162% of the wild type using a total cellular protein extract (Zheng *et al.*, 2006). The authors suggest disruption to the α-helix by proline may impair regulatory protein–protein interactions with the PHA synthase.

Although many beneficial mutations to PHA synthases of various classes have been characterized, the functional roles of the mutated residues have yet to be determined. Only mutations to the PHA synthase from *R. eutropha* have a suggested mechanism of action (Normi *et al.*, 2005; Zheng *et al.*, 2006). Ultimately, the resolution of three-dimensional structure will be a crucial step in understanding these structure–function relationships.

Substrate specificity

PHA synthases have different substrate specificities which determine the monomer composition of the polyester produced; this in turn determines the properties of the resulting material. Altering the substrate specificity of the enzyme is an important tool for the production of useful tailor-made biopolyesters. The substrate specificities of polyester synthases from *R. eutropha* and *A. vinosum* have been studied *in vitro* using analogues of PHB with varying chain lengths, branching, hydroxyl group chain position, and thioesters (Yuan *et al.*, 2001). These polyester synthases were seen to be highly specific *in vitro*.

Metabolic engineering to provide a range of 3-hydroxy fatty acid CoA thioesters has been used to study the *in vivo* substrate range of PHA synthases in recombinant *E. coli* (Antonio *et al.*, 2000; Dennis *et al.*, 1998; Qi *et al.*, 1997, 1998). In contrast to the specific *in vitro* specificity, *in vivo* the class I PHA synthase from *R. eutropha* displayed a broad substrate specificity, even accepting MCL 3-hydroxy fatty acid CoA esters as substrates (Antonio *et al.*, 2000; Dennis *et al.*, 1998). It has also been shown that growing *R. eutropha* on 3-mercaptopropionic acid as the carbon source produced a co-polymer of 3-hydroxybutyric acid and 3-mercaptopropionic acid linked by thioester bonds (Lütke-Eversloh *et al.*, 2001). Additionally, the increasing variety of PHA monomers detected in bacterially synthesized polyesters is further evidence that PHA synthases possess an extremely broad substrate specificity (Rehm, 2003).

Lastly, mutagenesis has been used to alter the substrate specificity of PHA synthases. Initially, the substrate specificity of class II PhaC from *P. putida* was altered by localized semi-random mutagenesis (Sheu and Lee, 2004). Later, alanine-510 of the *R. eutropha* polyester synthase was found to have a role in substrate specificity (Tsuge *et al.*, 2004), and altering the serine-477 residue of the *Pseudomonas* sp. 61-3 PHA synthase enhanced its incorporation of short-chain-length 3HA-CoA (Matsumoto *et al.*, 2006). Additionally, a chimeric synthase with 26% of the N-terminal region from *Aeromonas caviae* and 74% of the C-terminal region from *R. eutropha* maintained the advantageous properties of both enzymes: broad substrate specificity and high enzyme activity, respectively. The chimeric synthase accumulated a higher amount of PHA (50 wt%) which consisted of 2 mol% 3-hydroxyhexanoate (Matsumoto *et al.*, 2009).

In vivo substrate provision for polyester synthases

Formation of PHA inclusions is initiated by provision of 3-HA–CoA thioesters to a PHA synthase. Upon uptake of a suitable carbon source, anabolic and/or catabolic reactions convert the carbon compound into a HA–CoA thioester (Fig. 1.8). The synthase uses these thioesters as substrates, catalysing formation of the PHA polyester with concomitant release of CoA (Anderson and Dawes, 1990; Steinbüchel, 1991). Understanding the pathways that lead to biosynthesis of the CoA thioester precursors enables metabolic engineering of bacteria to produce tailor-made PHAs.

Provision of different carbon sources affects the type and composition of PHA, resulting in polyesters with varied properties including tensile strength and melting point (Steinbüchel *et al.*, 1998). However, *in vivo* substrate specificity can be determined experimentally only by analysing the chemical composition of PHAs produced through cultivation on various carbon sources. One such screen observed variation in the

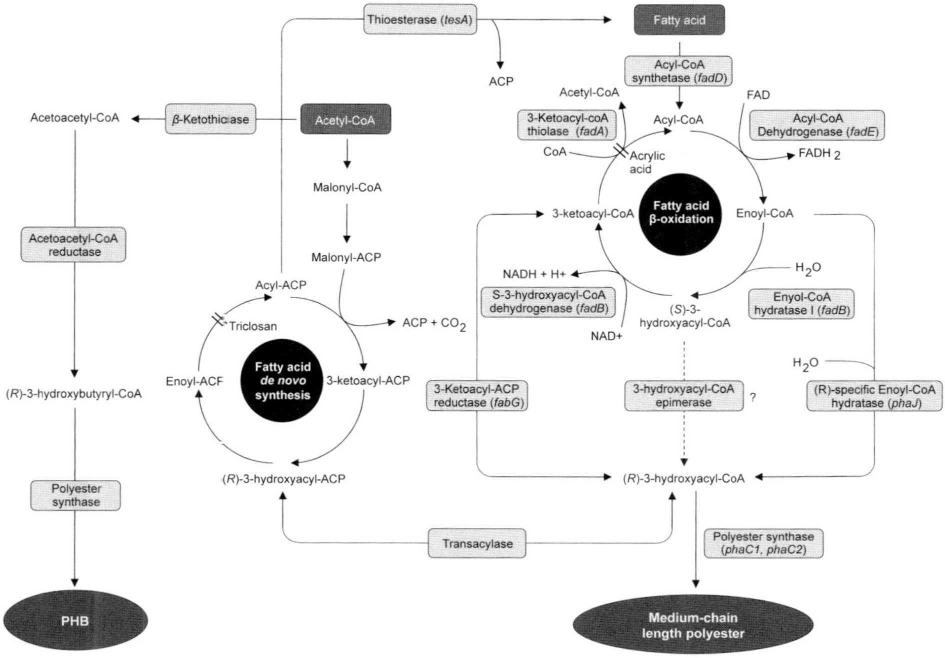

Figure 1.8 Metabolic path PHA production. (Modified from Rehm, 2007.)

composition of PHAs derived from 10 different carbon sources using the *Thiocapsa pfennigii* PHA synthase expressed recombinantly in *P. putida* (Valentin *et al.*, 1996).

Short-chain-length polyester biosynthesis

Production of PHA_{SCL} of 3–5 carbon atoms is best represented by *R. eutropha*'s synthesis of poly-3-hydroxybutyrate (P3HB) In *R. eutropha*, generation of thioester precursors from glucose is initiated by the enzyme PhaA, a β-ketothiolase, which catalyses the condensation of two acetyl-CoA molecules from the glycolysis pathway into acetoacetyl-CoA. The NADPH-dependent acetoacetyl-CoA reductase PhaB then reduces the acetoacetyl-CoA molecules to (*R*)-3-hydroxybutyrl-CoA, which the PHA synthase uses to produce poly-3-hydroxybutyrate (P3HB) (Peoples and Sinskey, 1989a.b; Schubert *et al.*, 1988; Slater *et al.*, 1988). However, use of octanoate and dodecanoate fatty acids as a carbon source instead of glucose results in alternative PHAs composed of 3-hydroxyoctanoate (3HO) and 3-hydroxydodecanoate (3HDD) in addition to P3HB (Antonio *et al.*, 2000).

Medium-chain-length polyester biosynthesis

In pseudomonads, synthesis of PHA_{MCL} of 6–14 carbon atoms involves the fatty acid *de novo* biosynthesis pathway. *P. putida* encodes the (*R*)-3-hydroxydecanoyl-ACP:CoA transacylase enzyme PhaG, which links fatty acid *de novo* biosynthesis to PHA biosynthesis. PhaG is only required for PHA synthesis when gluconate or other non-related carbon sources are provided (Fiedler *et al.*, 2000). Such carbon sources are oxidized to acetyl-CoA and enter the fatty acid *de novo* biosynthesis pathway, but PhaG diverts intermediates of the pathway towards PHA biosynthesis. *P. aeruginosa*'s PHA biosynthetic pathways are more diverse, possibly due to the presence of the protein Rh1A, which plays a role in surfactant synthesis and PHA production. Synthesis of rhamnolipids and PHA is closely related in *P. aeruginosa*, as fatty acid derivatives are used as substrates in both pathways (Soberón-Chávez *et al.*, 2005). However, it was shown that PhaC can only use CoA fatty acids as substrates, and thus activity of the PhaG transacylase is required for PHA_{MCL} production from sugar carbon sources (Rehm *et al.*, 1998). The sigma factor RpoN also appears to be involved in PHA biosynthesis gene regulation (Hoffmann and Rehm, 2004; Hoffmann and Rehm, 2005).

The fatty acid β-oxidation pathway provides precursors from fatty acids

Fatty acid β-oxidation is a general metabolic pathway in eukaryotes and prokaryotes involved in the catabolism of fatty acids. In *E. coli*, genes involved in β-oxidation fatty acid degradation are encoded by the *fad* operon, which is negatively regulated by *fadR*.

Free fatty acids enter the β-oxidation pathway upon ligation to CoA, generating acyl-CoA thioesters; this process is mediated by the acyl-CoA synthetase FadD (Hume *et al.*, 2009). The four-step fatty acid β-oxidation cycle generates various CoA thioester intermediates and ultimately produces acetyl-CoA (Fig. 1.8).

In the first step, the acyl-CoA thioesters produced by FadD are oxidized to enoyl-CoA by a FAD-dependent reaction with acyl-CoA dehydrogenase FadE; this is reportedly the rate-limiting step in the pathway (Lu *et al.*, 2003). The remaining three steps in the pathway are catalysed by the multienzyme complex FadBA, which is composed of two subunits in $\alpha_2\beta_2$ quaternary structure (Pawar and Schulz, 1981; Pramanik *et al.*, 1978).

The FadBA multienzyme complex encoded by the *fadBA* operon is thought to posses five enzymatic activities: enoyl-CoA hydratase (*fadB*), NADH-dependent (*S*)-3-hydroxyacyl-CoA dehydrogenase (*fadB*), *cis*-Δ^3-*trans*-Δ^2-enoyl-CoA isomerase, 3-hydroxyacyl-CoA epimerase and 3-ketoacyl-CoA thiolase (*fadA*) (Fiedler *et al.*, 2002). The crystal structure of the *fadAB* homologue *foaAB* from *Pseudomonas fragi* revealed that the α-dimer alone exhibits 2-enoyl-CoA hydratase and 3-hydroxyacyl-CoA dehydrogenase activity, involved in the catalytic conversion of enoyl-CoA and (*S*)-hydroxyacyl-CoA respectively. The 3-ketoacyl-CoA thiolase activity is mediated by contact between the α- and β-subunits, and is involved in the catalytic conversion of 3-ketoacyl-CoA to acetyl-CoA (Ishikawa *et al.*, 1997).

A relationship between β-oxidation and PHA_{MCL} synthesis is well known and extensively described in literature. Analysis of the PHA_{MCL} co-monomer composition from *P. putida* grown on fatty acids with ^{13}C-labelled substrates indicated the majority of the co-monomer was generated from β-oxidation (Eggink *et al.*, 1995). Additionally, genes encoding enzymes involved in β-oxidation and PHA_{MCL} biosynthesis in *P. putida* KT2440 were both highly expressed when grown on fatty acid (Wang and Nomura, 2010).

Acyl-CoA intermediates synthesized by the FadBA multienzyme complex, namely enoyl-CoA (*S*)-3-hydroxyacyl-CoA and 3-ketoacyl-CoA, serve as major precursors for PHA_{MCL} synthesis by class-II PHA synthases. However, due to the specificity of PHA synthases for (*R*)-3-hydroxyacyl-CoA thioesters, these intermediates cannot directly be utilized by the polyester synthase and must be converted to their (*R*)-3-hydroxyacyl-CoA moiety.

Enzymes involved in the conversion of β-oxidation intermediates have been extensively cloned and characterized in both PHA-producing and non- PHA-producing organisms. These enzymes include the (*R*)-specific enoyl-CoA hydratase (PhaJ) and the 3-ketoacyl-ACP reductase (FabG), involved in the conversion of enoyl-CoA and 3-ketoacyl-CoA respectively (Park *et al.*, 2005; Tsuge *et al.*, 2003b). The putative 3-hydroxyacyl-CoA epimerase encoded by the *fadBA* operon is thought to link the conversion of (*S*)-3-hydroxyacyl-CoA into (*R*)-3-hydroxyacyl-CoA; however, this has yet to be fully elucidated (Fiedler *et al.*, 2002).

Disruption of the β-oxidation pathway and/or amplification of genes involved with the conversion of β-oxidation intermediates have been employed to study the flux of carbon towards PHA synthesis (Chung *et al.*, 2009; Ren *et al.*, 2000; Snell *et al.*, 2002). For example, disruption of *fadA* in *P. putida* KCTC1639 increased the pool of available β-oxidation intermediates which could be channelled towards PHA synthesis, leading to an increase in PHA_{MCL} containing a higher percentage of longer-chain-length fatty acids (Vo *et al.*, 2007).

Alternative β-oxidation fatty acid enzymes

In eukaryotes, multiple enzymes exist for each step of the β-oxidation pathway, and different sets degrade short-, medium- and long-chain fatty acids (Brigham *et al.*, 2010). Similarly, some prokaryotes encode multiple enzyme sets. For example, *P. putida* has several sets of *fad* genes, contributing to its metabolic diversity (Escapa *et al.*, 2011). Likewise, the genome of *R. eutropha* H16 contains many genes potentially involved in β-oxidation, including 50 enoyl-CoA hydratases and 46 acyl-CoA dehydrogenases (Brigham *et al.*, 2010).

In *E. coli,* several β-oxidation pathway genes have been identified and characterized in literature. The *E. coli* genes *yfcX* and *yfcY* show significant homology to *fadB* and *fadA* respectively, and probably encode an alternative β-oxidation pathway utilized under anaerobic conditions (Park *et al.*, 2005). YfcX can substitute for FadB in the PHA_{MCL} production pathway, and displays multienzyme activity similar to FadB's hydratase and dehydrogenase activity (Park *et al.*, 2004).

In addition to YfcX, five additional homologues *paaG, paaF, bhbD, sceH* and *ydbU* displaying high homology to *fadB* have been identified in the *E. coli* genome and have been suggested to play a role in PHA_{MCL} biosynthesis in a *fadB* mutant background (Park *et al.*, 2004).

MaoC from *E. coli* is an enoyl-CoA hydratase showing 34% protein sequence homology to the *P. aeruginosa* (*R*)-specific hydratase PhaJ1. MaoC is suggested to provide a link between fatty acid β-oxidation and PHA synthesis in *E. coli* by the catalytic conversion of enoyl-CoA in a *fadB* mutant. MaoC was required to channel (*R*)-3-hydroxyacyl-CoA into PHA biosynthesis, as PHA_{MCL} accumulation was severely affected in a *maoC* knockout mutant. This was further confirmed by plasmid based complementation with MaoC (Park and Lee, 2003).

Inhibition of fatty acid β-oxidation

In both PHA-negative and PHA-producing organisms, FadB and/or FadA defective strains have been extensively used to study the link between the β-oxidation pathway and PHA synthesis. Disruption of the β-oxidation pathway results in a pool of β-oxidation intermediates that can be shunted towards PHA biosynthesis by specific hydratases, epimerases, and reductases. For example, production of PHA_{MCL} in PHA-negative *E. coli* was successfully established by heterologously expressing a class-II PHA synthase in a β-oxidation pathway defective background (Park *et al.*, 2004).

FadB and/or FadA defective strains are normally created by a genetic approach; however, chemical inhibition with acrylic acid or Cerulenin has also been effectively employed

(Park *et al.*, 2005). Acrylic acid has been proven to be a powerful tool for disruption of the β-oxidation pathway by inhibiting FadA. Acrylic acid exposure of recombinant *P. fragi* or PHA-negative *R. eutropha* expressing *P. aeruginosa phaC1* results in an increase in PHA_{MCL} accumulation (Fiedler *et al.*, 2002). Chemical inhibition of FadB with Cerulenin is also possible in recombinant *E. coli* (Park *et al.*, 2005).

Disruption of the β-oxidation pathway in PHA-producing organisms leads to synthesis of PHA_{MCL} consisting largely of long-chain-length fatty acids. *P. putida* KT2440 typically produces PHA_{MCL} consisting of 3-hydroxydecanoate (HHx), 3-hydroxyoctanoate (HO), 3-hydroxydecanoate (HD) and 3-hydroxydodecanoate (HDD) monomers when grown on dodecanoate as a sole carbon source, in which HHD will represent a small fraction of the overall composition. A significant shift in co-monomer composition towards the longer-chain-length fatty acid HHD was seen in a β-oxidation- impaired *P. putida* KT2440 *fadBA* mutant, validating the role of *fadA* and *fadB* in β-oxidation for the degradation of fatty acids (Ouyang *et al.*, 2007). Similar results were observed in a *P. putida* KCTC1639 *fadA* knockout (Vo *et al.*, 2007). Another study in *P. putida* KT2440 knockout mutants grown on dodecanoate resulted in co-monomers consisting of HD and HDD; it was suggested that these may be related to a secondary β-oxidation pathway or unknown isozymes or epimerases (Ma *et al.*, 2009).

Disruption of β-oxidation regulator genes can also cause accumulation of PHA_{MCL}. The *fadR* repressor protein in *E. coli* regulates *fadBA* genes and some *fadBA* homologues such as *yfcXY*, blocking their transcription during growth on fatty acids (Park *et al.*, 2005). Disruption of *fadR* results in the constitutive expression of *fad* genes, leading to a strong increase in accumulation of PHA_{MCL} (Qi *et al.*, 1998).

Putative 3-hydroxyacyl-CoA epimerase

A 3-hydroxyacyl-CoA epimerase enzyme is thought to link the β-oxidation and PHA synthesis pathways by converting the β-oxidation intermediate (*S*)-3-hydroxyacyl-CoA into its corresponding (*R*)-enantiomer. However, a specific epimerase involved in PHA_{MCL} biosynthesis has not yet been identified. The multienzyme complex FadBA has epimerase activity *in vitro* (Pramanik *et al.*, 1978) and may fill this role. However, a study investigating the putative epimerase function of recombinant *P. oleovorans* FadBA failed to demonstrate epimerase function *in vivo* (Fiedler *et al.*, 2002). Altogether, the source of 3-hydroxyacl-CoA epimerase activity linking the β-oxidation and PHA synthesis pathways remains unclear.

(*R*)-specific enoyl-CoA hydratases

The conversion of enoyl-CoA to (*R*)-3-hydroxyacyl-CoA is catalysed by (*R*)-specific enoyl-CoA hydratases. A monofunctional (*R*)-specific enoyl-CoA hydratase was first identified in *A. caviae* FA440 (Tsuge *et al.*, 2003b). Since then further hydratases have been identified in other PHA-producing organisms such as *P. aeruginosa, P. oleovorans, P. putida, R. rubrum,* and *R. eutropha* (Kawashima *et al.*, 2012). Homologous sequences have also been identified and characterized in non-PHA-producing organisms, such as MaoC in *E. coli* (Kawashima *et al.*, 2012; Park and Lee, 2003).

In *P. aeruginosa,* four (*R*)-specific enoyl-CoA hydratases (PhaJ1–PhaJ4) have been identified and characterized. All four PhaJs can supply (*R*)-3HA-CoA monomers for PHA synthesis when heterologously expressed with PhaC in *E. coli* (Tsuge *et al.*, 2003b). PhaJ1 demonstrates substrate specificity towards short-chain-length enoyl-CoAs of C_4–C_6, while PhaJ2–PhaJ4 exhibit a longer chain specificity of C_6–C_{12} (Tsuge *et al.*, 2003b).

PhaJ, like many other (*R*)-specific enoyl CoA hydratases, possesses a highly conserved hydratase 2 domain. The crystal structure of the *A. caviae* (*R*)-specific enoyl CoA hydratase revealed that the enzyme is a homodimer, with each monomer composed of five antiparallel β-sheets and five α-helices forming a «hot dog» fold (Hisano *et al.*, 2003; Tsuge *et al.*, 2003a). A binding pocket of highly conserved residues determines its specificity for C_4–C_6 substrates; however this specificity can be altered, as demonstrated by site directed mutagenesis (Tsuge *et al.*, 2003a). Many recently identified *PhaJ* homologues also contain an MaoC-like domain (pfam01575) (Kawashima *et al.*, 2012).

β-Ketoacyl-CoA reductase

FabG from *P. aeruginosa* is an NADH-dependent β-ketoacyl-CoA reductase which converts the last β-oxidation intermediate 3-ketoacyl-CoA into (*R*)-3-hydroxyacyl-CoA. It has been extensively studied in recombinant *E. coli* carrying the class II PHA synthase PhaC1 from *P. aeruginosa*. The function of FabG in β-oxidation was demonstrated by heterologous expression of *fab*G in *E. coli fabB* and *fabA* knockout mutants. Expression of FabG in an *E. coli fadR fadB* mutant did not significantly increase accumulation of PHA_{MCL} when grown on fatty acids as a sole carbon source; this was expected, as a *fabB* knockout would have disrupted the pathway generating the 3-ketoacyl-CoA substrate for FabG (Ren *et al.*, 2000). However, expression in a fadR *fadA* mutant had the opposite effect: the *fadA* knockout resulted in a larger pool of 3-ketoacyl-CoA for FabG to convert to (*R*)-3-hydroxyacyl-CoA, and a significant increase in PHA_{MCL} accumulation occurred (Ren *et al.*, 2000).

More recently, overexpression of *fadG* in *P. putida* KCTC1639 was found to reduce PHA accumulation, which is in contrast to previous results. This lower PHA accumulation was possibly due to FabG's ability to catalyse the reverse reaction of (*R*)-3-hydroxyacyl-CoA to 3-ketoacyl-CoA (Vo *et al.*, 2008).

Thioesterase

Accumulation of PHA_{MCL} in recombinant *E. coli* grown on fatty acids has been extensively demonstrated. However, PHA_{MCL} is not able to accumulate when sugars are the sole carbon source (Klinke *et al.*, 1999). Thioesterases play an important role in PHA_{MCL} production when carbon sources unrelated to the fatty acid β-oxidation pathway are used.

The thiolase protein family contains two classes: degradative and synthetic thiolases. Degradative thiolases convert ketoacyl-CoA into acyl-CoA in the β-oxidation pathway. Synthetic thiolases catalyse the conversion of the fatty acid *de* novo biosynthesis intermediate product acyl-ACP into its corresponding fatty acid. Synthetic thiolases thus provide a direct link between fatty acid *de novo* biosynthesis and β-oxidation (Hiltunen and Qin, 2000; Klinke *et al.*, 1999).

Overexpression of the native thioesterase-I gene *tesA* in a β-oxidation defective *fadR fadB E. coli* mutant harbouring *P. oleovorans phaC1* was capable of accumulating PHA_{MCL} when grown on gluconate with 3-hydroxyoctanoate as the main monomer constituent (Klinke *et al.*, 1999; Park *et al.*, 2005). In addition, an acyl-ACP thioesterase from the plant *Umbellularia californica* can mediate PHA_{MCL} production from gluconate in recombinant *E. coli* (Park *et al.*, 2005).

Polyester inclusions: self-assembly and structure

Self-assembly of polyester particles

Formation of PHA granules both *in vivo* and *in vitro* is mediated by the PHA synthase, which catalyses the synthesis of water-soluble (*R*)-3-hydroxyacyl-CoA monomers into water-insoluble high molecular weight polyesters. Throughout this synthesis process, the PHA synthase remains covalently attached to the growing polyester chain. Substrate is continually incorporated until metabolic or spatial constraints terminate polymerization (Grage *et al.*, 2009). Though the exact mechanism of PHA granule formation remains unknown, to date two models have been proposed: the micelle model and the budding model.

Micelle model

The micelle model is supported by the formation of PHA granules *in vitro* using only purified PHA synthase and substrate (Gerngross and Martin, 1995). In this model, the amphipathic nature of the synthase as it forms the hydrophobic polyester allows self-aggregation into a micelle-like structure (Stubbe and Tian, 2003) (Fig. 1.9). According to this model, nascent PHA granules formed *in vivo* are further coated with constituents other than the PHA synthase (Thomson and Sivaniah, 2010).

Budding model

The budding model considers the presence of the lipid membrane that has been observed surrounding isolated PHA granules. In this model, soluble PHA synthase exists close to the inner cell membrane, and the polymerization reaction occurs in the intra-membrane space, where the elongated PHA chains accumulate into a granule surrounded by a phospholipid monolayer, which eventually buds from the membrane into the cytoplasm (Rehm, 2007; Thomson and Sivaniah, 2010) (Fig. 1.9). However, since experimental evidence for the existence of a lipid layer *in vivo* is still lacking, it is conceivable that the lipids are a contaminant from the process of isolating the hydrophobic granules from lysed cells.

Location of granule formation

The primary evidence for the budding model of PHA inclusion formation is the observation of granules forming at cell poles. Jendorssek *et al.* (Jendrossek, 2005; Jendrossek *et al.*, 2007) examined early-stage PHA granule distribution in different strains using transmission electron microscopy (TEM) and confocal laser scanning fluorescence microscopy (CLSM) of cells stained with the fluorescent lipophilic dye Nile Red, or using CLSM coupled with fluorescence from a phasin–EGFP fusion. These experiments showed that the granules were located close to the cytoplasmic membrane, at or near the cell poles. Independently, Peters and Rehm demonstrated that newly emerging PHA granules formed at the cell poles via fluorescence microscopy (FM) using a PHA synthase–GFP fusion (Peters and Rehm, 2005).

However, Tian *et al.* observed via TEM of *R. eutropha* H16 cells that newly emerging granules arose only from the centre of the cell, at unknown dark-staining 'mediation elements', which might serve as scaffolding initiation sites for PHB granule formation (Tian *et al.*, 2005b). Mumtaz *et al.* (2011) also observed centrally located, dark staining mediation elements at early-stage PHBV inclusion formation in *Comamonas* sp. EB172.

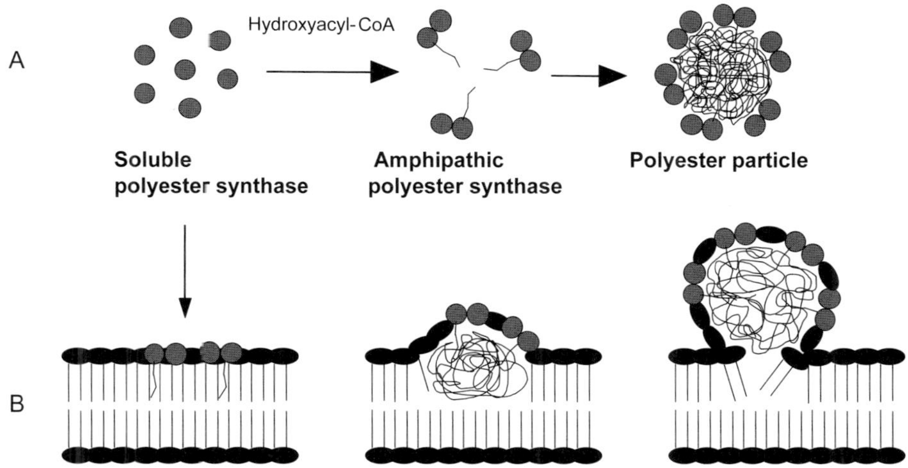

Figure 1.9 Models for polyester bead self-assembly. (a) Micelle model and (b) Budding model. (Reproduced from Rehm, 2007.)

Recent experiments with *Pseudomonas putida* KT2442 showed the nucleoid-associated phasin protein PhaF drives intracellular segregation of PHA granules, ensuring their equal distribution among daughter cells (Galán *et al.*, 2011). Later, Pfeiffer *et al.* discovered a similar phenomenon in *R. eutropha* H16: coordination between PhaM, the phasin PhaP5, and the PHB synthase PhaC1 determines the number, size, subcellular localization, and distribution of PHB granules to daughter cells (Pfeiffer and Jendrossek, 2011).

Taken together, along with the finding that PHA granule location at cell poles was dependent on nucleoid structure in recombinant *E. coli* (Peters and Rehm, 2005), these studies suggest that nucleoid structures may be the 'central mediation elements' to which PHA granules are anchored via phasin or phasin-like proteins. However, this anchoring is unnecessary for the initiation of PHA granule assembly, as granules can form not only in recombinant host strains lacking phasin proteins, but also in *in vitro* systems containing only PHA synthase and substrate (Gerngross and Martin, 1995).

Structure of PHA granules and the granule associated proteins

PHA granules consist of a hydrophobic polyester core coated with proteins and lipids. (However, as discussed above, the presence of lipids in native granules is currently unclear; for example, membrane composition and presence varied with PHA types, strains, and imaging approaches (Sudesh *et al.*, 2002).) It has been reported that water molecules exist in the PHA core *in vivo*, acting as plasticizers to prevent crystallization (Horowitz and Sanders, 1994). Major constituent proteins have been identified and designated to four categories: PHA synthases, depolymerases, phasins and regulatory proteins.

Depolymerases

The depolymerases attached to PHA granules are intracellular enzymes responsible for PHA degradation via thiolysis (Uchino *et al.*, 2008). Studies of PHA depolymerases are limited, but in *R. eutropha* several have been discovered and are designated PhaZ1-PhaZ (Saegusa *et al.*, 2001; York *et al.*, 2003). Presence of the PhaZ1 depolymerase on the surface of native PHA granules was demonstrated by Western blot with anti-PhaZ1 antibodies (Pötter and Steinbüchel, 2005). Another granule-associated protein, the 3-hydroxybutyrate oligomer hydrolase PhaY, is also involved in depolymerization (Haruhisa *et al.*, 2002).

Phasins

Phasins are low-molecular-weight proteins ranging from 11 to 25 kDa, and are the most abundantly produced granule-associated proteins, representing as much as 5% of total cellular protein (Hanley *et al.*, 1999; Wieczorek *et al.*, 1995). The four phasin homologues identified in *R. eutropha* are designated *phaP1–phaP4* (Pötter and Steinbüchel, 2005). PhaP1 is the most extensively characterized phasin protein and is known to form a triangular trimer (Neumann *et al.*, 2008).

Phasins are non-covalently associated with PHA granules through hydrophobic interactions with the polyester core and are nonessential for PHA accumulation. However, gene deletion experiments in *R. eutropha* showed absence of phasin lowers the rate of PHB production, reducing yield up to 50% compared with wild-type (York *et al.*, 2002). Phasins appear to play a role in controlling granule size and number, as cells defective in phasin production tend form a single large bead which occupies most cell space, rather than many small granules as seen in wild-type cells (Hanley *et al.*, 1999). Phasins in *P. putida* KT2442 and *R. eutropha* H16 have also been implicated in localizing PHA granules to the nucleoid, thus mediating distribution of granules among daughter cells (Galán *et al.*, 2011; Pfeiffer and Jendrossek, 2011).

Regulatory proteins

According to mutagenesis studies performed in *R. eutropha,* both granule synthesis and phasin (PhaP) production are tightly controlled by the transcriptional regulator PhaR (York *et al.*, 2002). Experiments in *R. eutropha* and *P. denitrificans* suggest the following autoregulation model. Under non-PHA accumulating conditions, cytoplasmic PhaR binds to the *phaP* promoter and inhibits its transcription. However, once PHA accumulation begins, PhaR binds to the hydrophobic surfaces of newly formed PHA granules, causing the cytoplasmic concentration of PhaR to drop. The *phaP* gene is thus derepressed, and PhaP protein begins to accumulate on the PHA granules. After the granules have reached maximum size, their surface is completely covered with PhaP phasins, preventing additional binding of PhaR. This results in an increase of cytoplasmic PhaR, which binds to the *phaP* promoter region and prevents transcription (Grage *et al.*, 2009; Pötter *et al.*, 2002; York *et al.*, 2002).

PhaR and other regulatory proteins bind non-covalently to the PHA granules (Rehm, 2006). It has been recently suggested PhaR may also play a role in other metabolic pathways, acting as a global PHA-repressor (Grage *et al.*, 2009). Other bead-associated proteins with regulatory functions have also been reported, such as PhaF from pseudomonads (Rehm, 2006).

Production of tailor-made functionalized biopolyester nanoparticles

The properties of biodegradability, biocompatibility, and production from renewable carbon sources make PHA a desirable alternative to petrochemical-based plastics. Imperial

Chemical Industries (ICI) first investigated *Azotobacter* spp., *Methylobacterium* spp. and *R. eutropha* as PHB production strains in the 1980s (Byrom, 1987; Luzier, 1992). High yields and low substrate cost led to commercial use of *R. eutropha* to produce 'Biopol' for shampoo bottles, films, coatings and packaging materials (Luzier, 1992). In the quest for cheaper production and varied material properties, PHA biosynthesis genes have been cloned and expressed in alternative hosts such as *E. coli* and *Pseudomonas* sp. bacteria (Schubert *et al.*, 1988; Slater *et al.*, 1988; Steinbüchel and Schubert, 1989), *Arabidopsis thaliana* plants (Nawrath *et al.*, 1994), *Spodopter frugiperda* insect cells (Williams *et al.*, 1996), and *Saccharomyces cerevisiae* yeasts (Leaf *et al.*, 1996). This expanded commercial interest led to an increased range of products and applications (Chen, 2009).

Isolation of PHA from host cells involves lysing the biomass, filtering the cell lysate to remove cellular debris, and precipitating the PHA with cold methanol or ethanol. PHA_{MCL} is soluble in acetone, allowing separation of PHA_{MCL} and PHA_{SCL} if required (Yao *et al.*, 1999). Cell lysis typically involves either enzymes such as lysozyme, or harsh solvents such as chloroform or methylene chloride; both methods can be used together to improve purity (Chen *et al.*, 2001; Choi and Lee, 1999; Furrer *et al.*, 2007). However, the aggressive chemicals used can cause polyester and protein degradation, and use of solvents can generate high production costs. Thus, the method chosen will depend on the purity and properties required of the end-product.

Despite much effort to develop economical large-scale production, PHAs are still more expensive to produce than oil-based plastics. Therefore, use of PHAs is primarily limited to specialty areas requiring their unique properties. For example, the biodegradability and biocompatibility of PHAs led to their use in a wide-range of medical applications such as cardiovascular patches and stents, bone and tissue scaffolds, drug delivery via implants, tablets, and microparticles (Williams and Martin, 2005; Xiong *et al.*, 2010). In 2007 the FDA approved suture material made of PH4B (TephaFLEX® Absorbable Suture) (Martin and Williams, 2003) and recently PHB polymers have been used for regeneration of nerve cells (Hazari *et al.*, 1999; Wang *et al.*, 2010).

One interesting feature of PHA is its natural production as nano-/micro-scale spherical granules. Nano-/micro-particles have a wide-range of applications in molecular biology and medicine, including drug and gene delivery, fluorescent labelling, separation of biological molecules and cells, and tissue engineering (Salata, 2004). Particles/beads can be made from a wide range of materials, including proteins, polymers, and inorganic compounds (Peek *et al.*, 2008). Proteins or chemicals can be incorporated into the particle to enable different functionalities, such as a drug core with a polymer coating for slow release properties (Huang *et al.*, 2012; Mailänder, 2009; Xiong *et al.*, 2010). The production of bioseparation resins consisting of recombinant proteins chemically conjugated to synthesized polymer beads is a multi-billion-dollar industry for biomedical research and pharmaceutical production.

In most cases, production of functionalized polymeric particles requires a minimum three-step process: (1) the polymer must be produced and purified; (2) the active ingredient, often a protein, must be produced and purified; and (3) the two components must be assembled, usually via chemical cross-linking or emulsification. Each of these processes can be costly, prone to variability, and require use and removal of toxic agents.

Recently, a method was developed that exploits the natural process of PHA granule formation to produce functionalized nano-/micro-beads. Genetic fusions of target proteins to

PHA-associated proteins such as phasins or synthases result in PHA particles naturally displaying the target protein on their surface (Fig. 1.10). This *in vivo* method for synthesizing functionalized particles has the clear advantage of one-step production, as the PHA beads and functional proteins are produced and cross-linked together in the same cell. However, the purification process for these beads needs to be gentler than the current commercial PHA extraction process to retain functionality (Atwood and Rehm, 2009; Jahns and Rehm, 2009; Peters and Rehm, 2008). The vast range of fusion proteins that can be produced in this way creates numerous applications for these PHA particles (Fig. 1.11). Some of these applications are discussed below.

Bioseparation

PHB beads displaying the ZZ domain of protein A from *Staphylococcus aureus* as the result of N-terminal fusion to PhaC have demonstrated a robust bioseparation power for purifying IgG from serum samples, with a binding capacity similar to commercial protein-A sepharose (Brockelbank *et al.*, 2006; Lewis and Rehm, 2009). Initially generated in recombinant *E. coli*, similar beads were recently produced in recombinant *Lactococcus lactis* (Mifune *et al.*, 2009). Intriguingly, the *L. lactis* produced beads were significantly smaller and had a higher binding capacity for IgG.

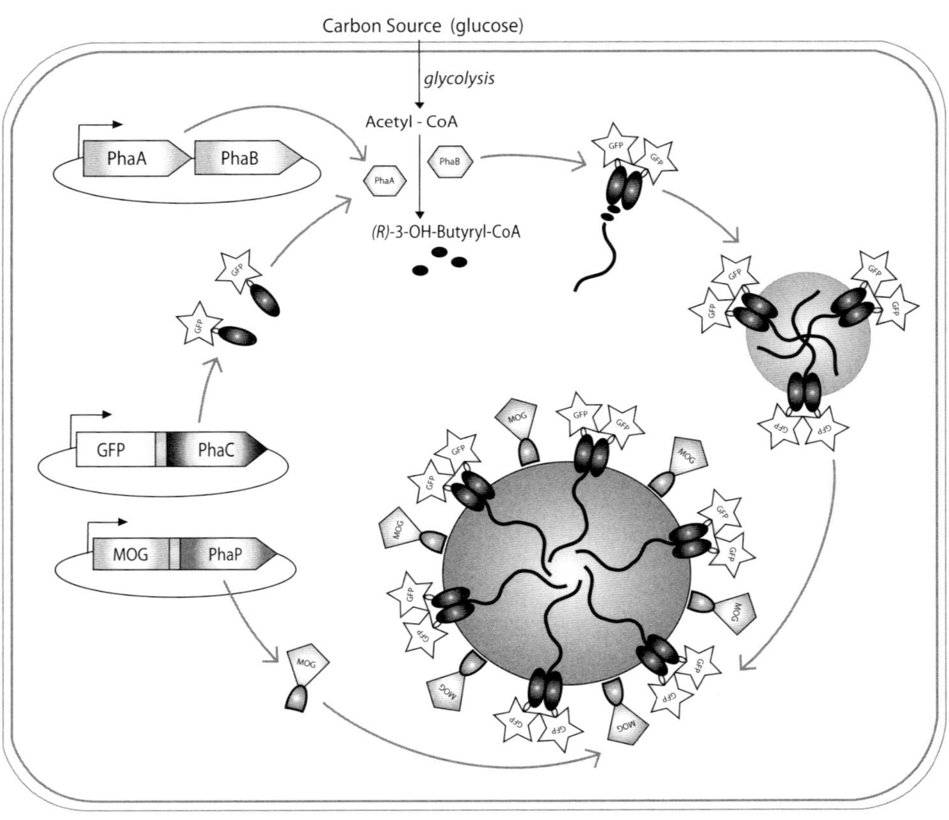

Figure 1.10 Formation of functionalized biopolyester granules from recombinant fusion proteins. (Reproduced from Draper and Rehm, 2012.)

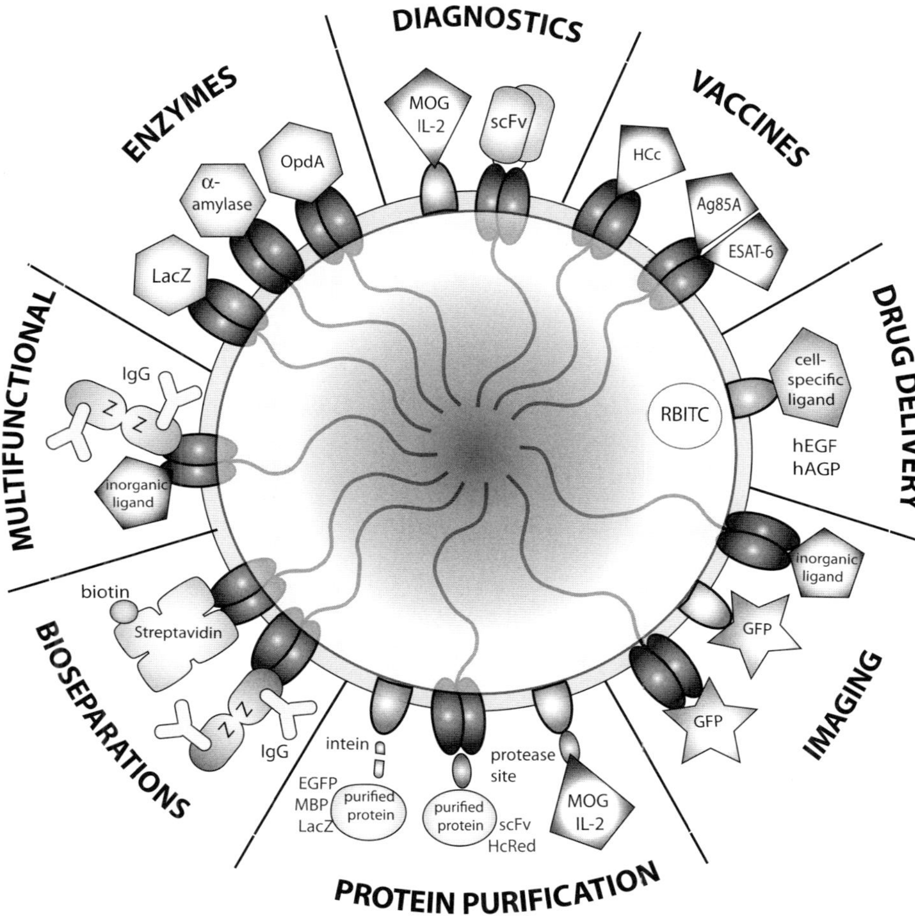

Figure 1.11 Experimentally demonstrated applications of functionalized biobeads. (Modified from Draper and Rehm, 2012.)

Similarly, a fusion of streptavidin to PhaC recombinantly expressed in *E. coli* generated PHB beads capable of binding biotin. The streptavidin-displaying beads were able to purify biotinylated antibodies and DNA (Peters and Rehm, 2008).

Recently, PHB beads capable of removing endotoxin from a liquid sample were developed. A fusion of human lipopolysaccharide binding protein (hLBP) to PhaP was produced in the yeast *Pichia pastoris* GS115, and the resulting rhLBP–PhaP fusion protein was immobilized *in vitro* on PHB particles via the natural hydrophobic interaction between PhaP and PHB. PHB beads thus coated could separate endotoxin from liquid samples with an efficiency of over 90% (Li *et al.*, 2011).

Drug delivery

While polymers such as polylactic acid (PLA) and poly(lactic-co-glycolic acid) (PLGA) have been used for drug delivery for many years, PHA has received less interest. Release of antibiotics from PHBV rods or microparticles has been investigated (Gurselt *et al.*; Sendil *et al.*, 1999), and recently an alternative production host, *Bacillus cereus* SPV, was used to produce PHA which was processed into microspheres containing gentamicin (Francis *et al.*, 2011).

Several experiments have demonstrated sustained chemical release from PHA particles. In one study, nanoparticles made of PHB, PHBHHx or PLA were loaded with rhodamine B isothiocyanate (RBITC) *in vitro* and targeted to macrophages; results showed RBITC was released for at least 20 days and macrophage viability was maintained (Xiong *et al.*, 2010). Likewise, PHA nanoparticles loaded with the phosphoinositide-3-kinases (PI3Ks) inhibitor TGX221 blocked proliferation of cancer cell lines, suggesting a potential application for PHB particles in cancer therapy (Lu *et al.*, 2011). A recent study showed promising drug release profiles using microspheres comprising PHB, cellulose acetate phthalate, and the anti-cancer drug 5-fluorouracil (5-FU) (Chaturvedi *et al.*, 2011). Animal studies have confirmed sustained drug delivery *in vivo*: PHBHHx nanoparticles loaded with an insulin phospholipid complex showed long-lasting insulin release in STZ diabetic rats (Peng *et al.*, 2012).

Targeted drug delivery has also been investigated. The cell-specific ligands human epidermal growth factor (hEGF) and mannosylated human 1-acid glycoprotein (hAGP) were recombinantly expressed as PhaP fusions and bound *in vitro* to separately generated rhodamine B isothiocyanate (RBITC)-loaded PHBHHx nanoparticles. These beads were taken up by macrophages and targeted to hepatocellular carcinoma cells both *in vitro* and *in vivo*, demonstrating the viability of functionalized PHA beads as a drug-delivery system (Yao *et al.*, 2008).

Protein purification

PHA bead-based protein purification strategies have utilized fusions of the protein of interest to the PHA bead-associated phasin, regulatory, or synthase proteins.

In the system developed by Banki *et al.* (2005), a phasin–intein–target protein fusion is expressed in PHA-producing *E. coli*. The phasin acts as a PHA affinity tag, anchoring the fusion protein to the PHA granule as it forms *in vivo*. The intein functions as self-cleaving tag to release target protein from the PHA granules after isolation. Maltose binding protein (MBP) and β-galactosidase (LacZ) were successfully purified by this method (Banki *et al.*, 2005). A similar strategy was used to purify β-galactosidase from *R. eutropha* (Barnard *et al.*, 2005).

A slightly different approach used PhaP as an *in vitro* PHA granule affinity tag: a PhaP-intein–target protein fusion recombinantly expressed in *E. coli* could be captured by manufactured PHA beads (Wang *et al.*, 2008). This method was recently shown to function using the regulatory protein PhaR instead of PhaP (Zhang *et al.*, 2010).

In 2011, Grage *et al.* developed an alternative method by fusing the target protein to PhaC, separated by an enterokinase cleavage site. This method was used to purify functional HcRed (a fluorescent protein), and anti-β-galactosidase single chain variable fragment (scFv) proteins. The use of PhaC synthase as affinity tag avoids the leakage shortcoming of previous approaches; the PhaC synthase remains covalently attached to the PHB particle, whereas the weaker attachment of PhaP/PhaR via hydrophobic interactions can result in contamination of the purified protein with the affinity partner (Grage *et al.*, 2011).

Enzyme immobilization

In the fields of pharmaceutical and fine chemical synthesis enzymes are becoming the catalytic agent of choice (Novick and Rozzell, 2005). Immobilization of enzymes onto a solid support facilitates separation from the product, often enhances stability, and allows for reuse of the expensive reaction catalyst (Katchalski-Katzir, 1993; Novick and Rozzell, 2005; Sheldon, 2007). The drawbacks to immobilization include some loss of enzyme activity and the high cost of enzyme purification and chemical immobilization (Katchalski-Katzir, 1993; Novick and Rozzell, 2005). The loss in activity is counterbalanced by increased stability and ease of reuse (Novick and Rozzell, 2005).

The first immobilization of a functional enzyme onto PHA granules was demonstrated by fusion of the β-galactosidase gene *lacZ* to the PHA synthase from *P. aeruginosa* PAO1 (Peters and Rehm, 2006). The enzyme beads were highly functional and stable in storage for several months, comparable to free β-galactosidase.

Enhanced enzymatic properties of thermostable α-amylase (BLA) were achieved by fusion to the N-terminus of PhaC from *R. eutropha* (Rasiah and Rehm, 2009). Immobilized BLA had a K_m of 5 M and a V_{max} of 506 mU/mg of bead protein, similar to the free enzyme. However, the temperature optimum for the immobilized enzyme increased from 55°C to 75°C and it remained stable up to 85°C. The pH optimum was also shifted from pH 5.6 to 8.0. These changes gave the PHA-immobilized BLA superior properties for use in starch liquefaction processes.

Recently, the OP hydrolase (OpdA) from *Agrobacterium radiobacter* was fused to the C-terminus of the *R. eutropha* PhaC via a designed linker (Blatchford *et al.*, 2011). OpdA PHA granules isolated from *E. coli* displayed functional enzyme with k_{cat} reduced 16.5 fold to 139/s and K_m increased 1.6 fold to 2.5×10^{-4} compared with free OpdA. Additionally, OpdA beads degraded organophosphorus pesticides (OPs) in the demanding environmental conditions of wool scour effluent, demonstrating their applicability in bioremediation.

Additionally, a single genetic fusion of the PhaA–PhaB–PhaC PHA synthesis pathway was shown to produce PHA granules in *E. coli*, demonstrating the ability to construct PHA beads recapitulating an entire enzymatic pathway (Mullaney and Rehm, 2010).

Although these examples show the ability of PHA granules to display functional enzymes, the potential relevance of enzyme immobilization to PHA granules for pharmaceutical and fine chemical synthesis is yet to be realized.

Diagnostics

PHB beads have been used for flow cytometry-based diagnostics. Hybrid genes encoding either the mouse interleukin 2 (IL-2) or the myelin oligodendrocyte glycoprotein (MOG) fused via an enterokinase site to the phasin PhaP were produced in recombinant PHB-producing *E. coli*. The resulting beads could be used to detect anti-IL-2 or anti-MOG antibodies using fluorescence-activated cell sorting (FACS); additionally, these beads retained functionality up to a year in storage (Bäckström *et al.*, 2007).

Another approach developed multifunctional beads: co-expression of PhaC-GFP and MOG–PhaP fusions generated fluorescent beads recognized by anti-MOG and anti-GFP antibodies (Atwood and Rehm, 2009). Such beads could be used for imaging and testing of diagnostic antibodies.

The recent success in producing PHB beads expressing a functional single-chain antibody variable fragment (scFv) in *E. coli* (Grage and Rehm, 2008) also has clear implications for diagnostics, suggesting that tailor-made diagnostic beads displaying antibody fragments against target antigens could be produced to order.

Imaging

Fluorescent PHB granules have been created by many groups, utilizing both PhaC and PhaP fusions to fluorescent proteins such as GFP and YFP (Banki *et al.*, 2005; Peters *et al.*, 2007; Peters and Rehm, 2005; Wang *et al.*, 2008).

Many of the PHB beads exhibiting bioseparation or diagnostic functions can also be used for imaging, especially if combined with an imaging agent such as fluorescence or inorganic contrast agents. Multifunctional beads displaying the MOG antigen combined with GFP have already been produced (Atwood and Rehm, 2009). Additionally, PHB beads displaying binding sites for both antibodies and inorganic gold or silica particles were recently developed (Jahns *et al.*, 2008). Antibody-mediated targeted delivery of an inorganic contrast agent by these beads could be used for bioimaging in medical applications.

Vaccine delivery

Over the past few decades, there has been increased interest in developing biodegradable nano-/micro-particles as vaccine delivery vehicles. Particulate vaccines target antigen-presenting cells and enhance cellular immune responses. Size and surface properties greatly influence the outcome of vaccination (Soppimath *et al.*, 2001). A range of polymers such as PLA and PLGA are often used, but the use of PHAs had not been investigated until recently.

PHB beads displaying hepatitis C core antigen (HCc) (Parlane *et al.*, 2011) and mycobacterial antigens antigen-85A (Ag85A) and 6 kDa early secretory antigenic target (ESAT6) (Parlane *et al.*, 2009, 2012) were recently demonstrated to act as safe and effective vaccines in mice. Initially, genetic fusions of these antigens to the PhaC synthase of *R. eutropha* were expressed recombinantly in *E. coli* (Parlane *et al.*, 2009), generating PHB vaccine particles naturally coated in the target antigen. However, the potential presence of endotoxin contaminants in *E. coli*-produced beads limited their potential as human vaccines. Thus, *L. lactis* was used as an alternative production host (Parlane *et al.*, 2011; Parlane *et al.*, 2012) due to its lack of endotoxin, previous use as a production host for recombinant proteins, and history of safe use for a range of human foods and products (Villatoro-Hernandez *et al.*, 2011). Antigen-specific cellular immune responses (Parlane *et al.*, 2011; Parlane *et al.*, 2012; Parlane *et al.*, 2009) and protection against tuberculosis (Parlane *et al.*, 2012) were observed in mice immunized with these PHB vaccine beads. These results suggest that PHB beads represent an emerging field for vaccinology.

Conclusion

The formation of biopolyesters in both naturally PHA-producing and recombinant bacteria has been extensively investigated. Many of the proteins involved in PHA synthesis, from synthases to regulators, have been cloned and characterized. The generation of PHA precursor molecules through multiple biosynthesis pathways has also been extensively investigated by many groups. Metabolic and genetic engineering approaches have demonstrated the ability to produce tailor-made biopolymers with unique material properties. However, despite much effort, biopolyesters remain more expensive to produce than conventional petroleum-based plastics. Thus, commercial use of biopolyesters is currently limited to high value niche applications which benefit from their unique properties, such as biomedical products requiring biocompatible and biodegradable materials.

Recently, the natural production of PHA granules as nano-/micro-scale beads coated with functional proteins has been exploited to produce functionalized beads. Similar functional beads are currently employed as bioseparation resins and represent a multi-billion-dollar industry; however their production is a costly multi-step process typically involving both chemical synthesis and recombinant protein production. In contrast, the one-step bacterial manufacture of functional PHA granules suitable for bioseparation, bioremediation, enzymatic catalysis, diagnostics, targeted drug delivery, and vaccine delivery provides an attractive alternative to conventional bead technologies. This new technology has great promise for cost-effective production of tailor-made functional nano-/micro-particles.

Acknowledgements

This work was supported by grants from Massey University and Polybatics Ltd.

References

Amara, A.A., and Rehm, B.H.A. (2003). Replacement of the catalytic nucleophile cysteine-296 by serine in class II polyhydroxyalkanoate synthase from *Pseudomonas aeruginosa*-mediated synthesis of a new polyester: identification of catalytic residues. Biochem. J. *374*, 413–421.

Amara, A.A., Steinbüchel, A., and Rehm, B.H.A. (2002). *In vivo* evolution of the *Aeromonas punctata* polyhydroxyalkanoate (PHA) synthase: isolation and characterization of modified PHA synthases with enhanced activity. Appl. Microbiol. Biotechnol. 59, 477–482.

Anderson, A.J., and Dawes, E.A. (1990). Occurrence, metabolism, metabolic role, and industrial uses of bacterial polyhydroxyalkanoates. Microbiol. Rev. *54*, 450–472.

Anderson, A.J., Haywood, G.W., and Dawes, E.A. (1990). Biosynthesis and composition of bacterial poly(hydroxyalkanoates). Int. J. Biol. Macromol. *12*, 102–105.

Antonio, R.V., Steinbüchel, A., and Rehm, B.H. (2000). Analysis of in vivo substrate specificity of the PHA synthase from *Ralstonia eutropha*: formation of novel copolyesters in recombinant *Escherichia coli*. FEMS Microbiol. Lett. *182*, 111–117.

Atwood, J.A., and Rehm, B.H.A. (2009). Protein engineering towards biotechnological production of bifunctional polyester beads. Biotechnol. Lett. *31*, 131–137.

Bäckström, B.T., Brockelbank, J.A., and Rehm, B.H.A. (2007). Recombinant *Escherichia coli* produces tailor-made biopolyester granules for applications in fluorescence activated cell sorting: functional display of the mouse interleukin-2 and myelin oligodendrocyte glycoprotein. BMC Biotechnol. 7, 3.

Banki, M.R., Gerngross, T.U., and Wood, D.W. (2005). Novel and economical purification of recombinant proteins: Intein-mediated protein purification using in vivo polyhydroxybutyrate (PHB) matrix association. Protein Sci. *14*, 1387–1395.

Barnard, G.C., McCool, J.D., Wood, D.W., and Gerngross, T.U. (2005). Integrated recombinant protein expression and purification platform based on *Ralstonia eutropha* integrated recombinant protein expression and purification platform based on *Ralstonia eutropha*. Society *71*, 5735–5742.

Bhubalan, K., Chuah, J.-A., Shozui, F., Brigham, C.J., Taguchi, S., Sinskey, A.J., Rha, C., and Sudesh, K. (2011). Characterization of the highly active polyhydroxyalkanoate synthase of *Chromobacterium* sp. strain USM2. Appl. Environ. Microbiol. *77*, 2926–2933.

Blatchford, P.A., Scott, C., French, N., and Rehm, B.H.A. (2011). Immobilization of organophosphohydrolase OpdA from *Agrobacterium radiobacter* by overproduction at the surface of polyester inclusions inside engineered *Escherichia coli*. Biotechnol. Bioeng. *109*, 1101–1108.

Brigham, C.J., Budde, C.F., Holder, J.W., Zeng, Q.D., Mahan, A.E., Rha, C., and Sinskey, A.J. (2010). Elucidation of beta-oxidation pathways in *Ralstonia eutropha* H16 by examination of global gene expression. J. Bacteriol. *192*, 5454–5464.

Brockelbank, J.A., Peters, V., and Rehm, B.H.A. (2006). Recombinant *Escherichia coli* strain produces a ZZ domain displaying biopolyester granules suitable for immunoglobulin G purification. Appl. Environ. Microbiol. *72*, 7394–7397.

Byrom, D. (1987). Polymer synthesis by micro-organisms: technology and economics. Trends Biotechnol. *5*, 246–250.

Chaturvedi, K., Kulkarni, A.R., and Aminabhavi, T.M. (2011). Blend microspheres of poly(3-hydroxybutyrate) and cellulose acetate phthalate for colon delivery of 5-fluorouracil. Ind. Eng. Chem. Res. *50*, 10414–10423.

Chen, G., Zhang, G., Park, S., and Lee, S. (2001). Industrial scale production of poly(3-hydroxybutyrate-co-3-hydroxyhexanoate). Appl. Microbiol. Biotechnol. *57*, 50–55.

Chen, G.-Q. (2009). A microbial polyhydroxyalkanoates (PHA) based bio- and materials industry. Chem. Soc. Rev. *38*, 2434–2446.

Choi, J.-i., and Lee, S.Y. (1999). Efficient and economical recovery of poly (3-Hydroxybutyrate) from recombinant *Escherichia coli* by simple digestion with chemicals. Biotechnol. Bioeng. *62*, 546–553.

Chung, A., Liu, Q., Ouyang, S.P., Wu, Q., and Chen, G.Q. (2009). Microbial production of 3-hydroxydodecanoic acid by pha operon and fadBA knockout mutant of *Pseudomonas putida* KT2442 harboring tesB gene. Appl. Microbiol. Biotechnol. *83*, 513–519.

Dennis, D., McCoy, M., Stangl, A., Valentin, H.E., and Wu, Z. (1998). Formation of poly(3-hydroxybutyrate-co-3-hydroxyhexanoate) by PHA synthase from *Ralstonia eutropha*. J. Biotechnol. *64*, 177–186.

Draper, J., and Rehm, B.H.A. (2012). Engineering bacteria to manufacture functionalized polyester beads. Bioengineered Bugs 3.

Eggink, G., Dewaard, P., and Huijberts, G.N.M. (1995). Formation of novel poly(hydroxyalkanoates) from long-chain fatty-acids. Can. J. Microbiol. *41*, 14–21.

Escapa, I.F., Morales, V., Martino, V.P., Pollet, E., Averous, L., Garcia, J.L., and Prieto, M.A. (2011). Disruption of beta-oxidation pathway in *Pseudomonas putida* KT2442 to produce new functionalized PHAs with thioester groups. Appl. Microbiol. Biotechnol. *89*, 1583–1598.

Fiedler, S., Steinbuchel, A., and Rehm, B.H. (2000). PhaG-mediated synthesis of Poly(3-hydroxyalkanoates) consisting of medium-chain-length constituents from nonrelated carbon sources in recombinant *Pseudomonas fragi*. Appl. Environ. Microbiol. *66*, 2117–2124.

Fiedler, S., Steinbüchel, A., and Rehm, B.H.A. (2002). The role of the fatty acid beta-oxidation multienzyme complex from *Pseudomonas oleovorans* in polyhydroxyalkanoate biosynthesis: molecular characterization of the fadBA operon from *P. oleovorans* and of the enoyl-CoA hydratase genes phaJ from *P. oleovorans* Archiv. Microbiol. *178*, 149–160.

Francis, L., Meng, D., Knowles, J., Keshavarz, T., Boccaccini, A.R., and Roy, I. (2011). Controlled delivery of gentamicin using poly(3-hydroxybutyrate) microspheres. Int. J. Mol. Sci. *12*, 4294–4314.

Füchtenbusch, B., Wullbrandt, D., and Steinbüchel, A. (2000). Production of polyhydroxyalkanoic acids by *Ralstonia eutropha* and *Pseudomonas oleovorans* from an oil remaining from biotechnological rhamnose production. Appl. Microbiol. Biotechnol. 53, 167–172.

Fukui, T., and Doi, Y. (1997). Cloning and analysis of the poly(3-hydroxybutyrate-co-3-hydroxyhexanoate) biosynthesis genes of *Aeromonas caviae*. J. Bacteriol. *179*, 4821–4830.

Furrer, P., Panke, S., and Zinn, M. (2007). Efficient recovery of low endotoxin medium-chain-length poly([R]-3-hydroxyalkanoate) from bacterial biomass. J. Microbiol. Methods *69*, 206–213.

Galán, B., Dinjaski, N., Maestro, B., de Eugenio, L.I., Escapa, I.F., Sanz, J.M., García, J.L., and Prieto, M.a. (2011). Nucleoid-associated PhaF phasin drives intracellular location and segregation of polyhydroxyalkanoate granules in *Pseudomonas putida* KT2442. Mol. Microbiol. *79*, 402–418.

Gerngross, T.U., and Martin, D.P. (1995). Enzyme-catalysed synthesis of poly[(R)-(–)-3-hydroxybutyrate]: formation of macroscopic granules in vitro. Proc. Natl. Acad. Sci. U.S.A. *92*, 6279–6283.

Gerngross, T.U., Reilly, P., Stubbe, J., Sinskey, A.J., and Peoples, O.P. (1993). Immunocytochemical analysis of poly-beta-hydroxybutyrate (PHB) synthase in *Alcaligenes eutrophus* H16: localization of the synthase enzyme at the surface of PHB granules. J. Bacteriol. *175*, 5289–5293.

Gerngross, T.U., Snell, K.D., Peoples, O.P., Sinskey, A.J., Csuhai, E., Masamune, S., and Stubbe, J. (1994). Overexpression and purification of the soluble polyhydroxyalkanoate synthase from *Alcaligenes eutrophus*: evidence for a required posttranslational modification for catalytic activity. Biochemistry 33, 9311–9320.

Grage, K., Jahns, A.C., Parlane, N., Palanisamy, R., Rasiah, I.A., Atwood, J.A., and Rehm, B.H.A. (2009). Bacterial polyhydroxyalkanoate granules: biogenesis, structure, and potential use as nano-/micro-beads in biotechnological and biomedical applications. Biomacromolecules *10*, 660–669.

Grage, K., Peters, V., and Rehm, B.H.A. (2011). Recombinant protein production by in vivo polymer inclusion display. Appl. Environ. Microbiol. 77, 6706–6709.

Grage, K., and Rehm, B.H.A. (2008). In vivo production of scFv-displaying biopolymer beads using a self–assembly-promoting fusion partner. Bioconj. Chem. *19*, 254–262.

Griebel, R.J., and Merrick, J.M. (1971). Metabolism of poly- -hydroxybutyrate: effect of mild alkaline extraction on native poly- -hydroxybutyrate granules. J. Bacteriol. *108*, 782–789.

Griebel, R.J., Smith, Z., and Merrick, J.M. (1968). Metabolism of poly-beta-hydroxybutyrate. I. Purification, composition, and properties of native poly-beta-hydroxybutyrate granules from *Bacillus megaterium*. Biochemistry *7*, 3676–3681.

Gurselt, I., Yagmurlu, F., Korkusuz, F., and Hasirci, V. (2002). In vitro antibiotic release from poly(3-hydroxybutyrate-co-3-hydroxyvalerate) rods. J. Microencaps. *19*, 153–164.

Hanley, S.Z., Pappin, D.J., Rahman, D., White, A.J., Elborough, K.M., and Slabas, A.R. (1999). Re-evaluation of the primary structure of *Ralstonia eutropha* phasin and implications for polyhydroxyalkanoic acid granule binding. FEBS Lett. *447*, 99–105.

Haruhisa, S., Mari, S., and Terumi, S. (2002). Cloning of an intracellular D(–)-3-hydroxybutyrate-oligomer hydrolase gene from *Ralstonia eutropha* H16 and identification of the active site serine residue by site-directed mutagenesis. J. Biosci. Bioeng. *94*, 106–112.

Hazari, A., Wiberg, M., Johansson-Rudén, G., Green, C., and Terenghi, G. (1999). A resorbable nerve conduit as an alternative to nerve autograft in nerve gap repair. Br. J. Plast. Surg. *52*, 653–657.

Hiltunen, J.K., and Qin, Y.M. (2000). Beta-oxidation – strategies for the metabolism of a wide variety of acyl-CoA esters. Biochim. Biophys. Acta. Mol. Cell Biol. Lipids *1484*, 117–128.

Hisano, T., Tsuge, T., Fukui, T., Iwata, T., Miki, K., and Doi, Y. (2003). Crystal structure of the (R)-specific enoyl-CoA hydratase from *Aeromonas caviae* involved in polyhydroxyalkanoate biosynthesis. J. Biol. Chem. *278*, 617–624.

Hoffmann, N., and Rehm, B.H.A. (2004). Regulation of polyhydroxyalkanoate biosynthesis in *Pseudomonas putida* and *Pseudomonas aeruginosa*. FEMS Microbiol. Lett. *237*, 1–7.

Hoffmann, N., and Rehm, B.H.A. (2005). Nitrogen-dependent regulation of medium-chain length polyhydroxyalkanoate biosynthesis genes in pseudomonads. Biotechnol. Lett. *27*, 279–282.

Holler, E., Angerer, B., Achhammer, G., Miller, S., and Windisch, C. (1992). Biological and biosynthetic properties of poly-L-malate. FEMS Microbiol. Lett. *103*, 109–118.

Hoppensack, A., Rehm, B.H., and Steinbüchel, A. (1999). Analysis of 4-phosphopantetheinylation of polyhydroxybutyrate synthase from *Ralstonia eutropha*: generation of beta-alanine auxotrophic Tn5 mutants and cloning of the panD gene region. J. Bacteriol. *181*, 1429–1435.

Horowitz, D.M., and Sanders, J.K.M. (1994). Amorphous, biomimetic granules of polyhydroxybutyrate: preparation, characterization, and biological implications. J. Am. Chem. Soc. *116*, 2695–2702.

Huang, Z.W., Laurent, V., Chetouani, G., Ljubimova, J.Y., Holler, E., Benvegnu, T., Loyer, P., and Cammas-Marion, S. (2012). New functional degradable and bio-compatible nanoparticles based on poly(malic acid) derivatives for site-specific anti-cancer drug delivery. Int. J. Pharmaceut. *423*, 84–92.

Hume, A.R., Nikodinovic-Runic, J., and O'Connor, K.E. (2009). FadD from *Pseudomonas putida* CA-3 Is a true long-chain fatty acyl coenzyme A synthetase that activates phenylalkanoic and alkanoic acids. J. Bacteriol. *191*, 7554–7565.

Hyakutake, M., Saito, Y., Tomizawa, S., Mizuno, K., and Tsuge, T. (2011). Polyhydroxyalkanoate (PHA) synthesis by class IV PHA synthases employing *Ralstonia eutropha* PHB(–)4 as host strain. Biosci. Biotechnol. Biochem. *75*, 1615–1617.

Ishikawa, M., Mikami, Y., Usukura, J., Iwasaki, H., Shinagawa, H., and Morikawa, K. (1997). Reconstitution, morphology and crystallization of a fatty acid beta-oxidation multienzyme complex from *Pseudomonas fragi*. Biochem. J. *328*, 815–820.

Jahns, A.C., Haverkamp, R.G., and Rehm, B.H.A. (2008). Multifunctional inorganic-binding beads self-assembled inside engineered bacteria. Bioconj. Chem. *19*, 2072–2080.

Jahns, A.C., and Rehm, B.H.A. (2009). Tolerance of the *Ralstonia eutropha* class I polyhydroxyalkanoate synthase for translational fusions to its C terminus reveals a new mode of functional display. Appl. Environ. Microbiol. *75*, 5461–5466.

Jendrossek, D. (2005). Fluorescence microscopical investigation of poly(3-hydroxybutyrate) granule formation in bacteria. Biomacromolecules *6*, 598–603.

Jendrossek, D., Selchow, O., and Hoppert, M. (2007). Poly(3-hydroxybutyrate) granules at the early stages of formation are localized close to the cytoplasmic membrane in Caryophanon latum. Appl. Environ. Microbiol. *73*, 586–593.

Jia, Y., Kappock, T.J., Frick, T., Sinskey, A.J., and Stubbe, J. (2000). Lipases provide a new mechanistic model for polyhydroxybutyrate (PHB) synthases: characterization of the functional residues in *Chromatium vinosum* PHB synthase. Biochemistry *39*, 3927–3936.

Jia, Y., Yuan, W., Wodzinska, J., Park, C., Sinskey, A.J., and Stubbe, J. (2001). Mechanistic studies on class I polyhydroxybutyrate (PHB) synthase from *Ralstonia eutropha*: class I and III synthases share a similar catalytic mechanism. Biochemistry *40*, 1011–1019.

John, M.E., and Keller, G. (1996). Metabolic pathway engineering in cotton: Biosynthesis of polyhydroxybutyrate in fibre cells. Proc. Natl. Acad. Sci. U.S.A. *93*, 12768–12773.

Katchalski-Katzir, E. (1993). Immobilized enzymes – learning from past successes and failures. Trends Biotechnol. *11*, 471–478.

Kawashima, Y., Cheng, W., Mifune, J., Orita, I., Nakamura, S., and Fukui, T. (2012). Characterization and functional analyses of R-specific enoyl coenzyme A hydratases in polyhydroxyalkanoate-producing *Ralstonia eutropha*. Appl. Environ. Microbiol. *78*, 493–502.

Kessler, B., Weusthuis, R., Witholt, B., and Eggink, G. (2001). Production of microbial polyesters: fermentation and downstream processes. Advanc. Biochem. Eng./Biotechnol. *71*, 159–182.

Kichise, T., Taguchi, S., and Doi, Y. (2002). Enhanced accumulation and changed monomer composition in polyhydroxyalkanoate (PHA) copolyester by in vitro evolution of *Aeromonas caviae* PHA synthase. Appl. Environ. Microbiol. *68*, 2411–2419.

Klinke, S., Ren, Q., Witholt, B., and Kessler, B. (1999). Production of medium-chain-length poly(3-hydroxyalkanoates) from gluconate by recombinant *Escherichia coli*. Appl. Environ. Microbiol. *65*, 540–548.

Kozhevnikov, I.V., Volova, T.G., Hai, T., and Steinbüchel, A. (2010). Cloning and molecular organization of the polyhydroxyalkanoic acid synthase gene (phaC) of *Ralstonia eutropha* strain B5786. Appl. Biochem. Microbiol. *46*, 140–147.

Lambov, N., Peikov, P., Belcheva, N., Gesheva, K., Zlatkov, A., and Danchev, N. (1997). Biodegradable cross-linked prodrug of bronchial dilator vephylline.1. Mechanism of the release. J. Controlled Release *45*, 193–197.

Leaf, T.A., Peterson, M.S., Stoup, S.K., Somers, D., and Srienc, F. (1996). *Saccharomyces cerevisiae* expressing bacterial polyhydroxybutyrate synthase produces poly-3-hydroxybutyrate. Microbiology *142*, 1169–1180.

Lemoigne, M. (1925). É'tudes sur l'autolyse microbienne acidification par formation d'acide-oxybutyrique. Annales de l'Institut Pasteur *39*, 144–173.

Lewis, J.G., and Rehm, B.H.A. (2009). ZZ polyester beads: an efficient and simple method for purifying IgG from mouse hybridoma supernatants. J. Immunol. Meth. *346*, 71–74.

Li, J., Shang, G., You, M., Peng, S., Wang, Z., Wu, H., and Chen, G.-Q. (2011). Endotoxin removing method based on lipopolysaccharide binding protein and polyhydroxyalkanoate binding protein PhaP. Biomacromolecules *12*, 602–608.

Li, P., Chakraborty, S., and Stubbe, J. (2009). Detection of covalent and noncovalent intermediates in the polymerization reaction catalysed by a C149S class III polyhydroxybutyrate synthase. Biochemistry *48*, 9202–9211.

Liebergesell, M., Schmidt, B., and Steinbüchel, A. (1992). Isolation and identification of granule-associated proteins relevant for poly(3-hydroxyalkanoic acid) biosynthesis in *Chromatium vinosum* D. FEMS Microbiol. Lett. *78*, 227–232.

Liebergesell, M., Sonomoto, K., Madkour, M., Mayer, F., and Steinbüchel, A. (1994). Purification and characterization of the poly(hydroxyalkanoic acid) synthase from *Chromatium vinosum* and localization of the enzyme at the surface of poly(hydroxyalkanoic acid) granules. Eur. J. Biochem. /FEBS *226*, 71–80.

Liebergesell, M., and Steinbüchel, A. (1992). Cloning and nucleotide sequences of genes relevant for biosynthesis of poly(3-hydroxybutyric acid) in *Chromatium vinosum* strain D. Eur. J. Biochem. /FEBS *209*, 135–150.

Liu, F., Li, W., Ridgway, D., Gu, T., and Shen, Z. (1998). Production of poly-β-hydroxybutyrate on molasses by recombinant *Escherichia coli*. Biotechnol. Lett. *20*, 345–348.

Lu, Q., Han, J., Zhou, L., Zhou, J., and Xiang, H. (2008). Genetic and biochemical characterization of the poly(3-hydroxybutyrate-co-3-hydroxyvalerate) synthase in *Haloferax mediterranei*. J. Bacteriol. *190*, 4173–4180.

Lu, X., Zhang, J., Wu, Q., and Chen, G.Q. (2003). Enhanced production of poly(3-hydroxybutyrate-co-3-hydroxyhexanoate) via manipulating the fatty acid beta-oxidation pathway in E-coli. FEMS Microbiol. Lett. *221*, 97–101.

Lu, X.-Y., Ciraolo, E., Stefenia, R., Chen, G.-Q., Zhang, Y., and Hirsch, E. (2011). Sustained release of PI3K inhibitor from PHA nanoparticles and in vitro growth inhibition of cancer cell lines. Appl. Microbiol. Biotechnol. *89*, 1423–1433.

Lütke-Eversloh, T., Bergander, K., Luftmann, H., and Steinbüchel, A. (2001). Identification of a new class of biopolymer: bacterial synthesis of a sulfur-containing polymer with thioester linkages. Microbiology (Reading, England) *147*, 11–19.

Luzier, W.D. (1992). Materials derived from biomass/biodegradable materials. Proc. Natl. Acad. Sci. U.S.A. *89*, 839–842.

Ma, L., Zhang, H.F., Liu, Q., Chen, J., Zhang, J., and Chen, G.Q. (2009). Production of two monomer structures containing medium-chain-length polyhydroxyalkanoates by beta-oxidation-impaired mutant of *Pseudomonas putida* KT2442. Biores. Technol. *100*, 4891–4894.

Mailänder, V., and Landfester, K. (2009). Interaction of nanoparticles with cells. Biomacromolecules *10*, 2379–2400.

Martin, D.P., and Williams, S.F. (2003). Medical applications of poly-4-hydroxybutyrate: a strong flexible absorbable biomaterial. Biochem. Eng. J. *16*, 97–105.

Matsumoto, K.i., Aoki, E., Takase, K., Doi, Y., and Taguchi, S. (2006). In vivo and in vitro characterization of Ser477X mutations in polyhydroxyalkanoate (PHA) synthase 1 from Pseudomonas sp. 61–3: effects of beneficial mutations on enzymatic activity, substrate specificity, and molecular weight of PHA. Biomacromolecules *7*, 2436–2442.

Matsumoto, K.i., Takase, K., Aoki, E., Doi, Y., and Taguchi, S. (2005). Synergistic effects of Glu130Asp substitution in the type II polyhydroxyalkanoate (PHA) synthase: enhancement of PHA production and alteration of polymer molecular weight. Biomacromolecules *6*, 99–104.

Matsumoto, K.i., Takase, K., Yamamoto, Y., Doi, Y., and Taguchi, S. (2009). Chimeric enzyme composed of polyhydroxyalkanoate (PHA) synthases from *Ralstonia eutropha* and *Aeromonas caviae* enhances production of PHAs in recombinant *Escherichia coli*. Biomacromolecules *10*, 682–685.

Matsusaki, H., Manji, S., Taguchi, K., Kato, M., Fukui, T., and Doi, Y. (1998). Cloning and molecular analysis of the Poly(3-hydroxybutyrate) and poly(3-hydroxybutyrate-co-3-hydroxyalkanoate) biosynthesis genes in *Pseudomonas* sp. strain 61-3. J. Bacteriol. *180*, 6459–6467.

McCool, G.J., and Cannon, M.C. (1999). Polyhydroxyalkanoate inclusion body-associated proteins and coding region in *Bacillus megaterium*. J. Bacteriol. *181*, 585–592.

Mifune, J., Grage, K., and Rehm, B.H. (2009). Production of functionalized biopolyester granules by recombinant *Lactococcus lactis*. Appl. Environ. Microbiol. *75*, 4668–4675.

Mullaney, J.A., and Rehm, B.H.A. (2010). Design of a single-chain multi-enzyme fusion protein establishing the polyhydroxybutyrate biosynthesis pathway. J. Biotechnol. *147*, 31–36.

Mumtaz, T., Abd-Aziz, S., Rahman, N.A.A., Yee, P.L., Wasoh, H., Shirai, Y., and Hassan, M.A. (2011). Visualization of core–shell PHBV granules of wild type Comamonas sp. EB172 in vivo under transmission electron microscope. Int. J. Polym. Anal. Charact. *16*, 228–238.

Nakashita, H., Arai, Y., Yoshioka, K., Fukui, T., Doi, Y., Usami, R., Horikoshi, K., and Yamaguchi, I. (1999). Production of biodegradable polyester by a transgenic tobacco. Biosci. Biotechnol. Biochem. *63*, 870–874.

Nawrath, C., Poirier, Y., and Somerville, C. (1994). Targeting of the polyhydroxybutyrate biosynthetic pathway to the plastids of *Arabidopsis* thaliana results in high levels of polymer accumulation. Proc. Natl. Acad. Sci. U.S.A. *91*, 12760–12764.

Neumann, L., Spinozzi, F., Sinibaldi, R., Rustichelli, F., Pötter, M., and Steinbüchel, A. (2008). Binding of the major phasin, PhaP1, from *Ralstonia eutropha* H16 to poly(3-hydroxybutyrate) granules. J. Bacteriol. *190*, 2911–2919.

Normi, Y.M., Hiraishi, T., Taguchi, S., Abe, H., Sudesh, K., Najimudin, N., and Doi, Y. (2005). Characterization and properties of G4X mutants of *Ralstonia eutropha* PHA synthase for poly(3-hydroxybutyrate) biosynthesis in *Escherichia coli*. Macromol. Biosci. *5*, 197–206.

Novick, S.J., and Rozzell, J.D. (2005). Immobilization of enzymes by covalent attachment. In Microbial Enzymes and Biotransformations, José Luis Barredo, ed. (New Jersey: Humana Press), pp. 247–271.

Ouyang, S.P., Luo, R.C., Chen, S.S., Liu, Q., Chung, A., Wu, Q., and Chen, G.Q. (2007). Production of polyhydroxyalkanoates with high 3-hydroxydodecanoate monomer content by fadB and fadA knockout mutant of *Pseudomonas putida* KT2442. Biomacromolecules *8*, 2504–2511.

Park, S.J., Choi, J.i., and Lee, S.Y. (2005). Engineering of *Escherichia coli* fatty acid metabolism for the production of polyhydroxyalkanoates. Enzyme Microbial Technol. *36*, 579–588.

Park, S.J., and Lee, S.Y. (2003). Identification and characterization of a new enoyl coenzyme A hydratase involved in biosynthesis of medium-chain-length polyhydroxyalkanoates in recombinant *Escherichia coli*. J. Bacteriol. *185*, 5391–5397.

Park, S.J., Lee, S.Y., and Yup Lee, S. (2004). New FadB homologous enzymes and their use in enhanced biosynthesis of medium-chain-length polyhydroxyalkanoates in FadB mutant *Escherichia coli*. Biotechnol. Bioeng. *86*, 681–686.

Parlane, N.A., Grage, K., Lee, J.W., Buddle, B.M., Denis, M., and Rehm, B.H.A. (2011). Production of a particulate hepatitis C vaccine candidate by an engineered *Lactococcus lactis* strain. Appl. Environ. Microbiol. *77*, 8516–8522.

Parlane, N.A., Grage, K., Mifune, J., Basaraba, R.J., Wedlock, D.N., Rehm, B.H.A., and Buddle, B.M. (2012). Vaccines displaying mycobacterial proteins on biopolyester beads stimulate cellular immunity and induce protection against tuberculosis. Clin. Vaccine. Immunol.*19*, 37–44.

Parlane, N.A., Wedlock, D.N., Buddle, B.M., and Rehm, B.H.A. (2009). Bacterial polyester inclusions engineered to display vaccine candidate antigens for use as a novel class of safe and efficient vaccine delivery agents. Appl. Environ. Microbiol. *75*, 7739–7744.

Pawar, S., and Schulz, H. (1981). The structure of the multienzyme complex of fatty acid oxidation from *Escherichia coli*. J. Biol. Chem. *256*, 3894–3899.

Peek, L.J., Middaugh, C.R., and Berkland, C. (2008). Nanotechnology in vaccine delivery. Advanced Drug Delivery Rev. *60*, 915–928.

Peng, Q., Zhang, Z.-R., Gong, T., Chen, G.-Q., and Sun, X. (2012). A rapid-acting, long-acting insulin formulation based on a phospholipid complex loaded PHBHHx nanoparticles. Biomaterials *33*, 1583–1588.

Peoples, O.P., and Sinskey, A.J. (1989a). Poly-beta-hydroxybutyrate (PHB) biosynthesis in *Alcaligenes eutrophus* H16. Identification and characterization of the PHB polymerase gene (phbC). J. Biol. Chem. *264*, 15298–15303.

Peoples, O.P., and Sinskey, A.J. (1989b). Poly-β-hydroxybutyrate biosynthesis in *Alcaligenes eutrophus* H16. Characterization of the genes encoding β-ketothiolase and acetoacetyl-CoA reductase. J. Biol. Chem. *264*, 15293–15297.

Peters, V., Becher, D., and Rehm, B.H.A. (2007). The inherent property of polyhydroxyalkanoate synthase to form spherical PHA granules at the cell poles: the core region is required for polar localization. J. Biotechnol. *132*, 238–245.

Peters, V., and Rehm, B.H.A. (2005). In vivo monitoring of PHA granule formation using GFP-labelled PHA synthases. FEMS Microbiol. Lett. *248*, 93–100.

Peters, V., and Rehm, B.H.A. (2006). In vivo enzyme immobilization by use of engineered polyhydroxyalkanoate synthase. Appl. Environ. Microbiol. *72*, 1777–1783.

Peters, V., and Rehm, B.H.A. (2008). Protein engineering of streptavidin for in vivo assembly of streptavidin beads. J. Biotechnol. *134*, 266–274.

Pfeiffer, D., and Jendrossek, D. (2011). Interaction between poly(3-hydroxybutyrate) granule-associated proteins as revealed by two-hybrid analysis and identification of a new phasin in *Ralstonia eutropha* H16. Microbiology (Reading, England) *157*, 2795–2807.

Portilla-Arias, J., Patil, R., Hu, J., Ding, H., Black, K.L., García-Alvarez, M., Muoz-Guerra, S., Ljubimova, J.Y., and Holler, E. (2010). Nanoconjugate platforms development based in poly(B,L-Malic Acid) methyl esters for tumour drug delivery. J. Nanomat. doi:10.1155/2010/825363.

Portilla-Arias, J.A., García-Alvarez, M., Martínez De Ilarduya, A., and Muñoz-Guerra, S. (2007). Ionic complexes of biosynthetic poly(malic acid) and poly(glutamic acid) as prospective drug-delivery systems. Macromol. Biosci. *7*, 897–906.

Pötter, M., Madkour, M.H., Mayer, F., and Steinbüchel, A. (2002). Regulation of phasin expression and polyhydroxyalkanoate (PHA) granule formation in *Ralstonia eutropha* H16. Microbiology (Reading, England) *148*, 2413–2426.

Pötter, M., and Steinbüchel, A. (2005). Poly(3-hydroxybutyrate) granule-associated proteins: impacts on poly(3-hydroxybutyrate) synthesis and degradation. Biomacromolecules *6*, 552–560.

Pramanik, A., Pawar, S., Antonian, E., and Schulz, H. (1978). Five different enzymatic activities are associated with the multienzyme complex of fatty acid oxidation from *Escherichia coli*. J. Bacteriol. *137*, 469–473.

Qi, Q., Rehm, B.H., and Steinbüchel, A. (1997). Synthesis of poly(3-hydroxyalkanoates) in *Escherichia coli* expressing the PHA synthase gene phaC2 from *Pseudomonas aeruginosa*: comparison of PhaC1 and PhaC2. FEMS Microbiol. Lett. *157*, 155–162.

Qi, Q., Steinbüchel, A., and Rehm, B.H. (1998). Metabolic routing towards polyhydroxyalkanoic acid synthesis in recombinant *Escherichia coli* (fadR): inhibition of fatty acid beta-oxidation by acrylic acid. FEMS Microbiol. Lett. *167*, 89–94.

Ramsay, J.A., Hassan, M.-C.A., and Ramsay, B.A. (1995). Hemicellulose as a potential substrate for production of poly(β-hydroxyalkanoates). Can. J. Microbiol. *41*, 262–266.

Rasiah, I.A., and Rehm, B.H. (2009). One-step production of immobilized alpha-amylase in recombinant *Escherichia coli*. Appl. Environ. Microbiol. *75*, 2012–2016.

Rathberger, K., Reisner, H., Willibald, B., Molitoris, H.-P., and Holler, E. (1999). Comparative synthesis and hydrolytic degradation of poly (L-malate) by myxomycetes and fungi. Mycological Research *103*, 513–520.

Rehm, B.H.A. (2003). Polyester synthases: natural catalysts for plastics. Biochem. J. *376*, 15–33.

Rehm, B.H.A. (2006). Genetics and biochemistry of polyhydroxyalkanoate granule self-assembly: The key role of polyester synthases. Biotechnol. Lett. *28*, 207–213.

Rehm, B.H.A. (2007). Biogenesis of microbial polyhydroxyalkanoate granules: a platform technology for the production of tailor-made bioparticles. Curr. Issues Mol. Biol. *9*, 41–62.

Rehm, B.H.A. (2010). Bacterial polymers: biosynthesis, modifications and applications. Nat. Rev. Microbiol. *8*, 578–592.

Rehm, B.H.A., Antonio, R.V., Spiekermann, P., Amara, A.A., and Steinbüchel, A. (2002). Molecular characterization of the poly(3-hydroxybutyrate) (PHB) synthase from *Ralstonia eutropha*: in vitro evolution, site-specific mutagenesis and development of a PHB synthase protein model. Biochim. biophys. Acta *1594*, 178–190.

Rehm, B.H.A., Krüger, N., and Steinbüchel, A. (1998). A new metabolic link between fatty acid de novo synthesis and polyhydroxyalkanoic acid synthesis. J. Biol. Chem. *273*, 24044–24051.

Ren, Q., Sierro, N., Witholt, B., and Kessler, B. (2000). FabG, an NADPH-dependent 3-ketoacyl reductase of *Pseudomonas aeruginosa*, provides precursors for medium-chain-length poly-3-hydroxyalkanoate biosynthesis in *Escherichia coli*. J. Bacteriol. *182*, 2978–2981.

Reusch, R., Hiske, T., and Sadoff, H. (1987). Cellular incorporation of poly-β-hydroxybutyrate into plasma membranes of *Escherichia coli* and Azotobacter vinelandii alters native membrane structure. Can. J. Microbiol./Rev. Can. Microbiol. *33*, 435–444.

Reusch, R.N. (1989). Poly-β-hydroxybutyrate/calcium polyphosphate complexes in eukaryotic membranes. Proc. Soc. Exp. Biol. Med. *191*, 377–381.

Reusch, R.N. (1992). Biological complexes of poly-β-hydroxybutyrate. FEMS Microbiol. Lett. *103*, 119–129.

Reusch, R.N., Bryant, E.M., and Henry, D.N. (2003). Increased poly-(R)-3-hydroxybutyrate concentrations in streptozotocin (STZ) diabetic rats. Acta Diabetol. *40*, 91–94.

Reusch, R.N., and Sadoff, H.L. (1988). Putative structure and function of a poly-β-hydroxybutyrate/calcium polyphosphate channel in bacterial plasma membranes. Proc. Natl. Acad. Sci. U.S.A. *85*, 4176–4180.

Reusch, R.N., Sparrow, A.W., and Gardiner, J. (1992). Transport of poly-β-hydroxybutyrate in human plasma. Biochim. Biophys. Acta Lipids Lipid Metab. *1123*, 33–40.

Ruth, K., Grubelnik, A., Hartmann, R., Egli, T., Zinn, M., and Ren, Q. (2006). Efficient production of (R)-3-hydroxycarboxylic acids by biotechnological conversion of polyhydroxyalkanoates and their purification. Biomacromolecules *8*, 279–286.

Saegusa, H., Shiraki, M., Kanai, C., and Saito, T. (2001). Cloning of an intracellular poly[D(−)-3-hydroxybutyrate] depolymerase gene from *Ralstonia eutropha* H16 and characterization of the gene product. J. Bacteriol. *183*, 94–100.

Salata, O.V. (2004). Applications of nanoparticles in biology and medicine. J. Nanobiotechnol. *2*, 3.

Schubert, P., Steinbüchel, A., and Schlegel, H.G. (1988). Cloning of the *Alcaligenes eutrophus* genes for synthesis of poly-beta-hydroxybutyric acid (PHB) and synthesis of PHB in *Escherichia coli*. J. Bacteriol. *170*, 5837–5847.

Seebach, D., and Fritz, M.G. (1999). Detection, synthesis, structure, and function of oligo(3-hydroxyalkanoates): Contributions by synthetic organic chemists. Int. J. Biol. Macromol. *25*, 217–236.

Sendil, D., Gürsel, I., Wise, D.L., and Hasirci, V. (1999). Antibiotic release from biodegradable PHBV microparticles. Journal of controlled release. Off. J. Controlled Rel. Soc. *59*, 207–217.

Sheldon, R.A. (2007). Enzyme immobilization: the quest for optimum performance. Advanc. Synth. Catal. *349*, 1289–1307.

Sheu, D.-S., and Lee, C.-Y. (2004). Altering the substrate specificity of polyhydroxyalkanoate synthase 1 derived from *Pseudomonas putida* GPo1 by localized semirandom mutagenesis. J. Bacteriol. *186*, 4177–4184.

Slater, S.C., Voige, W.H., and Dennis, D.E. (1988). Cloning and expression in *Escherichia coli* of the *Alcaligenes eutrophus* H16 poly-beta-hydroxybutyrate biosynthetic pathway. J. Bacteriol. *170*, 4431–4436.

Snell, K.D., Feng, F., Zhong, L., Martin, D., and Madison, L.L. (2002). YfcX enables medium-chain-length poly(3-hydroxyalkanoate) formation from fatty acids in recombinant *Escherichia coli* fadB strains. J. Bacteriol. *184*, 5696–5705.

Soberón-Chávez, G., Aguirre-Ramírez, M., and Sánchez, R. (2005). The *Pseudomonas aeruginosa* RhlA enzyme is involved in rhamnolipid and polyhydroxyalkanoate production. J. Ind. Microbiol. Biotechnol. *32*, 675–677.

Solaiman, D.K.Y., and Ashby, R.D. (2005). Genetic characterization of the poly(hydroxyalkanoate) synthases of various *Pseudomonas oleovorans* strains. Curr. Microbiol. *50*, 329–333.

Soppimath, K.S., Aminabhavi, T.M., Kulkarni, A.R., and Rudzinski, W.E. (2001). Biodegradable polymeric nanoparticles as drug delivery devices. Journal of controlled release: official journal of the Controlled Release Society *70*, 1–20.

Steinbüchel, A. (1991). Polyhydroxyalkanoic acids. In Biomaterials, D. Byrom, ed. (Macmillan), pp. 123–213.

Steinbüchel, A., Füchtenbusch, B., Gorenflo, V., Hein, S., Jossek, R., Langenbach, S., and Rehm, B.H.A. (1998). Biosynthesis of polyesters in bacteria and recombinant organisms. Polymer Degradation and Stability *59*, 177–182.

Steinbüchel, A., and Schubert, P. (1989). Expression of the *Alcaligenes eutrophus* poly(β-hydroxybutyric acid)-synthetic pathway in *Pseudomonas* sp. Archiv. Microbiol. *153*, 101–104.

Steinbüchel, A., and Valentin, H.E. (1995). Diversity of bacterial polyhydroxyalkanoic acids. FEMS Microbiol. Lett. *128*, 219–228.

Stubbe, J., and Tian, J. (2003). Polyhydroxyalkanoate (PHA) homeostasis: the role of the PHA synthase. Natural Product Reports *20*, 445.

Sudesh, K., Gan, Z., Maehara, A., and Doi, Y. (2002). Surface structure, morphology and stability of polyhydroxyalkanoate inclusions characterized by atomic force microscopy. Polymer Degradation and Stability *77*, 77–85.

Sudesh, K., and Iwata, T. (2008). Sustainability of biobased and biodegradable plastics. CLEAN – Soil, Air, Water *36*, 433–442.

Thomson, N., and Sivaniah, E. (2010). Synthesis, properties and uses of bacterial storage lipid granules as naturally occurring nanoparticles. Soft Matter *17*, 4045–4057.

Tian, J., Sinskey, A.J., and Stubbe, J. (2005a). Detection of intermediates from the polymerization reaction catalyzed by a D302A mutant of class III polyhydroxyalkanoate (PHA) synthase. Biochemistry *44*, 1495–1503.

Tian, J., Sinskey, A.J., and Stubbe, J. (2005b). Kinetic studies of polyhydroxybutyrate granule formation in *Wautersia eutropha* H16 by transmission electron microscop. Society *187*.

Tsuge, T., Hisano, T., Taguchi, S., and Doi, Y. (2003a). Alteration of chain length substrate specificity of *Aeromonas caviae* R-enantiomer-specific enoyl-coenzyme A hydratase through site-directed mutagenesis. Appl. Environ. Microbiol. *69*, 4830–4836.

Tsuge, T., Saito, Y., Narike, M., Muneta, K., Normi, Y.M., Kikkawa, Y., Hiraishi, T., and Doi, Y. (2004). Mutation effects of a conserved alanine (Ala510) in type I polyhydroxyalkanoate synthase from *Ralstonia eutropha* on polyester biosynthesis. Macromol. Biosci. *4*, 963–970.

Tsuge, T., Taguchi, K., Taguchi, S., and Doi, Y. (2003b). Molecular characterization and properties of (R)-specific enoyl-CoA hydratases from *Pseudomonas aeruginosa*: metabolic tools for synthesis of polyhydroxyalkanoates via fatty acid beta-oxidation. Int. J. Biol. Macromol. *31*, 195–205.

Tsuge, T., Watanabe, S., Shimada, D., Abe, H., Doi, Y., and Taguchi, S. (2007). Combination of N149S and D171G mutations in *Aeromonas caviae* polyhydroxyalkanoate synthase and impact on polyhydroxyalkanoate biosynthesis. FEMS Microbiol. Lett. *277*, 217–222.

Uchino, K., Saito, T., and Jendrossek, D. (2008). Poly(3-hydroxybutyrate) (PHB) depolymerase PhaZa1 is involved in mobilization of accumulated PHB in *Ralstonia eutropha* H16. Appl. Environ. Microbiol. *74*, 1058–1063.

Valentin, H., Schönebaum, A., and Steinbüchel, A. (1996). Identification of 5-hydroxyhexanoic acid, 4-hydroxyheptanoic acid and 4-hydroxyoctanoic acid as new constituents of bacterial polyhydroxyalkanoic acids. J. Bacteriol. *46*, 261–267.

Villatoro-Hernandez, J., Montes-de-Oca-Luna, R., and Kuipers, O.P. (2011). Exp. Opin. Biol. Ther. *11*, 261–267.

Vo, M.T., Lee, K.W., Jung, Y.M., and Lee, Y.H. (2008). Comparative effect of overexpressed phaJ and fabG genes supplementing (R)-3-hydroxyalkanoate monomer units on biosynthesis of mcl-polyhydroxyalkanoate in *Pseudomonas putida* KCTC1639. J. Biosci. Bioeng. *106*, 95–98.

Vo, M.T., Lee, K.W., Kim, T.K., and Lee, Y.H. (2007). Utilization of fadA knockout mutant *Pseudomonas putida* for overproduction of medium chain-length-polyhydroxyalkanoate. Biotechnol. Lett. *29*, 1915–1920.

Wang, L., Wang, Z.-H., Shen, C.-Y., You, M.-L., Xiao, J.-F., and Chen, G.-Q. (2010). Differentiation of human bone marrow mesenchymal stem cells grown in terpolyesters of 3-hydroxyalkanoates scaffolds into nerve cells. Biomaterials *31*, 1691–1698.

Wang, Q., and Nomura, C.T. (2010). Monitoring differences in gene expression levels and polyhydroxyalkanoate (PHA) production in *Pseudomonas putida* KT2440 grown on different carbon sources. J. Biosci. Bioeng. *110*, 653–659.

Wang, Z., Wu, H., Chen, J., Zhang, J., Yao, Y., and Chen, G.-Q. (2008). A novel self-cleaving phasin tag for purification of recombinant proteins based on hydrophobic polyhydroxyalkanoate nanoparticles. Lab on a Chip *8*, 1957–1962.

Wieczorek, R., Pries, A., Steinbüchel, A., Mayer, F., Wieczorek, R., Pries, A., and Steinbu, A. (1995). Analysis of a 24-kilodalton protein associated with the polyhydroxyalkanoic acid granules in *Alcaligenes eutrophus*. J. Bacteriol *177*, 2425–2435.

Williams, M.D., Rahn, J.A., and Sherman, D.H. (1996). Production of a polyhydroxyalkanoate biopolymer in insect cells with a modified eucaryotic fatty acid synthase. Appl. Environ. Microbiol. *62*, 2540–2546.

Williams, S.F., and Martin, D.P. (2005). Applications of PHAs in Medicine and Pharmacy. In Biopolymers Online (Wiley-VCH Verlag GmbH & Co. KGaA).

Wodzinska, J., Snell, K.D., Rhomberg, A., Sinskey, A.J., Biemann, K., and Stubbe, J. (1996). Polyhydroxybutyrate synthase: evidence for covalent catalysis. J. Am. Chem. Soc. *118*, 6319–6320.

Xiong, Y.-C., Yao, Y.-C., Zhan, X.-Y., and Chen, G.-Q. (2010). Application of polyhydroxyalkanoates nanoparticles as intracellular sustained drug-release vectors. J. Biomat. Sci. A *21*, 127–140.

Yao, J., Zhang, G., Wu, Q., Chen, G.Q., and Zhang, R. (1999). Production of polyhydroxyalkanoates by *Pseudomonas nitroreducens*. Antonie van Leeuwenhoek *75*, 345–349.

Yao, Y.-C., Zhan, X.-Y., Zhang, J., Zou, X.-H., Wang, Z.-H., Xiong, Y.-C., Chen, J., and Chen, G.-Q. (2008). A specific drug targeting system based on polyhydroxyalkanoate granule binding protein PhaP fused with targeted cell ligands. Biomaterials *29*, 4823–4830.

Yoncheva, K., Miloshev, S., Belcheva, N., and Lambov, N. (2001). Development and characterization of cross-linked poly(malate) microspheres with dipyridamole. Int. jJ Pharmaceut. *226*, 31–37.

York, G.M., Lupberger, J., Tian, J., Lawrence, A.G., Stubbe, J., and Sinskey, A.J. (2003). *Ralstonia eutropha* H16 encodes two and possibly three intracellular Poly[D-(–)-3-hydroxybutyrate] depolymerase genes. J. Bacteriol. *185*, 3788–3794.

York, G.M., Stubbe, J., and Sinskey, A.J. (2002). The *Ralstonia eutropha* PhaR protein couples synthesis of the PhaP phasin to the presence of polyhydroxybutyrate in cells and promotes polyhydroxybutyrate production. J. Bacteriol. *184*, 59–66.

Yuan, W., Jia, Y., Tian, J., Snell, K.D., Müh, U., Sinskey, A.J., Lambalot, R.H., Walsh, C.T., and Stubbe, J. (2001). Class I and III polyhydroxyalkanoate synthases from *Ralstonia eutropha* and *Allochromatium vinosum*: characterization and substrate specificity studies. Arch. Biochem. Biophys. *394*, 87–98.

Zhang, S., Wang, Z.H., and Chen, G.Q. (2010). Microbial polyhydroxyalkanote synthesis repression protein PhaR as an affinity tag for recombinant protein purification. Microbial Cell Factories *9*, 28.

Zheng, Z., Li, M., Xue, X.-J., Tian, H.-L., Li, D., Chen, G.-Q. (2006). Mutation on N-terminus of polyhydroxybutyrate synthase of *Ralstonia eutropha* enhaHB accumulation. Appl. Microbiol. Biotechnol. *72*, 896–905.

Self-assembly and Application of Cellulosomal Components

2

Daniel B. Fried, Sarah Moraïs, Qi Xu, Shi-You Ding, John O. Baker, Yannick J. Bomble, Michael E. Himmel and Edward A. Bayer

Abstract

Cellulosomes are modular, super-molecular enzyme systems secreted by anaerobic bacteria to degrade recalcitrant plant cell wall polysaccharides to simple sugars. The components of this molecular machine include a manifold of enzymatic units and carbohydrate-binding modules, as well as the cohesin and dockerin modules responsible for the system's unique self-assembly and connectivity. Known to be one of the highest affinity protein–protein interactions discovered to date, the cohesin–dockerin interaction is also the key to engineering the cellulosome. By mixing and matching the spare parts of the cellulosome and connecting them via cohesins and dockerins, biotechnologists have begun pursuing the concept of 'designer cellulosomes', which one day may be an important contributor to production of sustainable biomass-derived fuels. By incorporating molecules from other systems into the cellulosome paradigm, nanotechnologists have begun to harness the potential of this molecular construction kit to create diverse, self-assembling nanostructures for a broad variety of biotechnological applications.

Introduction

Cellulose is the dominant component of the plant cell wall and, along with hemicellulose, pectin, and lignin, it forms a highly heterogeneous, fibrous matrix, conferring mechanical strength and protection from enzymatic and chemical degradation. Despite cellulose's prevalence, few multicellular heterotrophs have the ability to digest it directly (Linton *et al.*, 2006), and most cellulose is degraded by microorganisms. So important are these microorganisms in returning cellulosic glucose to the food cycle, that many multicellular organisms have formed specialized symbiotic relationships, enlisting them as gut flora.

For some species, the metabolic challenges of the harsh anaerobic environment, along with the recalcitrance of insoluble substrates, have resulted in the evolution of particularly elaborate and efficient enzymatic ensembles for degrading cellulosic material. These modular, multi-enzyme structures, called cellulosomes (Lamed *et al.*, 1983; Bayer *et al.*, 1994), are either presented on the cell surface as protruding organelles or secreted into the environment. A defining feature of these enzyme systems is the high-affinity cohesin–dockerin interaction, which allows the assembly of enzymes and scaffoldin modules into an almost limitless number of arrangements. There is great potential for biotechnologists to utilize and expand on this molecular assembly motif. Reengineering the cellulosome itself may one day allow cellulosic waste to be utilized efficiently as a feedstock for biofuels; even more generally, applying the cellulosome's modularity and assembly features to new fields of study could allow the creation of new kinds of nanodevices.

Discovery

The discovery of the cellulosome occurred serendipitously during research on a putative bacterial cellulose-binding factor (then termed CBF). *Clostridium thermocellum*, a ubiquitous anaerobic, thermophilic bacterium, was known to possess the ability to bind crystalline cellulose, but the identity of the bacterium's CBF was unknown. To isolate the CBF, a cellulose-based affinity selection method was employed to isolate a spontaneous adhesion-deficient *C. thermocellum* mutant (Bayer *et al.*, 1983). Normally, *C. thermocellum* binds to insoluble cellulose and degrades it to the disaccharide cellobiose, which it then directly metabolizes. Growth medium containing cellobiose would permit the survival of *C. thermocellum* mutants that are deficient in binding or degrading cellulose. By pelleting the cellulose-bound wild-type bacteria, and selecting for free bacteria, cells containing only a disabled CBF were isolated. In this experiment, *C. thermocellum* was grown under anaerobic conditions with only cellobiose as a carbon source. An insoluble form of cellulose was then added to the suspension, which served to sediment most of the bacteria. The supernatant was then used to inoculate more cellobiose-only medium. This process was repeated six times, at which point the addition of cellulose failed to sediment the bacteria from the medium. The resulting cell suspension was dispersed on solid medium under anaerobic conditions, and single colonies were isolated. One of the adhesion-deficient mutants (AD2) discovered in this way was used for further experiments. Mutant AD2 was used to generate monospecific antibodies for the CBF by the following approach. First, antibodies were raised against the surface of the wild-type *C. thermocellum* cells, creating antibodies for all surface proteins, including the CBF. These antibodies were then adsorbed onto a mass of the CBF-deficient bacteria, precipitating the cells and purging the solution of all antibodies except those for the CBF. Using these purified, monotypic antibodies, the CBF of *C. thermocellum* was found to be part of a large protein complex with a molecular weight of 2.1 million daltons, and with a diameter of approximately 18 nm. Sodium dodecyl sulphate treatment of the complex yielded 14 protein components, of which only one large fragment (210,000 daltons) was active against the CBF antibody. Many of the complex's smaller fragments had cellulosic activity. These formative experiments led to the defining of the cellulosome as a discrete, multi-subunit complex, which both binds cellulose and has cellulolytic activity (Lamed *et al.*, 1983).

Structure of the cellulosome

Cellulose-targeting enzymes

Many organisms that degrade cellulose secrete or display cellulolytic enzymes that usually contain a carbohydrate-binding module (CBM) in addition to their catalytic module. The CBM targets the enzyme to the surface of cellulose, thereby increasing its effective concentration and compensating for the modest substrate binding affinity, characteristic of many of the catalytic modules of cellulases (Fig. 2.1).

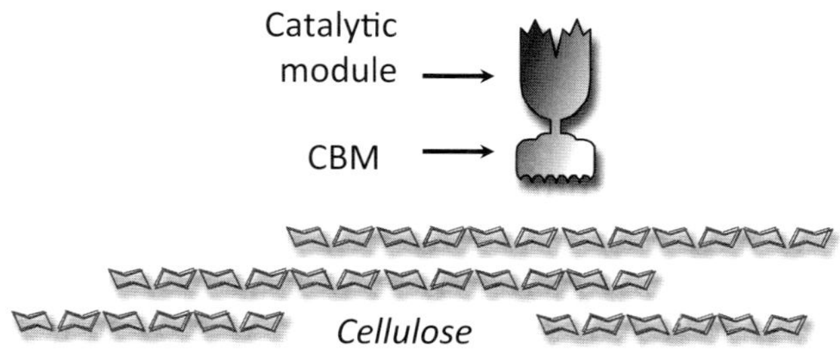

Figure 2.1 Representation of a CBM-containing modular cellulase interacting with the cellulose substrate.

Scaffoldin

The non-catalytic components of the cellulosome are called scaffoldin and contain, in single peptides, multiple copies of cohesins (and, in some cases, a single dockerin). These modules serve to organize and display the enzymatic subunits, which contain the complementary (dockerin or cohesin) binding partner. These scaffoldins include cellulase-integrating proteins (Cip) as a primary scaffoldin and, in some cases, cell-surface anchoring scaffoldins, depending on the species. The primary scaffoldin also contains in most cases a CBM, and, where appropriate, an anchoring component for binding to the cellulose and cell surfaces, respectively. Its modular components can serve as a rich source for nanotechnological application (Bayer *et al.*, 1994; Ding *et al.*, 2003b).

Dockerin

Many cohesin and dockerin sequences have been identified in microorganisms, but most originate from anaerobic bacteria. The cohesin-dockerin binding pair has an association constant among the highest in nature, and due to their different specificities, they represent a fertile source for nano-assembly components. Dockerins are about 70 amino acids in length and have structural similarity to the EF-hand loop–helix motif found in calcium binding proteins, such as calmodulin. Importantly, they contain a 22-amino acid duplicated segment.

Type I dockerins contain a pair of calcium-binding loops which stabilize the cohesin–dockerin interaction. A crystal structure of a heterodimeric complex of the two modules was revealed in 2003 (Carvalho *et al.*, 2003). Hydrophobic interactions involving these loops and portions of alpha helix-3 and alpha-helix-1 are responsible for the proper recognition between the dockerin and the cohesin. The 'recognition amino acids' that served as specificity determinants were first predicted from a combination of bioinformatics and affinity experiments (Pagés *et al.*, 1997) and then verified experimentally (Mechaly *et al.*, 2000; Schaeffer *et al.*, 2002). According to crystallographic data, serine-45 and threonine-46 on helix 3 provide an important hydrogen-bonding network between the two molecules. However, serine-10 and threonine-11, symmetrically located on the homologous face of helix 1, likely play a role in binding. Indeed, a dockerin with a disabled helix 3 permitted the crystallization of the 180 degree, reverse-binding cohesin–dockerin. In this case, helix-1 contributed primarily to the binding face (Carvalho *et al.*, 2007). Additionally, alanine scanning and truncation studies have shown that the first calcium-binding loop can be deleted with retention of specific cohesin binding properties (Karpol *et al.*, 2008). The ability of enzymes to switch from one binding mode to the other imparts structural flexibility to the cellulosome, which may be important for accessing the complex topology of crystalline cellulose. The dual binding mode of the dockerin/cohesin interaction, in addition to the dimensionally flexible nature of the scaffoldin itself, and the compositional and organizational complexity of the enzyme decoration of the scaffoldin, supports the notion that degradation of cell wall cellulose requires considerable plasticity in design (Gilbert, 2007).

Type II dockerins also require calcium for proper folding and bind with a 7 M affinity constant (Adams *et al.*, 2005). In the CipA scaffoldin, the type II dockerin at the C-terminus of the molecule is accompanied by an X-domain which stabilizes the dockerin conformation and promotes its solubility in aqueous solution. Calcium-induced folding causes the exposure of a hydrophobic face of the dockerin, which can either homodimerize with an affinity constant of 40 M, or bind to the corresponding type II cohesin.

Cohesin

Cohesins are about 140 amino acids in length, and fold into a typical jelly-roll topology consisting of a nine-stranded beta-sandwich (Shimon *et al.*, 1997a,b; Tavares *et al.*, 1997; Spinelli *et al.*, 2000). Type II cohesins have structures similar to those of type I cohesins, with the addition of an added alpha-helix between strands 6 and 7, as well as two beta-flaps that interrupt the progress of two of the beta strands (Noach *et al.*, 2005).

The cohesin–dockerin interaction

The cohesin–dockerin interaction is among the highest affinity protein–protein interactions in nature, with a K_a of greater than 10^9/M as measured by surface plasmon resonance (Fierobe *et al.*, 1999). Many different cohesins and dockerins have been identified throughout biology (Peer *et al.*, 2009). Type I and type II cohesin-dockerins have been known since the early days of cellulosome research (Leibovitz and Beguin, 1996). A more recent discovery involves the type III cohesin–dockerin interactions. These proteins are homologous to those from Sca A and Sca B of *Ruminococcus flavefaciens* (Ding *et al.*, 2001). For the most part, the cohesin–dockerin interaction is non-promiscuous, meaning they only bind one another if they are of the same type and originate from the same species. Conversely, cohesin–dockerin interactions are highly non-specific within a given type and species. For example, in *C. thermocellum*, all type I cohesin positions on the primary scaffoldin are equally likely to be filled by any *C. thermocellum* enzyme bearing a type I dockerin. This kind of binding behaviour presents biotechnologists with opportunities and challenges relating to binding specificity.

Investigation into the amino acid residues responsible for type- and species-specific recognition has employed various mutagenesis techniques. In one example, two repeated pairs of amino acids from the dockerin domains from *C. thermocellum* Cel48S and *C. cellulolyticum* Cel5A were swapped between species using site directed mutagenesis (Mechaly *et al.*, 2000). Located at the binding interface, four residues, located at positions 10 and 11 of the dockerin's duplicated calcium-binding sequence, were thought to account for species-specific cohesin-dockerin binding based on cross-species sequence alignment. In fact, the exchange of residues facilitated cross-species recognition, confirming the importance of those positions in recognizing the new cohesin. However, the self-species cohesin recognition was retained, indicating that additional binding interactions were involved in this recognition. Later, a non-rational homology-swapping approach was implemented, in which homologous stretches of amino acids from the dockerin of one species were exchanged for those of the other species (Nakar *et al.*, 2004). After analysing the binding activity between the hybrid dockerin with the cohesin of the competitive species, refinements to the original swapped domain were made, and the dockerin was tested again. After the fourth refinement, a chimeric *C. cellulolyticum* dockerin was obtained, in which mutation of only three residues succeeded in conferring 20% of the binding capacity for *C. thermocellum*. Similarly, dockerins from *C. thermocellum* have been shown to acquire cross-species binding affinity for *C. josui*, but both pairs of recognition amino acid residues must be swapped. Mutating only the first pair of recognition amino acids from those in *C. thermocellum* to those in *C. josui* did not interfere with species-specific binding; however, with the additional substitution of the second amino acid pair, the dockerin was able to recognize both species. This further demonstrated that the repeated recognition sequence is important, but not essential to species specificity (Sakka *et al.*, 2011).

A type I cohesin–dockerin complex was first crystallized in 2003 (Carvalho *et al.*, 2003), followed by a type–II complex (Adams *et al.*, 2006). The structure of the type I complex is shown in Fig. 2.2.

Cellulosomal enzymes

Cellulosomes employ an ensemble of xylanases, mannanases, pectinases, and cellulases. Most cellulosomal enzymes consist of glycoside hydrolases, which cleave the glycosidic linkage between sugar monomers, or between sugar monomers and other molecules. They can also contain polysaccharide lyases, which cleave polypeptides containing uronic acid; and carbohydrate esterases, which de-*N*-acetylate or de-*O*-acetylate sugars or cleave other acidic groups, such as ferulic acid from hemicelluloses, such as xylan. Cellulolytic enzymes can then be divided into enzymes which act on the polysaccharide termini (exo-acting) and

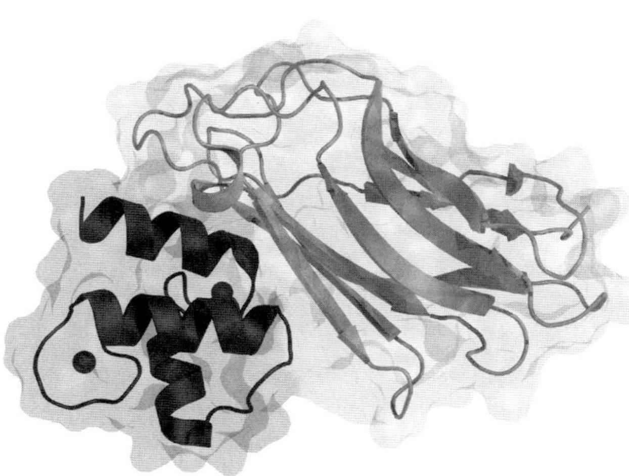

Figure 2.2 Crystal structure of a type I *C. thermocellum* cohesin–dockerin heterodimer, pdb accession code 1OHZ (Carvalho *et al.*, 2003). The dockerin is from xylanase 10B and the cohesin is the cohesin 2 from the CipA scaffoldin.

those which digest internal portions of the polysaccharide chain (endo-acting). Endo-acting enzymes tend to contain a grooved catalytic cleft, to facilitate entry of the internal portion of the polysaccharide chain, whereas exo-acting enzymes tend to contain a tunnel-shaped active site, into which the polysaccharide chain is threaded as the enzyme progresses. Cooperation between enzymes within the multienzyme complex leads to increased synergy, which is a key feature of the cellulosome function. Enzymes which are effective in the cellulosomal context are not necessarily functional in an isolated context. Cellulosomal bacteria themselves are sometimes obligated to function in concert with other bacteria. For example, despite its ability to degrade xylan, *C thermocellum* cannot utilize the sugars derived from it (Morag *et al.*, 1990). Cooperative relationships between bacteria are apparently essential components of such ecosystems. Additionally, some other non-glycoside-hydrolase proteins, such as serpins, may also be incorporated into cellulosomes, and may protect the complex against proteolysis (Zverlov *et al.*, 2005; Kang *et al.*, 2006).

CBM

Just as a wide range of cellulolytic enzymes is deployed by cellulosomes, in order to meet the challenges of their highly heterogeneous substrate, so is a wide range of CBMs. The CBMs are found on both enzymes and scaffoldins. CBMs of different families are known to bind a variety of polysaccharide substrates including: crystalline and non-crystalline types of cellulose, chitin, beta-1,2-glucans, beta-(1,3)-(1,4) mixed glucans, xylan, mannan, galactan, and starch (Boraston *et al.*, 2004). These proteins have substrate affinities in the low micromolar range, and often bind reversibly to cellulose (Families 1 and 4). Others bind essentially irreversibly (Families 2 and 3) and provide opportunities for affinity purification (Tomme *et al.*, 1998). The salient feature of the CBM binding face is the exposed aromatic side chains (Tormo *et al.*, 1996; McLean *et al.*, 2000). The CBM substrate-binding specificity depends on the content and arrangement of aromatic amino acid side chains located on the binding face. Modest binding affinities between the CBM and its substrate allow the cellulosome to process across its substrate. Of many CBM families, some recognize crystalline cellulose, as reflected in a planar binding surface composed of mostly aromatic side chains. Known as Type A CBMs, these bind to crystalline cellulose and/or chitin by employing a planar binding architecture (Lamed *et al.*, 1994; Tormo *et al.*, 1996; McLean *et al.*, 2000; Boraston *et al.*, 2004) and bind to the hydrophobic face of the cellulose chain. Other CBMs, known as Type B, contain a cleft binding surface in order to bind individual polysaccharide chains (Notenboom *et al.*, 2001; Simpson *et al.*, 2000). And other CBMs have are promiscuous binders, recognizing both crystalline and non-crystalline cellulose, or many forms of other polysaccharides (Boraston *et al.*, 2002). Both the presence of hydrophobic stacking between the CBM aromatic residues and the carbohydrate surface and water-mediated hydrogen bonding are essential for CBM binding (Pell *et al.*, 2003). The first crystal structure of a family-3 CBM (shown in Fig. 2.3) was obtained by Tormo *et al.* (1996). The mode of cellulose binding was demonstrated subsequently by electron microscopy (Lehtio *et al.*, 2003; Ding *et al.*, 2006a).

Because the active sites of cellulases are often partially hidden inside binding clefts or tunnels, they have a relatively low binding affinity for their substrate polysaccharides. To overcome this, many cellulases, whether non-cellulosomal or cellulosomal, employ CBMs in order to target their substrates. In aerobic cellulolytic organisms, the enzymes secreted to

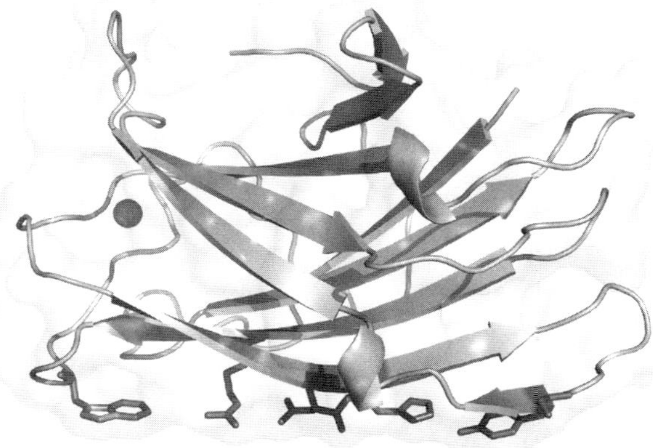

Figure 2.3 *C. thermocellum*'s Family-3 CBM, pdb accession code 1nbc (Tormo *et al.*, 1996). From left to right, residues important for cellulose binding, W118, R112, D56, H57, Y67, are shown. The CBM of *C. thermocellum* consists of a nine-stranded beta-sandwich and contains a cellulose-binding surface of aromatic and polar residues.

degrade cellulose always contain one or more catalytic modules fused to their own CBM in a single peptide. (Warren, 1996). However, in many anaerobic organisms, multiple enzymes within a cellulosome complex are targeted in concert to the cellulose surface via a common CBM on the scaffoldin subunit. In both cases, the CBM causes the enzymatic activity to be concentrated at the cellulose surface, increasing the effective concentration of the enzymes on the insoluble substrate that aids in catalysis. Some CBMs also appear to disrupt the crystalline cellulose surface, priming the cellulose surface for enzymatic activity (Bolam *et al.*, 1998). Some CBMs, in addition to targeting the cellulosome (directly) and the attached bacterium (secondarily and indirectly) to the cellulose surface, may actually help guide the carbohydrate polymer into the active site tunnel of the enzymatic units (Sakon *et al.*, 1997). This increases the catalytic potential of the enzyme by imparting to it a processive mode of action.

Cell-free cellulosomes

Cellulosomal enzymes contain dockerin modules that interact specifically with the cohesin modules of the scaffoldin subunit. Therefore, cellulosomes differ from simple cellulose binding enzymes in that they are composed of an array of enzymes, and contain multifaceted enzymatic activities. Soluble cellulosomes contain two kinds of subunits: enzymes and the scaffoldins. The enzymes are assembled on a scaffoldin subunit via the high-affinity cohesin–dockerin interactions. Although a given bacterial species employs numerous dockerins and cohesins, they display high homology within the type and species (see below). All the dockerins of a given 'type' produced by a given species can thus bind with any of the cohesins of the matching type that are on the scaffoldins produced by that species, allowing great complexity in enzyme display. However, it has been shown, using coarse grained modelling, that the specificity of cellulosomal enzymes for their cohesin partners on the scaffoldin can also be influenced by their shape and modularity (Bomble, 2011).

The scaffoldin subunit usually includes a CBM which permits substrate targeting of the multi-enzyme display system. After their release from the cell, soluble cellulosomes diffuse to the substrate surface and once bound, begin to degrade the insoluble material. In addition to the substrate-targeting CBMs, the scaffoldin contains multiple copies of the cohesins (Fig. 2.4), thereby promoting synergistic interaction among the enzymes due to their proximity on the scaffoldin subunit. However, this high-proximity may also limit the cellulase mobility within a cellulosome, and thus reduce its digestion efficiency. It has been proposed that the linkers of cellulosomal cellulases and scaffoldin provide plasticity to the whole cellulosome to solve this problem (Bayer *et al.*, 2004; Gilbert, 2007; Himmel *et al.*, 2010). To support this concept, recent predictions from bioinformatics studies confirmed the flexible nature of these linkers as well as those found in multi-modular cellulosomal enzymes (Bomble, 2011).

Cell-surface bound cellulosomes

Surface-displayed cellulosomes, which are perhaps the archetypical cellulosomes, target both the enzymes and the bacterium itself to the cellulosic substrate, allowing for tighter control of cellulose degradation and improved uptake of nutrients. These cellulosomes contain an S-layer homology module (SLH) and a CBM, which anchor the enzyme array, respectively, to the bacterial surface and to the cellulose substrate. The first cellulosome to be described was that of *C. thermocellum*, which is composed of three types of subunits: the scaffoldin, the enzymatic units, and the anchoring scaffoldin (see Fig. 2.5). Because of its added complexity, two types of self-specific cohesin–dockerin interactions (type I and type II) are needed to organize the cell-surface-bound assembly. The primary enzyme-integrating scaffoldin subunit (termed CipA) is composed of a single polypeptide chain which contains nine type I cohesin domains, a CBM, an X-domain of unknown function, and a type II dockerin (Leibovitz and Beguin, 1996). All the domains of the scaffoldin are connected by flexible linker peptides. The cellulosomal enzyme units all contain a type I dockerin, allowing each to bind to one of the type I cohesin modules on the scaffoldin. The

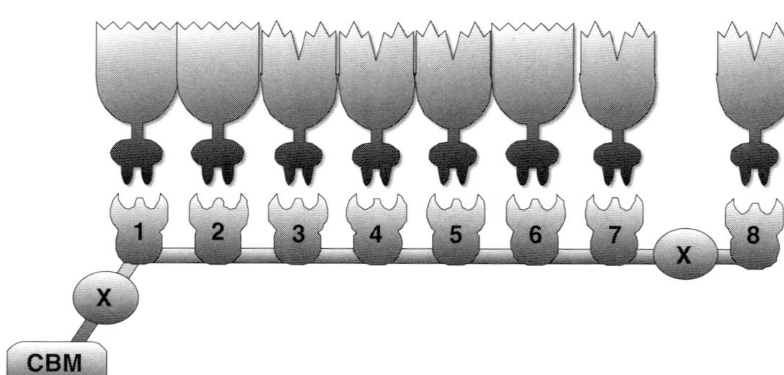

Figure 2.4 Schematic representation of a soluble cellulosome. The scaffoldin component is a single polypeptide chain, consisting of a CBM, X-modules of undetermined function, and a number of cohesins (enumerated). Dockerin-containing enzymes bind unselectively to the cohesin modules leading to a multiplicity of individual cellulosomes in a given population. This type of cellulosome is indicative of various mesophilic clostridia, e.g. *C. cellulolyticum*, *C. cellulovorans*, etc.

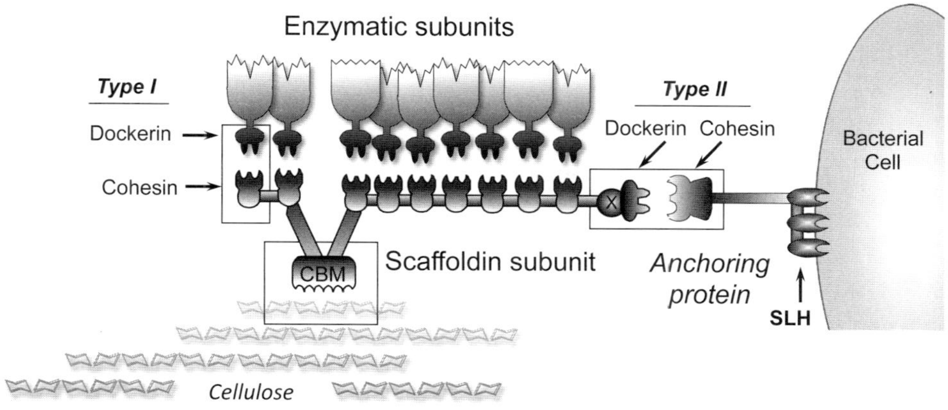

Figure 2.5 Simplified schematic view of the molecular disposition of the cellulosome and one of the associated anchoring proteins on the cell surface of *C. thermocellum*. Similar types of surface–bound complex cellulosomes are observed in other anaerobic bacteria, notably *Acetivibrio cellulolyticus*, *Bacteroides cellulosolvens* and *Ruminococcus flavefaciens*.

anchoring protein contains the SLH domains for connection to the cell surface, and a type II cohesin, which binds the terminal type II dockerin of the scaffoldin.

The architecture and assembly of the *C. thermocellum* cellulosome shown as an example in Fig. 2.5 is relatively simple, in that it contains a single enzyme-bearing scaffoldin. This bacterium, however, produces several types of anchoring scaffoldins, which contain one, two, or seven type II cohesins. The last could accommodate as many as 63 different cellulosomal enzymes. Such complex cellulosomes have been characterized from several different cellulolytic bacteria. These cellulosomes contain multiple secondary scaffoldins, all docked to a primary, cell-bound scaffoldin. For example, *Bacteroides cellulosolvens* contains a primary, cell-surface-anchored scaffoldin (termed ScaB) that contains 10 type I cohesins. Each of these cohesins can bind a secondary scaffoldin that contains a terminal type I dockerin, a CBM, and 11 type II cohesins (Xu *et al.*, 2004). This secondary scaffoldin (termed ScaA) arrays enzymes that contain type II dockerin domains (note that in *C. thermocellum*, the enzymes contain type I dockerins). Thus, if all cohesin positions were occupied, the cellulosome of *B. cellulosolvens* would contain 110 enzymes.

The even more complex cellulosomal system of *Ruminococcus flavefaciens* (Ding *et al.*, 2001; Rincon *et al.*, 2004, 2005; Alber *et al.*, 2008) contains four scaffoldin branches. The fact that these cellulosomes accommodate such large enzyme numbers and combinations of subunits gives us some sense of the enzymatic complexity needed for efficient degradation of the plant cell wall. Likewise, the cellulosome's modular and potentially complex architecture can also inspire nanotechnologists to design new kinds of useful artificial nanostructures.

The 3-D structure of the cellulosome

Small angle X-ray scattering has been used to show structural flexibility between the cohesin subunits. The flexibility imparted to the complex by the linkers in the scaffoldin, probably prevents cohesin domains, along with the enzymes they carry, from steric interference with one another and allow them to interact in a dynamic manner with selected portions of the plant cell wall polysaccharides (Hammel *et al.*, 2005). As observed by electron microscopy, the *C. thermocellum* cellulosome has a spherical shape and contains a cavity that includes a fibre-like scaffoldin and an outer shell containing the enzymes (Madkour and Mayer, 2003). It is thought that cellulosomes must be dynamic and amorphous in order for their enzymes to access their substrates within the complex topology of the plant cell wall (Mingardon *et al.*, 2007b). Larger portions of cellulosomes have been crystallized and their structures determined by X-ray diffraction studies (Bayer *et al.*, 2009; Adams *et al.*, 2010). Additionally, it was shown in a recent study using molecular modelling that the linkers in the cellulosome from *C. thermocellum* were extremely flexible and allowed a fully populated cellulosome to explore a very large number of conformations (Bomble, 2011).

Application of cellulosomal components in biotechnology

Early on, it was realized that the components of the cellulosome (cohesins, dockerins, CBMs, etc.) had potential applications outside the realm of designer cellulosomes (Bayer *et al.*, 1997). The modules from the cellulosome could be refashioned for new and divergent purposes to create novel kinds of affinity-based systems by combining them with other non-related systems (antibody-antigen, avidin-biotin, hormone-receptor, etc.) (Fig. 2.6). In this way, mediation between systems can be established, and their utility could be expanded. But, despite great advances in our knowledge about cellulosomes and their component protein modules, only a few of the proposed research areas have been explored and developed, and there is still a great deal of potential for innovation.

Cohesin-dockerin affinity assays

In order for the large and growing population of cohesins and dockerins to be tested, high throughput synthesis and binding-analysis methods have been developed. Isolated cohesins and dockerins have often suffered from poor stability, and it is now standard practice to link all the proteins of interest to another module (e.g. an enzyme or a CBM) using a fusion cassette system. This strategy improves the behaviour of the proteins and also facilitates high-throughput binding analysis. Cohesins are thus fused to a CBM and dockerins are fused to a xylanase. An affinity-based ELISA system (Barak *et al.*, 2005) was developed by which the cohesin–dockerin interactions can be determined by a standardized approach. In this technique, the CBM-containing cohesins are adsorbed onto microtitre plates, to which the xylanase–dockerin fusion proteins are applied. Xylanase concentration is then probed with antibodies in an ELISA format. Later, a similar method was developed in which the CBM-cohesin was more properly oriented for dockerin binding on cellulose microarrays (Haimovitz *et al.*, 2008). Using this technique, cohesins, each fused to a CBM, are deposited on a cellulose-coated chip, so that the cohesins are oriented into the solution, rather than adhering to the plate in random orientations. After incubation with xylanase-dockerins, the slide is then probed with fluorescently labelled anti-xylanase antibody.

Use in purification

Fusion proteins containing a CBM attached to a protein of interest can be used in a variety of applications, such as protein purification, microarrays, and cell sorting. A variety of affinity

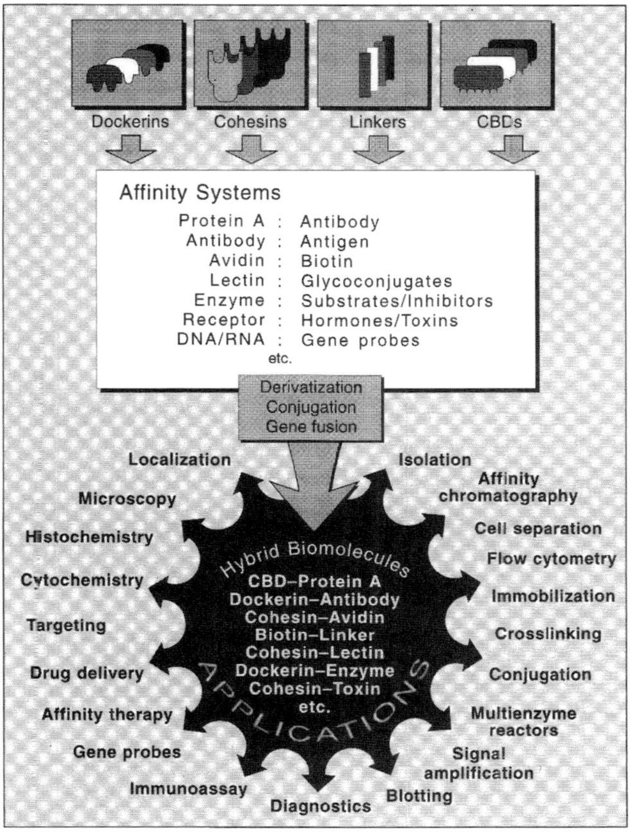

Figure 2.6 Scheme showing manifold potential applications of cellulosome components for affinity systems and other biotechnological techniques. Applications proposed in 1994 for cellulosomal modules in medical research and industrial fields (Bayer *et al.*, 1994). Cross-linking or fusing components of both the cellulosomal system and other affinity systems can be used to create hybrid biomolecules with a variety of applications.

purification strategies have been derived utilizing protein components of the cellulosome (Ong *et al.*, 2005; Shoseyov *et al.*, 2006; Haimovitz *et al.*, 2008).

Since the early days of cellulosome research, the potential for using cellulose modules for affinity purification was recognized. Application of the cohesin–dockerin interaction as a platform for affinity purification is an attractive alternative to metal chelation methods, since the components are non-toxic and transition to alkaline pHs is not required. Moreover, the cohesin and dockerin can serve as an alternative to small molecule-protein affinity methods, since all components are encodable, while still exhibiting very high affinity. The principle of this affinity purification method involves the calcium dependence of the cohesin–dockerin interaction and the susceptibility of the dockerin component to denaturation and unbinding with the addition of EDTA (Craig *et al.*, 2005, 2006). However, in practice this procedure did not allow for the elution to occur, even with the addition of high EDTA concentrations. To create a more reversible binding interaction, a truncated dockerin (Karpol *et al.*, 2009), whose *N*-terminal calcium binding loop was removed, was found to exhibit high affinity binding to the activated column while permitting EDTA-mediated elution. This approach resulted in reversible affinity purification columns, which used cellulose beads activated by the addition of a CBM-cohesin (Fig. 2.7). Proteins tagged with a matching dockerin could then be added and bound to the column and, in theory, eluted by the addition of EDTA. The EDTA would compete away dockerin-bound calcium and abolish the ability of the dockerin to bind the cohesin. By reducing cohesin-dockerin binding from the picomolar range to the nanomolar range, this strategy provides a cheap and reusable system for general protein purification (Demishtein *et al.*, 2010). The interaction between the cohesin and dockerin from *C. josui* has also been weakened so that it can be used for affinity purification (Kamezaki *et al.*, 2010). However, instead of a truncation, a single point-mutation of the dockerin at position 45, located in one of the repeated recognition epitopes, allowed for efficient dockerin-target-protein elution from the CBM-cohesin column (Dock tag system). Side-by-side testing with conventional poly-His nickel-bead purification showed higher purifications for the Dockerin tag system. Although high concentration of chelating agent will not allow it to function for proteins with bound metals, such a purification system could become a standard protein purification method for many other types of proteins

Designer cellulosomes

Already by 1994 (Bayer *et al.*, 1994), the field of designer cellulosomes was being anticipated, and a number of strategies for their synthesis were proposed (Fig. 2.8). The first type of hypothetical designer cellulosome utilized chemical cross-linking to the CBM modules to attach additional enzymes or other molecules, generating supercellulosomes from natural cellulosomes. Another designer cellulosome proposal called for creating heterocellulosomes

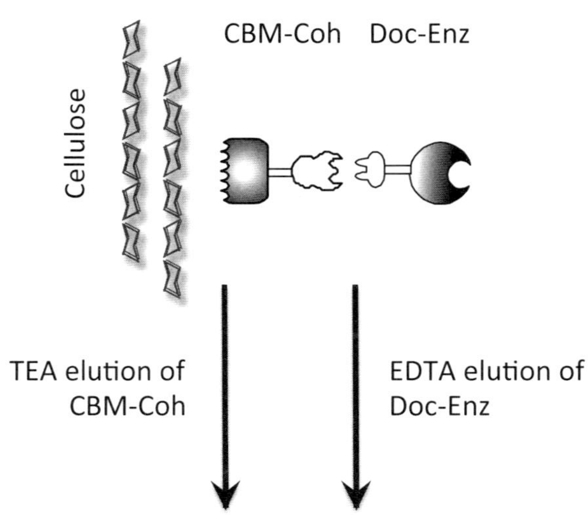

Figure 2.7 Scheme showing the cohesin-dockerin-based affinity purification strategy. The CBM-cohesin is first bound to beaded cellulose resin. The protein of interest, tagged with a matching dockerin module, is then applied to the cohesin-activated beads. After washing, the protein is then eluted using an EDTA gradient. The CBM-Coh can also be removed from the column using triethylamine (TEA).

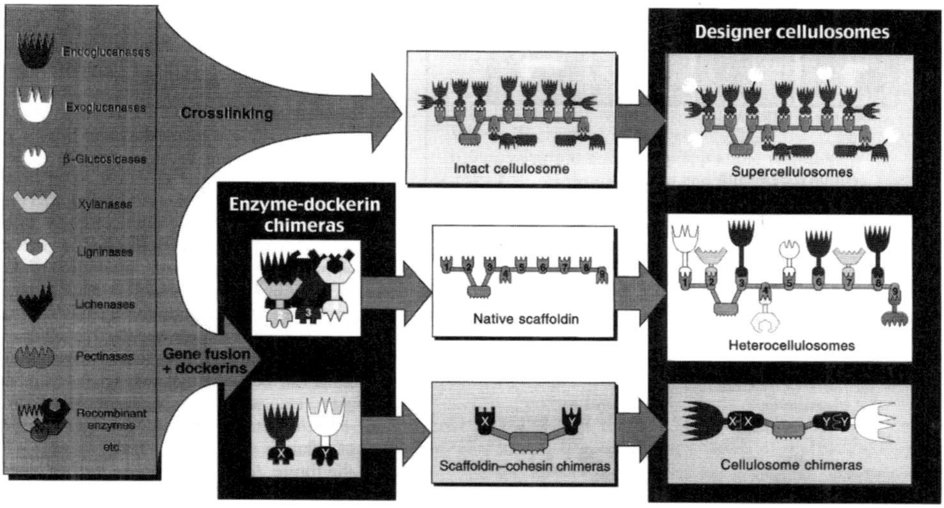

Figure 2.8 Outline of the future potential of designer cellulosomes as proposed in Bayer *et al.* (1994). Industrial-scale degradation of cellulosics could be accomplished by creating supercellulosomes, using enzymatic cross-linking, heterocellulosomes, using altered enzyme mixtures, and cellulosome chimeras, using recombinant DNA technology.

by doping the enzyme pool with specific enzyme-dockerins in order to create cellulosomes with altered enzyme display patterns. And finally, chimeras utilizing recombinant cellulosomal modules, most similar to the modern designer cellulosomes, was also proposed. Today, we are only beginning to harness the power of cellulosome modularity, and development of industrially relevant designer cellulosomes remains an important research area.

First attempts in the construction of small artificial cellulosomes for efficient degradation of specific cellulosic substrates were reported in 2001 (Fierobe *et al.* 2001), along with encouraging enhanced degradation of microcrystalline cellulose by *C. cellulolyticum* divalent designer cellulosomes (catalytic enhancement of 1.5 fold for some enzyme combinations). In this work, a chimeric scaffoldin was constructed to contain cohesins obtained from different species and exhibiting different dockerin-binding specificities. By constructing enzymes that contained matching dockerins, the authors were able to produce designer cellulosome complexes which were homogeneous in composition. The scaffoldin also contained a CBM for collective targeting of the enzymes to the cellulosic substrate, leading to increased synergy of up to 2- to 3-fold. Later studies defined two major factors that serve to enhance deconstruction of recalcitrant forms of cellulose (Fierobe *et al.* 2002): (1) the well-established effect of enzyme targeting to the substrate surface via the scaffoldin-borne CBM, and (2) the consequent proximity effect of the enzyme components. Additional chimeric scaffoldins and enzyme-dockerin modules were used to create an expanded library of 75 chimeric cellulosomes (Fierobe *et al.*, 2002), The catalytic activities of which were then tested on various cellulose sources. Different activities and synergistic effects were observed for various enzyme array/substrate combinations, and the role of the CBM as an important contributor to the enzymatic synergism was reinforced. With the advent of these chimeric systems, the repertoire of enzymes available for cellulosomal display could be greatly expanded, paving the way for realization of the designer cellulosome concept.

The recalcitrance of the cellulosic substrate was shown to be an important parameter for increased enzyme synergy in designer cellulosomes (Fierobe *et al.* 2005). In addition, this research demonstrated that the contribution of a greater spectrum of enzymes (from different families of glycoside hydrolases) specialized for different subcomponents is required for efficient degradation of complex lignocellulosic substrates.

In related studies, the mini-cellulosomes described were composed of a truncated scaffoldin from *C. cellulovorans* and copies of a recombinant cellulosomal endoglucanase; the complex exhibited synergistic activity on cellulosic substrates (Murashima *et al.* 2002). A *C. cellulovorans* chimeric cellulosome was reported by Cha *et al.* (2007) to exhibit enhanced activities (11- to 18-fold higher for the cellulosomal enzymes than in their free form) while integrating endoglucanase and xylanase on a scaffoldin with 1–4 cohesins. However, the scaffoldin contained cohesins with the same specificity, which were thus able to recognize only the dockerin of the same species. Consequently, unlike the designer scaffoldins produced by Fierobe *et al.* (2001, 2002, 2005), these mini-cellulosomes were heterogeneous in content. However, further research using the same scaffoldin indicated that in *C. cellulovorans,* the cohesin–dockerin interaction might be more selective than originally believed (Koukiekolo *et al.* 2005).

The designer cellulosome concept was used to design novel cellulosome geometries (Fig. 2.9), and their activities on crystalline cellulosic substrates were compared with those of designer cellulosomes presenting geometries found in natural cellulosomes (Mingardon *et*

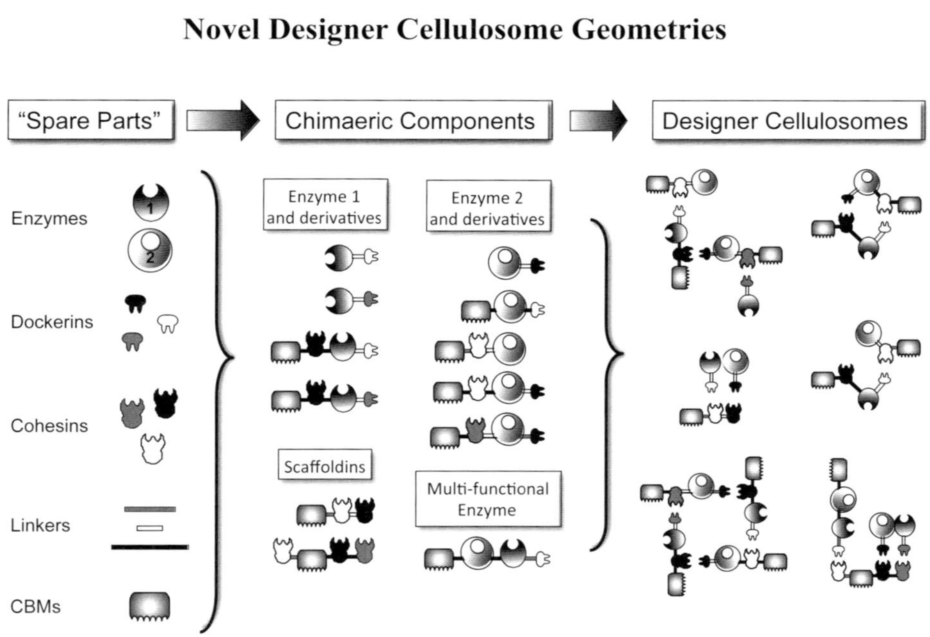

Figure 2.9 Schematic representations of the cellulosomal spare parts, recombinant components and possible designer cellulosomes complexes that can be formed. The designer cellulosomes present various geometries that were defined as polymeric, symmetrical, hybrid, asymmetrical, cyclic and mixed in a study by Mingardon and co-workers (Mingardon *et al.*, 2007). Only the hybrid and mixed cellulosomes are based on the native complexes.

al. 2007b). This intriguing work established the negative influence of multiple CBMs on cellulose hydrolysis activity of designer cellulosomes and indicated that enzyme mobility in designer cellulosomes is an important parameter. In one case, a fungus-derived cellulase from *Neocallimastix patriciarum* was incorporated into designer cellulosome modes together with standard cellulosomal enzymes (Mingardon *et al.* 2007a) and served to increase enzymatic activity 26-fold when scaffolded to a *C. cellulolyticum*-derived cellulosome. In this study, proximity effect and targeting effect were not observed simultaneously for the designer cellulosomes bearing enzymes from mixed sources. In that case, the origin of the enzymes from the different microbial systems could be responsible for the antagonism between the proximity and CBM targeting effects. It was then suggested that combined effects may occur in designer cellulosomes composed only of bacterial enzymes. Nevertheless, another study incorporated a fungal enzyme into the artificial cellulosome framework; in that case, the feruloyl esterase A from *Aspergillus niger* was fused to a dockerin from *C. thermocellum* (Levasseur *et al.*, 2004).

Research on designer cellulosomes demonstrates that the cellulosome modular structure can be broken into 'spare parts': i.e., enzymatic catalytic modules, dockerin modules, cohesin modules, linkers, and CBMs from different microorganisms can be separated and rearranged into chimeric components that are not found in nature (Fig. 2.9). This concept led to the design of novel atypical geometries, such as symmetrical, asymmetrical, or cyclic designer cellulosomes (Mingardon *et al.*, 2007b) and revealed that enzyme mobility is a critical parameter for designer cellulosome efficiency. Moreover, cellulosomal spare parts may in theory be used for the assembly of other unrelated multi-enzymatic systems in a single complex, i.e. systems, such as entire enzyme pathways, that are not necessarily produced as complexed systems in their native form. Such applications, however, have yet to be demonstrated.

In a recent article by Vazana and coworkers (2010), the designer cellulosome approach was used to examine the interplay of prominent cellulosomal and non-cellulosomal cellulases from *C. thermocellum* on crystalline cellulose. Two sets of related endoglucanases from *C. thermocellum*, one pair cellulosomal and one pair non-cellulosomal, were converted to the alternative form. The cellulosomal enzymes were converted to their free forms by replacing their dockerin domain with the CBM from their analogue, and the free enzymes were converted to the cellulosomal form by replacing their CBMs with dockerins from their analogues. Although both the free and cellulosomal enzymes act synergistically, the targeting effect was found to be the major factor responsible for the synergism among the enzyme combinations.

Later on, the conversion of the highly cellulolytic free-enzyme system of the aerobic thermophilic bacterium, *Thermobifida fusca*, into the cellulosomal mode was proposed. Initial research using the *T. fusca* system focused on the conversion of its free cellulase system to the cellulosomal mode and established that *T. fusca* enzymes could be converted to the cellulosomal mode and their integration into scaffoldin served to enhance cellulose-degrading activity compared with the free wild-type enzymes (Caspi *et al.*, 2006, 2008, 2009, 2010). This system also enabled examination of the significance of linker-length and dockerin position in enzyme design (Caspi *et al.*, 2009) and established that, unlike dockerin position, linker length between the catalytic module and the dockerin had apparently no influence on the activity of the chimeras. The combined action of cellulases and xylanases together in the same designer cellulosome complex was then considered on a natural complex wheat straw substrate (Moraïs *et al.*, 2010, 2011).

To date, the largest scaffoldin-based designer cellulosome to be reported was a tetravalent cellulosome incorporating complementary cellulases from *Thermobifida fusca*, the exoglucanase Cel48A, and the endoglucanase Cel5A, as well as two endoxylanases: Xyn10B and Xyn11A (Morais *et al.*, 2010a). This assembly gave a ~2.4-fold enzyme-activity enhancement on a wheat straw substrate. The specific proximity-induced synergy of the tetravalent designer cellulosome was confirmed by comparing the enzymatic activity with that of divalent designer cellulosomes containing the same enzymes. The tetravalent species exhibited 2-fold higher activity, demonstrating that enzymatic synergy is improved as the number of scaffoldin-bound enzymes increases. However, this proximity-induced synergy is not universal for all designer cellulosomes, and seems to likely depend on both the substrate and the enzymes involved.

In addition, a variation of the designer cellulosome approach was employed by fabricating a tetravalent complex that included the entire xylanolytic system of *T. fusca* (three endoxylanases and one β-xylosidase). The data demonstrated enhanced synergistic activities for the xylanolytic designer cellulosomes on the natural recalcitrant wheat straw substrate as compared to that of the free wild-type enzymes. The Xyl43A enzyme together with the three endoxylanases into designer cellulosome served to enhance the level of reducing sugars produced during wheat straw degradation and succeeded in degrading significant (~25%) levels of the total xylan component of the wheat straw substrate, suggesting that the incorporation of xylanases into cellulosome complexes is advantageous (Morais *et al.*, 2011).

The first *in vitro* assembly of engineered mini-cellulosomes was a mini-CipA, containing *C. thermocellum* CBM3a and two cohesin domains from *C. acetobutylicum* (Sabathe and Soucaille, 2003; Perret *et al.*, 2004), which resulted in a soluble artificial extracellular complex.

Challenges with designer cellulosome expression

Researchers have reported protein degradation problems when expressing designer cellulosome components in *E. coli*. This can be overcome by using a heterologous system where proteins can be produced in *Bacillus subtilis* WB800, a bacterium that is fast growing, is amenable to genetic manipulation, and can secrete proteins (Murashima *et al.*, 2002). Additionally, the WB800 strain is deficient in extracellular proteases (Cho *et al.*, 2004). This strain was used to co-express the EngB endoglucanase and an engineered mini-scaffoldin protein, mini-CbpA1, which contained a CBM and a cohesin. The construct was secreted by the bacterium, showing that *B. subtilis* can be used to produce difficult-to-express engineered cellulosomes.

Self-assembling cellulosome hybrids

The SP1 protein (stress-responsive homo-oligomeric protein) from quaking aspen (*Populus tremula*) is a self-assembling homo-dodecamer, forming an 11-nm ring-like assembly that is reminiscent of some proposed cellulosome structures. Although this structure is composed of distinct polypeptide chains, rather than a single, modular cellulosome scaffoldin, it has been used as a scaffold for cellulolytic enzymes, in a fashion analogous to the design of the cellulosome (Heyman *et al.*, 2007). In this construct, a cohesin molecule was fused, through a 23-residue linker peptide, to the SP1 monomer, and the chimeric monomer was allowed to self-assemble into the ring-like structure. The assembled cellulosome-like

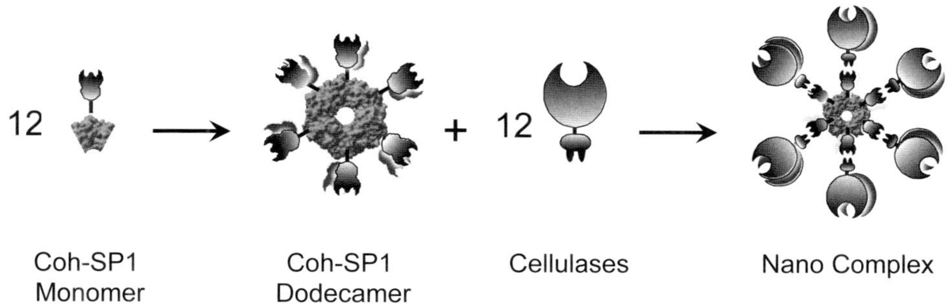

Figure 2.10 Formation of the endoglucanase-displaying SP1-based nano complex. The SP1 monomers, each containing a fused cohesin, first self-assemble into a ring-like docecamer. With the addition of the enzyme-linked dockerin, a cellulosome-like nano complex is formed.

structure reconstituted the dodecamer arrangement, and bound up to 10.5 *C. thermocellum* endoglucanase-dockerin enzymes (Fig. 2.10). As in the cellulosome, synergistic effects between closely packed enzymes were also detected. The chimera demonstrated two-fold enhanced activity compared to that of the free enzymes even though it contained no CBM to target the complex to the cellulose surface.

Unlike the cellulosomal scaffoldin, the SP1 scaffold lacks a straightforward method for incorporating different types of cohesins in an organized sequence. But even for the enzymatically homogeneous SP1 system, enzymatic synergism between different kinds of enzymes has been achieved. In a second study, the cohesin-containing SP1 scaffold was arrayed with 12 dockerin-bearing exoglucanases (Morais *et al.*, 2010b). In this case, SP1-scaffolded exoglucanases exhibited *reduced* cellulase activity when compared with free exoglucanases, which may be explained by the paucity of free chain-ends in crystalline cellulose. The addition of a small amount of free endoglucanase to the solution led to an increase in cellulolytic activity by providing a source of cellulose-chain ends, and resulted in significant synergistic effects for the SP1-bound exoglucanases.

In a similar strategy, a chaperonin from the extremophile archaeon, *Sulfolobus shibatae*, has been employed to mimic the cellulosomal scaffoldin's enzyme display, and produce similar synergistic effects (Mitsuzawa *et al.*, 2009). This chimera, termed the rosettazyme, consists of an 18-subunit complex which self-assembles into a double ring structure, with each ring consisting of 9 subunits. Each subunit is tethered to a cohesin, which in turn, can bind a dockerin-containing enzyme. Although the exact organization of the enzymes docked about the rosettazyme cannot be controlled, assembling rosettazymes with different ratios of enzymes allowed for the investigation of synergistic effects on cellulosic substrates. As more enzymes were added to the rosettazyme, and especially when exoglucanase and endoglucanase were both present, activity and synergism increased. Four complementary cellulolytic enzymes, bound to the rosettazyme in an optimized ratio, showed the greatest synergistic effects using this system.

Other recent examples utilized a non-cellulosomal scaffolding component that employs the avidin–biotin interaction to perform self-assembly (D.-M. Kim *et al.*, 2011, 2012). In these structures, an endoglucanase from *Aspergillus niger* and a CBM from *Cellulomonas fimi* were tagged with a biotin acceptor peptide and a polyhistidine tag. These components allowed for self-assembly on the avidin protein, and after enzyme–CBM ratios were optimized, increased catalytic activity over the free enzymes was observed. Clustering the biotin-tagged enzyme and CBM on streptavidin-conjugated cadmium selenide nanoparticles caused a 9.6-fold increase in activity for insoluble substrates, compared with the free enzyme.

Cell-surface display of enzymes and designer cellulosomes

In initial attempts, cellulases have been displayed in *E. coli* and *B. subtilis* by creating cellulase fusion proteins that are associated with the membrane (Francisco *et al.*, 1993; YS. Kim *et al.*, 2000; J.H. Kim *et al.*, 2005). *S. cerevisiae* has also been engineered to directly ferment cellulose to ethanol by functionalizing its surface with cellulolytic enzymes. This modified yeast contained endoglucanase and cellobiohydrolase from *T. reesei* and β-glucosidase from *Aspergillus aculeatus*. The β-glucosidase removes cellobiose generated by the cellulolytic enzymes and thus reduces product inhibition. This combination of displayed proteins on the yeast surface resulted in a yield of 0.45 g of ethanol per gram of cellulose (Fujita *et al.*, 2004). Other groups have constructed similar types of yeast strains that present bacterial enzymes on their surfaces to utilize xylose (Wahlbom *et al.*, 2003) and lignocellulose hydroxylate (Katahira *et al.*, 2006).

More recently, various research groups have created *S. cerevisiae* strains that display designer mini-cellulosomes using a covalent link to the β-1,6-glucan within the cell wall using a glycosyl phosphatidylinositol (GPI) signal motif to display the mini-scaffoldin on the yeast cell surface (Lilly *et al.*, 2009; Tsai *et al.*, 2009; Wen *et al.*, 2010). Mini-cellulosomes were assembled by incubating the cells with purified dockerin-containing cellulases (Lilly *et al.*, 2009; Tsai *et al.*, 2009) or *in vivo* by co-expressing the enzymes (Wen *et al.*, 2010). These research efforts established proof-of-concept to produce yeast strains that could produce ethanol from cellulose, a step towards the construction of a consolidated bioprocessing (CBP) microorganism that could produce high levels of ethanol directly from biomass. A recent article demonstrated that it is possible to use sortase enzymes to attach designer cellulosomes covalently to the surface of *B. subtilis* (Anderson *et al.*, 2011).

Nanoscience applications

The modular connectivity of the cellulosome has been expanded from that of an enzyme display system to that of a general nanostructure-fabricator (Ding *et al.*, 2006b). One such re-purposing is the use of cellulosomal components in assembling quantum dot-containing structures. Known for their unique optical and electric properties, quantum dots have many applications in diverse fields such as bioimaging, light harvesting, and nanoelectronics; however, precise positioning and organization is essential. The first application of cellulosomics to the quantum dot was the fabrication of a *C. thermocellum*–(CsSe)ZnS quantum dot conjugate. In this structure, the zinc-sulphate shell provides a coordination surface for a polyhistidine tag on the cohesin (Ding *et al.*, 2003a). The resultant protein-coated quantum dot assemblies formed various nanostructures spontaneously, including a major product that was trefoil-shaped, and exhibited light harvesting properties; as well as single quantum dots surrounded by bound cohesin. The trefoil nanostructures contain several quantum dots and measure ~10 nm in diameter. In order to better control the nano-assembly process of these quantum dot assemblies, a more complex quantum dot binding

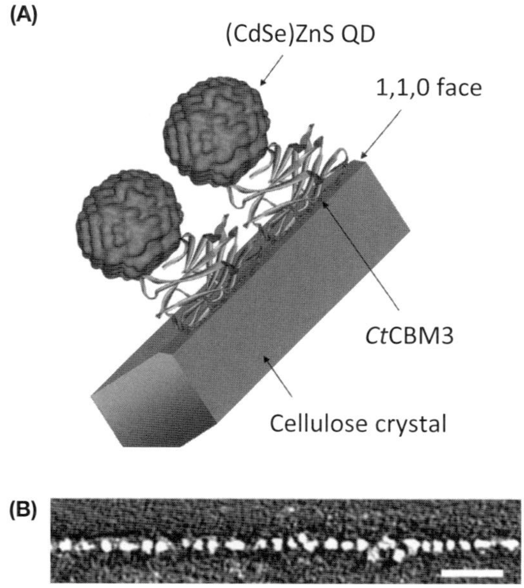

Figure 2.11 (A) Schematic model of QD/CBM/cellulose bio-assembly. The recombinant *C. thermocellum* CBM3 protein specifically binds to cellulose crystal (1,1,0 surface) and the QDs are anchored on CBM proteins by the His-tags fused at both *N*- and *C*-termini. QDs, CBMs, and cellulose are depicted in scale based on their published X-ray structures. (B) TEM images of QD/CBM/cellulose bio-assembles showing that QDs are linearly arrayed on the cellulose nanocrystal. QDs are shown as white spots. Bar = 50 nm. (From Xu *et al.*, Cellulose, 2009.)

protein chimera was constructed that incorporated cohesin-dockerin connectivity. This chimera consisted of a type II dockerin linked to a polyhistidine tag through an inert type I cohesin (Ding *et al.*, 2004). This set-up would allow the self-assembly of linear arrays of (CdSe)ZnS quantum dot-bearing units. After conjugation and fractionation, populations of quantum–dot-protein conjugates were found with a variety of stoichiometries. One population contained large particles with molecular weights over one million daltons. Another population contained smaller particles containing one, two, or three quantum dots, and bound to 10, 15, or 18 cohesin-dockerin monomers. Since each quantum dot is capable of binding multiple proteins, controlling the assembly of the particles needs refinement; nonetheless, these nanostructures do have interesting properties. Another related conjugate utilized a CBM to facilitate binding QDs to the surface to a microcrystalline cellulose surface (Zhang, 2006).

Xu *et al.* also demonstrated a method of creating linearly arrayed quantum dots using semiconductor (CdSe)ZnS quantum dot-CMB conjugates (Xu *et al.* 2009). By binding these conjugates to crystals of the macro algae *Valonia*, linear quantum dot arrays could be assembled on the planar face of the cellulose (see Fig. 2.11). Additionally, by modulating the size of the quantum dot and the identity of the CBMs, different labelling densities could be achieved. The tunable nature of this nano self-assembly technique demonstrates the utility of cellulosomal components for applications in nanoscale construction, and it also highlights the potential to create more detailed maps of the fine structure of the plant cell wall.

Conclusion

Since the initial discovery of the *C. thermocellum* cellulosome, we have discovered that the cellulosomal components, particularly the CBMs, cohesins and dockerins, are in fact prevalent throughout biology, and a multiplicity of cellulosomes, cellulosomal components, and their interactions have been characterized. This has allowed for advances towards creating designer cellulosomes, and also for the utilization of cellulosomal components for affinity purification methods. The cellulosome paradigm has also been used to make inroads into the field of nanomaterials and nanotechnology; however, much of the cellulosome's potential remains untapped. Although the original proposals for their use in bionanotechnology were described nearly two decades ago (Bayer *et al.*, 1994), most of the possible applications suggested at the time have not yet been pursued. A growing wealth of information about the cellulosome's structure and biology is waiting to be mobilized to create new structures that harness the modularity and diversity of this self-assembling protein machine.

Acknowledgements

This work was funded by the DOE Office of Science, Office of Biological and Environmental Research through the BioEnergy Science Center (BESC), a DOE Bioenergy Research Center. Grants from the following foundations are gratefully acknowledged: the United States–Israel Binational Science Foundation (BSF), Jerusalem, Israel, the Israel Science Foundation (grant nos 966/09 and 24/11), The Israeli Centers of Research Excellence (I-CORE) program (Center No. 152/11), The Alternative Energy Research Initiative (AERI) Bioenergy Consortium. E.A.B. is the incumbent of The Maynard I. and Elaine Wishner Chair of Bio-organic Chemistry.

References

Adams, J.J., Webb, B.A., Spencer, H.L., and Smith, S.P. (2005). Structural characterization of type II dockerin module from the cellulosome of *Clostridium thermocellum*: calcium-induced effects on conformation and target recognition. Biochemistry *44*, 2173–2182.

Adams, J.J., Pal, G., Jia, Z., and Smith, S.P. (2006). Mechanism of bacterial cell-surface attachment revealed by the structure of cellulosomal type II cohesin–dockerin complex. Proc. Natl. Acad. Sci. U.S.A. *103*, 305–310.

Adams, J.J., Currie, M.A., Ali, S., Bayer, E.A., Jia, Z., and Smith, S.P. (2010). Insights into higher-order organization of the cellulosome revealed by a dissect-and-build approach: Crystal structure of interacting *Clostridium thermocellum* multimodular components. J. Mol. Biol. *396*, 833–839.

Anderson, T.D., Robson, S.A., Jiang, X.W., Malmirchegini, G.R., Fierobe, H.P., Lazazzera, B.A., and Clubb, R.T. (2011). Assembly of minicellulosomes on the surface of *Bacillus subtilis*. Appl. Environ. Microbiol. *77*, 4849–4858.

Alber, O., Noach, I., Lamed, R., Shimon, L.J., Bayer, E.A., and Frolow, F. (2008). Preliminary X-ray characterization of a novel type of anchoring cohesin from the cellulosome of *Ruminococcus flavefaciens*. Acta Crystallograph. Section F, Struct. Biol. Crystalliz. Commun. *64*, 77–80.

Barak, Y., Handelsman, T., Nakar, D., Mechaly, A., Lamed, R., Shoham, Y., and Bayer, E.A. (2005). Matching fusion protein systems for affinity analysis of two interacting families of proteins: the cohesin–dockerin interaction. J. Mol. Recog. *18*, 491–501.

Bayer, E.A., Kenig, R., and Lamed, R. (1983). Adherence of *Clostridium thermocellum* to cellulose. J. Bacteriol. *156*, 818–827.

Bayer, E.A., Morag, E., and Lamed, R. (1994). The cellulosome – a treasure-trove for biotechnology. Trends Biotechnol. *12*, 379–386.

Bayer, E.A., Belaich, J.-P., Shoham, Y., and Lamed, R. (2004). The cellulosomes: multi-enzyme machines for degradation of plant cell wall polysaccharides. Annu. Rev. Microbiol. *58*, 521–554.

Bayer, E.A., Smith, S.P., Noach, I., Alber, O., Adams, J.J., Lamed, R., Shimon, L.J.W., and Frolow, F. (2009). Can we crystallize a cellulosome? In Biotechnology of Lignocellulose Degradation and Biomass

Utilization, K. Sakka, S. Karita, T. Kimura, M. Sakka, H. Matsui, H. Miyake, and A. Tanaka, eds. (Ito Print Publishing Division), pp. 183–205.

Bolam, D.N., Ciruela, A., McQueen-Mason, S., Simpson, P., Williamson, M.P., Rixon, J.E., Boraston, A., Hazlewood, G.P., and Gilbert, H.J. (1998). *Pseudomonas* cellulose-binding domains mediate their effects by increasing enzyme substrate proximity. Biochem. J. *331*, 775–781.

Bomble, Y.J., Beckham, G.T., Matthews, J.F., Nimlos, M.R., Himmel, M.E., and Crowley, M.F. (2011). A study of cellulosomal enzyme complex self-assembly. J. Biol. Chem. *286*, 5614–5623.

Boraston, A.B., Nurizzo, D., Notenboom, V., Ducros, V., Rose, D.R., Kilburn, D.G., and Davies, G.J. (2002). Differential oligosaccharide recognition by evolutionarily-related beta-1,4 and beta-1,3 glucan-binding modules. J. Mol. Biol. *319*, 1143–1156.

Boraston, A.B., Bolam, D.N., Gilbert, H.J., and Davies, G.J. (2004). Carbohydrate-binding modules: fine-tuning polysaccharide recognition. Biochem. J. *382*, 769–781.

Carvalho, A.L., Dias, F.M., Nagy, T., Prates, J.A., Proctor, M.R., Smith, N., Bayer, E.A., Davies, G.J., Ferreira, L.M., Romao, M.J., *et al.* (2007). Evidence for a dual binding mode of dockerin modules to cohesins. Proc. Natl. Acad. Sci. U.S.A. *104*, 3089–3094.

Carvalho, A.L., Dias, F.M.V., Prates, J.A.M., Nagy, T., Gilbert, H.J., Davies, G.J., Ferreira, L.M.A., Romao, M.J., and Fontes, C.M.G.A. (2003). Cellulosome assembly revealed by the crystal structure of the cohesin–dockerin complex. Proc. Natl. Acad. Sci. U.S.A. *100*, 13809–13814.

Cha, J., Matsuoka, S., Chan, H., Yukawa, H., Inui, M., and Doi, R.H. (2007). Effect of multiple copies of cohesins on cellulase and hemicellulase activities of *Clostridium cellulovorans* mini-cellulosomes. J. Mol. Biotechnol. *17*, 1782–1788.

Cho, H.Y., Yukawa, H., Inui, M., Doi, R.H., and Wong, S.L. (2004). Production of minicellulosomes from *Clostridium cellulovorans* in *Bacillus subtilis* WB800. Appl. Environ. Microbiol. *70*, 5704–5707.

Craig, P.O., Berguer, P.M., Ainciart, N., Zylberman, V., Thomas, M.G., Martinez Tosar, L.J., Bulloj, A., Boccaccio, G.L., and Goldbaum, F.A. (2005). Multiple display of a protein domain on a bacterial polymeric scaffold. Proteins *61*, 1089–1100.

Craig, S.J., Foong, F.C., and Nordon, R. (2006). Engineered proteins containing the cohesin and dockerin domains from *Clostridium thermocellum* provides a reversible, high affinity interaction for biotechnology applications. J. Biotechnol. *121*, 165–173.

Demishtein, A., Karpol, A., Barak, Y., Lamed, R., and Bayer, E.A. (2010). Characterization of a dockerin-based affinity tag: application for purification of a broad variety of target proteins. J. Mol. Recognit. *23*, 525–535.

Ding, S.Y., Rincon, M.T., Lamed, R., Martin, J.C., McCrae, S.I., Aurilia, V., Shoham, Y., Bayer, E.A., and Flint, H.J. (2001). Cellulosomal scaffoldin-like proteins from *Ruminococcus flavefaciens*. J. Bacteriol. *183*, 1945–1953.

Ding, S.Y., Jones, M., Tucker, M.P., Nedeljkovic, J.M., Wall, J., Simon, M.N., Rumbles, G., and Himmel, M.E. (2003a). Quantum dot molecules assembled with genetically engineered proteins. Nano Lett *3*, 1581–1585.

Ding, S.-Y., Lamed, R., Bayer, E.A., and Himmel, M.E. (2003b). The bacterial scaffoldin: Structure, function and potential applications in nanoscience. In Genetic Engineering: Principles and Methods, J.K. Setlow, ed. (New York, Kluwer Academic Publishers), pp. 209–226.

Ding, S.Y., Rumbles, G., Jones, M., Tucker, M.P., Nedeljkovic, J., Simon, M.N., Wall, J.S., and Himmel, M.E. (2004). Bioconjugation of (CdSe)ZnS quantum dots using a genetically engineered multiple polyhistidine tagged cohesin/dockerin protein polymer. Macromol Mater Eng *289*, 622–628.

Ding, S.-Y., Xu, Q., Mursheda, K.A., Baker, J.O., Bayer, E.A., Barak, Y., Lamed, R., Sugiyama, J., Rumbles, G., and Himmel, M.E. (2006a). Versatile derivatives of carbohydrate-binding modules for imaging of complex carbohydrates approaching the molecular level of resolution. BioTechniques *41*, 435–440.

Ding, S.-Y., Xu, Q., Bayer, E.A., Qian, X., Rumbles, G., and Himmel, M.E. (2006b). Bacterial protein complexes with potential applications in nanotechnology. In Microbial bionanotechnology: Biological self-assembly systems and biopolymer-based nanostructures, B. Rehm, ed. (Norfolk, UK, Horizon Scientific Press), pp. 269–305.Fierobe, H.P., Bayer, E.A., Tardif, C., Czjzek, M., Mechaly, A., Belaich, A., Lamed, R., Shoham, Y., and Belaich, J.P. (2002). Degradation of cellulose substrates by cellulosome chimeras. Substrate targeting versus proximity of enzyme components. The J. Biol. Chem. *277*, 49621–49630.

Fierobe, H.P., Mechaly, A., Tardif, C., Belaich, A., Lamed, R., Shoham, Y., Belaich, J.P., and Bayer, E.A. (2001). Design and production of active cellulosome chimeras – Selective incorporation of dockerin-containing enzymes into defined functional complexes. J. Biol. Chem. *276*, 21257–21261.

Fierobe, H.P., Pages, S., Belaich, A., Champ, S., Lexa, D., and Belaich, J.P. (1999). Cellulosome from *Clostridium cellulolyticum*: Molecular study of the dockerin/cohesin interaction. Biochemistry *38*, 12822–12832.

Francisco, J.A., Stathopoulos, C., Warren, R.A.J., Kilburn, D.G., and Georgiou, G. (1993). Specific adhesion and hydrolysis of cellulose by intact *Escherichia coli* cells expressing surface-anchored cellulase or cellulose-binding domains. Bio/technology *11*, 491–495.

Fujita, Y., Ito, J., Ueda, M., Fukuda, H., and Kondo, A. (2004). Synergistic saccharification, and direct fermentation to ethanol, of amorphous cellulose by use of an engineered yeast strain codisplaying three types of cellulolytic enzyme. Appl. Environ. Microbiol. *70*, 1207–1212.

Gilbert, H.J. (2007). Cellulosomes: microbial nanomachines that display plasticity in quaternary structure. Mol. Microbiol. *63*, 1568–1576.

Haimovitz, R., Barak, Y., Morag, E., Voronov-Goldman, M., Shoham, Y., Lamed, R., and Bayer, E.A. (2008). Cohesin-dockerin microarray: Diverse specificities between two complementary families of interacting protein modules. Proteomics *8*, 968–979.

Hammel, M., Fierobe, H.P., Czjzek, M., Kurkal, V., Smith, J.C., Bayer, E.A., Finet, S., and Receveur-Brechot, V. (2005). Structural basis of cellulosome efficiency explored by small angle X-ray scattering. The J. Biol. Chem. *280*, 38562–38568.

Han, W., Tang, P., Li, X., Qiu, F., Zhang, H., and Yang, Y. (2008). Self-assembly of star ABC triblock copolymer thin films: self-consistent field theory. J. Phys. Chem. B *112*, 13738–13748.

Heyman, A., Barak, Y., Caspi, J., Wilson, D.B., Altman, A., Bayer, E.A., and Shoseyov, O. (2007). Multiple display of catalytic modules on a protein scaffold: nano-fabrication of enzyme particles. J. Biotechnol. *131*, 433–439.

Himmel, M.E., Xu, Q., Luo, Y., Ding, S.-Y., Lamed, R., and Bayer, E.A. (2010). Microbial enzyme systems for biomass conversion: emerging paradigms. Biofuels *1*, 323–341.

Kamezaki, Y., Enomoto, C., Ishikawa, Y., Koyama, T., Naya, S., Suzuki, T., and Sakka, K. (2010). The Dock tag, an affinity tool for the purification of recombinant proteins, based on the interaction between dockerin and cohesin domains from *Clostridium josui* cellulosome. Prot. Express. Purif. *70*, 23–31.

Kang, S., Barak, Y., Lamed, R., Bayer, E.A., and Morrison, M. The functional repertoire of prokaryote cellulosomes includes the serpin superfamily of serine proteinase inhibitors. Mol. Microbiol. (2006).

Karpol, A., Barak, Y., Lamed, R., Shoham, Y., and Bayer, E.A. (2008). Functional asymmetry in cohesin binding belies inherent symmetry of the dockerin module: insight into cellulosome assembly revealed by systematic mutagenesis. Biochem. J. *410*, 331–338.

Karpol, A., Kantorovich, L., Demishtein, A., Barak, Y., Morag, E., Lamed, R., and Bayer, E.A. (2009). Engineering a reversible, high-affinity system for efficient protein purification based on the cohesin–dockerin interaction. J. Mol. Recognit. *22*, 91–98.

Katahira, S., Mizuike, A., Fukuda, H., and Kondo, A. (2006). Ethanol fermentation from lignocellulosic hydrolysate by a recombinant xylose- and cellooligosaccharide-assimilating yeast strain. Appl. Microbiol. Biotechnol. 72, 1136–1143.

Kim, J.H., Park, I.S., and Kim, B.G. (2005). Development and characterization of membrane surface display system using molecular chaperon, prsA, of *Bacillus subtilis*. Biochem Biophys Res Commun *334*, 1248–1253.

Kim, D.-M., Umetsu, M., Takai, K., Matsuyama, T., Ishida, N., Takahashi, H., Asano, R., and Kumagai, I. (2011). Enhancement of cellulolytic enzyme activity by clustering cellulose binding domains on nanoscaffolds. Small *7*, 656–664.

Kim, D.-M., Nakazawa, H., Umetsu, M., Matsuyama, T., and Ishida, N., Ikeuchi, A., Takahashi, H., Asano, R., and Kumagai, I. (2012). A nanocluster design for the construction of artificial cellulosomes. Cataly. Sci. Technol. *2*, 499–503.

Kim, Y.S., Jung, H.C., and Pan, J.G. (2000). Bacterial cell surface display of an enzyme library for selective screening of improved cellulase variants. Appl. Environ. Microbiol. *66*, 788–793.

Koukiekolo, R., Cho, H.Y., Kosugi, A., Inui, M., Yukawa, H., and Doi, R.H. (2005). Degradation of maize fibre by *Clostridium cellulovorans* cellulases and hemicellulases and contribution of scaffolding protein CbpA. Appl. Environ. Microbiol. *71*, 3504–3511.

Lamed, R., Setter, E., and Bayer, E.A. (1983). Characterization of a cellulose-binding, cellulase-containing complex in *Clostridium thermocellum*. J. Bacteriol. *156*, 828–836.

Lamed, R., Tormo, J., Chirino, A.J., Morag, E., and Bayer, E.A. (1994). Crystallization and preliminary X-ray analysis of the major cellulose-binding domain of the cellulosome from *Clostridium thermocellum*. J. Mol. Biol. *244*, 236–237.

Lehtio, J., Sugiyama, J., Gustavsson, M., Fransson, L., Linder, M., and Teeri, T.T. (2003). The binding specificity and affinity determinants of family 1 and family 3 cellulose binding modules. Proc. Natl. Acad. Sci. U.S.A. *100*, 484–489.

Leibovitz, E., and Beguin, P. (1996). A new type of cohesin domain that specifically binds the dockerin domain of the *Clostridium thermocellum* cellulosome-integrating protein CipA. J. Bacteriol. *178*, 3077–3084.

Levasseur, A., Pages, S., Fierobe, H.P., Navarro, D., Punt, P., Belaich, J.P., Asther, M., and Record, E. (2004). Design and production in *Aspergillus niger* of a chimeric protein associating a fungal feruloyl esterase and a clostridial dockerin domain. Appl. Environ. Microbiol. *70*, 6984–6991.

Lilly, M., Fierobe, H.P., van Zyl, W.H., and Volschenk, H. (2009). Heterologous expression of a *Clostridium* minicellulosome in *Saccharomyces cerevisiae*. FEMS Yeast Res. *9*, 1236–1249.

Linton, S.M., Greenaway, P., and Towle, D.W. (2006). Endogenous production of endo-beta-1,4-glucanase by decapod crustaceans. J. Comp. Physiol. B, Biochei. Syst. Environ. Physiol. *176*, 339–348.

Liu, C.B., Gong, C.Y., Huang, M.J., Wang, J.W., Pan, Y.F., Zhang, Y.D., Li, G.Z., Gou, M.L., Wang, K., Tu, M.J., *et al.* (2008). Thermoreversible gel-sol behaviour of biodegradable PCL-PEG-PCL triblock copolymer in aqueous solutions. J. Biomed. Mat. Res. Part B, Applied Biomat. *84*, 165–175.

Long, D., Lu, F., Zhang, R., Qiao, W., Zhan, L., Liang, X., and Ling, L. (2008). Suspension assisted synthesis of triblock copolymer-templated ordered mesoporous carbon spheres with controlled particle size. Chem. Commun. (Camb.), 2647–2649.

McLean, B.W., Bray, M.R., Boraston, A.B., Gilkes, N.R., Haynes, C.A., and Kilburn, D.G. (2000). Analysis of binding of the family 2a carbohydrate-binding module from *Cellulomonas fimi* xylanase 10A to cellulose: specificity and identification of functionally important amino acid residues. Protein Eng. *13*, 801–809.

Madkour, M., and Mayer, F. (2003). Structural organization of the intact bacterial cellulosome as revealed by electron microscopy. Cell Biol. Int. *27*, 831–836.

Mechaly, A., Yaron, S., Lamed, R., Fierobe, H.P., Belaich, A., Belaich, J.P., Shoham, Y., and Bayer, E.A. (2000). Cohesin-dockerin recognition in cellulosome assembly: experiment versus hypothesis. Proteins *39*, 170–177.

Mingardon, F., Chanal, A., Lopez-Contreras, A.M., Dray, C., Bayer, E.A., and Fierobe, H.P. (2007a). Incorporation of fungal cellulases in bacterial minicellulosomes yields viable, synergistically acting cellulolytic complexes. Appl. Environ. Microbiol. 73, 3822–3832.

Mingardon, F., Chanal, A., Tardif, C., Bayer, E.A., and Fierobe, H.P. (2007b). Exploration of new geometries in cellulosome-like chimeras. Appl. Environ. Microbiol. *73*, 7138–7149.

Mitsuzawa, S., Kagawa, H., Li, Y.F., Chan, S.L., Paavola, C.D., and Trent, J.D. (2009). The rosettazyme: A synthetic cellulosome. J. Biotechnol. *143*, 139–144.

Morag, E., Bayer, E.A., and Lamed, R. (1990). Relationship of cellulosomal and noncellulosomal xylanases of *Clostridium thermocellum* to cellulose-degrading enzymes. J. Bacteriol. *172*, 6098–6105.

Morais, S., Barak, Y., Caspi, J., Hadar, Y., Lamed, R., Shoham, Y., Wilson, D.B., and Bayer, E.A. (2010a). Cellulase-xylanase synergy in designer cellulosomes for enhanced degradation of a complex cellulosic substrate. Mbio *14*, 1.

Morais, S., Heyman, A., Barak, Y., Caspi, J., Wilson, D.B., Lamed, R., Shoseyov, O., and Bayer, E.A. (2010b). Enhanced cellulose degradation by nano-complexed enzymes: Synergism between a scaffold-linked exoglucanase and a free endoglucanase. J. Biotechnol. *147*, 205–211.

Morais, S., Barak, Y., Hadar, Y., Wilson, D.B., Shoham, Y., Lamed, R., and Bayer, E.A. (2011). Assembly of xylanases into designer cellulosomes promotes efficient hydrolysis of the xylan component of a natural recalcitrant cellulosic substrate. MBio 2.

Murashima, K., Chen, C.L., Kosugi, A., Tamaru, Y., Doi, R.H., and Wong, S.L. (2002). Heterologous production of *Clostridium cellulovorans* engB, using protease-deficient *Bacillus subtilis*, and preparation of active recombinant cellulosomes. J. Bacteriol. *184*, 76–81.

Nakar, D., Handelsman, T., Shoham, Y., Fierobe, H.P., Belaich, J.P., Morag, E., Lamed, R., and Bayer, E.A. (2004). Pinpoint mapping of recognition residues on the cohesin surface by progressive homologue swapping. J. Biol. Chem. *279*, 42881–42888.

Noach, I., Frolow, F. Jakoby, H., Rosenheck, S., Shimon, L.W., Lamed, R., and Bayer, E.A. (2005). Crystal structure of a type-II cohesin module from the *Bacteroides cellulosolvens* cellulosome reveals novel and distinctive secondary structural elements. J. Mol. Biol. *348*, 1–12.

Notenboom, V., Boraston, A.B., Kilburn, D.G., and Rose, D.R. (2001). Crystal structures of the family 9 carbohydrate-binding module from *Thermotoga maritima* xylanase 10A in native and ligand-bound forms. Biochemistry *40*, 6248–6256.

Ong, E., Alimonti, J.B., Greenwood, J.M., Miller, R.C., Jr., Warren, R.A., and Kilburn, D.G. (1995). Purification of human interleukin-2 using the cellulose-binding domain of a prokaryotic cellulase. Bioseparation *5*, 95–104.

Pagès, S., Belaich, A., Belaich, J.-P., Morag, E., Lamed, R., Shoham, Y., and Bayer, E.A. (1997). Species-specificity of the cohesin–dockerin interaction between *Clostridium thermocellum* and *Clostridium cellulolyticum*: prediction of specificity determinants of the dockerin domain. Proteins *29*, 517–527.

Peer, A., Smith, S.P., Bayer, E.A., Lamed, R., and Borovok, I. (2009). Non-cellulosomal cohesin and dockerin-like modules in the three domains of life. FEMS Microbiol. Lett. *291*, 1–16.

Pell, G., Williamson, M.P., Walters, C., Du, H., Gilbert, H.J., and Bolam, D.N. (2003). Importance of hydrophobic and polar residues in ligand binding in the family 15 carbohydrate-binding module from *Cellvibrio japonicus* Xyn10C. Biochemistry *42*, 9316–9323.

Perret, S., Belaich, A., Fierobe, H.P., Belaich, J.P., and Tardif, C. (2004). Towards designer cellulosomes in Clostridia: mannanase enrichment of the cellulosomes produced by *Clostridium cellulolyticum*. J. Bacteriol. *186*, 6544–6552.

Qi, M., O'Brien, J.P., and Yang, J. (2008). A recombinant triblock protein polymer with dispersant and binding properties for digital printing. Biopolymers *90*, 28–36.

Qiao, M., Chen, D., Hao, T., Zhao, X., Hu, H., and Ma, X. (2008). Injectable thermosensitive PLGA-PEG-PLGA triblock copolymers-based hydrogels as carriers for interleukin-2. Die Pharmazie *63*, 27–30.

Rincon, M.T., Martin, J.C., Aurilia, V., McCrae, S.I., Rucklidge, G., Reid, M., Bayer, E.A., Lamed, R., and Flint, H.J. (2004). ScaC, an adaptor protein carrying a novel cohesin that expands the dockerin-binding repertoire of the *Ruminococcus flavefaciens* 17 cellulosome. J. Bacteriol. *186*, 2576–2585.

Rincon, M.T., Cepeljnik, T., Martin, J.C., Lamed, R., Barak, Y., Bayer, E.A., and Flint, H.J. (2005). Unconventional mode of attachment of the *Ruminococcus flavefaciens* cellulosome to the cell surface. J. Bacteriol. *187*, 7569–7578.

Sabathe, F., and Soucaille, P. (2003). Characterization of the CipA scaffolding protein and in vivo production of a minicellulosome in *Clostridium acetobutylicum*. J. Bacteriol. *185*, 1092–1096.

Sakka, K., Sugihara, Y., Jindou, S., Sakka, M., Inagaki, M., Sakka, K., and Kimura, T. (2011). Analysis of cohesin–dockerin interactions using mutant dockerin proteins. FEMS Microbiol. Lett. *314*, 75–80.

Sakon, J., Irwin, D., Wilson, D.B., and Karplus, P.A. (1997). Structure and mechanism of endo/exocellulase E4 from *Thermomonospora fusca*. Nature Struct. Biol. *4*, 810–818.

Schaeffer, F., Matuschek, M., Guglielmi, G., Miras, I., Alzari, P.M., and Beguin, P. (2002). Duplicated dockerin subdomains of *Clostridium thermocellum* endoglucanase CelD bind to a cohesin domain of the scaffolding protein CipA with distinct thermodynamic parameters and a negative cooperativity. Biochemistry *41*, 2106–2114.

Shimon, L.J., Bayer, E.A., Morag, E., Lamed, R., Yaron, S., Shoham, Y., and Frolow, F. (1997a). A cohesin domain from *Clostridium thermocellum:* the crystal structure provides new insights into cellulosome assembly. Structure *5*, 381–390.

Shimon, L.J., Frolow, F., Yaron, S., Bayer, E.A., Lamed, R., Morag, E., and Shoham, Y. (1997b). Crystallization and preliminary X-ray analysis of a cohesin domain of the cellulosome from *Clostridium thermocellum.* Acta Crystallogr. Section D, Biol. Crystallogr. *53*, 114–115.

Shoseyov, O., Shani, Z., and Levy, I. (2006). Carbohydrate binding modules: biochemical properties and novel applications. Microbiol. Mol. Biol. Rev.*70*, 283–295.

Simpson, P.J., Xie, H.F., Bolam, D.N., Gilbert, H.J., and Williamson, M.P. (2000). The structural basis for the ligand specificity of family 2 carbohydrate-binding modules. J. Biol. Chem. *275*, 41137–41142.

Spinelli, S., Fierobe, H.P., Belaich, A., Belaich, J.P., Henrissat, B., and Cambillau, C. (2000). Crystal structure of a cohesin module from *Clostridium cellulolyticum:* implications for dockerin recognition. J. Mol. Biol. *304*, 189–200.

Sun, M., Wang, P., Qiu, F., Tang, P., Zhang, H., and Yang, Y. (2008). Morphology and phase diagram of ABC linear triblock copolymers: parallel real-space self-consistent-field-theory simulation. Phys. Rev. E, Stat., Nonlinear, Soft Matter Physi. *77*, 016701.

Tavares, G.A., Beguin, P., and Alzari, P.M. (1997). The crystal structure of a type I cohesin domain at 1.7 A resolution. J. Mol. Biol. *273*, 701–713.

Tomme, P., Boraston, A., McLean, B., Kormos, J., Creagh, A.L., Sturch, K., Gilkes, N.R., Haynes, C.A., Warren, R.A., and Kilburn, D.G. (1998). Characterization and affinity applications of cellulose-binding domains. J. Chromatogr. B, Biomed. Sci. Appl. *715*, 283–296.

Tormo, J., Lamed, R., Chirino, A.J., Morag, E., Bayer, E.A., Shoham, Y., and Steitz, T.A. (1996). Crystal structure of a bacterial family-III cellulose-binding domain: a general mechanism for attachment to cellulose. EMBO J. *15*, 5739–5751.

Tsai, S.L., Oh, J., Singh, S., Chen, R., and Chen, W. (2009). Functional assembly of minicellulosomes on the *Saccharomyces cerevisiae* cell surface for cellulose hydrolysis and ethanol production. Appl. Environ. Microbiol. *75*, 6087–6093.

Wahlbom, C.F., van Zyl, W.H., Jonsson, L.J., Hahn-Hagerdal, B., and Otero, R.R. (2003). Generation of the improved recombinant xylose-utilizing *Saccharomyces cerevisiae* TMB 3400 by random mutagenesis and physiological comparison with *Pichia stipitis* CBS 6054. FEMS Yeast Res. *3*, 319–326.

Vazana, Y., Moraïs, S., Barak, Y., Lamed, R., and Bayer, E.A. (2010). Interplay between *Clostridium thermocellum* family-48 and family-9 cellulases in the cellulosomal versus non-cellulosomal states. Appl. Environ. Microbiol. *76*, 3236–3243.

Warren, R.A. (1996). Microbial hydrolysis of polysaccharides. Annu. Rev. Microbiol. *50*, 183–212.

Wen, F., Sun, J., and Zhao, H. (2010). Yeast surface display of trifunctional minicellulosomes for simultaneous saccharification and fermentation of cellulose to ethanol. Appl. Environ. Microbiol. *76*, 1251–1260.

Xu, Q., Bayer, E.A., Goldman, M., Kenig, R., Shoham, Y., and Lamed, R. (2004). Architecture of the *Bacteroides cellulosolvens* cellulosome: description of a cell surface-anchoring scaffoldin and a family 48 cellulase. J. Bacteriol. *186*, 968–977.

Xu, Q., Tucker, M.P., Ahrenkiel, P., Ai, S., Rumbles, G., Sugiyama, J., Himmel, M.E., and Ding, S.-Y. (2009). Labelling the planar face of crystalline cellulose using quantum dots directed by type-i carbohydrate-binding modules. Cellulose *16*, 19–26.

Zhang, P. (2006). Investigation of novel quantum dots/proteins/cellulose bioconjugate using NSOM and fluorescence. J. Fluoresc. *16*, 349–353.

Zverlov, V.V., Kellermann, J., and Schwarz, W.H. (2005). Functional subgenomics of Clostridium thermocellum cellulosomal genes: identification of the major catalytic components in the extracellular complex and detection of three new enzymes. Proteomics *14*, 3646–3653.

Protein-aided Mineralization of Inorganic Nanostructures

3

Brandon L. Coyle, Weibin Zhou and François Baneyx

Abstract

Designer proteins combine the adhesive or synthesizing properties of solid-binding peptides (SBPs) selected by combinatorial techniques with the desirable characteristics of a host scaffold. Like natural biomineralizing proteins, these chimeric constructs are powerful tools to control the nucleation, growth, morphogenesis and crystallography of inorganic phases. They also hold great potential for the assembly of hybrid structures in which inorganic, biological and synthetic components are organized with the high degree of precision needed to take advantage of the unique properties of matter at the nanoscale. After briefly discussing common approaches for identifying SBPs, we discuss the mechanisms by which they modulate materialization, which variables influence the process, and review recent progress in the use of designer proteins to fabricate complex architectures.

Introduction

Biologically driven inorganic synthesis has evolved in organisms whose survival is only possible within a limited range of temperature, pressure and pH. Within these operational constraints, living organisms manufacture inorganic materials with exceptional mechanical, transport, optical, and magnetic properties by exerting nanoscale control over position, composition and crystal structure (Baeuerlein, 2004; Bradbury and Marks, 2004). Silica mineralization is best exemplified by diatoms which build intricate amorphous SiO_2 cell walls (or frustules) with regular patterning at both the nano- and microscales (Kröger, 2007). The Gram-negative bacterium *Magnetospirillum magnetotacticum* has evolved acidic vesicles called magnetosomes to mineralize nanometer-sized, single-domain crystals of iron oxide that it uses to align itself with earth's magnetic field (Maratea and Blakemore, 1981; Schleifer *et al.*, 1991). The bones and teeth of vertebrates are both made of calcium phosphate. However, differences in the mineral's composition, shape and crystallography lead to unique properties in each hard tissue (Gorski, 1998). Finally, although mollusc shells are made of crystalline aragonite platelets, this form of calcium carbonate is nominally not thermodynamically stable under the conditions where it is formed (Mayer and Sarikaya, 2002).

A limitation of these otherwise remarkable natural processes is that the palette of inorganics compounds produced is limited. Expanding the compositional universe, biomimetics focuses on emulating Nature's manufacturing prowess with an emphasis on valuable (nano) materials and structures (Cheng *et al.*, 2006; Lu and Lieber, 2007; Mann, 2002; Sun and Rogers, 2007). Traditional routes to inorganic materials fabrication revolve around the use of organic solvents and high temperatures which are under scrutiny in a society yearning for 'green' and cost-effective processes. Proteins offer an alternative for synthesizing, patterning and manipulating inorganic materials. Not only are they in the correct size range to build structures whose unique properties originate from nanoscale effects, but they can be modified to bind to virtually any material. More importantly, their shape, chemistry and function can be tailored by genetic engineering. Here, we review commonly used approaches for the discovery of peptides capable of binding and/or mineralizing inorganic phases and focus on the use of genetically engineered proteins and organisms in the synthesis, organization and functionalization of hybrid nanomaterials. For more information on peptide-based mineralization and assembly schemes, the reader is referred to a number of recent reviews (Dickerson *et al.*, 2008; Papapostolou and Howorka, 2009; Tamerler *et al.*, 2010).

Selection of solid-binding peptides

Because nature produces only a limited amount of inorganics, alternative approaches have been developed to build 'designer' proteins suitable for mineralizing, binding and organizing materials of engineering interest. Combinatorial display has been a major tool in enabling this molecular biomimetic approach to materials synthesis (Sarikaya *et al.*, 2003, 2004).

In 1985, George Smith modified the genome of the M13 filamentous bacteriophage by fusing an antibody epitope to one of the viruses' minor coat proteins and showed that the modified bacteriophage was capable of binding to the antibody (Smith, 1985). This seminal work laid the foundation for the development of the Ph.D. M13 bacteriophage display system (Parmley and Smith, 1988) which has become the most widely used technology for the identification of solid binding peptides (SBPs). In M13 display, random sequences of amino acids are expressed as fusions to the N-terminus of the p3 minor coat protein to yield a library of virions that can be screened for candidates capable of binding to a particular antibody, protein or even solid material (Fig. 3.1A). Three libraries are commercially available from New England Biolabs (Beverly, MA, USA): linear heptapeptide (Ph.D.-7) and dodecapeptide (Ph.D.-12) libraries, and the Ph.D.-C7C library in which random heptapeptides are constrained by a disulphide bond between flanking cysteine residues to restrict the conformational space sampled by the inserts (Fig. 3.1B). Each random sequence is displayed at 3–7 copies (usually five) at the end of the bacteriophage that normally binds to the conjugative F pilus of *E. coli* (Fig. 3.1A). Typical Ph.D. libraries contain about 10^9 primary clones but phages are usually amplified so that up to 55 copies of the same sequence might be present in the working library. Non-commercial systems in which random peptides are fused to the p7 or p9 minor coat proteins (Fig. 3.1A) have also been built (Endemann and Model, 1995). However, they have not been used as extensively as the Ph.D. system. It should also be noted that while it is possible to fuse peptides to the p8 major coat protein to display them at nearly 2700 copies on the virion surface, this approach is not well suited for combinatorial screens because certain basic sequences inhibit phage biogenesis (Kehoe and Kay, 2005).

Cell surface display is an alternative to phage display in which random peptides are exposed on the cell surface by virtue of being fused to (or within) a membrane-anchored protein. While a number of such technologies have been developed over the years (Boder and Wittrup, 1997; Daugherty, 2007; Georgiou *et al.*, 1997), the prevailing cell surface display tools in the materials science community are the FliTrx system (which used to be – but

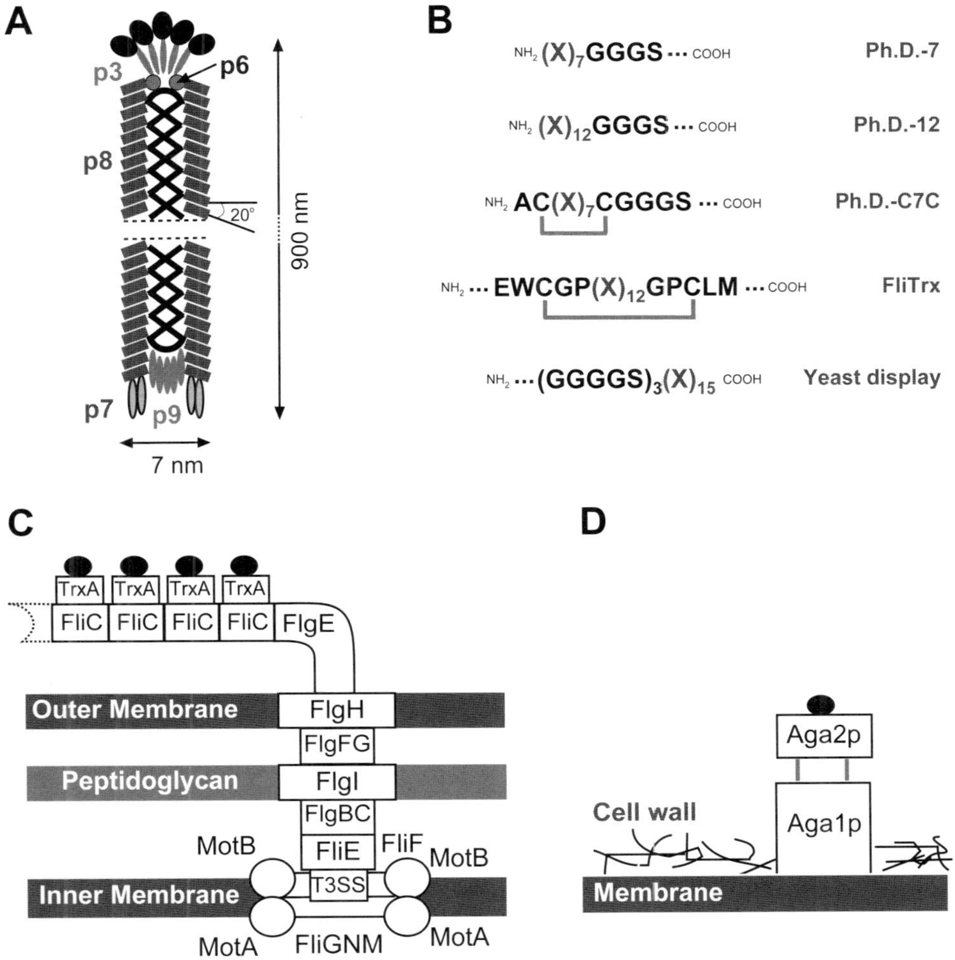

Figure 3.1 (A) Schematic representation of an M13 bacteriophage identifying capsid proteins and geometric parameters. (B) Structure of random peptides displayed as N-terminal fusions to the p3 protein in the Ph.D.7, Ph.D.–12 and Ph.D.-C7C systems, as internal fusions in the FliTrx system, and as C-terminal fusions in the yeast display system. Random amino acids are denoted × and grey lines correspond to disulphide bonds. (C) Schematic structure of the *E. coli* cell envelope and defective flagella architecture used in in FliTrx display. Black ovals corresponds to random peptides inserted within TrxA which is itself fused within FliC. (D) Schematic representation of the *S. cerevisiae* cell envelope in the yeast display system. Aga2p which bears the random peptide (black oval) as a C-terminal fusion is connected to Aga1p through two disulphide bridges (grey lines).

is no longer – commercialized by Invitrogen, Carlsbad, CA) and yeast surface display. In FliTrx display (Fig. 3.1C), disulfide-bonded dodecapeptides are displayed at multiple copies within defective *E. coli* flagella. This is done by replacing the protruding CGPC active site of the cysteine oxidoreductase thioredoxin 1 (TrxA) by random peptides and inserting the resulting fusions within a permissive site of the major flagellar protein FliC (Fig. 3.1B) (Lu *et al.*, 1995). Yeast surface display operates under similar principles, but relies on C-terminal fusions to the *Saccharomyces cerevisiae* Aga2p protein which projects from the cell wall (Fig. 3.1D) (Boder and Wittrup, 1997; Boder and Wittrup, 1998). As a result, random sequences can efficiently interact with their targets. The size of both libraries is limited by the efficiency of transformation. The FliTrx system has a diversity of about 10^8, which is only a small fraction of the 4.1×10^{15} possible sequence space. Yeast surface display libraries can be built to a diversity of 10^9 primary clones, but the process is labour-intensive and it is more common to use libraries of $\approx 10^7$ members that can be produced in a single day (Boder and Wittrup, 1997).

Ribosomal and mRNA display technologies do not rely on a host organism to connect genotype and phenotype. As a result, transformation is not needed and libraries can contain as many as 10^{13} to 10^{14} unique members (Gold, 1995). Such a degree of diversity is sometimes required to identify strong consensus motifs, e.g. when searching for high-affinity protein ligands. However, inorganic surfaces are inhomogeneous in terms of roughness, oxidation state, and crystallography. As a result, SBPs obtained after 3–5 rounds of biopanning typically do not converge towards a consensus sequence of amino acids. This is not to say that they do not share materials-specific physicochemical characteristics, as exemplified graphically in Fig. 3.2 for five different inorganic compounds (Chiu *et al.*, 2012; Thai *et al.*, 2004; Zhou *et al.*, 2010; B.L. Coyle, unpublished).

To summarize, library diversity is generally not an issue when looking for solid-binding peptides (SBPs), peptides that bind to the same material share physicochemical characteristics, and most of the sequences reported in the literature have been isolated using the few *in vivo* display systems that were both robust and commercially available.

Mechanisms of materialization

One of the major challenges in molecular biomimetics is the construction of 'designer' proteins that retain their original biological function while becoming endowed with the ability to modulate inorganic nucleation and growth due to the incorporation of an SBP. In materialization processes, such proteins can act as traditional surfactants, reductants and ligands, or they may function as selective surface active agents that stabilize nascent nuclei or bind to specific crystal faces. This classification is affected by the concentration needed to enact the desired effect (Fig. 3.3). In the vicinity of the equilibrium disassociation constant (K_d) for protein–inorganic interactions – a value that corresponds to half-monolayer protein

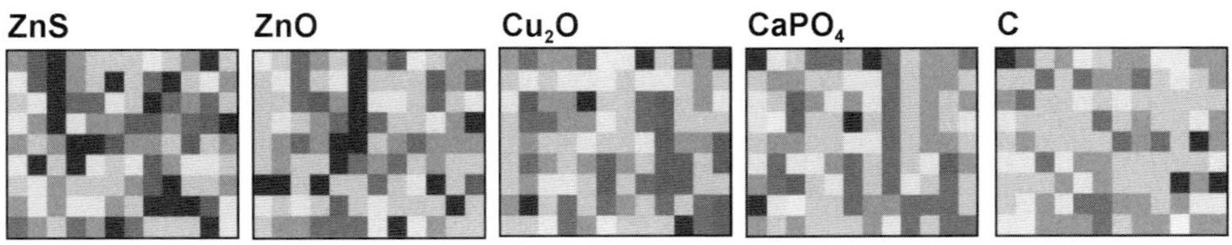

Figure 3.2 SBPs do not converge towards a consensus sequence but share materials-specific characteristics. Patterns are graphical representation of the physicochemical properties of nine solid binding dodecapeptides selected on zinc sulphide (ZnS), zinc oxide (ZnO), cuprous oxide (Cu_2O), calcium phosphate ($CaPO_4$) and carbon (C) substrates via FliTrx display system. Shades of grey correspond to different side-chains chemistries (from dark to light: acidic, amide, hydroxyl, basic, hydrophobic, and neutral).

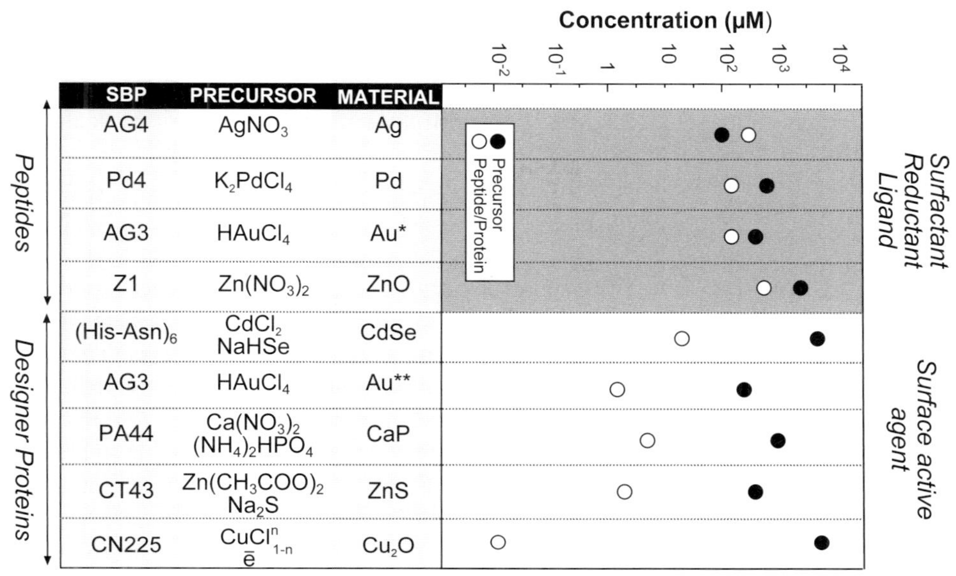

Figure 3.3 Mechanism of action of SBPs and designer proteins. When used at comparable concentrations as the precursor salts, SBPs function as surfactants, reductants and ligands, much like conventional additives (e.g. polyols, CTAB, ascorbic acid, etc.) do. When engineered within a protein scaffold, SBPs can be used at concentrations much lower than that of the precursors and function as surface active agents. Note for example the difference in concentrations when the gold binding peptide AG3 (AYSSGAPPMPPF) (Slocik *et al.*, 2005) was used in a synthetic form (Au*) (Chen *et al.*, 2008) or as a TrxA fusion protein (Au**) to produce gold nanoparticles. Data was extracted from the following references: Ag (Naik *et al.*, 2002); Pd (Pacardo *et al.*, 2009); Au* (Chen *et al.*, 2008); ZnO (Tomczak *et al.*, 2009); CdSe (Aryal and Benson, 2007); Au** (Shimojo *et al.*, 2012); CaP (Chiu *et al.*, 2012); ZnS (Zhou *et al.*, 2010); Cu_2O (Dai *et al.*, 2005).

coverage and typically falls in the nanomolar to micromolar range – the dominant mechanism is selective surface activity. In this regime, a few protein molecules can drastically alter the final composition, morphology, or crystal structure of an inorganic compound. This highly sought after capability is one of the major advantages of biomineralization. However, if morphogenetic effects only become apparent at millimolar concentrations, it is likely that the designer protein (or synthetic SBP) functions as traditional material growth modifiers do, for instance, by acting on the ions in solution through chelation or reduction (Fig. 3.3).

Designer proteins as selective surface active agents

Determining the precise mechanisms by which a protein influences inorganic mineralization can be a difficult task unless the experiment is designed to ask this very question. In one such example, we inserted the cuprous oxide (Cu_2O) binding peptide CN225 (RHTDGLRRIAAR) towards the C-terminus of TraI, a 192-kDa DNA-binding protein encoded by the *E. coli* F plasmid (Dai *et al.*, 2005; Thai *et al.*, 2004). The resulting protein called TraIi1753::CN225 remained capable of binding DNA and was found to have an equilibrium dissociation constant $K_d = 12$ nM for Cu_2O by quartz crystal microbalance (QCM) measurements. The ability of TraIi1753::CN225 to influence cuprous oxide mineralization was next compared to that of the TraIi1753 wild type at a concentration of 12 nM ($1 \times K_d$) using a copper chloride solution specifically formulated to thermodynamically favour the formation of atacamite ($Cu_2Cl(OH)_3$), a secondary copper mineral. In the absence of additive, or upon addition of TraIi1753, large (100–300 nm) particles of the expected solid were obtained. However, when the solution was supplied with TraIi1753::CN225 at a molar ratio of protein to Cu^{2+} of 1:500,000, we obtained 2 nm diameter Cu_2O particles surrounded by a protein corona (Dai *et al.*, 2005). These results can be explained by TraIi1753::CN225 binding to and stabilizing Cu_2O nuclei that transiently form under thermodynamically unfavourable conditions before they redissolve. Indeed, an equilibrium dissociation constant of 12 nM corresponds to a stabilizing Gibbs adsorption energy $\Delta G_{ads} = RT \ln(K_d)$ of −45 kJ/mol per protein. Capping of the growing particles by TraIi1753::CN225 once they have reached a critical diameter of about 2 nm accounts for the formation of core-shell architectures (Dai *et al.*, 2005).

The knowledge of the dissociation constant between a protein engineered for inorganic binding and its cognate solid is a valuable piece of information. It may also be difficult to obtain by QCM or surface plasmon resonance (SPR) techniques due to the challenge of depositing a compositionally pure and physically stable thin film of target material onto a QCM crystal of SPR chip (see Sarikaya *et al.* (2004) for a discussion of these analytical techniques and Kitayaporn *et al.* (2012) for an example of engineering thin film stability). Fortunately, K_d is not required to determine if an inorganic-binding protein functions as a surface active agent, as this can be inferred from the knowledge of the protein to ion molar ratio required for morphogenetic effects (Fig. 3.3) and/or from the appearance of an inorganic phase of unexpected composition or crystal structure.

Consider, as an example, the protein-aided fabrication of zinc sulphide (ZnS) nanocrystals (Zhou *et al.*, 2010). As an alternative to traditional organometallic precursors and high temperature schemes for quantum dot synthesis, we used FliTrx display to identify ZnS binding peptides, employed standard molecular biology approaches to engineer a compositionally diverse set of such binders in the active site of TrxA, and tested the resulting proteins for their ability to mineralize luminescent ZnS nanocrystals from an aqueous solution of zinc salts and sodium sulphide. While bulk ZnS precipitation occurred in the absence of protein or when 2 μM of wild-type TrxA and several of its ZnS-binding variants were added to the solution, one derivative incorporating the CT43 sequence (AGDSSGVDSRSV) led to the production of optically active and physically stable particles consisting of a ≈4 nm wurtzite (hexagonal) ZnS core capped by a protein shell. Considering that this was achieved at a protein to Zn^{2+} ratio of 1:200 and under conditions that should favour the mineralization of the ZnS zinc blende crystal phase, TrxA::CT43 clearly functions as a surface active agent (Zhou *et al.*, 2010).

Solid-binding peptides as growth modifiers

Most studies exploring the mineralizing or morphogenetic properties of SBPs use synthetic peptides at concentrations comparable to that of the inorganic precursors. Under such conditions, SBPs function like conventional growth modifiers: they ligate or reduce ionic species or adsorb to the growing solid, sometimes with crystallographic specificity (Fig. 3.3). As an example, Naik and coworkers used phage display to identify silver-binding dodecapeptides (Naik *et al.*, 2002). One of these peptides called AG4 (NPSSLFRYLPSD) supported the mineralization of silver nanoparticles when added at a concentration of about

0.3 mM to a 0.1 mM solution of silver nitrate. The particles produced were often faceted but also heterogeneous in terms of shape and size. At a peptide to Ag^+ ratio of nearly 3:1, AG4 cannot function as a selective agent. Rather, it was proposed to provide a reducing environment around silver nuclei and to exhibit some degree of specificity for the (111) face of silver (Naik *et al.*, 2002).

Not all SBPs are capable of inducing inorganic mineralization. In fact, most of them merely adhere to inorganic particles after they have been formed through (electro)chemical reduction or precipitation. Such capping mechanisms can be used not only to stabilize particles, but also to control their size or alter their properties (see below). Knecht and coworkers provide a good example of this approach by using a palladium binding peptide called Pd4 (TSNAVHPTLRHL) to produce Pd nanoparticles (Pacardo *et al.*, 2009). In their synthesis scheme, Pd4 is mixed in a solution of K_2PdCl_4 at a peptide to Pd^{2+} ratio of approximately 1:4 and the solution is supplemented with the strong reducing agent sodium borohydride ($NaBH_4$). Owing to the presence of high concentrations of Pd4, no bulk precipitation occurs. Rather, palladium nanoparticles that are monodisperse and 2 nm in diameter are obtained, presumably due to peptide-mediated coordination of Pd^{2+} cations before reduction and capping of the newly formed particles by Pd4 (Pacardo *et al.*, 2009).

Morphogenetic effects

Precise control of particle size and shape is critical to take full advantage of nanoscale effects. Designer proteins that incorporate SBPs offer an opportunity to do so in an environmentally friendly manner. However, if a predictive, rather than 'cook and look' approach is needed, the task is not simple. In fact, even seemingly unimportant variables such as the carry-over concentration of biological buffer and the host protein scaffold can have a profound influence on the size and shape of the particles produced (Kitayaporn *et al.*, 2012). Once these parameters are well controlled, the influence of other variables can be studied, as described below using examples from our laboratory.

Concentration effects

One of the most straightforward ways to modify the shape of a crystalline particle is to quench the growth of one of its faces by adsorbing a molecule to it. For designer proteins, this means using an SBP that can discriminate between the various crystal faces of a material. Based on morphological observations, Naik *et al.* (2002) suggested that the silver mineralizing peptide AG4 preferentially bound to the thermodynamically favourable (111) faces of silver (Tao *et al.*, 2008). To examine this possibility in more detail, we fused the AG4 sequence to the C-terminus of *E. coli* maltose binding protein (MBP) and found that the fusion protein exhibited a K_d of 180 nM for polycrystalline silver, a material that is mostly (111) (Sengupta *et al.*, 2008). To determine how MBP-AG4 would influence silver morphogenesis, we generated isolated and multifaceted silver nuclei on amorphous carbon and grew the particles electrochemically in the presence of control MBP or increasing amounts of MBP-AG4 (Grosh *et al.*, 2009). While particles produced in the absence of additive or with 180 nM ($1 \times K_d$) control protein had similar pentagonal morphology (Fig. 3.4A), addition of the same concentration of MBP-AG4 led to the appearance of branched, rosette-like structures (Fig. 3.4C). This drastic morphological change is likely due to the passivation of (111) terraces by adsorbed MBP-AG4 accompanied by the rapid growth of new faces at protein-free defect regions that now experience high local current densities (Grosh *et*

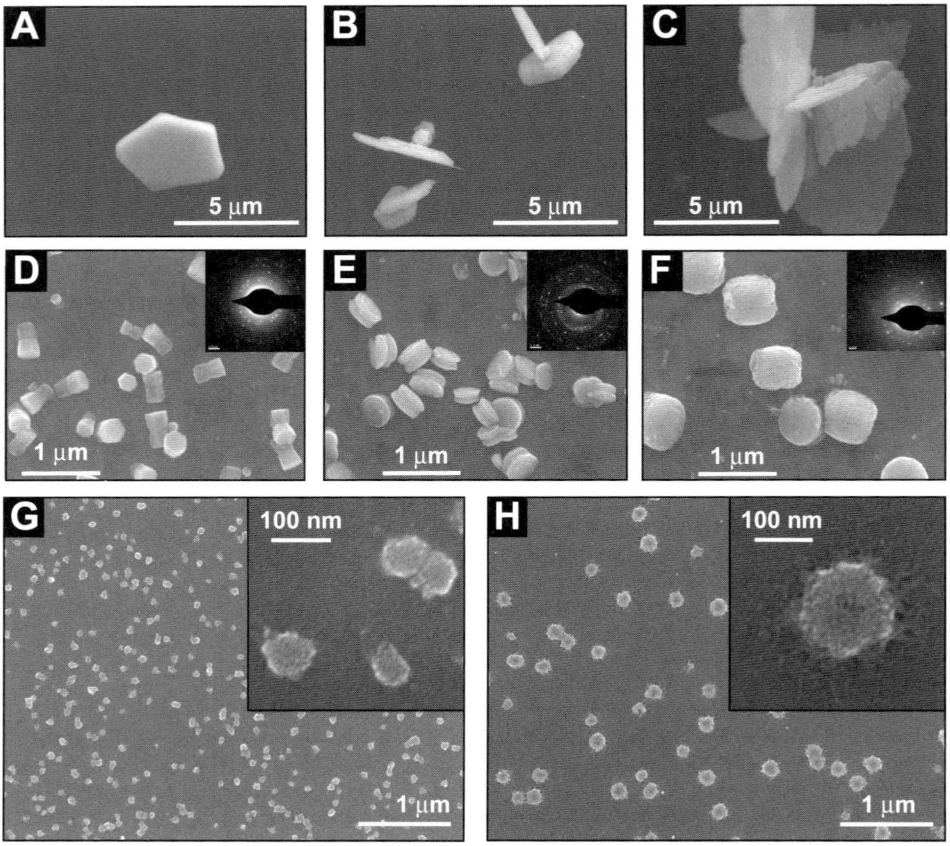

Figure 3.4 Three approaches to control inorganic morphology using designer proteins engineered with silver (A–C), zinc oxide (D–F) and calcium phosphate (G and H) binding peptides. In A–C, a fusion protein between *E. coli* maltose binding protein (MBP) and a linear silver binding peptide controls silver in a concentration-dependent manner: no additive or wild type MBP (A), 90 nM (B) and 180 nM (C) of MBP-Ag4 (Grosh *et al.*, 2009). In D and E, derivative of *E. coli* thioredoxin A (TrxA) in which disulphide-bonded ZnO binding dodecapeptides replace the native CGPC active site change zinc oxide morphology (Kitayaporn *et al.*, 2012). While hexagonal rods are obtained without protein or the wild type (D), 1 μM of derivatives containing different ZnO binders lead to thin (E), or cylindrical (F), laminated plates. In G–H, 5 μM of a TrxA derivative containing a calcium phosphate binding peptide in a disulphide-constrained loop (G) or not (H) gives rise to nanoparticles that differ in size by a ratio of 3 (Chiu *et al.*, 2012).

al., 2009). Fig. 3.4B and C also shows that the number of projecting facets increased with protein concentration presumably due to higher surface coverage of the original silver platelet and the formation of a greater numbers electrical hot spots on a now mostly insulating substrate.

It is important to point out here that *all* proteins will adsorb to high-energy surfaces (and in particular to noble metals such as gold and silver) whether or not they have been engineered to do so. In fact, the K_d of wild type MBP for silver is comparable to that of MBP-AG4 (Grosh *et al.*, 2009). What AG4 confers is an ability to more tightly pack proteins onto specific faces and in an oriented manner (Sengupta *et al.*, 2008). Thus, at concentration that exceed full monolayer coverage ($2 \times K_d$), both designer and control proteins completely passivate the surface and morphogenetic changes, let alone particle growth, can no longer be electrically driven.

Compositional effects

Because SBPs do not converge towards a consensus sequence, morphogenetic effects are typically evaluated using as little as one and rarely more than five different peptides or designer proteins. Even with such small populations, significant differences may be observed in the shape and size of mineralized particles. For example, we studied how TrxA derivatives incorporating four different disulphide-constrained ZnO binding peptides within their active would affect zinc oxide morphogenesis at 70°C (Kitayaporn *et al.*, 2012). While this temperature is high for a typical protein-aided fabrication process, it is required to form pure ZnO crystals. Fortunately, TrxA is a highly thermostable protein that allows one to access this temperature range. The four ZnO-binding peptides used, CN107 (VRAADSRSAHLA), CN111 (PAGLQVGFAVEV), CN179 (RIGHGRQIRKPL), and CN185 (RTDDGVAGRTWL), were chosen for their diverse physicochemical characteristics and range from acidic and hydrophobic (CN111) to basic and hydrophilic (CN179) (Thai *et al.*, 2004). Fig. 3.4D, shows that whereas addition of 100 nM of TrxA::CN107 led to the formation of hexagonal ZnO rods similar to those obtained without additive or in the presence of wild type TrxA (Kitayaporn *et al.*, 2012), TrxA::CN179 produced stacked plates (Fig. 3.4E), and cylindrical multilayered particles lacking distinct faceting were obtained with TrxA::CN185 (Fig. 3.4F). These morphological changes can be explained by the fact that the four designer proteins have different affinities for the (002) face of ZnO (Kitayaporn *et al.*, 2012). They also provide as a good illustration of how much variation in particle shape and size one might expect by making use of SBPs that bind to the same material but differ in amino acid composition.

Conformational effects

Both the Ph.D.-C7C and FliTrx systems yield disulphide-bonded SBPs that are less flexible than their linear counterparts. How this decrease in conformational degrees of freedom impacts inorganic affinity appears to depend on the nature of the SBP. For instance, we have previously shown that presentation in a disulphide-bonded loop is essential for the CN225 Cu_2O binding peptide to adhere to cuprous oxide surfaces (Choe *et al.*, 2007). Others have reported that the affinity of Pt- and Au-binding peptides for their cognate materials increases or remains unchanged when they are disulphide-constrained (Hnilova *et al.*, 2008; Seker *et al.*, 2007). On the other hand, a TiO_2/SiO_2 binder was found to interact with titanium surfaces more efficiently when linear compared to disulphide-bonded (Chen *et al.*, 2009b).

To explore the influence of SBP conformation on mineralization, we made use of a TrxA derivative containing a disulphide-bonded calcium phosphate binder called PA44 (KDVVVGVPGGQD) and constructed the TrxA::PA44-C32S mutant in which one of the flanking cysteine residues had been converted to a serine (Chiu *et al.*, 2012). When an ammonium phosphate solution containing 5 μM of TrxA::PA44 was supplemented with calcium nitrate, we obtained 70 ± 15 nm diameter particles consisting of an amorphous calcium phosphate core capped by a protein shell (Fig. 3.4G). Repeating the experiment with the mutant protein (which no longer presents PA44 into a loop) yielded particles that were less abundant and nearly 3-fold larger (Fig. 3.4H). This morphological change is likely to be associated with TrxA::PA44-C32S having a lower affinity for calcium phosphate than TrxA::PA44 and to the rapid kinetics of formation of amorphous calcium phosphate. Under this scenario, larger particles would have time to grow before being capped by TrxA::PA44-C32S compared to TrxA::PA44 (Chiu *et al.*, 2012). Thus, by modulating the efficiency of inorganic recognition, the conformation of an SBP can affect the size or shape of a nascent inorganic phase.

Ferritin: a versatile biomineralization scaffold

Proteins extracted from hard tissues or mineralizing organisms can be powerful tools to control the nucleation and growth of inorganic phases. Unfortunately, their usefulness is generally limited to the compound with which they are normally associated. An exception to this rule are ferritins, a family of proteins that store iron within their central core, but whose internal cavity has been used to produce nanoparticles of many other compositions.

Ferritins are oligomeric iron storage proteins present in a large number of species (Harrison and Arosio, 1996). They consist of 24 subunits that assemble to define an approximately 8-nm-diameter internal pocket used for iron mineralization. Channels allow metal ions to diffuse in and out of the central cavity which contains nucleation sites for the formation of ferric oxyhydroxide which allows for nontoxic storage of essential iron (Lawson *et al.*, 1991). When ferritin is empty, it is referred to as apoferritin and can be used as a reaction vessel to template the mineralization of inorganics nanomaterials (Uchida *et al.*, 2010).

Horse spleen apoferritin has been extensively used to synthesize nanoparticles of Mn(O)OH, Co(O)OH, Mn_3O_4, Fe_3O_4, Fe_2S_3 and CdS (Douglas *et al.*, 1995; Harrison and Arosio, 1996; Meldrum *et al.*, 1991, 1992; Wong and Mann, 1996). This early work relied on the ability of the oligomer to accumulate transition metal ions at supersaturating concentrations within its core while excluding counter-ions on its surface (Douglas *et al.*, 2005). Klem and coworkers expanded the scheme to high-oxidation-state ions by using a xenon lamp to drive the photochemical reduction of Eu^{3+}, Fe^{3+} or Ti^{4+} and form metal oxyhydroxides within the cage (Klem *et al.*, 2008). In a more robust process, Ueno *et al.* (2004) chelated Pd^{2+} ions within the apoferritin cavity and relied on sodium borohydride reduction to produce nanoparticles of palladium that were 2 nm in diameter and highly monodisperse. Gálvez and coworkers refined the scheme by including a dialysis step to remove free metal ions from solution before adding the reducing agent, thus preventing bulk precipitation of copper and its adsorption to the protein surface. In this fashion, they were able to produce 3-nm-diameter copper particles within the apoferritin cavity (Gálvez *et al.*, 2005).

One of the advantages of apoferritin is that it confers solubility to the solids that are encapsulated within its core. For instance, particles of CdS produced by exposing apoferritin to Cd^{2+} and reacting the complex with a sulphur solution remained soluble for weeks although they had irregular morphologies and poor optical properties (Wong and Mann, 1996). High-quality luminescent nanocrystals of CdS, CdSe, ZnSe could nevertheless be templated by using a slow chemical reaction to control the flux of selenium or sulphide ions to Cd^{2+}- or Zn^{2+}-bound apoferritin (Iwahori *et al.*, 2005, 2006, 2007; Yamashita *et al.*, 2004).

A logical next step in apoferritin-mediated mineralization was to modify the protein with SBPs specific for desirable inorganic compounds. Taking advantage of the fact that human light chain ferritin (LCF) forms a ferritin-like cage with C-termini projecting towards the inside of the cavity (Sano *et al.*, 2006), Naik and coworkers constructed and purified a fusion protein between LCF and the silver-mineralizing peptide AG4. The resulting protein,

LCF-AG4, self-assembled into the expected oligomeric structure and mineralized 5-nm-diameter silver nanoparticles within its core when incubated with a silver nitrate solution (Kramer *et al.*, 2004).

Sano *et al.* (2005) chose to engineer the exterior of the ferritin cage with a TiO_2 binding peptide to generate a protein capable of interacting with two different materials. This was achieved by fusing a minimized TiO_2 binder called MinTBP-1 (RKLPDA) to the N-terminus of LCF, which conferred the protein a K_d of 4 nM for titanium when displayed at 24 copies on the surface of the recombinant oligomer. As a result, the chimeric protein became capable of binding titanium surfaces with high avidity while retaining an ability to mineralize ferrihydrite and Co_3O_4 in its core (Sano *et al.*, 2005; Yamashita *et al.*, 2006)

Mineralization and assembly of complex architectures

How the natural ability of apoferritin to induce inorganic mineralization within its cavity has been exploited and modified over the years makes a compelling story in the field of protein-aided nanofabrication. However, what can be achieved with a single protein (or family of proteins) is limited. The most powerful aspect of molecular biomimetics is that it allows for combining virtually any useful protein attribute (e.g. structural, regulatory, binding, transport, catalytic, etc.) with the inorganic binding or synthesizing properties of an SBP. What can be achieved with the resulting chimeric proteins is only limited by the imagination of the designer. The following paragraphs touch upon design issues and describe a few examples that are but the first steps in the creation of increasingly complex systems and devices.

Construction of designer proteins

Clearly, any useful SBP must remain capable of binding to its cognate solid (or of mineralizing a desirable inorganic phase) once it has been transferred to a larger protein scaffold. In addition to this portability requirement, SBPs should not, or only minimally interfere, with host protein function. This means that the sites used for their insertion should be both solvent-accessible and permissive for folding.

Because most proteins tolerate the addition of short sequences to their N- or C-termini, fusing an SBP to these locations is the simplest way to build a designer protein. The addition of an inorganic binding sequence isolated by M13 Ph.D. display to a protein's N-terminus should not affect solid recognition since those peptides were originally selected as an N-terminal fusion to the p3 coat protein (Fig. 3.1A). Similarly, SBPs identified by yeast display are normally fused to the C-terminus of Aga2p (Fig. 3.1D) and should retain solid binding ability when fused to the carboxyl end of other proteins. Provided that they are fairly hydrophilic and that they are connected through flexible linkers (e.g. GGGS; Fig. 3.1B), such extensions will typically not interfere with the function of the host scaffold unless they lie close to the active/binding site or affect its structure via long-range effects (Sastry *et al.*, 2009).

More desirably from a design standpoint, SBPs can be inserted within internal permissive sites that have been identified by subjecting the host scaffold to transposon or linker mutagenesis (Manoil and Traxler, 2000). One such permissive location is the CGPC active site of *E. coli* TrxA and it is rather straightforward to transfer a disulphide-bonded dodecapeptide identified by FliTrx display to this scaffold since the flanking DNA sequences are identical (Chiu *et al.*, 2012; Kitayaporn *et al.*, 2012; Zhou *et al.*, 2010). TrxA has a number of additional advantages: its active site protrudes from the protein core which maximizes SBP availability (Holmgren, 1985); other proteins can be fused to its N- and C-termini and it tends to act as a solubilizing fusion partner (LaVallie *et al.*, 2000); it can be expressed at high level in *E. coli* and is easy to purify by thermal or osmotic shock (LaVallie *et al.*, 1993; Zhou *et al.*, 2010); and, as previously mentioned, it is thermostable and allows one to conduct biofabrication schemes at up to 85°C, a temperature at which most mesophilic proteins unfold (Kitayaporn *et al.*, 2012). Other internal permissive sites that have been used to build solid-binding designer proteins include those present towards the N-terminus of alkaline phosphatase (Brown, 1997, 2001; Kacar *et al.*, 2009), between the two structural domains of TraI (Hall Sedlak *et al.*, 2010), and towards the C-terminus of the same protein (Dai *et al.*, 2005). For both internal and end fusions, one should keep in mind that a highly hydrophobic SBP might cause the chimeric protein to aggregate and that certain amino acid sequences will increase proteolytic susceptibility because they destabilize the fold or contain protease recognition sites (B.L. Coyle, unpublished data).

Biomimetic layer-by-layer mineralization

In traditional layer-by-layer (LbL) assembly processes, layers of positively and negatively charged molecules or particles are alternatively deposited onto a surface with intermediate wash steps to build composite thin films that are usually about 1 µm thick. By taking advantage of the MinTBP-1-modified light chain ferritin variant (MinTBP-1-LCF; see above), Shiba and coworkers expanded LbL to the biomimetic world (Sano *et al.*, 2006). In the so-called BioLbL approach (Fig. 3.5A), the ability of minTBP-1 to bind to titanium is first used to form a base apoferritin monolayer. Because minTBP-1 also binds to Si, those SBPs that do not contact the titanium surface are used to mineralize a capping silica layer from a tetramethylorthosilicate (TMOS) precursor. A fresh monolayer of MinTBP-1-LCF is then assembled on the silica substrate by relying on minTBP-1 ability to bind to Si. The process is repeated to stack multiple protein layers on intervening silica layers (Fig. 3.5A). One of the most attractive aspects of BioLbL is that a broad range of materials can be mineralized within the apoferritin cage (see above). Thus, Sano and coworkers were able to stack thin layers of alternative chemistries (Fe, Co, CdSe) separated by silica layers (Fig. 3.5A) (Sano *et al.*, 2006). This ingenious process was later expanded to intervening titania (rather than silica) layers, which allowed the group to pattern a Pt base layer with stripes of titania via traditional lithography while conserving their ability to build multilayer structures through BioLBL (Fig. 3.5B). The resulting approach (called DP-BioLBL for direct patterning biomimetic layer by layer) combines control of chemistry in the z-direction with positional control in the x–y plane (Sano *et al.*, 2007).

Assembling nanoparticles on DNA guides with designer proteins

With recent progress in design algorithms and lower costs for nucleic acid synthesis, DNA is becoming an increasingly useful building block for the construction of intricate nanostructures and the organization of inorganic nanoparticles into well-defined crystallographic symmetries (Cutler *et al.*, 2012; Yang *et al.*, 2010). The many proteins that bind to single- or double-stranded nucleic acids (and do so with or without sequence specificity) provide an orthogonal chemistry. If these DNA-binding proteins are further engineered with one or more SBPs, an additional inorganic binding or mineralizing capability becomes available.

In one such example, a TiO_2 binding peptide was fused to the C-terminus of the *E. coli*

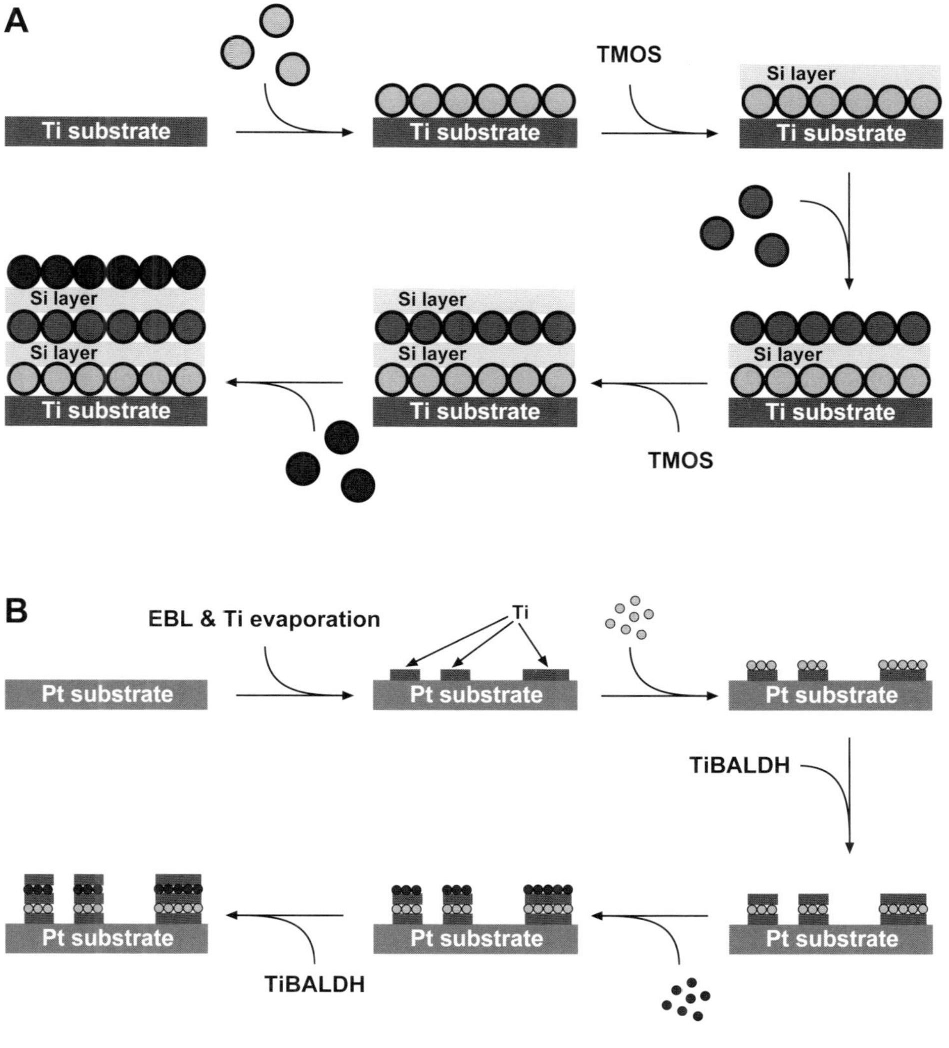

Figure 3.5 Biomimetic layer-by-layer assembly. (A) The original BioLbL process exploits the dual ability of the MinTBP-1-LCF ferritin variant to mineralize nanodots of various chemical compositions within its cavity (different colour circles) while binding Ti and mineralizing silica through the 24 surface-exposed SBPs. Prehydrolysed TMOS serves as a precursor for silica layer mineralization on top of each ferritin layer. (B) In the DB-BioLBL process, a traditional electron beam lithography step followed by evaporation of Ti and developing is used to create Ti patterns on a Pt substrate. The ability of MinTBP-1-LCF containing nanodots of different composition to both bind Ti and mineralize titania from a Ti(IV) bis-(ammonium lactato)-dihydroxide (TiBALDH) precursor is used to build in the *z* direction. See text for details.

lactose operon LacI repressor protein and the resulting fusion protein was used to tether TiO_2 nanoparticles to plasmid DNA containing one or two *lacO* binding sequences (Chen *et al.*, 2009a). TraI has also proven valuable to build composite architectures (Dai *et al.*, 2005; Hall Sedlak *et al.*, 2010). This protein is unique in that its relaxase domain exhibits sequence-specific single-stranded-DNA-binding activity while its helicase domain binds single and double-stranded-DNA non-specifically. More importantly, several permissive sites suitable for the insertion of SBPs have been identified (Haft *et al.*, 2006). One of these sites (at codon 369) lies in the linker region connecting the protein's functional domains and insertion of a gold binding peptide at this location yielded a chimeric protein capable of coupling gold nanoparticles to single-stranded DNA although particle coverage was sparse (Hall Sedlak *et al.*, 2010).

The double-stranded DNA binding ability of TraI has also been used to assemble mineralized products onto circular DNA guides. As discussed above, TraIi1753::CN225, a TraI variant containing a cuprous oxide binder near its C-terminus, is suitable for producing nanoparticles consisting of a 2 nm Cu_2O core capped by a protein shell (Dai *et al.*, 2005). When these particles were mixed with the double-stranded relaxed form II DNA of bacteriophage *f*X174 (which adopts a circular conformation when one of its two strands is nicked), they spontaneously assembled to form smooth circles that were 100 nm diameter circles and consisted of tightly packed particles with evenly separated inorganic cores (Dai *et al.*, 2005).

Self-assembled one-dimensional nanostructures

A number of proteins including amyloid proteins, keratin, collagen, tubulin and actin assemble into high aspect ratio fibrils that are suitable for mineralizing or organizing inorganic materials into functional nanostructures. Such protein fibres are stiffer than DNA and remain stable over the broad range of temperatures and aggressive chemistries that are generally needed to process a nanomaterial (e.g. via electroplating). In a seminal demonstration of the usefulness of protein fibrils in nanoelectronics, Lindquist and coworkers exploited the ability of the N-terminal and middle regions (NM) of the *S. cerevisiae* Sup35p prion protein to self-assemble into β-sheet-rich amyloid fibres to build nanowires of gold and silver (Scheibel *et al.*, 2003). Short (60–200 nm long) recombinant NM fibres were produced by vigorous agitation and could be used at different ratios to seed the polymerization of longer fibres ranging in length from 500 nm to 5 μm. Because those structures were electrically insulating, Scheibel and coworkers modified NM by engineering a surface-exposed cysteine that was used for coupling maleimide-derivatized gold nanoparticles 1.4 nm in diameter. A continuous sheath of silver or gold was finally obtained by electroless deposition of gold or silver. The resulting metallic fibres were 100 nm in width and conducted electricity with low resistance (Scheibel *et al.*, 2003).

In an attempt to bypass the metallization process, Ostrov and Gazit (2010) fused the silver-mineralizing peptide AG4 at the N-terminus of *E. coli* FtsZ, a prokaryotic homologue of tubulin that forms fibres upon addition of GTP or cations. When AG4-FtsZ polymerization was induced with GTP in a solution of silver nitrate, filaments decorated with mineralized Ag nanoparticles were visualized by TEM. However, the biomineralized particles were heterogeneous in size and there were gaps in coverage. Replacing the AG4 moiety with the gold-binding peptide p8#9 (VSGSSPDS) (Huang *et al.*, 2005) and supplementing the GTP polymerization mixture with synthetic 2.5 nm Au particles led to the organization of colloidal gold onto the protein fibre, but once again coverage was incomplete (Ostrov and Gazit, 2010). This study underscores two important points. First, any biomineralization event that does not involve a capping step or controls the flux of precursors will likely lead to the formation of polydisperse particles. Second, how SBPs will be positioned with respect to one another in a higher order designer protein structure is an important consideration for practical applications.

Complex and multifunctional nanostructures

In addition to providing an environmentally friendly route to nanofabrication through biomineralization, one of the main advantages of designer proteins is that they can be modified by genetic engineering to build complex architectures that are suitable for harnessing unique nanoscale properties or imparting useful functionalities to hybrid nanostructures. In an early demonstration of protein-driven multi-material assembly, Brown (2001) constructed 6GB-bio, an artificial protein consisting of six tandem repeats of a gold binding peptide (QATSIGVEKLAGMAESKPTKT) fused to the biotin acceptor peptide GGLNDIFEAQKIEWHEDS (Beckett *et al.*, 1999). Following *in vitro* biotinylation of the C-terminal tail with the enzyme biotin holoenzyme synthetase (BHS), 6GB-bio was immobilized onto streptavidin-coated polystyrene spheres. These conjugates were next assembled onto larger gold particles by relying on interactions between the fusion protein's gold binding domains and the Au surface (Fig. 3.6A) (Brown, 2001). More recently, an FtsZ variant containing an N-terminal biotin acceptor peptide was used to couple avidin-coated magnetic beads through GTP-induced polymerization of the biotinylated protein (Fig. 3.6B). The protein filaments were 2–4 nm thick and 200–800 nm in length and often aligned into larger sheets (Ostrov and Gazit, 2010).

Metal-enhanced fluorescence (MEF) is a near-field phenomenon in which the photoluminescence of a fluorophore (e.g. a quantum dot) is increased through interactions with the local electric field of metal nanoparticles exposed to incident illumination (Aslan *et al.*, 2005). Although it is a very attractive means of increasing brightness, MEF requires precise control of the distance separating quantum dot and metallic particles as loss of enhancement occurs when the separation is too large while fluorescence is quenched if the metal is too close (Haridas *et al.*, 2011). Taking advantage that proteins are in the correct size range to harness field enhancement effects, Leong *et al.* (2010) made use of a biotinylated gold binding peptide called *l*-AuBP2 (WALRRSIRRQSY), gold nanoparticles and a fabricated gold nanoarray to tune the distance separating biotinylated quantum dots and gold nanostructures. In the optimal architecture (consisting of the fabricated gold array, and successive layers of biotinylated *l*-AuBP2, streptavidin, biotinylated quantum dots, streptavidin, biotinylated *l*-AuBP2, and gold nanoparticles; Fig. 3.6C) a separation of 17.5 nm between Au nanostructures and quantum dots creates a plasmonic 'hot spot' around the nanocrystal that leads to a 15-fold increase in emission brightness (Leong *et al.*, 2010).

In addition to the biotin-streptavidin system, other protein–protein interactions can be exploited to build complex architectures. Fig. 3.6D provides a good illustration of the process. Starting with a TrxA derivative incorporating the CT43 ZnS-mineralizing peptide, we fused tandem repeats of the B from *Staphylococcus aureus* protein A to the N-terminus of the protein to endow it with the ability to bind to the Fc fragment of several classes of IgG. The resulting chimera not only remained suitable for the production of luminescent ZnS nanocrystals but its BB domain could bind antibodies with a K_d of 60 nM (Zhou *et al.*, 2010). Furthermore, addition of a small amount of transition metal ions during the biomineralization process changed the maximum emission wavelength from 430 nm (undoped) to 480 nm (Cu^{2+}) or 590 nm (Mn^{2+}), and the amount of antibodies bound by the quantum dot protein shell could be controlled by modifying the molar ratio of BB-TrxA::CT43 to IgG (Zhou and Baneyx, 2011).

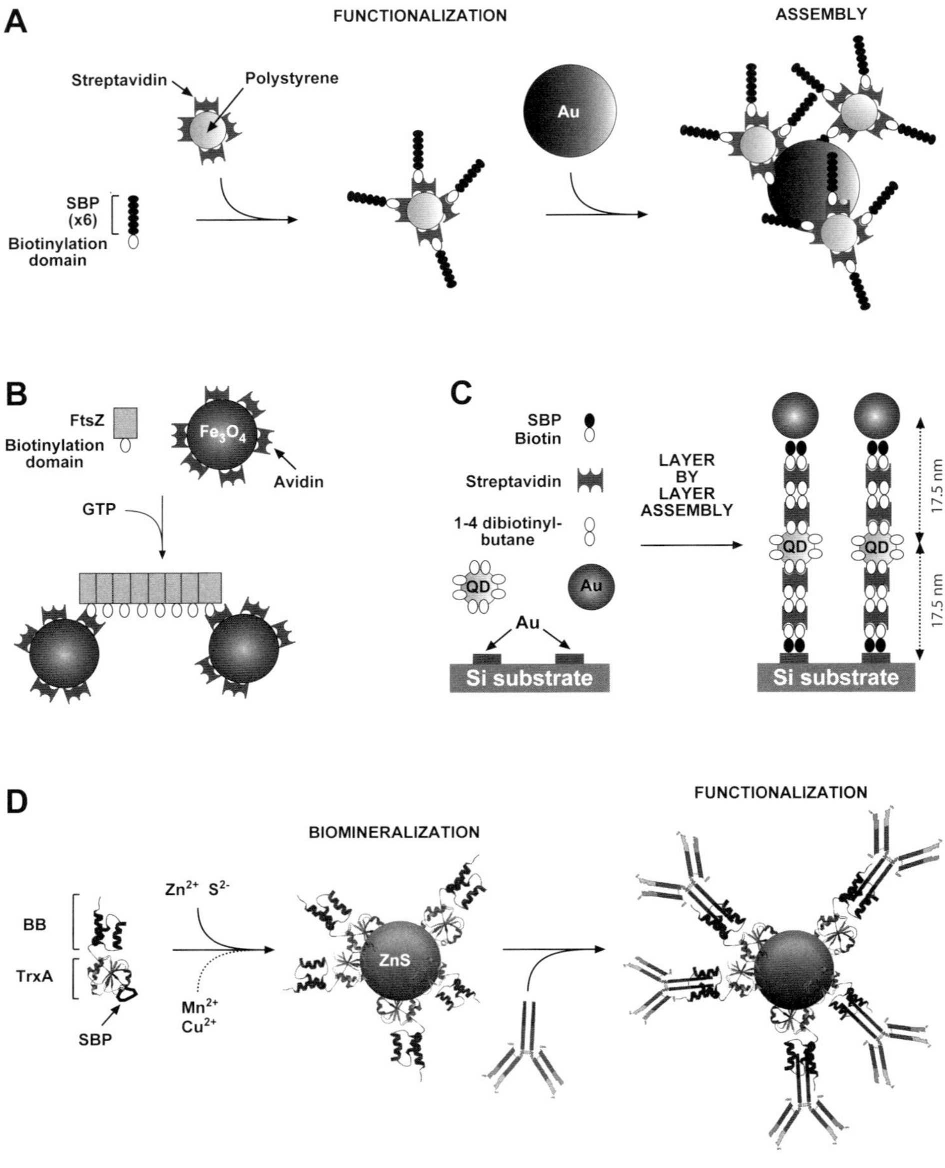

Figure 3.6 Using designer proteins to build complex architectures. (A) A fusion between six repeats of a gold binding peptide and a biotinylation domain is used to assemble streptavidin-coated polystyrene spheres onto larger gold particles. (B) A fusion between a biotinylation domain and the cell division protein FtsZ is used to couple avidin-coated magnetic particles via GTP-induced polymerization. (C) A biotinylated gold binding peptide, biotin-coated quantum dots (QDs), 1,4-dibiotinyl-butane and streptavidin are used to precisely separate patterned gold structures and gold nanaparticles so that a plasmonic hot spot is generated around the QD. (D) A fusion between an antibody-binding domain (BB) and a ZnS-binding TrxA derivative is used to mineralize and cap ZnS nanocrystals that can be derivatized with antibodies.

Conclusion

Proteins hold great potential in nanotechnology due to their ability to precisely control the shape, size, crystal structure, stability and activity of inorganic materials. Inspiration from organisms that mineralize and assemble inorganic phases for improved structure and function quickly led to the modification of natural proteins for the synthesis of technologically relevant materials, as exemplified in this chapter with the apoferritin case study. By harnessing the power of combinatorial display, molecular biomimetics has made it possible to endow a variety of proteins with inorganic binding or synthesizing activity without disrupting their original function. These designer proteins are suitable for building complex hybrid architectures that take full advantage of nanoscale effects (e.g. quantum confinement and plasmonic enhancement).

Yet, building with proteins can be expensive and the application space remains so far limited to high value-added structures for nanoelectronics, biology, diagnostics and nanomedicine. While this chapter focused on protein-aided nanofabrication, it is important to note that SBPs can also be incorporated within the coat of viruses or on the surface of microorganisms to cheaply mineralize inorganic nanoparticles for commodity products ranging from batteries to catalysts (Bansal *et al.*, 2006; Lee *et al.*, 2009; Manocchi *et al.*, 2010; Nam *et al.*, 2006; Neltner *et al.*, 2010). These approaches have been reviewed exhaustively elsewhere (Hyman, 2012; Lee *et al.*, 2012; Mohanpuria *et al.*, 2007; Narayanan and Sakthivel, 2010).

The next major advance in molecular biomimetics will be to rationally design proteins for inorganic synthesis, multifunctionality, and complex structure assembly. This remains a difficult task since *de novo* scaffold design is challenging and because small changes in SBP concentration, identity, and structure can result in large changes in the material produced. Progress in the computational design of artificial proteins and in elucidating the mechanisms underpinning protein–inorganic interactions will undoubtedly drive this exciting field forward.

References

Aryal, B.P., and Benson, D.E. (2007). Polyhistidine fusion proteins can nucleate the growth of CdSe nanoparticles. Bioconjug. Chem. *18*, 585–589.

Aslan, K., Gryczynski, I., Malicka, J., Matveeva, E., Lakowicz, J.R., and Geddes, C.D. (2005). Metal-enhanced fluorescence: an emerging tool in biotechnology. Curr. Opin. Biotechnol. *16*, 55–62.

Baeuerlein, E. (2004). Biomineralization: Progress in Biology, Molecular Biology and Application (Wiley-VCH).

Bansal, V., Poddar, P., Ahmad, A., and Sastry, M. (2006). Room-temperature biosynthesis of ferroelectric barium titanate nanoparticles. J. Am. Chem. Soc. *128*, 11958–11963.

Beckett, D., Kovaleva, E., and Schatz, P.J. (1999). A minimal peptide substrate in biotin holoenzyme synthetase-catalyzed biotinylation. Protein science: a publication of the Protein Society *8*, 921–929.

Boder, E.T., and Wittrup, K.D. (1997). Yeast surface display for screening combinatorial polypeptide libraries. Nat. Biotechnol. *15*, 553–557.

Boder, E.T., and Wittrup, K.D. (1998). Optimal screening of surface-displayed polypeptide libraries. Biotechnol. Prog. *14*, 55–62.

Bradbury, A.R., and Marks, J.D. (2004). Antibodies from phage antibody libraries. J. Immunol. Meth. *290*, 20.

Brown, S. (1997). Metal-recognition by repeating polypeptides. Nat. Biotechnol. *15*, 269–272.

Brown, S. (2001). Protein-mediated particle assembly. Nano. Lett. *1*, 391–394.

Chen, C.L., Zhang, P., and Rosi, N.L. (2008). A new peptide-based method for the design and synthesis of nanoparticle superstructures: construction of highly ordered gold nanoparticle double helices. J. Am. Chem. Soc. *130*, 13555–13557.

Chen, H., Su, X., Neoh, K.-G., and Choe, W.S. (2009a). Engineering LacI for self-assembly of inorganic nanoparticles on DNA scaffold through the understanding of LacI binding to solid surfaces. Adv. Func. Mater. *19* 1186–1192.

Chen, H.B., Su, X.D., Neoh, K.-G., and Choe, W.S. (2009b). Context-dependent adsorption behavior of cyclic and linear peptides on metal oxide. Langmuir *25*, 1588–1593.

Cheng, M.M.-C., Cuda, G., Bunimovich, Y.L., Gaspari, M., Heath, J.R., Hill, H.D., Mirkin, C.A., Nijdam, A.J., Terracciano, R., Thundat, T., *et al.* (2006). Nanotechnologies for biomolecular detection and medical diagnostics. Curr. Opin. Chem. Biol. *10*, 11–19.

Chiu, D., Zhou, W., Kitayaporn, S., Schwartz, D.T., Murali-Krishna, K., Kavanagh, T.J., and Baneyx, F. (2012). Biomineralization and size control of stable calcium phosphate core–protein shell nanoparticles: potential for vaccine applications. Bioconjugate *Chem. 23*, 610–617.

Choe, W.S., Sastry, M.S.R., Thai, C.K., Dai, H., Schwartz, D.T., and Baneyx, F. (2007). Conformational control of inorganic adhesion in a designer protein engineered for cuprous oxide binding. Langmuir *23*, 11347–11350.

Cutler, J.I., Auyeung, E., and Mirkin, C.A. (2012). Spherical nucleic acids. J. Am. Chem. Soc. *134*, 1376–1391.

Dai, H., Choe, W.-S., Thai, C.K., Sarikaya, M., Traxler, B.A., Baneyx, F., and Schwartz, D.T. (2005). Nonequilibrium Synthesis and Assembly of Hybrid Inorganic–Protein Nanostructures Using an Engineered DNA Binding Protein. J. Am. Chem. Soc. *127*, 15637–15643.

Daugherty, P.S. (2007). Protein engineering with bacterial display. Curr. Opin. Struct. Biol. *17*, 474–480.

Dickerson, M.B., Sandhage, K.H., and Naik, R.R. (2008). Protein- and Peptide-Directed Syntheses of Inorganic Materials. Chem Rev *108*, 4935–4978.

Douglas, T., Allen, M., and Young, M. (2005). Self-assembling Protein Cage Systems and Applications in Nanotechnology. In Biopolymers Online (Wiley-VCH Verlag GmbH & Co. KGaA).

Douglas, T., Dickson, D.P.E., Betteridge, S., Charnock, J., Garner, C.D., and Mann, S. (1995). Synthesis and structure of an iron(III) sulfide-ferritin bioinorganic nanocomposite. Science *269*, 54–57.

Endemann, H., and Model, P. (1995). Location of filamentous phage minor coat proteins in phage and in infected cells. J. Mol. Biol. *250*, 496–506.

Gálvez, N., Sánchez, P., and Domínguez-Vera, J.M. (2005). Preparation of Cu and CuFe Prussian Blue derivative nanoparticles using the apoferritin cavity as nanoreactor. Dalton Trans. 2492–2494.

Georgiou, G., Stathopoulos, C., Daugherty, P.S., Nayak, A.R., Iverson, B.L., and Curtiss, R., 3rd (1997). Display of heterologous proteins on the surface of microorganisms: from the screening of combinatorial libraries to live recombinant vaccines. Nat. Biotechnol. *15*, 29–34.

Gold, L. (1995). Oligonucleotides as research, diagnostic, and therapeutic agents. J. Biol. Chem. *270*, 13581–13584.

Gorski, J.P. (1998). Is all bone the same? distinctive distributions and properties of non-collagenous matrix proteins in lamellar vs. woven bone imply the existence of different underlying osteogenic mechanisms. Crit. Rev. Oral Biol. Med. *9*, 201–223.

Grosh, C.D., Schwartz, D.T., and Baneyx, F. (2009). Protein-based control of silver growth habit using electrochemical deposition. Crystal Growth & Design *9*, 4401–4406.

Haft, R.J., Palacios, G., Nguyen, T., Mally, M., Gachelet, E.G., Zechner, E.L., and Traxler, B. (2006). General mutagenesis of F plasmid TraI reveals its role in conjugative regulation. J. Bacteriol. *188*, 6346–6353.

Hall Sedlak, R., Hnilova, M., Gachelet, E., Przybyla, L., Dranow, D., Gonen, T., Sarikaya, M., Tamerler, C., and Traxler, B. (2010). An engineered DNA-binding protein self-assembles metallic nanostructures. ChemBioChem *11*, 2108–2112.

Haridas, M., Tripathi, L.N., and Basu, J.K. (2011). Photoluminescence enhancement and quenching in metal-semiconductor quantum dot hybrid arrays. Appl. Phys. Lett. *98*, 063305–063305.

Harrison, P.M., and Arosio, P. (1996). The ferritins: molecular properties, iron storage function and cellular regulation. Biochim. Biophys. Acta (BBA) – Bioenergetics *1275*, 161–203.

Hnilova, M., Oren, E.E., Seker, U.O.S., Wilson, B.R., Collino, S., Evans, J.S., Tamerler, C., and Sarikaya, M. (2008). Effect of molecular conformations on the adsorption behaviour of gold-binding peptides. Langmuir *24*, 12440–12445.

Holmgren, A. (1985). Thioredoxin. Annu. Rev. Biochem. *54*, 237–271.

Huang, Y., Chiang, C.-Y., Lee, S.K., Gao, Y., Hu, E.L., Yoreo, J.D., and Belcher, A.M. (2005). Programmable assembly of nanoarchitectures using genetically engineered viruses. Nano Lett. *5*, 1429–1434.

Hyman, P. (2012). Bacteriophages and nanostructured materials. Adv. Appl. Microbiol. *78*, 55–73.

Iwahori, K., Enomoto, T., Furusho, H., Miura, A., Nishio, K., Mishima, Y., and Yamashita, I. (2007). Cadmium sulfide nanoparticle synthesis in Dps protein from *Listeria innocua*. Chem. Mater. *19*, 3105–3111.

Iwahori, K., Morioka, T., and Yamashita, I. (2006). The optimization of CdSe nanoparticles synthesis in the apoferritin cavity. Physica Status Solidi (A) Appl. Res. *203*, 2658–2661.

Iwahori, K., Yoshizawa, K., Muraoka, M., and Yamashita, I. (2005). Fabrication of ZnSe nanoparticles in the apoferritin cavity by designing a slow chemical reaction system. Inorg. Chem. *44*, 6393–6400.

Kacar, T., Zin, M.T., So, C., Wilson, B., Ma, H., Gul-Karaguler, N., Jen, A.K., Sarikaya, M., and Tamerler, C. (2009). Directed self-immobilization of alkaline phosphatase on micro-patterned substrates via genetically fused metal-binding peptides. Biotechnol. Bioeng. *103*, 696–705.

Kehoe, J.W., and Kay, B.K. (2005). Filamentous phage display in the new millennium. Chem. Rev. *105*, 4056–4072.

Kitayaporn, S., Zhou, W., Schwartz, D.T., and Baneyx, F. (2012). Laying out ground rules for protein-aided nanofabrication: ZnO synthesis at 70°C as a case study. Biotechnol. Bioeng. *109*, 1912–1918.

Klem, M.T., Mosolf, J., Young, M., and Douglas, T. (2008). Photochemical mineralization of europium, titanium, and iron oxyhydroxide nanoparticles in the ferritin protein cage. Inorg. Chem. *47*, 2237–2239.

Kramer, R.M., Li, C., Carter, D.C., Stone, M.O., and Naik, R.R. (2004). Engineered protein cages for nanomaterial synthesis. J. Am. Chem. Soc. *126*, 13282–13286.

Kröger, N. (2007). Prescribing diatom morphology: toward genetic engineering of biological nanomaterials. Curr. Opin. Chem. Biol. *11*, 662–669.

LaVallie, E.R., DiBlasio, E.A., Kovacic, S., Grant, K.L., Schendel, P.F., and McCoy, J.M. (1993). A thioredoxin gene fusion expression system that circumvents inclusion body formation in the *E. coli* cytoplasm. Nat. Biotechnol. *11*, 187–193.

LaVallie, E.R., Lu, Z., Diblasio-Smith, E.A., Collins-Racie, L.A., and McCoy, J.M. (2000). Thioredoxin as a fusion partner for production of soluble recombinant proteins in *Escherichia coli*. Methods Enzymol. *326*, 322–340.

Lawson, D.M., Artymiuk, P.J., Yewdall, S.J., Smith, J.M., Livingstone, J.C., Treffry, A., Luzzago, A., Levi, S., Arosio, P., and Cesareni, G. (1991). Solving the structure of human H ferritin by genetically engineering intermolecular crystal contacts. Nature *349*, 541–544.

Lee, S.Y., Lim, J.S., and Harris, M.T. (2012). Synthesis and application of virus-based hybrid nanomaterials. Biotechnol. Bioeng. *109*, 16–30.

Lee, Y.J., Yi, H., Kim, W.-J., Kang, K., Yun, D.S., Strano, M.S., Ceder, G., and Belcher, A.M. (2009). fabricating genetically engineered high-power lithium-ion batteries using multiple virus genes. Science *324*, 1051–1055.

Leong, K., Chen, Y., Masiello, D.J., Zin, M.T., Hnilova, M., Ma, H., Tamerler, C., Sarikaya, M., Ginger, D.S., and Jen, A.K.Y. (2010). Cooperative near-field surface plasmon enhanced quantum dot nanoarrays. Advanc. Funct. Mater. *20*, 2675–2682.

Lu, W., and Lieber, C.M. (2007). Nanoelectronics from the bottom up. Nat. Mater. *6*, 841–850.

Lu, Z., Murray, K.S., Van Cleave, V., LaVallie, E.R., Stahl, M.L., and McCoy, J.M. (1995). Expression of thioredoxin random peptide libraries on the *Escherichia coli* cell surface as functional fusions to flagellin: a system designed for exploring protein–protein interactions. BioTechnology *13*, 366–372.

Mann, S. (2002). Biomineralization (Oxford: Oxford University Press).

Manocchi, A.K., Horelik, N.E., Lee, B., and Yi, H. (2010). Simple, readily controllable palladium nanoparticle formation on surface-assembled viral nanotemplates. Langmuir *26*, 3670–3677.

Manoil, C., and Traxler, B. (2000). Insertion of in-frame sequence tags into proteins using transposons. Methods *20*, 55–61.

Maratea, D., and Blakemore, R.P. (1981). *Aquaspirillum magnetotacticum* sp. nov., a magnetic spirillum. Int. J. Syst. Bacteriol. *31*, 452–455.

Mayer, G., and Sarikaya, M. (2002). Rigid biological composite materials: structural examples for biomimetic design. Exp. Mech. *42*, 395–403.

Meldrum, F.C., Heywood, B.R., and Mann, S. (1992). Magnetoferritin: in vitro synthesis of a novel magnetic protein. Science *257*, 522–523.

Meldrum, F.C., Wade, V.J., Nimmo, D.L., Heywood, B.R., and Mann, S. (1991). Synthesis of inorganic nanophase materials in supramolecular protein cages. Nature *349*, 684–687.

Mohanpuria, P., Rana, N.K., and Yadav, S.K. (2007). Biosynthesis of nanoparticles: technological concepts and future applications. Journal of Nanoparticle Research *10*, 507–517.

Naik, R.R., Stringer, S.J., Agarwal, G., Jones, S.E., and Stone, M.O. (2002). Biomimetic synthesis and patterning of silver nanoparticles. Nat Mater *1*, 169–172.

Nam, K.T., Kim, D.-W., Yoo, P.J., Chiang, C.-Y., Meethong, N., Hammond, P.T., Chiang, Y.-M., and Belcher, A.M. (2006). Virus-enabled synthesis and assembly of nanowires for lithium ion battery electrodes. Science *312*, 885–888.

Narayanan, K.B., and Sakthivel, N. (2010). Biological synthesis of metal nanoparticles by microbes. Advances in Colloid and Interface Science *156*, 1–13.

Neltner, B., Peddie, B., Xu, A., Doenlen, W., Durand, K., Yun, D.S., Speakman, S., Peterson, A., and Belcher, A. (2010). Production of hydrogen using nanocrystalline protein-templated catalysts on M13 phage. ACS Nano. *4*, 3227–3235.

Ostrov, N., and Gazit, E. (2010). Genetic engineering of biomolecular scaffolds for the fabrication of organic and metallic nanowires. Angew Chem. Int. Ed. Engl. *49*, 3018–3021.

Pacardo, D.B., Sethi, M., Jones, S.E., Naik, R.R., and Knecht, M.R. (2009). Biomimetic Synthesis of Pd Nanocatalysts for the Stille Coupling Reaction. ACS Nano. *3*, 1288–1296.

Papapostolou, D., and Howorka, S. (2009). Engineering and exploiting protein assemblies in synthetic biology. Mol. Biosys. *5*, 723–732.

Parmley, S.F., and Smith, G.P. (1988). Antibody-selectable filamentous fd phage vectors: affinity purification of target genes. Gene *73*, 305–318.

Sano, K., Ajima, K., Iwahori, K., Yudasaka, M., Iijima, S., Yamashita, I., and Shiba, K. (2005). Endowing a ferritin-like cage protein with high affinity and selectivity for certain inorganic materials. Small *1*, 826–832.

Sano, K., Yoshii, S., Yamashita, I., and Shiba, K. (2007). In aqua structuralization of a three-dimensional configuration using biomolecules. Nano Lett *7*, 3200–3202.

Sano, K.-I., Sasaki, H., and Shiba, K. (2006). Utilization of the Pleiotropy of a Peptidic Aptamer To Fabricate Heterogeneous Nanodot-Containing Multilayer Nanostructures. J. Am. Chem. Soc. *128*, 1717–1722.

Sarikaya, M., Tamerler, C., Jen, A.K.Y., Schulten, K., and Baneyx, F. (2003). Molecular biomimetics: nanotechnology through biology. Nat. Mater. *2*, 577–585.

Sarikaya, M., Tamerler, C., Schwartz, D.T., and Baneyx, F. (2004). Materials assembly and formation using engineered polypeptides Annu. Rev. Mater. Res. *34*, 373–408.

Sastry, M.S., Zhou, W., and Baneyx, F. (2009). Integrity of N- and C-termini is important for E. coli Hsp31 chaperone activity. Protein Sci. *18* 1439–1447.

Scheibel, T., Parthasarathy, R., Sawicki, G., Lin, X.M., Jaeger, H., and Lindquist, S.L. (2003). Conducting nanowires built by controlled self-assembly of amyloid fibres and selective metal deposition. Proc. Natl. Acad. Sci. U.S.A. *100*, 4527–4532.

Schleifer, K.H., Schueler, D., Spring, S., Weizenegger, M., Amann, R., Lduwig, W., and Koehler, M. (1991). The genus magnetospirillum gen. nov. description of Magnetospirillum gryphiswaldense sp. nov. and transfer of *Aquaspirillum magnetotacticum* to *Magnetospirillum magnetotacticum* comb. nov. Syst. Appl. Microbiol. *14*, 379–385.

Seker, U.O.S., Wilson, B., Dincer, S., Kim, I.W., Oren, E.E., Evans, J.S., Tamerler, C., and Sarikaya, M. (2007). Adsorption behaviour of linear and cyclic genetically engineered platinum binding peptides. Langmuir *23*, 7895–7900.

Sengupta, A., Thai, C.K., Sastry, M.S.R., Matthaei, J.F., Schwartz, D.T., Davis, E.J., and Baneyx, F. (2008). A genetic approach for controlling the binding and orientation of proteins on nanoparticles. Langmuir *24*, 2000–2008.

Shimojo, K., Niide, T., Taguchi, T., Naganawa, H., Kamiya, N., and Goto, M. (2012). Facile, rapid and efficient fabrication of gold nanoparticles decorated with functional proteins. Analyst *137*, 2300–2303.

Slocik, J.M., Stone, M.O., and Naik, R.R. (2005). Synthesis of gold nanoparticles using multifunctional peptides. Small 1,1048–1052.

Smith, G.P. (1985). Filamentous fusion phage: novel expression vectors that display cloned antigens on virion surface. Science *228*, 3.

Sun, Y., and Rogers, J.A. (2007). Inorganic semiconductors for flexible electronics. Advanc. Mater. *19*, 1897–1916.

Tamerler, C., Khatayevich, D., Gungormus, M., Kacar, T., Oren, E.E., Hnilova, M., and Sarikaya, M. (2010). Molecular biomimetics: GEPI-based biological routes to technology. Biopolymers *94*, 78–94.

Tao, A.R., Habas, S., and Yang, P. (2008). Shape control of colloidal metal nanocrystals. Small *4*, 310–325.

Thai, C.K., Dai, H., Sastry, M.S.R., Sarikaya, M., Schwartz, D.T., and Baneyx, F. (2004). Identification and characterization of Cu(2)O- and ZnO-binding polypeptides by *Escherichia coli* cell surface display: towards an understanding of metal oxide binding. Biotechnol. Bioeng. *87*, 129–137.

Tomczak, M.M., Gupta, M.K., Drummy, L.F., Rozenzhak, S.M., and Naik, R.R. (2009). Morphological control and assembly of zinc oxide using a biotemplate. Acta Biomater. *5*, 876–882.

Uchida, M., Kang, S., Reichhardt, C., Harlen, K., and Douglas, T. (2010). The ferritin superfamily: Supramolecular templates for materials synthesis. Biochim. Biophys. Acta *1800*, 834–845.
Ueno, T., Suzuki, M., Goto, T., Matsumoto, T., Nagayama, K., and Watanabe, Y. (2004). Size-selective olefin hydrogenation by a Pd nanocluster provided in an apo-ferritin cage. Angewandte Chem. Int. Ed. *43*, 2527–2530.
Wong, K.K.W., and Mann, S. (1996). Biomimetic synthesis of cadmium sulfide-ferritin nanocomposites. Advanc. Mater. *8*, 928–932.
Yamashita, I., Hayashi, J. and Hara, M. (2004). Bio-template synthesis of uniform CdSe nanoparticles using cage-shaped protein, apoferritin. Chem. Lett. *33*, 1158–1159.
Yamashita, I., Kirimura, H., Okuda, M., Nishio, K., Sano, K., Shiba, K., Hayashi, T., Hara, M., and Mishima, Y. (2006). Selective nanoscale positioning of ferritin and nanoparticles by means of target-specific peptides. Small 2, 1148–1152.
Yang, D., Campolongo, M.J., Nhi Tran, T.N., Ruiz, R.C., Kahn, J.S., and Luo, D. (2010). Novel DNA materials and their applications. Wiley Interdiscip. Rev. Nanomed. Nanobiotechnol. *2*, 648–669.
Zhou, W., and Baneyx, F. (2011). Aqueous, protein-driven synthesis of transition metal-doped ZnS immuno-quantum dots. ACS Nano. *5*, 8013–8018.
Zhou, W., Schwartz, D.T., and Baneyx, F.o. (2010). Single-Pot biofabrication of zinc sulfide immuno-quantum dots. J. Am. Chem. Soc. *132*, 4731–4738.

Amyloid Fibrils as Bionanomaterials

4

Jared K. Raynes and Juliet A. Gerrard

Abstract

It is becoming increasingly clear that nature employs amyloid fibrils in a functional role for a range of processes, from immune responses, to aiding in the colonization of bacteria. These functional amyloid fibrils have inspired researchers to investigate the potential of amyloid fibrils as novel bionanomaterials. The amyloid fibril structure possesses many features that make it an ideal candidate for use in bionanomaterials. These include: their nanometer size, which gives rise to a high surface-to-volume ratio enabling high loading capacities of decorations on their surface; the ability to self-assemble, which affords a bottom-up approach to material design; the potential to be manufactured from waste materials; and their diverse chemical functionality, arising from their amino acid composition, which allows for decoration with chemicals and biomolecules *via* amino acid moieties such as amino and sulphur groups. This chapter focuses on the assembly of amyloid fibrils and how these features are enabling their emerging uses as novel bionanomaterials.

Introduction

Amyloid fibrils are a type of highly ordered, insoluble, self-assembling protein nanofibre, first associated with protein misfolding diseases such as Creutzfeldt–Jakob, Huntington's, Parkinson's, Alzheimer's and type II diabetes (Chiti and Dobson, 2006; Harrison *et al.*, 2007; Pedersen and Otzen, 2008). Although amyloid fibrils have generally been associated with disease, it is thought that all proteins share the ability to form amyloid fibrils (Chiti *et al.*, 1999). The lack of similarity between the protein structures and polypeptide sequences of the amyloid forming proteins lends strength to this assertion. Further evidence in support of this hypothesis emerged when a systematic genome wide search for amyloid forming peptide sequences was undertaken. The genome search discovered that almost all proteins contain self-complementary sequences required for amyloid formation (Goldschmidt *et al.*, 2010). The relatively recent discovery of functional amyloid fibrils that do not cause disease, but are in fact beneficial to the host organism, also supports the idea that the amyloid fibril fold is a universal protein fold, not just one that occurs from the misfolding of proteins (Otzen, 2010).

The fact that nature utilizes functional amyloid fibrils, and the increasing evidence that it is the soluble pre-fibrillar oligomers that are the toxic disease causing species (Kirkitadze *et al.*, 2002; Laganowsky *et al.*, 2012), not the mature amyloid fibrils, validates the use of amyloid fibrils for many applications. Amyloid fibrils also possess many desirable features making them ideal for use in bionanotechnology applications. They typically measure 7–20 nm in diameter and up to several μm in length (Goldsbury *et al.*, 2011), providing a very large aspect ratio that is thought of as one dimensional (Xu *et al.*, 2010). This aspect ratio provides a surface-to-volume ratio much greater than an equivalent amount of a conventional material, which allows a higher loading of decorations on their surface. They can also provide a large increase in the surface area of a material with which they are associated (Ping, 2006). The building blocks of amyloid fibrils are amino acids, which provide chemical functionality in the form of the side chain residues, allowing for simple functionalization, as seen with non-fibrillar proteins (Fágáin, 1995). These building blocks also illicit the self-assembling nature of amyloid fibrils, a highly desirable property in the design and manufacture of 'bottom-up' materials (Gras, 2007).

In this chapter, the discussion is restricted to the use of proteins as the starting materials for amyloid fibril formation. The use of self-assembling peptides to create amyloid-like fibril structures has also been widely explored, the interested reader is pointed to excellent recent reviews of the structure (Valéry *et al.*, 2012) and uses (Cherney and Gazit, 2008) of self-assembling peptides in this context.

In addition to their size and ability to be functionalized, the strength of amyloid fibrils is also highly desirable. For example, the strength of insulin amyloid fibrils has been shown to be ~0.6 GPa, which is comparable to steel, and their stiffness (~3.3 GPa) is equivalent to that of natural β-sheet protein silk (Smith *et al.*, 2006). The rigidity of amyloid fibrils arises from the extensive inter-backbone hydrogen bonding network (Knowles *et al.*, 2007). The extensive hydrogen bonding network gives a high level of order and also confers stability at extremes of pH, temperature, pressure, and usually in both aqueous and organic solvents (Dirix *et al.*, 2005). The strength and stability derived from the high level of order within amyloid fibrils makes them highly suited for generation of new materials (Mankar *et al.*, 2011).

The self-assembling nature, nanometer dimensions, ease of functionality, strength, and stability make amyloid fibrils an ideal candidate for use in the manufacture of novel bionanomaterials. Shown in Fig. 4.1 is an overview of some of the many potential uses of amyloid fibrils. One of the main driving forces in the use of amyloid fibrils is their intrinsic self-assembling ability, allowing exquisite fine tuning of the hierarchical assembly process, and therefore material design.

Protein folding and misfolding

Proteins are synthesized at the ribosome where the polypeptide chain elongates and starts folding into a thermodynamically stable structure as defined by the amino acid sequence and the particular environment of the protein (Luheshi and Dobson, 2009). Protein folding is tightly controlled by the cellular protein quality control systems which include chaperones and proteolytic systems (Koga *et al.*, 2011). Chaperones help to increase the rate of folding for proteins and ensure correct folding (Hartl and Hayer-Hartl, 2002). The ubiquitin system on the other hand, works to degrade surplus or damaged proteins by breakdown in the proteasome to limit the misfolding of proteins (Hershko and Ciechanover, 1998; Glickman and Ciechanover, 2002; Ron and Walter, 2007). The energy that the cell puts into ensuring correctly folded and functional proteins illustrates how important protein folding is, as misfolded proteins can cause disease (Koga *et al.*, 2011).

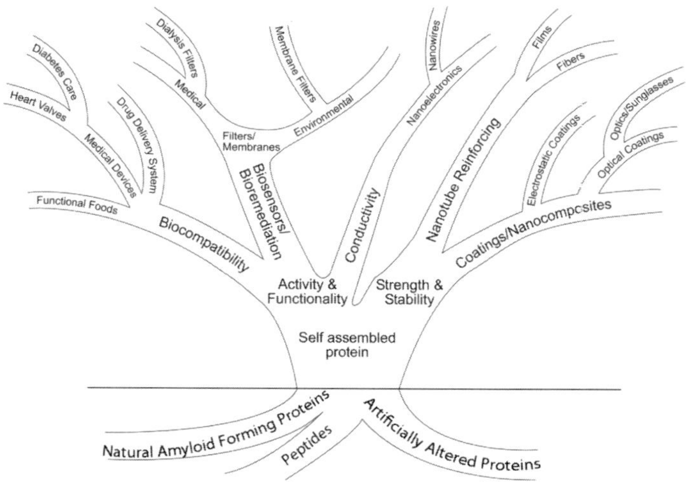

Figure 4.1 Potential uses of amyloid fibrils.

For a protein to fold into its native conformation it must pass through a number of intermediately folded states as shown in Fig. 4.2 (Tyedmers *et al.*, 2010). The kinetics of this process are governed by the environment (protein concentration, pH, temperature, pressure, ionic composition) in which the protein is found (Hamada and Dobson, 2002). For amyloid fibrils to form from natively folded globular proteins, a partial unfolding event needs to occur (Dobson, 1999; Uversky *et al.*, 2002). Partial unfolding can be induced outside the cell by high temperatures, high pressure, low pH or organic solvents (Brange *et al.*, 1997).

The different conformational states that a protein can adopt are driven not only by the environmental conditions in which the protein is found, but also by the entropy of the system (Jahn and Radford, 2008; Mishra and Winter, 2008). This process can be illustrated by an energy landscape (Fig. 4.3), which conceptualizes the folding process by taking into account the energy and entropy of all of the protein conformations. Energy landscapes are generally funnel like, where at the high energy, high entropy part of the funnel there are numerous unfolded conformations which the polypeptide can adopt. The polypeptide moves down the energy gradient via a series of partially folded intermediates to reach an energetically favourable native state (Onuchic and Wolynes, 2004; Wolynes, 2005).

Changes in environmental conditions or polypeptide sequence can influence the topography of the funnel, creating an increase or decrease in the number of energy minima (Mishra and Winter, 2008). Changes in the topography can result in the accumulation of partially folded intermediates at the additional energy minima. Under normal protein folding conditions, these states would be transient; however, upon perturbation of the physiochemical conditions, these partially folded intermediates can accumulate, resulting in non-natural intermolecular interactions. This can result in folding pathways in which the polypeptide chain has access to entropically lower energy states such as amorphous aggregates or amyloid fibrils (Jahn and Radford, 2008).

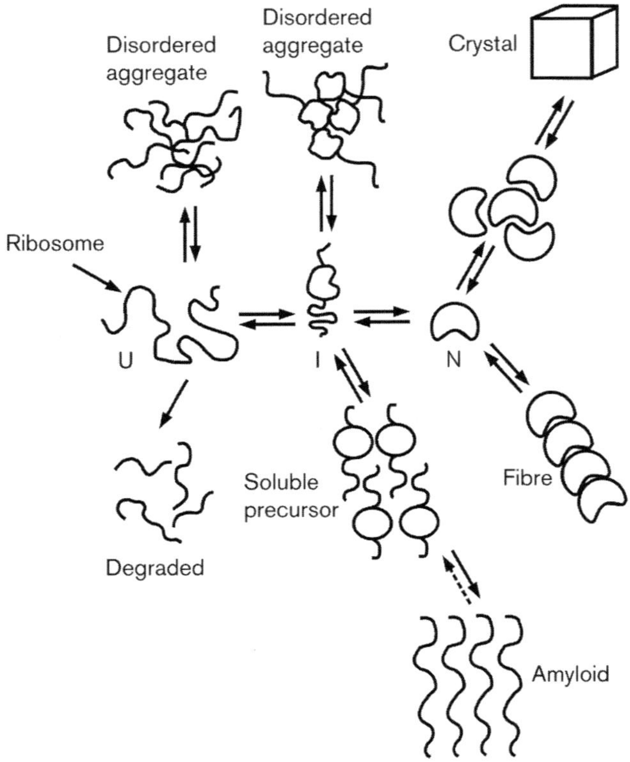

Figure 4.2 The process of protein folding. Reprinted from Dobson and Karplus (1999), with permission from Elsevier.

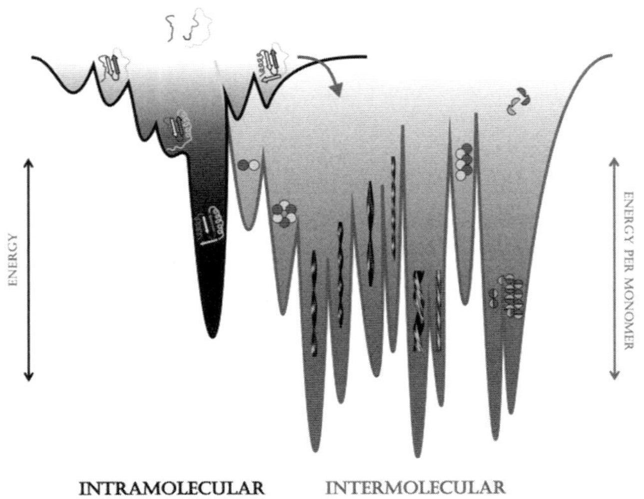

Figure 4.3 A schematic energy landscape of protein folding and aggregation/amyloid fibril formation. Reprinted from Eichner and Radford (2011), with permission from Elsevier. A colour version of this figure is located in the plate section at the back of the book.

Amyloid fibrils – a generic protein structure?

It is widely thought that amyloid fibril formation is a universal property of most, if not all proteins (Chiti *et al.*, 1999). The lack of sequence similarity of the known amyloid-forming proteins and the existence of non-disease forming amyloid proteins which are beneficial to organisms suggests that the amyloid structure is possibly a generic protein fold of all polypeptide chains (Nerelius *et al.*, 2010). However, specific sequences are also known to promote or reduce the tendency of the protein to adopt the amyloid form (Giraldo, 2007; Nerelius *et al.*, 2010). The polypeptide sequences responsible for contributing to amyloid fibril formation have been found to be those that can form a complementary interface with other, usually identical sequences, to form steric zippers (Goldschmidt *et al.*, 2010). Fig. 4.4 highlights the large tertiary structural diversity of the amyloid precursor proteins, illustrating that the amyloid fibril fold is a universally adoptable protein fold (Eichner and Radford, 2011).

Amyloid fibril formation

Amyloid fibril formation is thought to occur through a nucleated-polymerization mechanism (Liang *et al.*, 2010). Examining the time-course of the transition of peptides and proteins into amyloid fibrils using techniques such as thioflavin T (ThT) fluorescence (Fig. 4.5) and light scattering, generally shows a pronounced lag phase, followed by an exponential growth phase (Naiki *et al.*, 1997; Serio *et al.*, 2000; Uversky *et al.*, 2002; Beck *et al.*, 2004). The observed lag phase is thought to be due to the formation of nuclei. Once formed, fibril growth proceeds rapidly through the association of monomers or oligomers.

Schmit *et al.* (2011) have developed a theory to describe the nucleation process. They hypothesized that there are three states of equilibrium for amyloid peptides: the monomer, oligomer, and the fibril (Fig. 4.6). As with other nucleation dependent mechanisms this equilibrium can be shifted with the addition of preformed fibrils or 'seeds' where the lag phase can be shortened or eliminated completely (Naiki *et al.*, 1997; Serio *et al.*, 2000).

Amyloid fibril structure

Identifying the atomic-level structure of amyloid fibrils has been a goal of numerous research groups worldwide. These elusive atomic structures have been extremely difficult to determine because of the non-crystalline and insoluble nature of amyloid fibrils. The first atomic-level structural model of an amyloid–like cross-β spine, which is a signature of all amyloid fibrils, was published in 2005 (Nelson *et al.*, 2005). This X-ray crystal structure came from a seven-residue peptide sequence of the Sup35 protein, which is a yeast prion-like forming protein. The crystal structure of the entire amyloid forming protein has yet to be achieved. Other techniques have been used to gain an understanding of the structure of amyloid fibrils including cryo-electron microscopy (cryo-EM), X-ray fibre diffraction, transmission electron microscopy (TEM), atomic force microscopy (AFM), and most recently, solid-state nuclear magnetic resonance (ssNMR).

The first structures of amyloid fibrils came from TEM, X-ray fibre diffraction, and later AFM (Chiti and Dobson, 2006). These techniques gave vital information about the size, unbranched morphology, the presence of a core structure made up of protofilaments and the repeating nature of the characteristic cross-β spine (Sunde and Blake, 1997; Serpell and Smith, 2000; Serpell *et al.*, 2000). More recently cryo-EM, together with 3D reconstruction

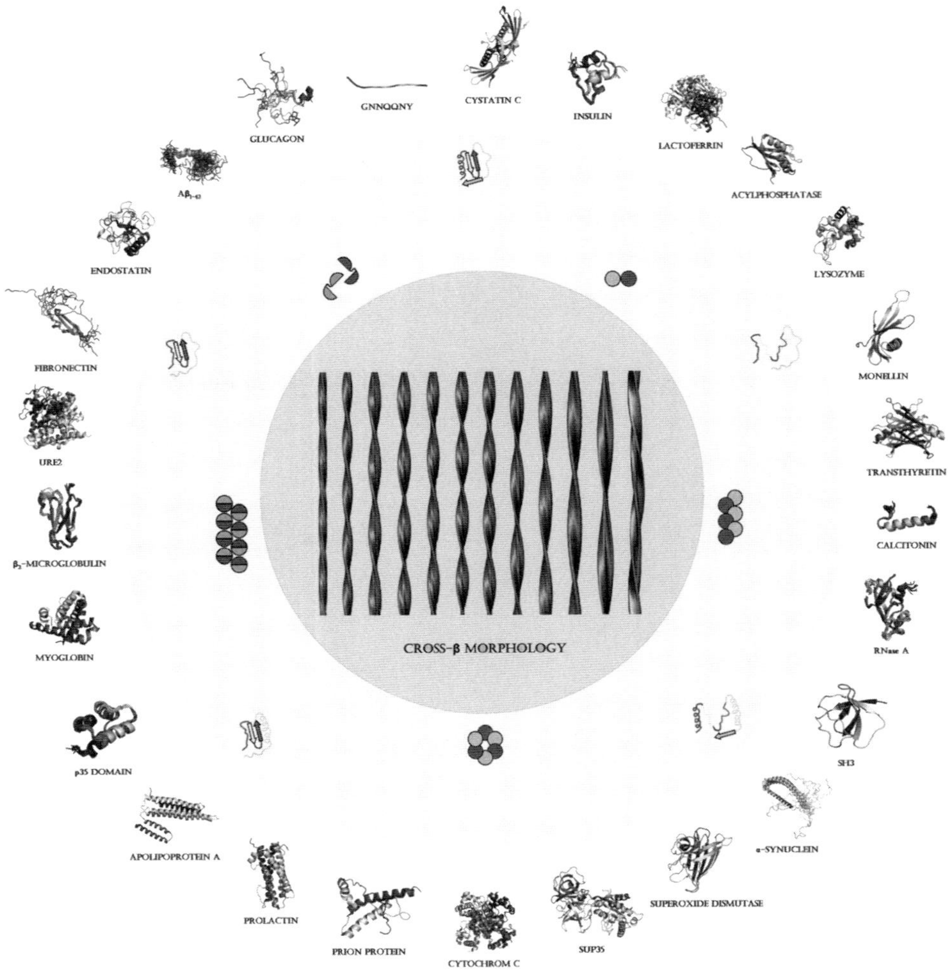

Figure 4.4 The diversity of amyloid precursor protein tertiary structures (reprinted from Eichner and Radford, 2011, with permission from Elsevier.) derived from X-ray crystallography and solution NMR are shown in rainbow colours. The centre (purple area) shows 12 different fibrillar polymorphs of Aβ_{1-40} determined by cryo-EM (reprinted from Meinhardt and Fändrich (2009), with permission from Elsevier). The green and yellow areas depict the monomers and oligomers that may be accessible by a polypeptide chain en route to the amyloid fibril fold. A colour version of this figure is located in the plate section at the back of the book.

software, has provided highly detailed models of amyloid fibrils at resolutions of up to 8 Å (Fig. 4.4) (Meinhardt and Fändrich, 2009; Schmidt *et al.* (2009). Mass-per-length measurements have allowed for the number of peptides (~ 2.5) per cross-β repeat per protofilament for the Aβ(1–42) and Aβ(1–40) peptide fibrils to be modelled (Schmidt *et al.*, 2009).

Since the first atomic level structure of amyloid-like fibrils in 2005 (Nelson *et al.*, 2005), X-ray crystallography has provided a great deal of knowledge about the structure of peptide derived amyloid fibrils. When these structures are used in combination with ssNMR (Tycko, 2011), a detailed understanding of amyloid structures becomes clear. Supplementation of ssNMR data with constraints from lower resolution structures, such as those from

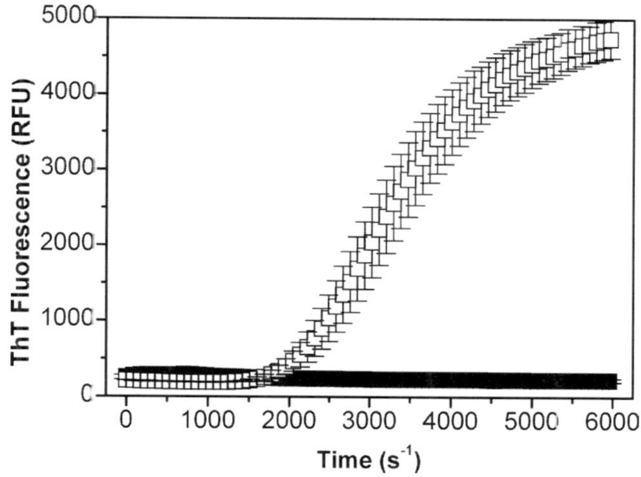

Figure 4.5 Time course profile of bovine insulin (□) amyloid fibril formation and a buffer control (■) at 60°C, as monitored by ThT fluorescence. Data are from the authors' laboratory. Each sample was measured in triplicate, and the error bars are the standard deviation of the mean.

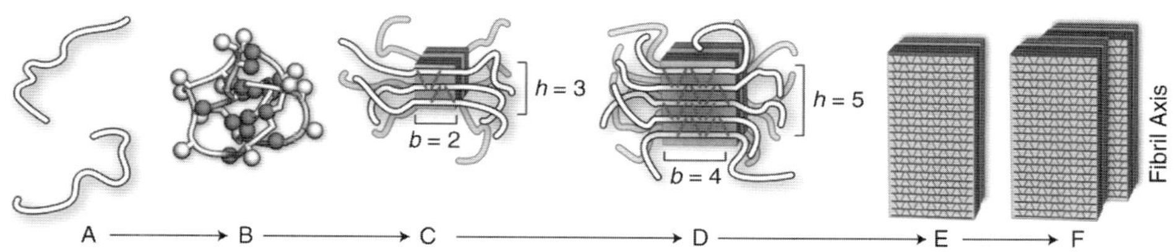

Figure 4.6 A model process of amyloid fibril formation via a nucleated growth mechanism. Reprinted from Schmit *et al.* (2011), with permission from Elsevier. '(State A) Isolated peptide monomers in solution. (State B) Oligomeric assembly of a few peptide chains. (State C) Nucleus of β-sheet structure. The peptide backbone runs perpendicular to the fibre axis. (State D) Postcritical nucleus structure showing more β-structure. (State E) A protofilament is a single long thread of β-structure consisting of a β-sandwich and two face-to-face β-sheet planes. (State F) The full fibril, a bundle of protofilaments shown here to contain p=2 protofilament threads.' A colour version of this figure is located in the plate section at the back of the book.

electron microscopy, makes it possible to develop full molecular models for amyloid fibrils (Tycko, 2011).

ssNMR has shown that whilst amyloid fibrils and amyloid-like fibrils all share the common cross-β structure, the way in which polypeptide chains adopt the β-sheet structure can be variable. For example, amylin fibrils are composed of parallel β-sheets, whilst the $A\beta_{11-25}$ fibrils comprise antiparallel β-sheets (Tycko, 2006). A new high-resolution structural model of mature $A\beta_{1-40}$ amyloid fibrils (Fig. 4.7) obtained using ssNMR, highlights the polymorphism of amyloid fibrils by having a different fold for the mature fibrils (Bertini *et al.*, 2011). This model was obtained using highly homogeneous isotopically labelled amyloid fibrils which generated a very accurate model compared to other models which used heterogeneous samples.

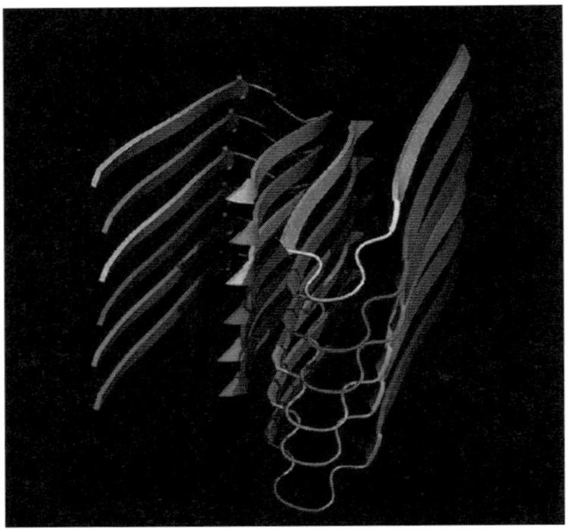

Figure 4.7 Structural model of mature $A\beta_{1-40}$ amyloid fibrils. Reprinted from Bertini *et al.* (2011), with permission from the American Chemical Society. A colour version of this figure is located in the plate section at the back of the book.

Structural polymorphisms occur not only between different amyloid forming proteins, but also within the same amyloid forming protein. This is highlighted by the 12 polymorphs of $A\beta_{1-40}$ amyloid fibrils, as seen in Fig. 4.4. A single structural model is therefore unlikely to be accurate for all amyloid forming proteins. This needs to be taken into account when producing amyloid structures for use in nanotechnology. Changing the conditions under which the amyloid fibrils are formed can result in structural diversity and therefore non-reproducibility. However, polymorphism could also be advantageous in that the morphology of amyloid forming proteins could be tailored to suit particular applications (Rao *et al.*, 2011). Glucagon amyloid fibril formation is a good example of the possibility of generating reproducible amyloid fibril morphologies by subtle changes in protein concentration, and buffer conditions, which has yielded at least five different fibrillar morphologies (Pedersen *et al.*, 2010).

Recently, the Eisenberg group published the crystal structure of a segment of the amyloid-forming protein αB crystallin in which the peptide segment formed an oligomeric complex displaying characteristic features of other amyloid oligomers: β-sheet-rich structure, cytotoxicity, and recognition by an oligomer-specific antibody (Laganowsky *et al.*, 2012). This crystal structure adds an important piece of evidence supporting the view that it is the pre-fibrillar oligomers that are the toxic species, and not the mature amyloid fibrils.

Defining characteristics

The generally recognized criteria defining amyloid or amyloid like fibrils include: the binding of the histological dyes Congo red (CR) and ThT; fibrillar morphology by AFM, and TEM; and β-sheet secondary structure as measured by Fourier transform infrared (FTIR) spectroscopy and X-ray fibre diffraction (Nilsson, 2004; Gras and Squires, 2011; Gras *et al.*, 2011; Hatters and Griffin, 2011). X-ray fibre diffraction is the 'gold standard' method for amyloid fibril characterization, where the amyloid structures give characteristic diffraction patterns (Fig. 4.8) with ~4.7 Å meridional reflections and ~10 Å equatorial reflections

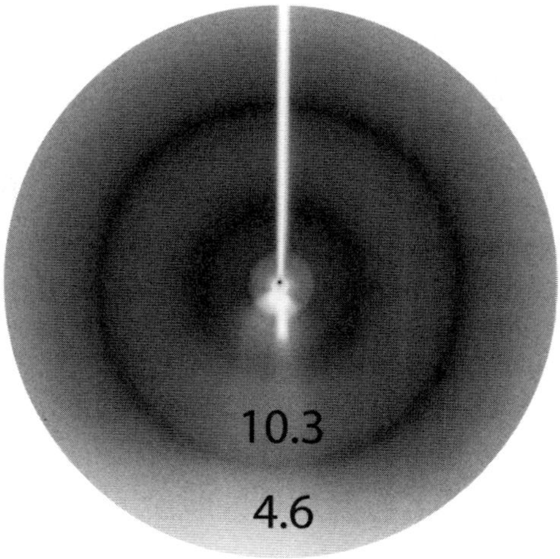

Figure 4.8 X-ray fibre diffraction pattern of crystallin amyloid fibrils. Reprinted from Healy *et al.* (2012), with permission from John Wiley and Sons. The meridional (outer band) reflection of 4.6 Å and the equatorial (inner band) reflection of 10.3 Å are shown.

(Eanes and Glenner, 1968). This diffraction pattern reflects the highly ordered repeating structural units within the fibrils, corresponding to a structure containing a cross-β-spine, with β-strands perpendicular to the fibrils axis (Kirschner *et al.*, 1986; Sunde and Blake, 1997; Sunde *et al.*, 1997).

CR is a widely used diazo dye used to test for the presence of amyloid fibrils (Howie and Brewer, 2009). The structure of CR is shown in Fig. 4.9. Once stained with CR, the dye binds perpendicular to the fibril axis and so when viewed under cross-polarized light an apple-green birefringence is observed (Westermark *et al.*, 1999; Nilsson, 2004; Hatters and Griffin, 2011).

ThT is a benzothiazole dye that selectively binds to amyloid fibrils and has a structure as shown in Fig. 4.10. Amyloid fibrils can be detected using ThT through excitation at 450 nm

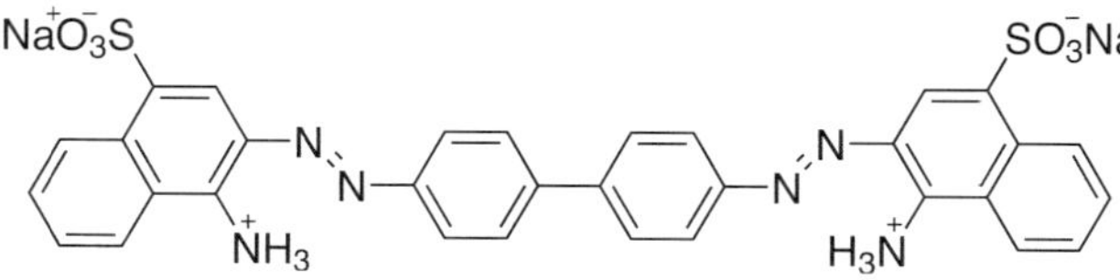

Figure 4.9 The structure of Congo red (Nilsson, 2004).

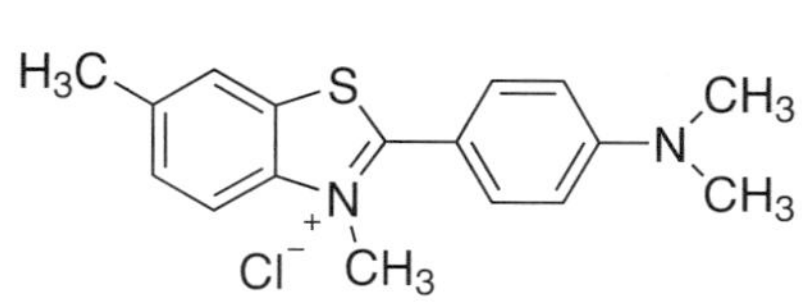

Figure 4.10 The structure of Thioflavin T (Biancalana and Koide, 2010).

and recording the emission at 485 nm (LeVine III, 1999). ThT is thought to bind to the cavities running parallel to the fibril axis, between the protofilaments of the amyloid fibrils (Groenning, 2010). Other fluorescent dyes such as amino derivatives of benzanthrone are being investigated as amyloid-binding dyes because they can potentially offer greater sensitivity and lower background protein fluorescence (Gorbenko *et al.*, 2010).

TEM is one of the most effective methods of distinguishing amyloid fibrils from other types of protein aggregates because of their fibrillar morphology (Nilsson, 2004). Negative staining of amyloid fibrils with uranyl acetate is used to view the amyloid fibrils with TEM, which appear as an unbranched twisted fibrillar morphology with dimensions of 6–20 nm wide and up to several microns in length (Fig. 4.11) (Goldsbury *et al.*, 2011).

AFM is another microscopy tool for investigating amyloid fibril morphology. AFM is a type of scanning probe microscopy that uses a sharp tip connected to a cantilever to scan a surface (Giessibl, 2003). When the tip comes into contact with, for example, an amyloid fibril, the tip is raised and a height measurement is recorded. An example of an AFM image of bovine insulin amyloid fibrils is shown in Fig. 4.12.

Functional amyloids in nature

Whilst amyloid fibrils are associated with twenty-five human diseases, functional non-disease related amyloid fibrils are seen throughout nature in all domains of life (Coustou *et al.*, 1997; Vidal *et al.*, 1998; Iconomidou *et al.*, 2000; Chapman *et al.*, 2002; Fowler *et al.*, 2005; Sabate *et al.*, 2010). Nature utilizes amyloid fibrils for many different applications, demonstrating that amyloid fibrils can be assembled without causing disease. The cellular processes employed to control the assembly of functional amyloid fibrils are tightly controlled to limit the possibility of uncontrolled protein folding. For example, Pmel17 amyloid fibril formation, involved in the deposition of melanin (Fowler *et al.*, 2005) is controlled by cleavage

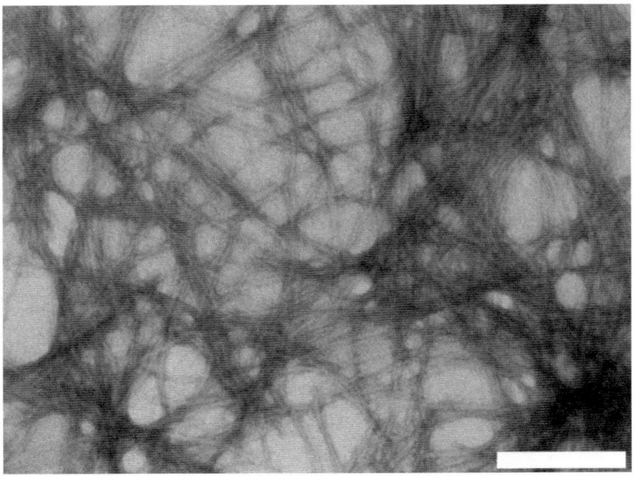

Figure 4.11 Representative TEM micrograph of bovine insulin amyloid fibrils from the authors' laboratory. The scale bar = 200 nm. Reprinted from Raynes *et al.* (2011), with permission from John Wiley and Sons.

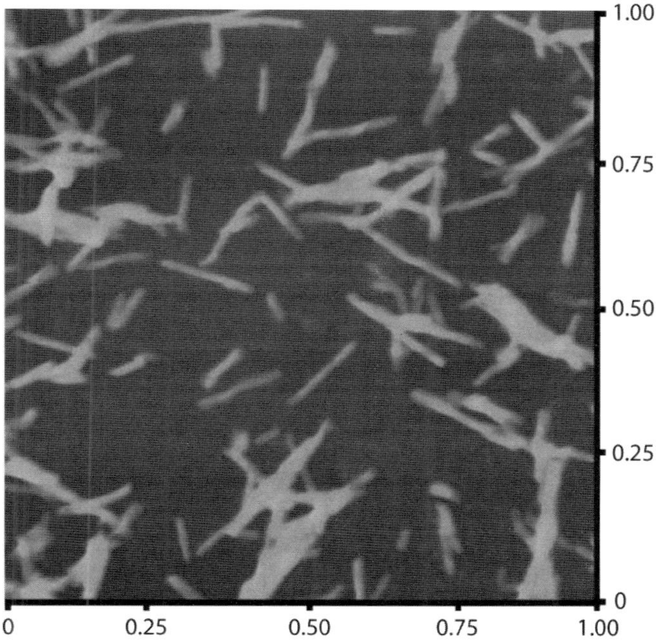

Figure 4.12 Representative AFM image of bovine insulin amyloid fibrils. The size is recorded in µm. Reprinted from Domigan *et al.* (2012), with permission from John Wiley and Sons.

of the full-length protein, releasing the Mα fragment, which undergoes further proteolysis and glycosylation before forming functional amyloid fibrils in the acidic conditions of the melanosomes (McGlinchey *et al.*, 2011). This highly controlled environment allows for the delicate assembly of functional amyloid fibrils without the risk of forming disease, or leading to the aggregation of other proteins. Nature has clearly evolved the ability to utilize the desirable properties of amyloid fibrils, whilst controlling undesirable protein misfolding, for the benefit of the organisms in which they are formed. The recent interest in methods to create self-assembling materials in a controlled and bottom-up approach has inevitably lead researchers to look at nature for inspiration. To begin investigating the potential of amyloid fibrils in the creation of bionanomaterials, an examination of nature's use of functional amyloid fibrils is highly relevant.

One of the best studied functional amyloid systems is that of *Escherichia coli,* which produce surface proteins called curli (Chapman *et al.*, 2002). Curli are a component of the extracellular matrix that is essential for biofilm formation, host cell invasion, and adhesion (Vidal *et al.*, 1998; Bian *et al.*, 2000; Gophna *et al.*, 2001; Zogaj *et al.*, 2001). Filamentous fungi such as *Neurospora crassa* produce hydrophobin proteins which can form amyloid-like structures called rodlets (Mackay *et al.*, 2001). The hydrophobins self-assemble into amphipathic monolayers on the surface of fungal spores and fruiting bodies at air–water interfaces (Wösten *et al.*, 1999). This allows hyphae to penetrate the growth media and gives spores a water-repellent coating to enable aerial dispersion (Wösten *et al.*, 1999). Silkmoths produce chorion proteins which can self-assemble into amyloid fibrils (Iconomidou *et al.*, 2000). These amyloid fibrils act as a protective shell around the oocyte and embryo (Iconomidou *et al.*, 2011). Melanosomes are cellular organelles found in the mammalian cells, melanocytes, and retinal pigment epithelium, located in the skin and eyes, respectively (Hearing, 2000; Marks and Seabra, 2001). As discussed above, an amyloid-like forming protein, Pmel17, is found in melanosomes and is required to polymerize into amyloid structures for the production of mature melanin (Fowler *et al.*, 2005). Melanin protects cells against toxins, UV radiation, and is required in pigmentation. Recently amyloid fibrils have been associated with the protection of *Heliothis virescens* larvae due to their release upon an immune challenge (Falabella *et al.*, 2012). The released amyloid fibrils form a coating on foreign objects entering the haemocoel, promoting melanin synthesis and adhesion of immune cells, aiding in defence against the foreign intruder.

The widespread use of functional amyloid fibrils in nature for many different processes illustrates the usefulness of this protein fold. Biomimicry of not only the amyloid fibril structure, but the processes that allow the controlled assembly and use of amyloid fibrils without causing disease has the potential to be exploited for applications in material design.

Amyloid fibrils in bionanotechnology

Amyloid fibril-containing bionanomaterials have potential applications in a wide range of fields, including biosensing, bioremediation, tissue engineering, and cell growth and adhesion (Waterhouse and Gerrard, 2004; Gras *et al.*, 2008). Amyloid fibrils are highly suited to incorporation within biomaterials because of their ability to self-assemble, their high surface to volume ratio, and their nanoscaffold capability (Gras *et al.*, 2008; Scanlon and Aggeli, 2008). The self-assembling nature of amyloid fibrils allows for the potential of bottom-up approaches to the design of novel materials (Gras, 2007; Paparcone and Buehler, 2011). Bottom-up material design offers potential for the delicate fine-tuning of a material, and in the case of protein structures such as amyloid fibrils, there is the potential to make single amino acid substitutions one at a time via mutagenesis until the material has the desired properties (Hauser and Zhang, 2010). Protein engineering is already widely used to alter the properties of enzymes to suit a combination of altered catalytic function, increased stability and optimal catalytic activity (Quin and Schmidt-Dannert, 2011). Recent advances in the knowledge of amyloid fibril assembly, and the increased capabilities of amyloid predicting algorithms, means that it may soon be possible to use protein engineering to create amyloid fibrils with desired strength, stability, and assembly processes (Woolfson and Mahmoud, 2010; Maji, 2011).

Nature employs the bottom-up design approach to solve the problems of assembling highly sophisticated molecular machines and structures in the form of enzymes and protein structures such as microtubules and microfilaments (Seeman, 2002). Mimicking nature on the nanoscale may provide a means of designing 'smart' self-assembling materials, and amyloid fibrils fulfil these requirements (Zhang, 2003; Tu and Tirrell, 2004; Knowles and Buehler, 2011). The potential use of amyloid fibrils in bionanotechnology has only recently been recognized (MacPhee and Dobson, 2000; MacPhee and Woolfson, 2004; Waterhouse and Gerrard, 2004; Gras, 2007). In the past ten years amyloid fibrils have been explored for a multitude of different bionanotechnology applications, some of which are summarized by Table 4.1.

As can be seen in Table 4.1, insulin is a commonly used amyloid forming protein for investigating potential bionanotechnology applications. Bovine insulin is commonly used as a model amyloid-forming protein because it readily forms amyloid fibrils *in vitro* at high

Table 4.1 Summary of papers utilizing amyloid fibrils in bionanotechnology

Amyloid fibril-forming protein	Technology	Reference
ApoA1	Glutathione S-transferase fusion	Guglielmi *et al.* (2009)
Insulin	Cell adhesion	Sakono *et al.* (2011)
Insulin and α-lactalbumin	Coaggregation with hydrolases	Kim *et al.* (2012)
Insulin	Hybrid bioinorganic fibrils	Tang *et al.* (2010)
Insulin	Nanocomposite	Rao *et al.* (2012)
Insulin	Nanoscaffold for organophosphate hydrolase immobilization	Raynes *et al.* (2011)
Insulin	Nanoscaffold for glucose oxidase immobilization	Pilkington *et al.* (2010)
Insulin	Phosphorescent amyloid-like protein fibrils	Rizzo *et al.* (2010b)
Insulin	Surface assembly	Ha and Park (2005)
N-terminal & middle region of Sup35p	Conducting nanowires	Scheibel *et al.* (2003)
α-synuclein	Amyloid hydrogel for enzyme entrapment	Bhak *et al.* (2010)
Transthyretin (TTR)	Biological cell adhesion sequence fusion	Gras *et al.* (2008)
$TTR_{105-115}$-cycloRGDfK	Cell adhesion	Bongiovanni *et al.* (2011)
Tandem repeat of the SH3 domain	*b*-Type cytochrome fusion	Baldwin *et al.* (2006)
Ure2 prion protein	Barnase fusion	Baxa *et al.* (2002)
Ure2 prion protein	Carbonic anhydrase fusion	Baxa *et al.* (2002)
Ure2 prion protein	Green fluorescent protein fusion	Baxa *et al.* (2002)
Ure2 prion protein	GST fusion	Baxa *et al.* (2002)

temperatures and low pH, but also because it is extremely well characterized (Waugh, 1946; Ortiz *et al.*, 2007).

Early work exploring amyloid fibrils as nanomaterial components was based around fusing the gene of a particular enzyme with an amyloid forming component peptide to create enzymatically active amyloid fibrils. The amyloid forming peptide component can either be derived from the sequence of an amyloid forming protein such as transthyretin (Gras *et al.*, 2008), or can be a synthesized peptide (Channon and MacPhee, 2008). For example, the fusing of the Ure2 prion domain with four proteins, namely carbonic anhydrase, barnase, glutathione S-transferase (GST), and the green fluorescent protein (GFP) (Baxa *et al.*, 2002) was carried out, with the four proteins chosen because of their small substrates, to minimize the possible steric hindrance from the aggregates formed. Subsequent activity assays revealed that only the GST and GFP amyloid fusion aggregates retained significant activity, ~80% and ~130%, respectively. Whereas, the barnase and carbonic anhydrase amyloid fusion aggregates had minimal activity. The reduction in barnase and carbonic anhydrase activity was surmised to be from diffusional limitations because both of these enzymes are near-diffusion-limited enzymes (Jönsson and Wennerström, 1978; Day *et al.*, 1992).

Guglielmi *et al.* (2009) created a similar amyloid fusion protein with GST, but, instead of the URE2 prion domain, they used a 93 residue polypeptide derived from the amyloidogenic domain of apolipoprotein A-I. In this research, the amyloid fusion enzyme had to be incubated for 3 weeks to generate amyloid fibrils. This is an extremely long time if this technology were to be used industrially. The resulting GST amyloid fusion aggregates were again shown to be active and were able to be stored for up to two weeks without loss of activity.

Baldwin *et al.* (2006) generated a *b*-type cytochrome amyloid fusion by constructing a gene which has a cytochrome domain and an amyloid forming domain derived from a tandem repeat of a SH3 domain. The fusion amyloid fibrils were shown to be able to bind metalloporphyrins and from TEM experiments it was estimated that the density of metalloporphyrins was 2.6 per nm. The authors reason that at this concentration rapid electron transfer would be able to occur and that this technology could be used to create self-assembling molecular wires.

Gras *et al.* (2008) and Bongiovanni *et al.* (2011) use an 11 amino acid sequence corresponding to residues 105–115 from the amyloidogenic protein transthyretin as the amyloid forming component of their fusion peptides for the creation of functional amyloid fibrils for cell adhesion. Gras *et al.* (2008) fused a biological cell adhesion sequence (RGD) to the amyloid forming component and created amyloid fibrils capable of promoting cell adhesion, whereas, Bongiovanni *et al.* (2011) fused a cycloRGDfk peptide to target mammalian cell surface $\alpha v\beta 3$ integrin receptors, and again showed that the fusion amyloid fibrils were capable of promoting cell adhesion and spreading. Peptide synthesis was used in the creation of both the fusion peptides, demonstrating the applications of being able to modify one amino acid at a time to create a desired sequence.

The above examples create fusion proteins, either at the gene level, or synthetically using standard solid phase peptide synthesis chemistry. The fusion method could have great potential when creating small peptide fusion amyloid fibrils such as the TTR105–115-RGD amyloid fibrils because they can be completely synthetically made. The major obstacle for the gene based fusions could be in the protein growth and purification because of the intrinsic aggregation potential of the gene product, therefore, the bacterial host may produce low yields. Generating gene based fusions could also be time consuming if the active enzyme component of the fusion amyloid fibrils turns out to have a low activity or be inactivated, as was the case for the barnase and carbonic anhydrase (Baxa *et al.*, 2002).

Recently there has been increased interest in utilizing nanosupports to immobilize enzymes due to their large surface area, allowing for a higher specific activity, less diffusional resistance, and increases in stability by immobilization on or in a support of similar dimensions to enzymes (Ansari and Husain 2012). Therefore, the potential uses of novel nanosupports such as amyloid fibrils have gathered interest.

Immobilizing enzymes post amyloid formation is an alternative method of creating nanostructures with enzymatic activity. The two methods for manufacturing enzymatically active amyloid fibrils are shown in Fig. 4.13. Pilkington *et al.* (2010) proved this to be a valid approach when immobilizing glucose oxidase to insulin amyloid fibrils by cross-linking with glutaraldehyde. Immobilization of organophosphate hydrolase (OPH) to insulin amyloid fibrils using glutaraldehyde was also successfully achieved (Raynes *et al.*,

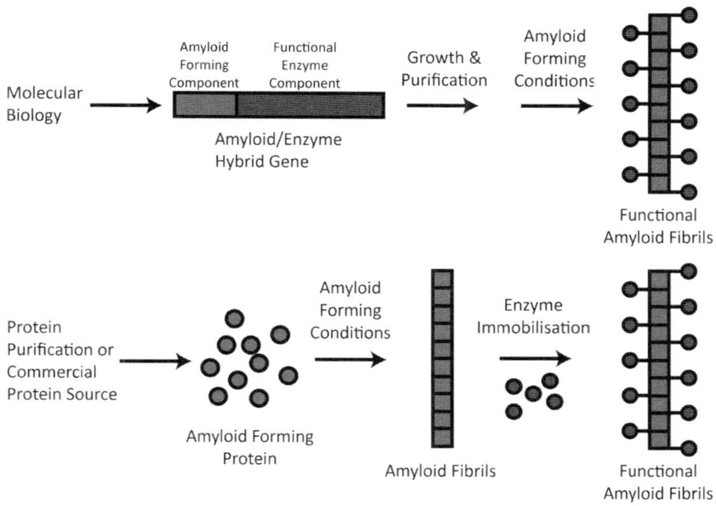

Figure 4.13 Formation of functional amyloid fibrils by (top) gene fusion of an amyloid forming peptide/protein sequence with a functional enzyme/protein sequence or (bottom) enzyme immobilization post amyloid formation. A colour version of this figure is located in the plate section at the back of the book.

2011). The immobilized OPH exhibited a significant increase in thermostability at 45 and 50°C, compared to free OPH. This result demonstrates the ability of insulin amyloid fibrils to stabilize the function of OPH, an enzyme that is not very stable outside of the cellular environment. Recently, Kim *et al.* (2012) immobilized an SGNH esterase to insulin and α-lactalbumin amyloid fibrils by first coaggregating the amyloid fibrils and enzyme using high concentrations of ammonium sulphate, followed by cross-linking with glutaraldehyde. This method also proved successful, but adds an extra step to the immobilization procedure. The advantages of post-assembly modification include the opportunity to create a more general material that can be readily modified with any desired enzyme or other decoration (Woolfson and Mahmoud, 2010).

Amyloid fibrils have also been used as a scaffold for decoration with inorganic molecules such as iridium, apatite and platinum complexes. The integration of the iridium complexes with bovine insulin amyloid fibrils was achieved by simply grinding the iridium complexes with lyophilized bovine insulin, followed by amyloid fibril formation at 65°C and low pH (Rizzo *et al.*, 2010a). The resulting hybrid amyloid fibrils were successfully used to create a white organic light-emitting diode (Rizzo *et al.*, 2010b). In the case of the platinum hybrid insulin amyloid fibrils, the counter ion of the positively charged amyloid fibrils was simply exchanged with the negatively charged hexachloroplatinate (IV) anion. The mineralization of apatite on the amyloid fibril surface was achieved by subsequently incubating platinum amyloid fibrils with phosphate and calcium ions. These hybrid inorganic amyloid fibrils could have potential uses in tissue engineering (Tang *et al.*, 2010).

Amyloid fibrils have been investigated as a nanocomposite to enhance the properties of the material in which they are incorporated. For example, Oppenheim *et al.* (2010) created a lysozyme amyloid fibril-elastomer composite containing a 10% amyloid fibril filling ratio which was shown to be at least twice as stiff as a carbon nanotube-elastomer composition with the same fill ratio. Rao *et al.* (2012) incorporated pre-manufactured insulin amyloid fibrils into poly (vinyl alcohol) (PVOH), creating a bionanocomposite. The mechanical properties of the resulting bionanocomposite were evaluated, and a PVOH film containing 0.6% w/w insulin amyloid fibrils was 15% stiffer than the control. Pilkington *et al.* (2010) extended the creation of bovine insulin amyloid fibril-PVOH bionanocomposites by incorporating glucose oxidase functionalized bovine insulin amyloid fibrils into PVOH. The bionanocomposite was shown to retain GOD activity, and be able to function as an antimicrobial material by inhibiting the growth of *E. coli* when supplemented with glucose. The creation of bionanocomposite materials with amyloid fibrils has shown that amyloid fibrils can provide benefits to the materials in terms of increases in mechanical strength and by acting as a scaffold for enzyme immobilization.

Amyloid fibrils have also been used as the sole component in the creation of nanostructured films (Knowles *et al.*, 2010). The films were produced from hen egg white lysozyme and were shown to be highly rigid, having a Young's modulus of ~5–7 GPa. The amyloid fibrils were found to be aligned within the films by binding ThT and measuring the fluorescence of film through a polarizing filter. The manufacture of a highly aligned and rigid nanomaterial film by the hierarchical self-assembly of amyloid fibrils illustrates the potential of this proteinaceous system for the production of bionanomaterials.

An alternative method for generating bionanomaterials from the amyloid fibril self-assembly process is that of Ha and Park (2005), who assembled bovine insulin amyloid fibrils on the surface of glass. The method to assemble the amyloid fibrils relied on the fact that amyloid fibril formation occurs *via* nucleated growth mechanism (Schmit *et al.*, 2011). The assembly process was split into two parts; firstly, amyloid seeds were covalently bound to the surface of the glass through amine coupling chemistry; secondly, the seeded glass was placed into a fresh solution of insulin at low pH and elevated temperature for five hours, causing mature amyloid fibrils to assemble from the surface bound amyloid seeds. The authors extended this method to produce functional bionanomaterials with glass beads by functionalizing surface assembled bovine insulin amyloid fibrils with immobilized GFP, GOD, and tobacco etch virus protease (Raynes, 2012). An overview of this process is shown in Fig. 4.14. Immobilizing enzymes to the surface assembled amyloid fibrils confirmed that the amyloid fibrils were increasing the surface area available for enzyme immobilization. Many surfaces require some type of modification before enzymes can be immobilized to them, and amyloid fibrils show the ability to not only enable enzyme immobilization on a surface, but also increase the surface area of the chosen surface.

The ability to control the self-assembly process and the dimensions of amyloid fibrils is recognized to be a key requirement in the future application of amyloid fibrils in bionanomaterials (Giraldo, 2010). Domigan *et al.* (2012) recently showed the ability to control the dimensions of amyloid fibrils formed from bovine insulin by carefully controlling the amyloid fibril forming conditions. Another recognized barrier to the use of amyloid fibrils is a cheap and renewable source of an amyloid forming protein. In many of the above bionanomaterial examples, the amyloid fibrils used for the creation of the bionanomaterials were formed from a highly purified, and expensive starting material. It has previously been shown that amyloid fibrils can be produced from a crude protein waste product of the agriculture industry, namely, bovine eye lenses (Garvey *et al.*, 2009). A recent advance on the use of crude crystallin proteins from fish eye lenses has also shown that the morphology of the resulting amyloid fibrils can be controlled to generate higher ordered structures by careful

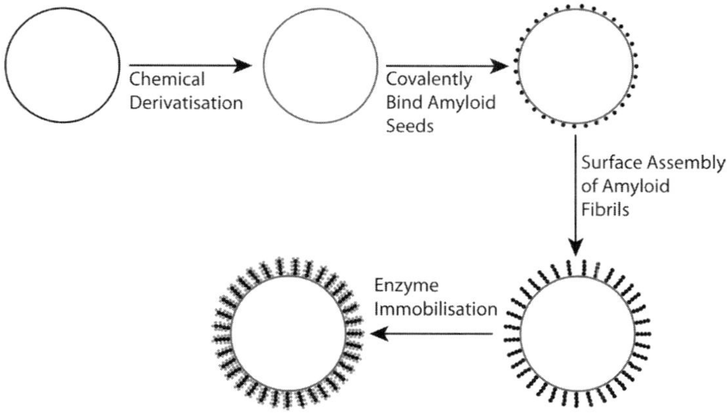

Figure 4.14 Overview of the functional bionanomaterial manufacturing process. Firstly, glass beads are chemically derivatized with aminopropyltriethoxysilane and disuccinimidyl carbonate to yield an activated surface that can then covalently bind amyloid seeds. Surface assembly of mature amyloid fibrils can then occur by immersion in native amyloid forming protein and heating at low pH. The surface assembled amyloid fibrils are then decorated with biomolecules by either physical adsorption or covalent coupling, depending on the biomolecule.

alteration of the amyloid fibril forming conditions (Healy *et al.*, 2012). Thus, renewable crude protein sources for the production of amyloid fibrils are now available.

Future prospects

The initial groundwork for the use of amyloid fibrils in novel bionanomaterials has been laid. One can envisage a future where smart, designer materials that self-assemble into intricate nanostructures become common place. To realize the full potential of amyloid fibrils, more fundamental research on the assembly mechanism needs to be undertaken to elucidate the entire process. In achieving this, the amyloid fibril assembly process could be harnessed for the creation of designer self-assembling bionanomaterials. These future designer bionanomaterials could be used to release a drug at a specified target, such as a cancer cell, or as a self-assembly switch that triggers upon a chosen environmental condition. Novel bionanomaterials created from amyloid fibrils have the potential to extend the repertoire of the material designer in a bottom-up approach, enabling complete control of all aspects of the material.

The biggest road block to the widespread use of amyloid fibrils in the future could be from their historical association with disease (Maji, 2011). The ever increasing knowledge about the mechanism of amyloid formation, the widespread existence of functional, non-disease related amyloid fibrils, and mounting evidence that it is in fact the pre-fibrillar soluble oligomers that are the toxic amyloid species, not the mature amyloid fibrils, should all serve to increase the future use of amyloid fibrils in bionanotechnology applications.

Conclusion

The advent of a better fundamental understanding of the amyloid forming process, as well as new technologies to design and create *de novo* polypeptide sequences with selected features, indicates the use of a self-assembling system such as amyloid fibrils for novel bionanomaterial design is only going to increase. The potential uses of amyloid fibrils are extremely diverse, and the growing number of applications in the literature suggests that amyloid fibrils will be useful components of bionanomaterials.

References

Baldwin, A.J., Bader, R., Christodoulou, J., MacPhee, C.E., Dobson, C.M., and Barker, P.D. (2006). Cytochrome display on amyloid fibrils. J. Am. Chem. Soc. *128*, 2162–2163.

Baxa, U., Speransky, V., Steven, A.C., and Wickner, R.B. (2002). Mechanism of inactivation on prion conversion of the *Saccharomyces cerevisiae* Ure2 protein. Proc. Natl. Acad. Sci. USA *99*, 5253–5260.

Beck, J.S., Christensen, G., and Otzen, D.E. (2004). Modulation of S6 fibrillation by unfolding rates and gatekeeper residues. J. Mol. Biol. *341*, 575–588.

Bertini, I., Gonnelli, L., Luchinat, C., Mao, J., and Nesi, A. (2011). A new structural model of Aβ40 fibrils. J. Am. Chem. Soc. *133*, 16013–16022.

Bhak, G., Lee, S., Park, J.W., Cho, S., and Paik, S.R. (2010). Amyloid hydrogel derived from curly protein fibrils of α-synuclein. Biomaterials *31*, 5986–5995.

Bian, Z., Brauner, A., Li, Y., and Normark, S. (2000). Expression of and cytokine activation by *Escherichia coli* curli fibres in human sepsis. J. Infect. Dis. *181*, 602–612.

Biancalana, M., and Koide, S. (2010). Molecular mechanism of Thioflavin-T binding to amyloid fibrils. BBA-Proteins Proteom. *1804*, 1405–1412.

Bongiovanni, M.N., Scanlon, D.B., and Gras, S.L. (2011). Functional fibrils derived from the peptide TTR1-cycloRGDfK that target cell adhesion and spreading. Biomaterials 32, 6099–6110.

Brange, J., Andersen, L., Laursen, E.D., Meyn, G., and Rasmussen, E. (1997). Towards understanding insulin fibrillation. J. Pharm. Sci. *86*, 517–525.

Channon, K., and MacPhee, C.E. (2008). Possibilities for 'smart' materials exploiting the self-assembly of polypeptides into fibrils. Soft Matter *4*, 647–652.

Chapman, M.R., Robinson, L.S., Pinkner, J.S., Roth, R., Heuser, J., Hammar, M., Normark, S., and Hultgren, S.J. (2002). Role of *Escherichia coli* curli operons in directing amyloid fibre formation. Science *295*, 851–855.

Cherny, I., and Gazit, E. (2008). Amyloids: Not only pathological reagents but also ordered nanomaterials. Angew. Chem. Int. Ed. *47*, 4062–4069.

Chiti, F., and Dobson, C.M. (2006). Protein misfolding, functional amyloid, and human disease. Annu. Rev. Biochem. *75*, 333–366.

Chiti, F., Webster, P., Taddei, N., Clark, A., Stefani, M., Ramponi, G., and Dobson, C.M. (1999). Designing conditions for in vitro formation of amyloid protofilaments and fibrils. Proc. Natl. Acad. Sci. U.S.A. *96*, 3590–3594.

Coustou, V., Deleu, C., Saupe, S., and Begueret, J. (1997). The protein product of the het-s heterokaryon incompatibility gene of the fungus *Podospora anserina* behaves as a prion analogue. Proc. Natl. Acad. Sci. U.S.A. *94*, 9773–9778.

Day, A.G., Parsonage, D., Ebel, S., Brown, T., and Fersht, A.R. (1992). Barnase has subsites that give rise to large rate enhancements. Biochemistry *31*, 6390–6395.

Dirix, C., Meersman, F., MacPhee, C.E., Dobson, C.M., and Heremans, K. (2005). High hydrostatic pressure dissociates early aggregates of TTR105–115, but not the mature amyloid fibrils. J. Mol. Biol. *347*, 903–909.

Dobson, C.M. (1999). Protein misfolding, evolution and disease. Trends Biochem. Sci. *24*, 329–332.

Dobson, C.M., and Karplus, M. (1999). The fundamentals of protein folding: bringing together theory and experiment. Curr. Opin. Struct. Biol.*9*, 92–101.

Domigan, L.J., Healy, J.P., Meade, S.J., Blaikie, R.J., and Gerrard, J.A. (2012). Controlling the dimensions of amyloid fibrils: toward homogenous components for bionanotechnology. Biopolymers *97*, 123–133.

Eanes, E.D., and Glenner, G.G. (1968). X-ray diffraction studies on amyloid filaments. J. Histochem. Cytochem. *16*, 673–677.

Eichner, T., and Radford, S.E. (2011). A diversity of assembly mechanisms of a generic amyloid fold. Mol. Cell *43*, 8–18.

Fágáin, C.Ó. (1995). Understanding and increasing protein stability. BBA-Protein Struct. M. *1252*, 1–14.

Falabella, P., Riviello, L., Pascale, M., Di Lelio, I., Tettamanti, G., Grimaldi, A., Iannone, C., Monti, M., Pucci, P., Tamburro, A.M., *et al.* (2012). Functional amyloids in insect immune response. Insect Biochem. Mol. Biol. *43*, 203–2011.

Fowler, D.M., Koulov, A.V., Alory-Jost, C., Marks, M.S., Balch, W.E., and Kelly, J.W. (2005). Functional amyloid formation within mammalian tissue. PLoS Biol. *4*, 100–107.

Garvey, M., Gras, S.L., Meehan, S., Meade, S.J., Carver, J.A., and Gerrard, J.A. (2009). Protein nanofibres of defined morphology prepared from mixtures of crude crystallins. Int. J. Nanotechnol. *6*, 258–273.

Giessibl, F.J. (2003). Advances in atomic force microscopy. Rev. Mod. Phys. *75*, 949–983.

Giraldo, R. (2007). Defined DNA sequences promote the assembly of a bacterial protein into distinct amyloid nanostructures. Proc. Natl. Acad. Sci. USA *104*, 17388–17393.

Giraldo, R. (2010). Amyloid assemblies: Protein legos at a crossroads in bottom-up synthetic biology. ChemBioChem *11*, 2347–2357.

Glickman, M.H., and Ciechanover, A. (2002). The ubiquitin-proteasome proteolytic pathway: destruction for the sake of construction. Physiol. Rev. *82*, 373–428.

Goldsbury, C., Baxa, U., Simon, M.N., Steven, A.C., Engel, A., Wall, J.S., Aebi, U., and Müller, S.A. (2011). Amyloid structure and assembly: Insights from scanning transmission electron microscopy. J. Struct. Biol. *173*, 1–13.

Goldschmidt, L., Teng, P.K., Riek, R., and Eisenberg, D. (2010). Identifying the amylome, proteins capable of forming amyloid-like fibrils. Proc. Natl. Acad. Sci. U.S.A. *107*, 3487–3492.

Gophna, U., Barlev, M., Seijffers, R., Oelschlager, T.A., Hacker, J., and Ron, E.Z. (2001). Curli fibres mediate internalization of *Escherichia coli* by eukaryotic cells. Infect. Immun. *69*, 2659–2665.

Gorbenko, G., Trusova, V., Kirilova, E., Kirilov, G., Kalnina, I., Vasilev, A., Kaloyanova, S., and Deligeorgiev, T. (2010). New fluorescent probes for detection and characterization of amyloid fibrils. Chem. Phys. Lett. *495*, 275–279.

Gras, S.L. (2007). Amyloid fibrils: From disease to design. New biomaterial applications for self-assembling cross-β fibrils. Aust. J. Chem. *60*, 333–342.

Gras, S.L., and Squires, A.M. (2011). Dried and hydrated X-ray scattering analysis of amyloid fibrils. In Protein Folding, Misfolding, and Disease, A.F. Hill, K.J. Barnham, S.P. Bottomley, and R. Cappai, eds. (Totowa, NJ: Humana Press), pp. 147–163.

Gras, S.L., Tickler, A.K., Squires, A.M., Devlin, G.L., Horton, M.A., Dobson, C.M., and MacPhee, C.E. (2008). Functionalised amyloid fibrils for roles in cell adhesion. Biomaterials *29*, 1553–1562.

Gras, S.L., Waddington, L.J., and Goldie, K.N. (2011). Transmission electron microscopy of amyloid fibrils. In Protein Folding, Misfolding, and Disease, A.F. Hill, K.J. Barnham, S.P. Bottomley, and R. Cappai, eds. (Totowa, NJ: Humana Press), pp. 197–214.

Groenning, M. (2010). Binding mode of Thioflavin T and other molecular probes in the context of amyloid fibrils-current status. J. Chem. Biol. *3*, 1–18.

Guglielmi, F., Monti, D.M., Arciello, A., Torrassa, S., Cozzolino, F., Pucci, P., Relini, A., and Piccoli, R. (2009). Enzymatically active fibrils generated by the self-assembly of the ApoA-I fibrillogenic domain functionalized with a catalytic moiety. Biomaterials *30*, 829–835.

Ha, C., and Park, C.B. (2005). Template-directed self-assembly and growth of insulin amyloid fibrils. Biotechnol. Bioeng. *90*, 848–855.

Hamada, D., and Dobson, C.M. (2002). A kinetic study of β-lactoglobulin amyloid fibril formation promoted by urea. Protein Sci. *11*, 2417–2426.

Harrison, R.S., Sharpe, P.C., Singh, Y., and Fairlie, D.P. (2007). Amyloid peptides and proteins in review. Rev. Physiol. Biochem. Pharmacol. *159*, 1–77.

Hartl, F.U., and Hayer-Hartl, M. (2002). Protein folding. Molecular chaperones in the cytosol: from nascent chain to folded protein. Science *295*, 1852–1858.

Hatters, D.M., and Griffin, M.D.W. (2011). Diagnostics for amyloid fibril formation: Where to begin? In Protein Folding, Misfolding, and Disease, A.F. Hill, K.J. Barnham, S.P. Bottomley, and R. Cappai, eds. (Totowa, NJ: Humana Press), pp. 121–136.

Hauser, C.A.E., and Zhang, S. (2010). Designer self-assembling peptide nanofibre biological materials. Chem. Soc. Rev. *39*, 2780–2790.

Healy, J., Wong, K., Roux, C., Domigan, L., Sunde, M., Gerrard, J.A., and Vasudevamurthy, M. (2012). Polymorphism and higher ordered structures of protein nanofibres from crude mixtures of fish lens crystallins: Towards useful materials. Biopolymers *97*, 595–606.

Hearing, V.J. (2000). The melanosome: The perfect model for cellular responses to the environment. Pigment Cell. Res. *13*, 23–34.

Hershko, A., and Ciechanover, A. (1998). The ubiquitin system. Annu. Rev. Biochem. *67*, 425–479.

Howie, A.J., and Brewer, D.B. (2009). Optical properties of amyloid stained by Congo red: History and mechanisms. Micron *40,* 285–301.
Iconomidou, V.A., Cordopatis, P., Hoenger, A., and Hamodrakas, S.J. (2011). The silkmoth eggshell as a natural amyloid shield for the safe development of insect oocyte and embryo: Insights from studies of silkmoth chorion protein peptide-analogues of the B family. Peptide Sci. *96,* 723–733.
Iconomidou, V.A., Vriend, G., and Hamodrakas, S.J. (2000). Amyloids protect the silkmoth oocyte and embryo. FEBS Lett. *479,* 141–145.
Jahn, T.R., and Radford, S.E. (2008). Folding versus aggregation: Polypeptide conformations on competing pathways. Arch. Biochem. Biophys. *469,* 100–117.
Jönsson, B., and Wennerström, H. (1978). Diffusion control in reversible enzyme reactions. Applications to carbonic anhydrase. Biophys. Chem. *7,* 285–292.
Kim, S., Bae, S.Y., Lee, B.Y., and Kim, T.D. (2012). Coaggregation of amyloid fibrils for the preparation of stable and immobilized enzymes. Anal. Biochem. *421,* 776–778.
Kirkitadze, M.D., Bitan, G., and Teplow, D.B. (2002). Paradigm shifts in Alzheimer's disease and other neurodegenerative disorders: the emerging role of oligomeric assemblies. J. Neurosci. Res. *69,* 567–577.
Kirschner, D.A., Abraham, C., and Selkoe, D.J. (1986). X-ray diffraction from intraneuronal paired helical filaments and extraneuronal amyloid fibres in Alzheimer disease indicates cross-β conformation. Proc. Natl. Acad. Sci. USA *83,* 503–507.
Knowles, T.P., Fitzpatrick, A.W., Meehan, S., Mott, H.R., Vendruscolo, M., Dobson, C.M., and Welland, M.E. (2007). Role of intermolecular forces in defining material properties of protein nanofibrils. Science *318,* 1900–1903.
Knowles, T.P.J., and Buehler, M.J. (2011). Nanomechanics of functional and pathological amyloid materials. Nat. Nano. *6,* 469–479.
Knowles, T.P.J., Oppenheim, T.W., Buell, A.K., Chirgadze, D.Y., and Welland, M.E. (2010). Nanostructured films from hierarchical self-assembly of amyloidogenic proteins. Nat. Nano. *5,* 204–207.
Koga, H., Kaushik, S., and Cuervo, A.M. (2011). Protein homeostasis and ageing: The importance of exquisite quality control. Ageing Res. Rev. *10,* 205–215.
LeVine III, H. (1999). Quantification of β-sheet amyloid fibril structures with thioflavin T. Methods Enzymol. *309,* 274–284.
Liang, Y., Lynn, D.G., and Berland, K.M. (2010). Direct observation of nucleation and growth in amyloid self-assembly. J. Am. Chem. Soc. *132,* 6306–6308.
Luheshi, L.M., and Dobson, C.M. (2009). Bridging the gap: From protein misfolding to protein misfolding diseases. FEBS Letters *583,* 2581–2586.
Mackay, J.P., Matthews, J.M., Winefield, R.D., Mackay, L.G., Haverkamp, R.G., and Templeton, M.D. (2001). The hydrophobin EAS is largely unstructured in solution and functions by forming amyloid-like structures. Structure *9,* 83–91.
MacPhee, C.E., and Dobson, C.M. (2000). Formation of mixed fibrils demonstrates the generic nature and potential utility of amyloid nanostructures. J. Am. Chem. Soc. *122,* 12707–12713.
MacPhee, C.E., and Woolfson, D.N. (2004). Engineered and designed peptide-based fibrous biomaterials. Curr. Opin. Solid St. M. *8,* 141–149.
Maji, S.K. (2011). Amytloid: A natural nanomaterial. Int. J. Nanosci. *10,* 909–917.
Mankar, S., Anoop, A., Sen, S., and Maji, S.K. (2011). Nanomaterials: amyloids reflect their brighter side. Nano Rev.s *2,* 1–12.
Marks, M.S., and Seabra, M.C. (2001). The melanosome: Membrane dynamics in black and white. Nat. Rev. Mol. Cell Biol. *2,* 738–748.
McGlinchey, R.P., Yap, T.L., and Lee, J.C. (2011). The yin and yang of amyloid: Insights from α-synuclein and repeat domain of Pmel17. Phys. Chem. Chem. Phys. *13,* 20066–20075.
Meinhardt, J., and Fändrich, M. (2009). Structure of amyloid fibrils. Pathologe *30,* 175–181.
Mishra, R., and Winter, R. (2008). Cold- and pressure-induced dissociation of protein aggregates and amyloid fibrils. Angew. Chem. Int. Ed. *47,* 6518–6521.
Naiki, H., Hashimoto, N., Suzuki, S., Kimura, H., Nakakuki, K., and Gejyo, F. (1997). Establishment of a kinetic model of dialysis-related amyloid fibril extension in vitro. Amyloid *4,* 223–232.
Nelson, R., Sawaya, M.R., Balbirnie, M., Madsen, A.O., Riekel, C., Grothe, R., and Eisenberg, D. (2005). Structure of the cross-β spine of amyloid-like fibrils. Nature *435,* 773–778.
Nerelius, C., Fitzen, M., and Johansson, J. (2010). Amino acid sequence determinants and molecular chaperones in amyloid fibril formation. Biochem. Biophys. Res. Commun. *396,* 2–6.
Nilsson, M.R. (2004). Techniques to study amyloid fibril formation *in vitro.* Methods *34,* 151–160.
Onuchic, J.N., and Wolynes, P.G. (2004). Theory of protein folding. Curr. Opin. Struc. Biol. *14,* 70–75.

Oppenheim, T., Knowles, T.P.J., Lacour, S.P., and Welland, M.E. (2010). Fabrication and characterization of protein fibril–elastomer composites. Acta Biomater. *6*, 1337–1341.

Otzen, D. (2010). Functional amyloid: turning swords into plowshares. Prion *4*, 256–264.

Paparcone, R., and Buehler, M.J. (2011). Failure of Aβ(1–40) amyloid fibrils under tensile loading. Biomaterials *32*, 3367–3374.

Pedersen, J.S., Andersen, C.B., and Otzen, D.E. (2010). Amyloid structure – one but not the same: The many levels of fibrillar polymorphism. FEBS J. *277*, 4591–4601.

Pedersen, J.S., and Otzen, D.E. (2008). Amyloid – A state in many guises: survival of the fittest fibril fold. Protein Sci. *17*, 2–10.

Pilkington, S.M., Roberts, S.J., Meade, S.J., and Gerrard, J.A. (2010). Amyloid fibrils as a nanoscaffold for enzyme immobilization. Biotechnol. Prog. *26*, 93–100.

Ping, W. (2006). Nanoscale biocatalyst systems. Curr. Opin. Biotech. *17*, 574–579.

Quin, M.B., and Schmidt-Dannert, C. (2011). Engineering of biocatalysts: From evolution to creation. Catalysis *1*, 1017–1021.

Rao, S.P., Meade, S.J., Healy, J.P., Sutton, K.H., Larsen, N.G., Staiger, M.P., and Gerrard, J.A. (2012). Amyloid fibrils as functionalizable components of nanocomposite materials. Biotechnol. Prog. *28*, 248–256.

Rao, S.P., Meade, S.J., Joyce, N.I., Healy, J.P., Sutton, K.H., Larsen, N.G., and Gerrard, J.A. (2011). Amyloid fibril formation from crude protein mixtures. Biotechnol. Prog. *27*, 1768–1776.

Raynes, J.K. (2012). Immobilising biomolecules on amyloid fibrils for biotechnology applications. PhD thesis, University of Canterbury.

Raynes, J.K., Pearce, F.G., Meade, S.J., and Gerrard, J.A. (2011). Immobilization of organophosphate hydrolase on an amyloid fibril nanoscaffold: Towards bioremediation and chemical detoxification. Biotechnol. Prog. *27*, 360–367.

Rizzo, A., Inganäs, O., and Solin, N. (2010a). Preparation of phosphorescent amyloid-like protein fibrils. Chem. Eur. J. *16*, 4190–4195.

Rizzo, A., Solin, N., Lindgren, L.J., Andersson, M.R., and Inganäs, O. (2010b). White light with phosphorescent protein fibrils in OLEDs. Nano. Lett. *10*, 2225–2230.

Ron, D., and Walter, P. (2007). Signal integration in the endoplasmic reticulum unfolded protein response. Nat. Rev. Mol. Cell. Biol. *8*, 519–529.

Sabate, R., De Groot, N.S., and Ventura, S. (2010). Protein folding and aggregation in bacteria. Cell. Mol. Life Sci. *67*, 2695–2715.

Sakono, M., Akiyama, S., Zako, T., Sakaki, S., Waku, T., Tanaka, N., and Maeda, M. (2011). Application of two morphologically different fibrillar and filamentous insulin amyloids as a biomaterial for cell culture surfaces. React. Funct. Polym. *71*, 324–328.

Scanlon, S., and Aggeli, A. (2008). Self-assembling peptide nanotubes. Nano Today 3, 22–30.

Scheibel, T., Parthasarathy, R., Sawicki, G., Lin, X.-M., Jaeger, H., and Lindquist, S.L. (2003). Conducting nanowires built by controlled self-assembly of amyloid fibres and selective metal deposition. Proc. Natl. Acad. Sci. U.S.A. *100*, 4527–4532.

Schmidt, M., Sachse, C., Richter, W., Xu, C., Fändrich, M., and Grigorieff, N. (2009). Comparison of Alzheimer Aβ(1–40) and Aβ(1–42) amyloid fibrils reveals similar protofilament structures. Proc. Natl. Acad. Sci. U.S.A. *106*, 19813–19818.

Schmit, J.D., Ghosh, K., and Dill, K. (2011). What drives amyloid molecules to assemble into oligomers and fibrils? Biophys. J. *100*, 450–458.

Seeman, N.C. (2002). Emulating biology: Building nanostructures from the bottom up. Proc. Natl. Acad. Sci. U.S.A. *99*, 6451–6455.

Serio, T.R., Cashikar, A.G., Kowal, A.S., Sawicki, G.J., Moslehi, J.J., Serpell, L.C., Arnsdorf, M.F., and Lindquist, S.L. (2000). Nucleated conformational conversion and the replication of conformational information by a prion determinant. Science *289*, 1317–1321.

Serpell, L.C., and Smith, J.M. (2000). Direct visualization of the β-sheet structure of synthetic Alzheimer's amyloid. J. Mol. Biol. *299*, 225–231.

Serpell, L.C., Sunde, M., Benson, M.D., Tennent, G.A., Pepys, M.B., and Fraser, P.E. (2000). The protofilament substructure of amyloid fibrils. J. Mol. Biol. *300*, 1033–1039.

Smith, J.F., Knowles, T.P.J., Dobson, C.M., MacPhee, C.E., and Welland, M.E. (2006). Characterization of the nanoscale properties of individual amyloid fibrils. Proc. Natl. Acad. Sci. U.S.A. *103*, 15806–15811.

Sunde, M., and Blake, C.C.F. (1997). The structure of amyloid fibrils by electron microscopy and X-ray diffraction. Adv. Protein Chem. *50*, 123–159.

Sunde, M., Serpell, L.C., Bartlam, M., Fraser, P.E., Pepys, M.B., and Blake, C.C.F. (1997). Common core structure of amyloid fibrils by synchrotron X-ray diffraction. J. Mol. Biol. *273*, 729–739.

Tang, Q., Solin, N., Lu, J., and Inganäs, O. (2010). Hybrid bioinorganic insulin amyloid fibrils. Chem. Commun. *46*, 4157–4159.

Tu, R.S., and Tirrell, M. (2004). Bottom-up design of biomimetic assemblies. Adv. Drug Deliver. Rev. *56*, 1537–1563.

Tycko, R. (2006). Molecular structure of amyloid fibrils: Insights from solid-state NMR. Q. Rev. Biophys. *39*, 1–55.

Tycko, R. (2011). Solid-state NMR studies of amyloid fibril structure. Annu. Rev. Phys. Chem. *62*, 279–299.

Tyedmers, J., Mogk, A., and Bukau, B. (2010). Cellular strategies for controlling protein aggregation. Nat. Rev. Mol. Cell. Biol. *11*, 777–788.

Uversky, V.N., Li, J., Souillac, P., Millett, I.S., Doniach, S., Jakes, R., Goedert, M., and Fink, A.L. (2002). Biophysical properties of the synucleins and their propensities to fibrillate. J. Biol. Chem. *277*, 11970–11978.

Valéry, C., Artzner, F., and Paternostre, M. (2012). Peptide nanotubes: molecular organizations, self-assembly mechanisms and applications. Soft Matter *7*, 9583–9594.

Vidal, O., Longin, R., Prigent-Combaret, C., Dorel, C., Hooreman, M., and Lejeune, P. (1998). Isolation of an *Escherichia coli* K-12 mutant strain able to form biofilms on inert surfaces: Involvement of a new ompR allele that increases curli expression. J. Bacteriol. *180*, 2442–2449.

Waterhouse, S.H., and Gerrard, J.A. (2004). Amyloid fibrils in bionanotechnology. Aust. J. Chem. *57*, 519–523.

Westermark, G., Johnson, K.H., and Westermark, P. (1999). Staining methods for identification of amyloid in tissue. Methods Enzymol. *309*, 3–25.

Wolynes, P.G. (2005). Energy landscapes and solved protein-folding problems. Phil. Transact. A. Math. Phys. Eng. Sci. *363*, 453–464.

Woolfson, D.N., and Mahmoud, Z.N. (2010). More than just bare scaffolds: Towards multi-component and decorated fibrous biomaterials. Chem. Soc. Rev. *39*, 3464–3479.

Wösten, H.A.B., van Wetter, M.-A., Lugones, L.G., van der Mei, H.C., Busscher, H.J., and Wessels, J.G.H. (1999). How a fungus escapes the water to grow into the air. Curr. Biol. *9*, 85–88.

Xu, Z., Paparcone, R., and Buehler, M.J. (2010). Alzheimer's Aβ(1–40) Amyloid fibrils feature size-dependent mechanical properties. Biophys. J. *98*, 2053–2062.

Zhang, S. (2003). Fabrication of novel biomaterials through molecular self-assembly. Nat. Biotech. *21*, 1171–1178.

Zogaj, X., Nimtz, M., Rohde, M., Bokranz, W., and Römling, U. (2001). The multicellular morphotypes of *Salmonella typhimurium* and *Escherichia coli* produce cellulose as the second component of the extracellular matrix. Mol. Microbiol. *39*, 1452–1463.

Bacteriophages: Self-assembly and Applications

5

Jasna Rakonjac and James F. Conway

Abstract

Bacteriophage biology ushered in the era of modern molecular and structural biology. The accumulated wealth of knowledge on phage assembly, structure and the life cycle permitted their utilization in a broad range of applications, from basic molecular biology to nanotechnology and pharmaceutical industry. This chapter reviews the current state of knowledge of bacteriophage assembly and structure represented by two morphologically different types, headed and filamentous bacteriophages. The principles of phage display are further presented, followed by a wide range of applications, including the most recent applications in nanotechnology.

Introduction

In the early days of molecular genetics, phage assembly was an object of intense research efforts. These simple biological entities provided the foundations of our knowledge of basic biological processes at the molecular level. Although tools for dissection of molecular mechanisms in higher organisms are available today, bacteriophage still fascinate new generations of researchers and continue to provide important insights into the structure of protein assemblies and their functions as molecular complexes and machines. The structural components of phage are being dissected at high resolution by a combination of X-ray crystallography and cryo-electron microscopy, to give unprecedented detail of phage structure and the processes of assembly and infection. In parallel to the structural studies, phages have been the major force in modern combinatorial biochemistry. The potential of phages in this area of applied research is based on the following characteristics:

1. Virion proteins are relatively tolerant to fusion of peptides and proteins, allowing display of foreign peptides and proteins on the virion surface.
2. Self-assembly of virion subunits fused with exogenous proteins into stable phage particles enables their *in vitro* manipulation and selection of variants with desired properties.
3. The physical link of the fusion to the respective encoding sequence allows any selected variant of displayed peptide or protein to be amplified, which in turn allows researcher-directed genetic change to be preserved and propagated.
4. The physical stability of the particles makes a phage not only a means to isolate a variant of interest, but also an easily tracked nanoparticle carrier.
5. Finally, the highly ordered structure of bacteriophages provides a template for assembly of organic and inorganic nanostructures.

There are some thousands of publications on phage assembly and applications which are impossible to cover in a single review. Our focus will be on illustrating this area by the use of several examples close to our expertise.

Phage self-assembly

Morphologies of viruses

Viruses are parasitic, depending on a host cell for maintenance and expression. Just as cells have many roles, shapes and states, so too have viruses a large variety of morphologies according to their lifecycle (Ruigrok *et al.*, 2005). In general, a protein shell (capsid) encloses the viral genome providing a protective environment and a platform on which factors involved in cell-recognition and entry are arrayed either directly or indirectly. These factors may include extended fibre molecules, glycoproteins embedded in an external lipid bilayer, or a receptor-binding motif displayed on the capsid itself. The capsid is also more or less actively involved in assembly of the virus, including packaging of the genome and attachment of external accessory proteins, and may itself undergo large-scale conformational changes as part of the control for sequencing the assembly pathway. One well-studied assembly pathway is that of bacteriophage HK97, which is discussed further below. Morphology is then an important descriptive tool for identifying and comparing viruses, but to be complete it must also encompass the dynamic aspect of structural evolution during assembly and maturation.

Morphology is not the only way to characterize viruses. Other attributes are related to distinctive strategies for infection, reproduction, egress, and so on. For example, viral genomes may be DNA or RNA, determining the cell compartment to which it must be delivered; positive-sense or negative-sense single-stranded, or double-stranded, requiring appropriate replication machinery that may be fully provided by the host or packaged in part with the genome; the protein capsid may acquire an internal lipid bilayer, or one or two external bilayers into which receptor-binding molecules are embedded; cell recognition and entry may be through an elaborate 'baseplate' attached to a protein tube extending a long distance from the capsid, or by displaying a binding site on the surface of the capsid itself. However, despite the seemingly endless variation exhibited by viruses, morphology is an important descriptor of character and is a primary basis for classification by family or finer. Much of this character is evident in the capsid, and if present, the tail. General features revealed by electron microscopy, such as capsid shape (spherical, prolate, helical, asymmetric) and size (which is more limited for RNA genomes due to the higher rate of replication errors), tail length and diameter, are useful starting points for classification.

The capsid is the main structural component of a virus, harbouring the genome and often displaying all or some of the cell recognition/entry apparatus. The size of the capsid is clearly related to the size of the genome, and varies from the small, such as the simple 60-subunit icosahedra of parvovirus (250 Å diameter), to the giant, as exemplified by the 4000-Å diameter Mimivirus family (Arslan *et al.*, 2011; Boyer *et al.*, 2009; Colson *et al.*, 2011; La Scola *et al.*, 2003; Raoult *et al.*, 2004) or the 8600-Å-long Ff filamentous phage, the length of which depends on the size of the encapsidated DNA. Some form of regular packing is often employed so as to build a large container from many copies of a few different subunits, thereby minimizing the proportion of the genome that must be dedicated to the structural components while maximizing the volume of the container and thereby the size of the genome. Symmetrical arrangements include the helix which forms the basis for filamentous phage capsid structure and other extended capsids such as tobacco mosaic virus. Another ubiquitous symmetry is icosahedral, which appears to be the most common architectural form for virus capsids. Estimates based on soil and water samples put the number of bacteriophage at 10^{31} particles (Hatfull, 2008; Hendrix, 2003), and about 90–95% of those characterized, albeit a tiny fraction of the total, have double-stranded DNA genomes and icosahedral capsids. Further, the few for which high resolution structures have been solved appear to share a common fold that was first characterized for the major capsid protein of phage HK97 (Helgstrand *et al.*, 2003; Wikoff *et al.*, 2000). Geometrically, the icosahedral capsids can be regarded as a network of hexavalent sites that is curved and closed by incorporation of pentameric subunits at specific locations according to size and complexity (Johnson and Speir, 1997). Capsid structures that are morphologically in between the extended helix and spherical icosahedron are also observed which are essentially helical tubes of defined length capped by icosahedral ends. These are called prolate icosahedra, and a well-known example is phage T4.

The relationship between bacteriophages – viruses that specifically target bacteria – and plant, insect and animal viruses has become clearer through structural studies in particular, as protein sequence homology does not often show any relationship between viral proteins with common folds. One striking example is the unusual trimeric capsomer common to adenovirus and phage PRD1, both of which have capsids with pseudo-T=25 lattice geometry (Belnap and Steven, 2000). Although the subunits – the hexon protein of adenovirus and the P3 protein of PRD1 – are of quite different sizes, the core fold has been revealed by crystallography to be similar (Benson *et al.*, 2002). A further branch of this story was revealed when the capsid and subunit structures of archaeal virus STIV were found to employ the same double-β-barrel fold (Khayat *et al.*, 2005; Rice *et al.*, 2004). This connection between three such vastly separated viruses is suggestive of common ancestry albeit quite removed. Recently this group has been expanded to include the major capsid protein, vp54, of an algae virus PBCV-1 (*Paramecium bursaria* Chlorella virus type 1), which also forms a trimeric capsomer (Yan *et al.*, 2000) and shares the same core fold (Nandhagopal *et al.*, 2002) although the T=169 capsid is much larger than those of adenovirus and phage PRD1. An even larger member of this structural family, PpV01 (*Phaeocystis pouchetii* virus) that infects a phytoplankton host, was subsequently described with capsid geometry of T=219 and a maximum diameter of 2200 Å (Fokine *et al.*, 2005).

More evidence for common ancestry is mounting as structural data accumulate. Recent work on herpes simplex virus has demonstrated a procapsid form that embarks on a maturation pathway of structural modifications yielding the mature, DNA-filled particle (Heymann *et al.*, 2003; Trus *et al.*, 1996). This is analogous to the prohead-expansion-head pathway of some bacteriophages, such as T4, P22 and HK97. In addition, the HSV-1 dsDNA genome is packaged through a specialized vertex (Cardone *et al.*, 2007; Newcomb *et al.*, 2001; Trus *et al.*, 2004), as for tailed phages although the nature of the packaging mechanism has yet to be elaborated. HSV-1 also employs minor capsid proteins (VP19c, VP23, VP26) whose roles include stabilization of the procapsid and/or capsid (Booy *et al.*, 1994; Newcomb *et al.*, 1993; Spencer *et al.*, 1998), much like various phage decoration proteins such as T4's hoc and soc (Steven *et al.*, 1992) and λ's gpD. As for adenovirus, a structural connection between herpesvirus and a bacteriophage, SPO1, has recently been proposed based on a common capsid architecture including the so-far unique arrangement of capsomers with triangulation number $T=16$, and the observation of asymmetric hetero-trimers located between capsomers on the exterior surfaces (Duda *et al.*, 2006). Members of this family also exhibit large capsids, such as the 1800-Å-diameter phage G (Ageno *et al.*, 1973; Murphy and Philipson, 1962), and evidence is emerging of the HK97 capsid protein fold in viruses of archaea that connects them with the dsDNA icosahedral tailed phages and eukaryotic herpesviruses (Akita *et al.*, 2007). Additionally, the HK97 fold has been discovered in capsid-like compartments, such as the encapsulins that sequester enzymatic reactions in bacteria and archaea (Sutter *et al.*, 2008). Whether this fold originated in a cellular component that learned to package DNA, or from a virus that changed to enclose protein, the various ways it has been exploited in nature suggest opportunities for biotechnological use.

These examples of shared morphology and apparent evolutionary connections between eukaryotic and prokaryotic viruses serve to highlight the relative simplicity of the phage capsid, which generally has fewer and smaller proteins and is easier to manipulate and study. But even the simplest capsid is assembled by polymerization of many copies of the constituent subunit or subunits, in some cases by self-assembly and in others with the aid of a scaffold structure, and in most cases through a series of irreversible steps leading to a unique end-product. This process is of quite remarkable fidelity, dependent on in-built control mechanisms that are poorly understood but which must direct assembly towards the final structure while avoiding non-productive aggregation. In addition, the assembly pathway often leads from a precursor capsid that undergoes large-scale modifications such as proteolysis and expansion, cross-linking or the acquisition of decoration molecules, as well as packaging of the viral genome and docking of the tail structure if relevant, before becoming a mature and infectious particle. Virus assembly is thus a rather complex process involving many players, including a coerced host, and its study can teach us much about protein structure, interactions between proteins and with DNA/RNA, protein subunit dynamics and, more operatively, protein complexes as molecular machinery. In addition, since proteins and viruses are such an important part of our lives, the lessons of phage assembly can impact human health and commercial activity as well as offering opportunity for exploitation in research or technological settings.

Assembly of icosahedral capsids

Bacteriophages with icosahedral or prolate capsids are included in our term 'headed' phages. The majority of these – estimated at ~95% – have tails and are of the order Caudovirales, which is divided into three families. Well-known examples that cover the three families can be found among the seven T or 'Type' phages, collected from pre-existing stocks and renamed by Delbrück and co-workers in the 1940s to provide a focus for work on bacteriophages. Several of these have been extensively studied, particularly T4, a member of the Myoviridae family that is characterized by an extended tubular tail which contracts during DNA injection into the host cell. The so-called T-even phages, T2, T4 and T6, are structurally similar, with prolate icosahedral capsids, several sets of fibres involved in host recognition and binding, and an elaborate multi-protein baseplate. Phages with extended but non-contractile tails belong in the Siphoviridae family and include T1 and T5, for which limited structural information is available, as well as the better known λ, HK97 and SPP1 (Subtilis Phage Pavia-1) that have icosahedral heads comprising 415 copies of their major capsid proteins arranged with $T=7$ lattice geometry. The third family, the Podoviridae, is distinguished by a minimal tail and includes the similar T3 and T7 as well as P22 and the prolate φ29. The boundaries within this system of classification are sometimes blurred by phages sharing features of several families, such as the gene structure of one but the morphology of another (for example, see Chibani-Chennoufi *et al.*, 2004) and perhaps should not be considered to have the deepest biological significance.

Many variations on phage capsid assembly have been observed and perhaps many more are yet to be discovered. This section will focus on tailed phages; in particular one of the better studied cases, HK97, will be described in detail here since extensive structural information is available for the various stages of the capsid assembly pathway. In addition, the HK97 capsid protein, gp5, represents a novel fold that has since been identified in capsid proteins of other phages representing the three families of the Caudovirales: the T4 capsid proteins gp23 and gp24 (Fokine *et al.*, 2005), P22 (Jiang *et al.*, 2003), ε15 (Jiang *et al.*, 2008), φ29 (Morais *et al.*, 2005) and, by morphological similarities at 20 Å resolution, probably also in T5 (Effantin *et al.*, 2006), amongst others. If the HK97 gp5 fold is indeed common within the Caudovirales, it may represent one of the more numerous folds in nature.

More information is known on the capsid and portal structures of phages than on their tails. Crystallography of portal-less capsids and of portals has provided immense detail of the structures, albeit lacking in the dynamic aspects such as the action of the DNA-packaging machinery or capsid maturation. Cryo-electron microscopy has been more informative of the latter and the combination of structures from both methodologies has yielded pseudo-atomic resolution of the capsid–portal complex in the case of φ29 (Morais *et al.*, 2005; Morais *et al.*, 2001) and the maturation dynamics of the HK97 capsid (Conway *et al.*, 2001; Gan *et al.*, 2006; Helgstrand *et al.*, 2003; Lata *et al.*, 2000; Ross *et al.*, 2005, 2006; Wikoff *et al.*, 2006) amongst others. The tail of phage T4 has been under intensive crystallographic scrutiny, revealing the complex organization of the multi-protein baseplate to which six long tail fibres and six short tail fibres are attached (Kanamaru *et al.*, 2002; Kostyuchenko *et al.*, 2003; Leiman *et al.*, 2010) as well as changes during infection (Aksyuk *et al.*, 2009; Leiman *et al.*, 2004, 2010). We may consider these to be samples of a variety of structural forms employed by bacteriophages (and viruses) for effecting their life cycle. As more structural data become available, general principles underlying the bewildering diversity of the phage universe will doubtlessly be revealed.

A generalized assembly pathway for icosahedral phage capsids is outlined in Fig. 5.1. The nucleation for assembly *in vivo* is thought to be the portal, which is a 12-subunit ring that occupies one of the pentavalent sites in the complete capsid. This combination involves a curious symmetry mismatch – 12 to 5 – for which no definitive explanation is yet available. Current opinion holds that symmetry mismatches may be related to rotational motion, such as for the protein degradation machine, ClpAP (Beuron *et al.*, 1998), and the F0-F1 rotary proton pump (Junge *et al.*, 2001). However, another example where a trimeric fibre molecule binds to pentameric capsomers in phages P22 and PRD1, and also adenovirus, presents no obvious basis for invoking rotation. In the case of the phage capsid–portal mismatch, motion is proposed to be implicated in an ATP-driven pump that packages the genome into the pre-formed capsid (Simpson *et al.*, 2001). For T7 a further symmetry mismatch is expressed by a structure called the 'core' that assembles on to the 12-subunit portal on the interior of the procapsid and is essential for infection – this 'core' has 8-fold symmetry and is likely an octamer of gp15 (Cerritelli *et al.*, 2003b) and while it is essential for infectivity, its function is only now becoming understood (Chang *et al.*, 2010).

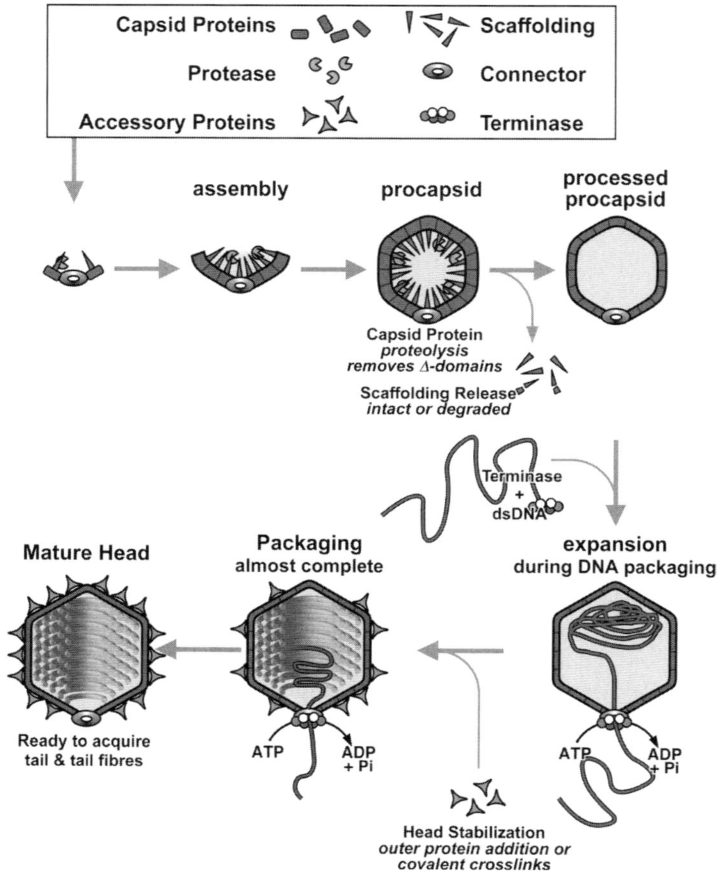

Figure 5.1 Schematic view of a generic assembly and maturation pathway for icosahedral capsids. Figure adapted from Steven *et al.* (2005).

The mechanism of nucleating icosahedral capsid assembly by accretion of subunits into a particular symmetry arrangement is not understood. Polymerization could take place subunit-by-subunit, but a more hierarchical accumulation involving capsomers or capsomer groupings seems likely for larger capsids, but has not been demonstrated. Several different strategies are employed by phages (and viruses in general) to aid in capsid assembly. One is to construct a preformed core, or scaffold, around which the capsid subunits will be recruited. The scaffold protein gene is usually part of the structural set that codes for the portal, scaffold, protease, decoration and major capsid proteins (not necessarily in that order). Presumably the core binds the capsid protein as monomers or capsomers (hexamers and pentamers) enforcing the correct curvature and interactions between the subunits. Alternatively, polymerization can take place by docking complexes of the capsid protein and scaffold onto the growing structure, again enforcing the correct geometry. Interestingly, the scaffold core is not necessarily icosahedral, and it is also possibly not uniform from particle to particle, despite yielding uniform capsids. A third mechanism is to dispense with a separate scaffold protein altogether – and indeed smaller capsids may polymerize without any such aid, while larger capsids such as that of HK97 have an extension to the major capsid protein that appears to be involved in pre-forming pentamers and hexamers in solution for incorporation into the nascent capsid (Duda *et al.*, 1995). In this case, the scaffold function appears to have fused with the capsid protein gene, or alternatively was never separated. The HK97 capsid was the first to be observed with this scaffold–capsid fusion, and was the largest such at ~500 Å diameter, but recent work on phage T5 shows that it is similar to HK97 in this and other respects but has a higher triangulation number of 13 and consequently a larger capsid of ~950 Å diameter (Effantin *et al.*, 2006; Huet *et al.*, 2010). Yet a fourth mechanism has been observed for phage φX174, which employs both an internal scaffold (protein B) and, unusually, an external scaffold (protein D) (Dokland *et al.*, 1997; Uchiyama and Fane, 2005).

Precursor capsids embark upon the maturation process by ejecting any internal scaffold and in some cases by removal also of some portion of the major capsid protein or proteins. Proteolysis is commonly employed to dismantle and reduce the sizes of proteins and domains so that they might be expelled through small holes in the procapsid wall. For HK97 (see Fig. 5.2), a separately encoded viral protease, gp4, is incorporated in the procapsid at a copy number of 60–70 per capsid and becomes activated post-assembly by some unknown mechanism. The major capsid protein, gp5, is trimmed from 384 residues by removal of the 102 N-terminal domain, which includes four or five potential cleavage sites (Duda *et al.*, 1995). The protease is also auto-proteolytic, thus inactivating and fragmenting itself. Mass measurements by scanning transmission electron microscopy (STEM) indicate that the capsid protein N-terminal domains and the protease fragments are removed from the capsid interior (Conway *et al.*, 1995), although how this is achieved is not known. It is possible that during the packaging step, the incoming genome effectively squeezes the fragments out through small pores in the capsid. Scaffold cores are also removed by proteolysis in some cases, but in others such as phage P22 the scaffold disassembles and subunits exit through large pores in the capsid for subsequent re-use (Greene and King, 1994).

The processed procapsid is primed for incorporation of the viral genome and, in some cases, for undergoing an expansion transformation that will exchange one set of subunit interactions for another as well as exposing motifs for binding decoration/stabilization proteins or for additional capsid protein subunit interactions. The binding of terminases and the dsDNA genome to the portal signals the beginning of packaging, which is powered by ATP hydrolysis and results in a nucleotide concentration of up to 500 mg/ml in the mature capsids of, for example, phages T7 and T4, and with inter-duplex spacings of between 22–28 Å on average. Although the genome cannot be organized with icosahedral symmetry, an onion layer appearance is often evident in reconstructions from electron microscopy images, and certain cryoEM images of DNA-filled capsids reveal a ring-like appearance in the projected density. For T7, the fortuitous pairing portal-to-portal of mutant capsids allowed the packaging axis to be distinguished in cryo-electron micrographs, and additionally certain views along a 5-fold axis could be assigned as coincident with the packaging axis, both kinds of image suggesting that the dsDNA is arranged in a spool about that axis (Cerritelli *et al.*, 2003a). Larger capsids such as for T4 may instead have local ordering rather than a global spooling of DNA (Booy *et al.*, 1991).

The pressure of the internally packaged DNA together with the rigors of the extracellular environment requires the capsid to be a strong container. Unlike eukaryotic

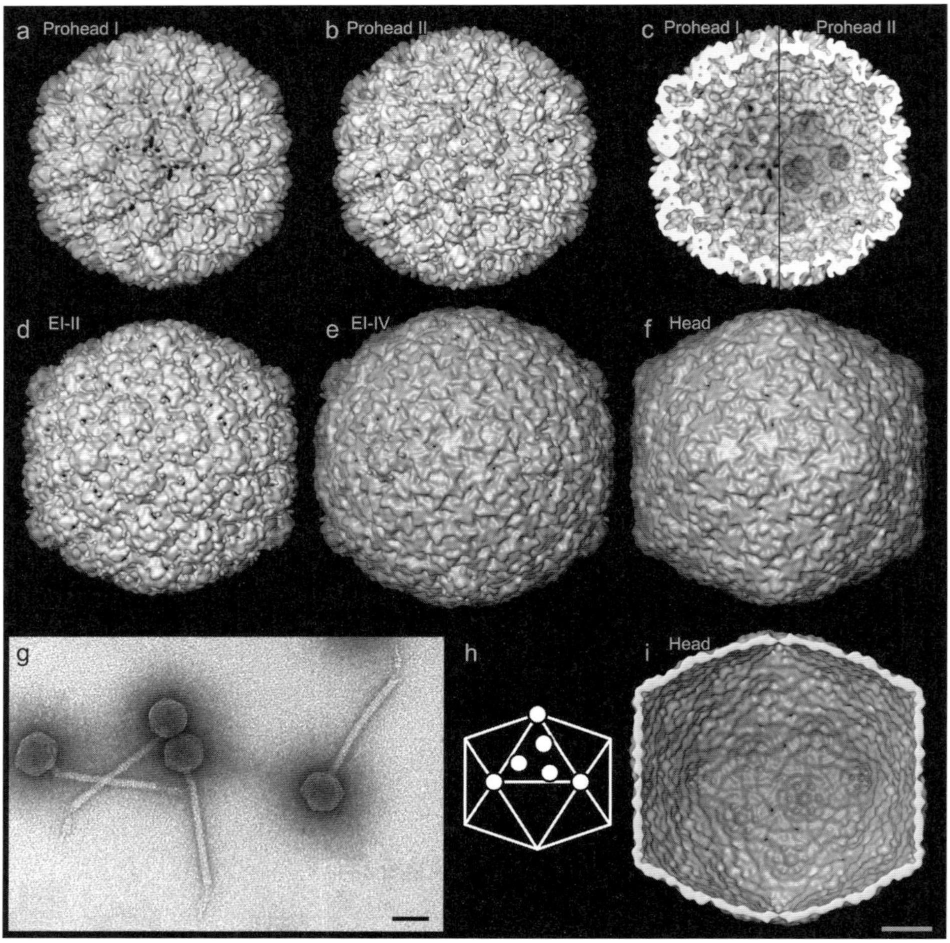

Figure 5.2 Phage HK97 capsid assembly intermediates. Surface renditions of structures solved by cryoEM to a resolution of 14Å, except where indicated. (a) Prohead I, the first complete capsid, is built from pentamers of the major capsid protein, gp5, at the icosahedral vertices and hexamers elsewhere arranged with triangulation number T=7*laevo* (see panel h). Prohead I is proteolytically processed to form Prohead II (b) by removal of the 102-amino acid N-terminal domains located on the interior of the capsid, as revealed in views of the back-half of the capsids (c). Several intermediates in the expansion step have been characterized, including EI-II (d) and EI-IV (e) shown here. These are progressively larger, rounder and smoother, as well as following a pattern of formation of covalent cross-links between subunits. The final mature capsid (f) is rendered from a crystal structure of the capsid to the same resolution as the cryoEM density maps. This form is very planar with straight facets and thin walls. The sectioned view (i) may be compared with those of the Proheads (c) to reveal the differences in the capsid walls. Bar = 100Å. (g) Negative-stain electron micrographs of mature phage particles, whose capsids are consistent with the polyhedral appearance of the mature capsid (f). Bar = 500Å. (h) Schematic layout of capsomers (circles) according to *T* = 7*l* lattice geometry. Figure adapted from (Steven *et al.*, 2005) and including panel (g) with kind permission of Drs R.L. Duda and R.W. Hendrix (University of Pittsburgh).

viruses that enter the host cell and disassemble to deliver their genome (for example, DNA viruses such as adenovirus and polyomavirus) the tailed phage capsid does not disassemble, allowing the use of irreversible covalent modifications or tightly binding decoration proteins for stabilization. Binding sites for stabilizing molecules are generated or exposed on the exterior surface by expansion of the capsid during genome packing for phages such as λ (gpD trimers for λ) and T4 (trimers of soc and monomers of hoc for T4). HK97 uses a different stabilizing mechanism – the formation of a peptidyl bond between an asparagine residue on one capsid protein subunit with a lysine on another (Duda *et al.*, 1995). This is a remarkable reaction, autocatalysed by a nearby glutamic residue on a third subunit and resulting in branched protein chains that form catameric rings topologically linked like chainmail or the Olympic rings (Duda, 1998; Helgstrand *et al.*, 2003; Wikoff *et al.*, 2000). The subunit motions have been modelled for the precursor capsid from the final crystallographically determined positions in the mature capsid by fitting subunit atomic models into a 12 Å cryoEM-based reconstruction (Conway *et al.*, 2001) and expansion intermediates (Gan *et al.*, 2006; Wikoff *et al.*, 2006). On average, each capsid protein subunit is displaced ~40 Å outwardly, and rotates ~40° during capsid expansion, and in addition, an extended loop, the E-loop containing one of the two cross-linking residues, hinges at its base by ~15° from an upright position towards the capsid surface much as the drawbridge of a castle descends. Cross-links cannot form prematurely since the amino acids involved are not in position until after expansion has begun. However, cross-linking does follow a distinctive and processive pattern across the capsid surface, possibly preventing back-sliding as the expansion process progresses (Gan *et al.*, 2004; Lee *et al.*, 2008). Whether such a pattern of stabilization also holds for capsids that bind decoration molecules remains to be seen.

The morphological result of expansion for the HK97 capsids is doubling of the internal volume, and reshaping of the capsid from the highly contoured and round precursor with knobby capsomers into a smooth, thin-shelled and polyhedral mature capsid (Conway *et al.*, 1995; Lata *et al.*, 2000). The large subunit motions indicate a complete rearrangement of the network of interactions holding the capsid together. An estimate of the change of buried surface area suggests that it almost doubles from precursor to the mature capsid, despite the concurrent thinning of the capsid wall (Conway *et al.*, 2001), although this additional contact area does not appear to contribute to stability according to physical measurements by scanning calorimetry (Ross *et al.*, 2005) but may instead be involved in the control mechanism guiding 415 copies of the capsid protein through the elaborate expansion transformation.

The packaged, mature capsid then acquires a tail, which is assembled through a parallel pathway and may involve as much or more complexity than head assembly given the numerous proteins involved in more elaborate baseplates, contractile tails, and multiple sets of fibres. Few structural details are available for this process and it will not be described here. The end result is an infectious particle that will be released from the cell, in some cases by lysis, primed for injecting its genome down the tail tube when its tail fibres encounter one or more receptor molecules on a host bacterium and the tail tip has gained entry. Assembly of the head has taken place within the environment of the host, and folding of capsid proteins and others may involve host factors such as groE (named for its role in folding the phage λ gpE capsid protein). However, polymerization of capsids from properly folded protein is often autonomous and may proceed thereafter *in vitro*. Maturation without genome incorporation may also be forced by very mild treatment, as is the case with HK97, resulting in uniform particles of a defined size and architecture.

The production of infectious phages, mutants or empty capsids is relatively easy and is potentially of some medical and biotechnological importance. Interest in phage therapy for combating bacterial infections in humans has yet to translate into an effective medical application, but the possibility of phage use in some topical infection (e.g. acne) does not seem farfetched. Treatment of food for bacterial contamination has been demonstrated in cheese (Guenther and Loessner, 2011), fresh and prepared food (Guenther *et al.*, 2009; Leverentz *et al.*, 2006), and meats (Anany *et al.*, 2011). A recently recognized adoption of phage tails by the bacterial hosts is the pyocin family that may have applications as antibiotics (Nakayama *et al.*, 2000; Scholl *et al.*, 2012). Phage display, allowing high-throughput screening of protein interactions, was originally developed in filamentous phage but icosahedral capsid and tail proteins have also been exploited in this way (see the λ and T7 phage display sections below). Phage particles are also being tested for use in constructing nano-materials based on the uniform sizes of the particles that may lend themselves to well-defined properties at the nano-scale for medicine (Souza *et al.*, 2006) or materials (Nam *et al.*, 2006), although it is early days yet. These promising avenues of phage biotechnology are explored further below.

Assembly of filamentous bacteriophages

Filamentous bacteriophages are not only different from the headed phages in their morphology, but also in their assembly (reviewed by Marvin, 1998; Model and Russel, 1988; Webster, 1996, 2001). The hallmark of filamentous phage assembly is its location: virion proteins are integrated into the membrane prior to assembly, and the assembly itself is carried out at the membranes of their hosts (Endemann and Model, 1995; Russel, 1995). In contrast to the autonomous and ordered assembly of the headed phage in the cytoplasm, filamentous phage assembly is a secretion-like process, carried out by phage-encoded trans-envelope assembly machinery.

The secretory fashion of assembly, during which the virion is simultaneously extruded through the host cell envelope, is the reason why the host cell survives virion production. In fact, the secretory assembly is much more efficient than the one-shot burst of the headed phage. The exemplary F-specific phages of *E. coli* (Ff phages), which are used as cloning vectors, are assembled/secreted at a rate as high as 1000 phages per cell per hour. As a result, the titres of the Ff phages can be as high as 10^{12}–10^{13} per ml of culture, as opposed to phage λ, whose titre is at most 10^{9}–10^{10} per ml.

The filamentous phages infect Gram-negative bacteria (with only two exceptions; Day, 2011). With only 70 filamentous phages isolated to date, they represent a minor fraction of bacterial viruses found in nature. However, it is possible that the number of filamentous bacteriophage species is significantly underestimated, because many of them do not show any visible plaques when plated on their hosts. (Traditionally, plaque-forming is the chief assay for discovery of phages.) Incidentally, the first identified filamentous bacteriophages, the F-specific phages (Ff) of *Escherichia coli*, replicate as episomes at a high rate, slowing the host growth rate by about 50% and allowing for the formation of turbid plaques. This feature permitted the discovery and analysis of the filamentous phages. In contrast, a very distantly related filamentous phage of *Vibrio cholerae*, CTXΦ, whose genome is integrated into the host chromosome and which encodes the deadly cholera toxin, is produced at a rate of only 1 per 1000 cells per generation time and therefore cannot form plaques. The genome and the genes of this 'lysogenic' phage have been studied for years as 'auxiliary' toxins until Waldor and Mekalanos (1996) showed, by transduction of an engineered antibiotic marker, that the set of genes around the *ctxAB* are in fact coding for the virion proteins of a filamentous phage! Since this original discovery, a number of *V. cholerae* lysogenic filamentous phages have been discovered and their life cycles, including very unusual type of lysogeny, have been investigated. A number of lysogenic filamentous phages have also been discovered in genera *Neisseria, Pseudomonas, Xanthomonas, Yersinia* and *Escherichia* (reviewed in Day, 2011; McLeod *et al.*, 2005; Rakonjac *et al.*, 2011).

Most of the structural information about virion structure and assembly comes from episomally replicating *P. aeruginosa* phages Pf1 and Pf3, as well as F-pilus-specific *E. coli* phages. These phages, almost identical in DNA sequence and used interchangeably, are f1, fd and M13, and will be referred to in this review as Ff. The Ff phages are the focus of this review not only because most of what we know about filamentous phage assembly is revealed using these phages as model system, but also because they are the most commonly used nanoparticle carriers for display of proteins of interest in phage display technology.

Filamentous phage structure

As their name implies, filamentous phages are long thin filaments, very much like the bacterial surface structures called pili. Although filamentous phages and pili are not genetically related, they are both built by helical arrangement of many copies of a single subunit, and their morphology at the resolution of the transmission electron microscopy (TEM) is indistinguishable. Moreover, if the release of the filamentous phages is prevented by deleting the minor virion protein that is required for termination of assembly and release from the host envelope, phage filaments become permanently anchored to the cell surface as DNA-containing pili (Rakonjac and Model, 1998). Since the virion protein (pIII) that terminates the assembly is also the protein which mediates infection, it is possible that filamentous phages may have evolved from conjugative pili by gaining a protein that provided an efficient mode of spread in bacterial populations, without a need for direct contact between the donor and the recipient.

The Ff virion is 6 nm in diameter and 860 nm in length. The length of the virion depends on the size of the DNA packaged into the filament. The filamentous phage genome and virion can accommodate DNA inserts of up to 12,000 nt, almost twice the size of the phage genome itself (6407 nt). Moreover, in the wild-type Ff population, 5% of the virions are twice as long as expected, and presumably carry two sequentially packaged phage genomes. Fig. 5.3A shows an atomic force microscope image of two filamentous phage particles that differ in size due to packaging of the two 'genomes': the longer carries the phage ssDNA genome, whereas the shorter carries the ssDNA of a phagemid vector generated by replication from the f1 origin engineered into the phagemid, using proteins encoded by the helper phage.

The tube of wild-type Ff phages is composed of about 2700 copies of the major coat protein (pVIII, also referred to as gp8 or g8p or p8). The ends of the virion are formed by two different pairs of minor virion proteins: pVII/pIX and pIII/pVI. Each of the minor proteins is present in the virion in 3–5 copies (Fig. 5.3B).

The major coat protein is 50 residues in length. X-ray fibre diffraction was used to determine the structure of the major coat protein and the arrangement of the monomers in the virion of several filamentous phages (reviewed by Marvin, 1998). Each pVIII subunit is an α-helix, with the axis slightly tilted relative to the axis of the filament, the N-terminus pointing outwards, and the C terminus inwards (Fig. 5.3C; Marvin *et al.*, 2006). The pVIII

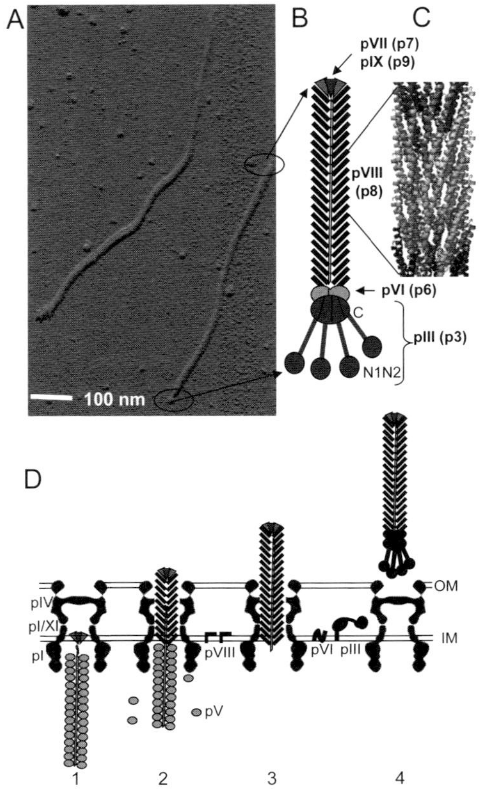

Figure 5.3 Organization and assembly of Ff phages. (A) Atomic force microscope image of two filamentous phage particles. The longer and the shorter particle carry the ssDNA of the helper phage R408 and a phagemid, respectively. (B) Schematic representation of the Ff virion. N1N2. (C) The domains of pIII, respectively. C. Ribbon representation of pVIII structure and arrangement within the virion, derived from the RCSB PDB database accession number 2cOw (Marvin *et al.*, 2006). (D) Ff assembly. 1, initiation of assembly; 2 and 3, elongation; 4, termination. Proteins pVIII, pVI and pIII in the inner membrane prior to assembly are shown. Outline of the pIV, the outer membrane channel is based on coordinates from Opalka *et al.* (2003). The outline of the inner membrane complex is based on the type III secretion system inner ring complex (Marlovits *et al.*, 2004); the actual structure of the inner membrane pI/pXI complex has not been determined.

subunits pack with two different types of symmetries, corresponding to two different structural classes. In *E. coli* phage Ff, IKe and If1 (class I), pVIII subunits are arranged on a five-start helix with 2-fold screw axis (C_5S_2 symmetry) of about 32 Å pitch, with five subunits in each 'ring' of the virion tube, related by a 5-fold rotation axis (Day *et al.*, 1988; Marvin *et al.*, 2006). In the phage of Pseudomonas and Xanthomonas (class II) pVIII subunits are arranged on a simple one-start helix of ~ 15 Å pitch, with 5.4 subunits per turn ($C_1S_{5.4}$) (Goldbourt *et al.*, 2007).

The ratio of subunits per nucleotide is specific to each phage. In Ff it is 0.42 ± 0.01, whereas in Pf1 it is 1 (Day *et al.*, 1988). The DNA is arranged as a right-handed helix; however the conformations in the two classes differ greatly. In Ff, the bases are stacked at the centre with the phosphates on the outside interacting with the capsid; the bases from opposing sides of the single-stranded DNA circle are H-bonded to one another, but only about 25% of the H-bonding can be of the Watson–Crick type. In Pf1, where there is one subunit per nucleotide, strands are highly stretched and twisted, resulting in an unusual conformation in which the sugar phosphate back-bones are in contact with each other at the centre and the bases are exposed on the outside for interaction with the capsid; because of wide separation between the bases of opposing DNA strands, H-bonding is not possible (Liu and Day, 1994; Tsuboi *et al.*, 2010).

As opposed to the wealth of detailed knowledge of the structure and arrangement of the major coat protein subunits along the filament, no structural information is available at a resolution higher than that of transmission electron microscopy (TEM) about the ends of the virion. It is known that the pVII/pIX end is blunt, whereas the pIII/pVI end is pointy. Proteins pVII and pIX are extremely short and hydrophobic, 33 and 32 amino acids (aa), respectively. The structure of the pVII/pIX end of the virion has not been determined. These two proteins are the first to be assembled and extruded during phage assembly.

At the opposite (pIII/pVI) end of the filament, the minor protein pVI is relatively larger (112 aa), but is mostly hydrophobic. Protein pIII is an exception among the virion proteins in that it is large (406 aa) and mostly hydrophilic, except for a C-terminal hydrophobic helix. This protein has a modular organization. Three knobs-on-a-strings are sometimes discernible at the pIII/pVI end; the knobs are removed by subtilisin treatment, which also eliminates the N1 and N2 domains of pIII, suggesting that these two domains are separated from the body of the virion by a flexible linker (Grey *et al.*, 1979). The structure of two N-terminal domains, N1 and N2 which are required for binding to Ff receptors on the host cells to initiate the infection has been solved at a high resolution using X-ray crystallography and NMR and their functionality in binding the host receptors has been shown (Holliger *et al.*, 1999; Lubkowski *et al.*, 1998, 1999; Martin and Schmid, 2003; Reichmann and Holliger, 1997). The pIII C domain structure has not been determined. It has been shown, however, that together with pVI and a small number of the major coat protein pVIII subunits, the C domain of pIII forms a tight complex (Gailus and Rasched, 1994). Steps in the infection that follow receptor binding and that result in the insertion of the virion coat protein into the inner membrane and entry of phage DNA into the cytoplasm require the C domain of pIII (Bennett *et al.*, 2011). Furthermore, the C domain of pIII is required for the termination of phage assembly and release of the virions from the host cell into the medium (Crissman and Smith, 1984; Rakonjac *et al.*, 1999; Stengele *et al.*, 1990).

Assembly

The single-stranded circular DNA of filamentous phages replicates by a rolling circle mechanism which is very different from the originally proposed 'toilet roll' model of the λ phage in that it is carried out one strand at a time and results in production of unit-length single stranded circular DNA products. The replication is initiated after the positive strand ssDNA from the virion enters the host cell during infection. The negative strand is synthesized by the host proteins; the RNA polymerase first forms a primer at the negative origin, and the DNA polymerase holoenzyme uses that plasmid as a primer to synthesize the negative strand (Higashitani *et al.*, 1996). The double-stranded or replicative form (RF) this produced serves both as a template for replication of the negative strand and as a template for transcription of the phage genes required for proceeding of the life cycle. Replication of the positive strand starts from the positive strand origin and is initiated by the phage-encoded replication protein pII. As the new positive strand is synthesized the old one is displaced (Asano *et al.*, 1999; Higashitani *et al.*, 1994). Positive- and negative-strand origins of replication and the packaging signal are all located in the intergenic region of the phage genome, within a few hundred base pairs. This fragment was engineered into many plasmid vectors to generate phagemid vectors, which allow replication and packaging of plasmid ssDNA in the presence of a helper phage (Dente *et al.*, 1983). Early in the infection, positive strands produced by replication are used as templates to create more copies of the double-stranded RF form of the phage. Later in the infection, phage-encoded ssDNA-binding protein pV accumulates in the cytoplasm, down-regulating replication and coating the ssDNA to form a packaging substrate for phage assembly (Fig. 5.3D; Michel and Zinder, 1989). The packaging signal (a hairpin loop) is bare while the rest of the molecule is coated by pV, a phage-encoded ssDNA binding protein. The packaging signal, as its name implies, is recognized by a trans-envelope assembly-secretion machinery (Brissette and Russel, 1990; Rapoza and Webster, 1995).

Most of what is known about the mechanism of assembly (schematically depicted in Fig. 5.3D) comes from genetic analyses. The only structurally analysed component is the outer membrane protein pIV, an outer membrane protein of the secretin family, which was shown by cryo-EM and electrophysiological analyses to be a gated channel with an inner diameter of 8 nm, sufficient to export the phage particle, which is 6 nm in diameter (Linderoth *et al.*, 1997; Marciano *et al.*, 1999, 2001; Opalka *et al.*, 2003). The inner membrane component of the assembly machinery is composed of two proteins, pI and pXI, of which the latter corresponds to a C-terminal portion of pI and is produced by translation from an internal ATG codon within gI. Both pI and pXI are required for assembly (Haigh and Webster, 1999; Rapoza and Webster, 1995). The pI protein contains a putative ATP-binding domain necessary for assembly. It is thought that pI/pXI complex forms a channel and it has been shown to have an ATPase activity (Feng, 2000; Horabin and Webster, 1988). A semi-permeable system for testing the requirements for assembly has shown that both ATP and proton motive force are required for assembly (Feng *et al.*, 1997). Experiments testing interchangeability between pIV of Ff and a related phage Ike showed that pI and pIV from the same phage are required for assembly and that the periplasmic domain of the pIV interacts with the C-terminal portion of pI which is located in the periplasm (Daefler *et al.*, 1997; Feng *et al.*, 1999; Russel, 1993).

Prior to assembly, all virion proteins are integrated into the inner membrane (Endemann and Model, 1995). Therefore, the inner membrane component of the assembly machinery, pI/pXI, is likely to serve as an assembler, taking the subunits up from the surrounding lipid bilayer and feeding them onto the ssDNA substrate which is simultaneously being translocated across the membrane. For the assembly to be initiated, the packaging signal has to interact with the pI/pXI inner membrane complex and minor virion proteins pVII and pIX and pI (Russel and Model, 1989).

The elongation phase ensues, during which the virion is assembled as it is extruded from the cells. As the ssDNA traverses the membrane, pV dissociates and is replaced by the major coat protein pVIII. During assembly, the virion proteins are transferred from the phospholipid bilayer of membrane to the lipid-free filament (Newman *et al.*, 1977) where, at least in the case of the major coat protein pVIII, former phospholipid-interacting hydrophobic helices now interact with other subunits of the virion (McDonnell *et al.*, 1993; Papavoine *et al.*, 1998; Stopar *et al.*, 2002).

In the membrane, prior to assembly into the virion, pVIII is a dimer; preservation of dimer is required for assembly (Haigh and Webster, 1998). Each minor protein associates with pVIII in the membrane, suggesting that it guides incorporation of the minor proteins into the virion (Endemann and Model, 1995). Genetic data suggest that pVIII may directly interact with pI (Russel, 1993).

When DNA is completely coated with pVIII, assembly is terminated and virions are released from the cells by excision from the membrane and insertion into the virion of two minor proteins, pIII and pVI, which form a detergent- and heat-resistant complex or 'cap' at the end of the filament. If either of these two proteins is absent, the phages remain tethered to the membrane (Rakonjac *et al.*, 1999; Rakonjac and Model, 1998). Interestingly, rather than stopping the assembly altogether, the filaments keep elongating by adding new genomes and pVIII monomers, resulting in a bizarre appearance of the infected cell in electron micrographs: after a 90 min infection 200–300 several micrometer long pili-like structures emanate from the cell surface.

Filamentous phage replication has been a topic of intense research in the 1980s and the knowledge about it was applied to construct filamentous phage-derived cloning and phage display vectors, including those that are used as templates for the assembly of nanoparticles.

Applications

Bacteriophages and phage display

Bacteriophages were the original model systems instrumental in establishing basic tenets of molecular biology, providing most of the fundamental knowledge about of the processes of replication, transcription and translation. In parallel, because of their easy production in large amounts and the knowledge about their replication and assembly, bacteriophage have been adapted and widely used as cloning vectors for the recombinant DNA technology (Sambrook and Russell, 2001).

In the mid-1980s and early 1990s, phages took the centre stage in development of combinatorial or display methods. The basic idea of these methods is to create libraries of peptide or antibody variants, each variant bound to a nanoparticle, and to affinity select and identify variants that bind to a target of interest. Although non-biological nanoparticle libraries fulfil these conditions, their limitation is inability to amplify the selected variants (Chen *et al.*, 1996). If a large library (e.g. 10 million variants) is screened for the peptides of interest, each variant is present in a very low copy number; therefore the number of particles of a single variant can be too small to allow identification. To overcome this problem, biological *in vitro* and *in vivo* systems, in which the DNA encoding for the peptide or antibody is physically linked to the displayed peptide, have been developed (Cull *et al.*, 1992; Cwirla *et al.*, 1990; Devlin *et al.*, 1990; Mattheakis *et al.*, 1994; Scott and Smith, 1990). The link of displayed protein or peptide to genetic material allows amplification of the selected variant, which in turn permits several rounds of affinity enrichment that eventually results in a pure population of a single or small number of biding variants. This in turn allows for easy identification of selected variants by sequence analysis of the nucleic acid coding for the peptide/antibody. In the 'synthetic' display systems, the physical link between the peptide and encoding sequence is provided by trapping the protein product and encoding mRNA to a single ribosome. Another approach is to fuse the peptides or proteins to a transcriptional regulator bound to DNA which encodes the fusion. Templates of the affinity selected proteins can be amplified by PCR. By far the most versatile system proved to be display on the surface of bacteriophage, by fusion to the coat (capsid) proteins. The advantage of phage display over other methods is the stability of the protein–DNA interaction and ease of amplification. Among dozens of *E. coli*-infecting phages, the functions of which have been studied intensively over several decades, Ff, λ and T7 phages are in common use today as phage display vectors, Ff being used predominantly.

The aim of all phage display library screenings is finding a rare variant. As a consequence, the strategy is to construct a library large enough to accommodate all possible variations. For example, a library of random hexapeptides will require a diversity of $20^6 = 6.4 \times 10^7$ in order for every variant to be represented. Therefore, the first important characteristic of a phage display library is its size. In theory, it should be possible to express all combinations by inserting into a phage display vector a random coding sequence 18 nucleotides in length. However, in order for the library to be displayed on the phage and amplified, it has to be transformed into *E. coli*. The limit of transformation depends on many factors, but it can be taken that, for a phagemid, it would be up to 10^{10} primary transformants per μg of DNA; therefore limiting the length of the randomized peptide. Very often, longer peptides are desired and in that case the library is not expected to be saturated, therefore decreasing the chance to isolate a particular variant.

Once constructed, the library has to be screened to isolate the variant of interest. First the phages have to be assembled and released from the host cells in sufficient number to achieve high enough odds for a particular variant to be selected. Success will depend on the affinity of the peptide for the ligand and the number of copies of the peptide per virion. For a library of 10^{10} variants, at least 10^{11} virions displaying peptides need to be used in the first round of affinity selection to have a chance of isolating the binding peptide. This is generally not a problem, because phages used in phage display technology are produced at the concentration (or titre) of 10^9 (λ) or 10^{12} (Ff) per ml of culture. Increasing the culture volume and concentrating the phage particles further increases this number.

The basic method for affinity screening of the phage display libraries is often referred to as 'biopanning' (Fig. 5.4; Parmley and Smith, 1988). The ligand (or bait) of interest is bound to a solid phase, and the phage display library (in the form of purified virions) in solution is applied to it to allow binding of the specific variants. Multiple rounds of washing are applied to eliminate the non-binders, and what remains is eluted. If no specific eluent is available, low or high pH are commonly used to disrupt interactions of the displayed peptide with the solid phase-attached ligand, although it is very important that the elution conditions do not interfere with the infectivity of the viral particle since disrupting the structure and infectivity of the virion would prevent the amplification and consequently identification of the variant of interest. Because of a degree of unspecific binding of phages to the matrix surrounding the ligand, at least three rounds of panning are performed in order to amplify the binding variants and eliminate the 'background' of non-specific binders. The readers are referred to dedicated phage display manuals for detailed phage display protocols (Barbas *et al.*, 2001; Kay *et al.*, 2002).

The original phage display publication (Smith, 1985) described mapping the epitope of an antibody by displaying a library of random fragments of the antigen on the surface of the phage in fusion with pIII. However, the real potential of this technique was realized by construction and screening of phage display libraries of both peptides and antibodies,

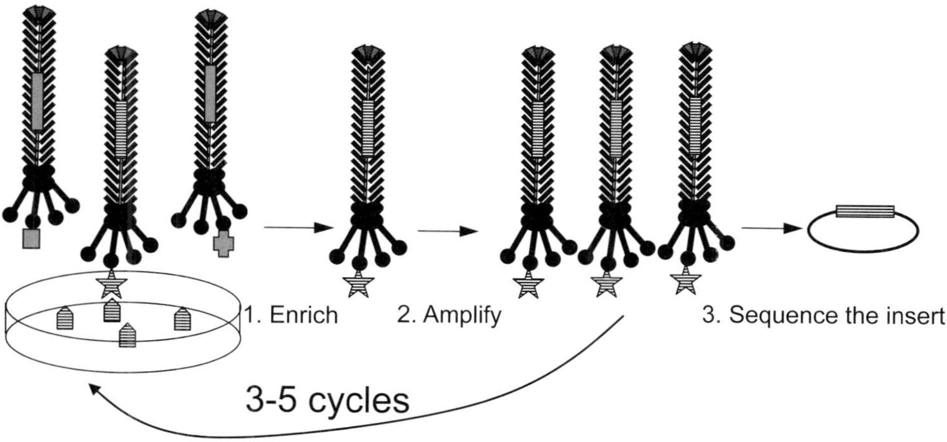

Figure 5.4 Schematic representation of the biopanning protocol. The library is symbolized by three phagemid particles. Different shapes associated with pIII represent different variants of displayed protein. The rectangles in the centre of the particles represent the coding sequences for the displayed proteins. The phage or transducing particle interacting with the shapes bound to the solid support is enriched for by washing off the non-binding phage, followed by amplification. This cycle typically needs to be repeated 3–5 times to obtain a pure population of binding variant(s). The selected variant is identified by determining the sequence of the insert.

followed by an explosion of phage display use in protein and antibody engineering (Barbas *et al.*, 1991; Clackson *et al.*, 1991; Cwirla *et al.*, 1990; Devlin *et al.*, 1990; Scott and Smith, 1990). In March 2012, 8713 publications were retrieved from the ISI Web of Science database using 'phage display' as a keyword, of which 789 were reviews.

The Ff phage display

Because of exceptional resistance to a broad range of pH or temperatures, and even high pressure expected during large-scale fermentation (Branston *et al.*, 2011), Ff filamentous phages are prevalently used in phage display technology and nanotechnology applications. Each of the five Ff virion proteins have been used as platforms for phage display, but most commonly used are the minor protein pIII present in 3–5 copies at the end of the virion, and the major protein pVIII which forms the tube of the virion and is present in a few thousand copies per virion (Fig. 5.5B–F, H and I). Both proteins in Ff phages have an essential N-terminally located signal sequence which is required for targeting of the virion to the inner membrane via the SecYEG translocon prior to assembly into the virion. The accessory proteins that are involved in membrane targeting along with the translocon, however, differ for the two coat proteins: it is YidC for pVIII and SecA/B for pIII (Rapoza and Webster, 1993; Samuelson *et al.*, 2001). During translocation, the signal sequence is cleaved by the signal peptidase, and the N terminus of the mature protein is located in the periplasm. Both pIII and pVIII are retained in the inner membrane by the C-terminally located hydrophobic α-helix. After assembly, their N termini are exposed on the surface of the virion. Therefore, for display on the phage surface, coding sequences have to be inserted, in frame, between the signal sequence and the mature portion of these two proteins.

Proteins pVII and pIX do not have a signal sequence; the short length and hydrophobicity guides their insertion into the membrane. Initially, it was thought that these two proteins

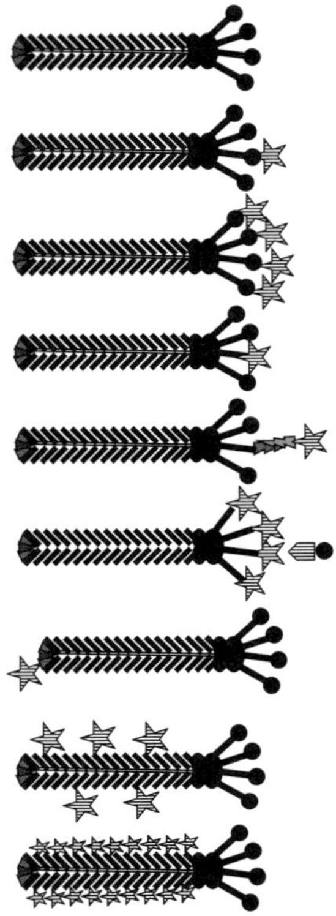

Figure 5.5 Types of phage display. (A) Wild-type phage, B-F, types of pIII display. (B) Monovalent display, fusion with the full length pIII. (C) Polyvalent display, fusion with the full length pIII. (D) monovalent display, fusion to the C domain of pIII. (E) Indirect phage display. The connecting shapes between the star-shaped displayed protein and truncated pIII represent the leucine zipper dimerization domains of transcription factors Fos and Jun. (F) SAP (SIP). Displayed proteins are fused to the C domain of pIII. These particles are non-infectious because they lack the N1N2 domain of pIII, required for receptor binding. The ligand is fused to the N domain of pIII; the interaction of the phage displayed protein with the ligand restores infectivity. The pIII polyvalent display is obtained by infection with the ΔgIII helper phage. (G) pVII and/or pIX display; H, mosaic pVIII display: wild-type pVIII is co-incorporated into the virion together with the pVIII fusion. I, uniform pVIII display. All pVIII subunits are fused to peptides.

could not be used in display; however Gao *et al.* (1999) have shown that, if a signal sequence is provided at the N-terminus of the insert, efficient display of the fused protein can be achieved (Fig. 5.5G). Moreover, two subunits of a heterodimeric protein can be displayed in pairs when fused to pVII and pIX expressed in the same host cell, resulting in display of the heterodimeric protein on the surface of the phage.

The topological organization of pVI has not been investigated. One report showed that fusion of protein sequence to the C-terminus of pVI results in display (Jespers *et al.*, 1996); however this site had not been used frequently.

The pVIII display

In pVIII display, the number of copies of the foreign insert per virion depends on the size of the displayed peptide. The cut-off length for display on every pVIII copy (Fig. 5.5I) is six or seven residues (Iannolo *et al.*, 1995). In order to display larger peptides and proteins, a mix of the wild-type pVIII and the display fusions have to be assembled into the phage (Fig. 5.5H). An interesting modification of pVIII display used two different fusions which include a cysteine-containing linker. Because pVIII assembles as a dimer, cysteines of the two linkers become covalently linked by an S–S bridge, thereby forming a dimer upon assembly into the virion (Zwick *et al.*, 2000).

The pIII display

Protein pIII is most commonly used for display. Both the full-length (Fig. 5.5B and C) and the truncated pIII, containing only the C-domain, are used as a platform for display (e.g.

(Barbas *et al.*, 1991; Griffiths *et al.*, 1993). If truncated pIII is used for display, a wild-type (full length) pIII must be provided from the helper phage to allow infection of the *E. coli* host to allow amplification of selected clones (Fig. 5.5D).

The pIII fusions are most commonly expressed from phagemids vectors, which carry both plasmid and f1 origins of replication and also the f1 packaging signal. Upon helper phage infection, phagemids replicate from the f1 origin, resulting in production of ssDNA with the packaging signal that is assembled into the virion. Phage display phagemid vectors carry the cloning site between the signal sequence-coding segment and the mature full length or truncated pIII. This allows proteins encoded by the inserts to be displayed on the surface of the phage in fusion with pIII; at the same time, the DNA encoding the fusion is packaged inside the particle, usually called a phagemid particle or transducing particle (TDP). To amplify a phagemid, the F^+ host cells are mixed with the particles. The phagemid DNA is introduced into the cell by infection, resulting in expression of the antibiotic resistance encoded by the phagemid. Therefore, the phagemid is transfected into the host cell. Except for the mode of phagemid DNA introduction into the cell, transfection is equivalent to transformation: in the absence of the helper phage, phagemid replicates from the plasmid origin of replication.

The phagemid offers flexibility in that the pIII fusion for display can be expressed from an inducible promoter. This is desirable because many pIII fusions are relatively toxic to the host cell. Furthermore, the size of the phagemid vector is generally smaller than that of the phage genome, therefore increasing the efficiency of transformation, which is the bottleneck limiting the diversity (or primary size) of phage display libraries. Phagemid phage display vectors also offer flexibility in the number of copies of pIII per virion. If the wild-type helper phage is used, there is on average less than one display fusion per virion, due to preferential incorporation of the wild-type pIII encoded by the helper phage. However, when helper phage that carries a deletion of pIII is used, all pIII copies in the virion display the pIII fusion encoded by the phagemid (Fig. 5.5C; de Wildt *et al.*, 2002; Griffiths *et al.*, 1993; Rakonjac *et al.*, 1997).

There are several other variants of pIII display. One system was developed to exploit the requirement of N1N2 domains in the infection. In this system, called selectively infective phage (SIP) or selection and amplification of phage (SAP), protein–protein interaction screens were based on restoration of infectivity through interaction of two fusions: one soluble, with the N1N2 domain fragment of pIII, and the other, with the C domain of pIII incorporated into the phage particle (Duenas and Borrebaeck, 1994; Krebber *et al.*, 1995). The interaction between the proteins fused to two parts of pIII link the N1N2 domains of pIII to the virion, thereby allowing infection (Fig. 5.5F). A similar system was used for selection of catalytic antibodies carrying out a chemical reaction which covalently links N1N2 domains with the phage via an organic molecular bridge (Gao *et al.*, 1997).

The λ phage display

The λ phage was adapted for phage display about 10 years after the filamentous phage (Dunn, 1995; Maruyama *et al.*, 1994; Sternberg and Hoess, 1995). The main advantage of λ over the Ff phage is that its site of assembly is in the cytoplasm; therefore the proteins for display do not have to cross the inner membrane of *E. coli* into the periplasm. As stated in the previous section, many proteins from the cytosol that contain multiple cysteines are misfolded in the periplasm due to formation of disulphide bridges; therefore assembling them into the λ phage particle in the cytosol is an advantage. Two λ phage proteins have been used as platforms for phage display, gpD and gpV. GpD is a 110-residue head accessory protein which binds as a trimer to the capsid during expansion and is essential for packaging the complete genome. Furthermore, this protein structurally stabilizes the phage head; in its absence, high Mg^{2+} concentration is required for maintaining the head structure; exposure of phage lacking GpD to EDTA results in disintegration of the phage head. Fusions to both the N and C terminus of gpD are assembled into the phage and displayed on the surface. The structure of the free gpD trimer, excluding the N-terminal 15 residues, has been determined by X-ray diffraction to a resolution of 1.1 Å, and it was fitted into the cryo-EM-derived structure of the phage (Yang *et al.*, 2000). Interestingly, both N and C termini are located on the face of gpD proximal to the capsid; therefore they are not directly exposed on the surface. The success of display can be explained by the use of long linkers engineered between gpD and the fused insert. The second λ protein used as a display platform is gpV, the tail shaft protein. Both proteins are capable of displaying relatively large fusion partners. Display using gpV is achieved at a low copy number for large proteins and high copy number for short peptides (Dunn, 1996b). In contrast, the gpD display of large proteins has been achieved at a high copy number per virion (Gupta *et al.*, 2003). This is desirable for selection of low affinity interactions, which are common e.g. among many signal-transduction proteins. Despite these advantages, λ phage display vectors are used significantly less frequently than Ff vectors, mainly because cloning into λ is not as easy as it is into phagemid or Ff-derived vectors. A second disadvantage is lower stability of the λ phage, which has a much narrower pH tolerance range than does Ff. It is inactivated at pH 2.2, the most common elution condition in affinity-purification or biopanning. To overcome this limitation, some λ display cloning constructs include protease target sites between the gpD and the inserted protein (Maruyama *et al.*, 1994; Mikawa *et al.*, 1996). The λ phage display cloning vectors are of two types: (i) λ vectors, adapted for creating either gpD or gpV fusions (Dunn, 1995; Maruyama *et al.*, 1994); and (ii) systems consisting of *in vivo* recombining plasmid and phage, where the plasmid is manipulated *in vitro* to create a gpD fusion, then it is recombined *in vivo* with the λ 'acceptor' phage (Gupta *et al.*, 2003; Santi *et al.*, 2000; Sternberg and Hoess, 1995). The site-specific recombination system consists of *lox* sites and corresponding *lox*-specific Cre recombinase which mediate the integration of the fusion-encoding segment of the plasmid into the acceptor phage. The plasmid system is used to facilitate *in vitro* manipulations which, when λ vectors are used, are rather cumbersome because the genome size is nearly 50 kbp. Moreover, the efficiency of plasmid transformation is up to 100 times as high as that of transfection of *in vitro* packaged the λ genome. The λ phage display system has successfully been used to display enzymes, peptides, and cDNA libraries (e.g. Cicchini *et al.*, 2002; Dunn, 1996a; Hoess, 2002).

The T7 phage display

T7 phage display (Rosenberg *et al.*, 1996) has been available from the company Novagen and has been used relatively frequently in the recent several years. The advantages of T7 over λ phage are that the virions are more stable and can withstand harsh treatments such as high salt, urea, guanidine, EDTA, SDS and alkaline pH (up to 10), which allows for easy and inexpensive elution of phages in the biopanning procedure. Another advantage of T7 over λ is a tenfold higher efficiency of *in vitro* packaging. The platform for display is the major capsid protein gp10. The gene codes for two variants, gp10A (344 aa) and gp10B

(397 aa), of which gp10B is a frame-shift product present at a proportion of ~10% in the head (Condron *et al.*, 1991). Because of its increased length and lack of conservation of the C-terminus with gp10A, the probability that gp10B C-terminus is exposed is higher than that of the C-terminus of gp10A. Therefore, phage display vectors for T7 are designed so that the insert is fused to the C terminus of gp10B. Peptides of up to 50 residues can be displayed at a high density (up to 415 per virion) in the vectors in which the gene 10 was modified to code only for gp10B variant (without shifting the frame). The limitation of the T7 system is its inability to display larger proteins at a high copy number. Larger proteins, up to 1200 aa in length, can be displayed only at a density of 0.1 to 1 per virion, from a T7 phage display vector in which the expression of gp10B is low and the wild-type gene 10 is provided from a plasmid. This lowers the chance of isolating low affinity interactions, which is important in interaction screens of cDNA libraries. Currently there are about 150 reports in the database using T7 display, and besides peptide and antibody display, successful applications for isolating proteins from cDNA libraries for binding to, for example, specific RNA or DNA sequence motifs, chromatin or toxins have been reported; e.g. (Danner and Belasco, 2001; Gearhart *et al.*, 2002; Kurakin *et al.*, 2004; Wang *et al.*, 2004). Furthermore, it was shown that the peptide libraries displayed on T7 are less biased than those on Ff; hence the T7 peptide libraries have been used with an increased frequency in recent years (Krumpe *et al.*, 2006; Larman *et al.*, 2011; Teesalu *et al.*, 2012).

Phage display of antibodies

Antibodies are irreplaceable reagents used in diagnostics and detection of trace amount of substances, from pesticides and hormones to viruses. The monoclonal antibody technology has been successfully used to isolate uniform populations or clones of the immortalized mouse B cells which produce a single antibody that binds to a specific target (antigen) of interest (Kohler and Milstein, 1975).

In the early 1990s, phage display technology empowered researchers to use phage and *in vitro* affinity selection instead of immunization and hybridomas to successfully isolate monoclonal recombinant antibodies that recognize the target or antigen of interest. These initial reports showed that the variable domains of the heavy (V_H) and light (V_L and V_K) chains can fold correctly and recognize the cognate target (antigen) when expressed in *E. coli* and assembled into the Ff phages fused to pIII or pVIII (Barbas *et al.*, 1991; McCafferty *et al.*, 1990). Two types of display were designed, one in which the heavy and light chain variable domains were expressed as separate polypeptides from the same vector and became associated in the *E. coli* periplasm prior to phage assembly (Fab; Barbas *et al.*, 1991), and the other where they were expressed as the same peptide in which a long flexible linker connected them (scFv; Clackson *et al.*, 1991). These first antibody libraries were amplified from B cell mRNA of immunized animals or volunteers. However, antibodies against human (self) target proteins, to which humans are naturally tolerant, are crucial targets for therapy. Phage display, by virtue of being an *in vitro* affinity selection protocol, allowed isolation of antibodies constructed from the mRNA obtained from the pre-immune repertoire, derived from immature or inactivated B-cells which circulate in the human peripheral blood (Griffiths *et al.*, 1993). Pioneered in 1991 and improved over more than a decade, the phage display protocols for isolating human antibodies against human antigens have already been used for development of therapeutic recombinant antibodies (Chames *et al.*, 2002; de Haard *et al.*, 1999; Marks *et al.*, 1991; Schier *et al.*, 1996) and exploration of use for gene delivery into the human cells (Poul *et al.*, 2000). The unique advantage of phage display is that it provides human antibodies, which, unlike the animal antibodies, are tolerated by the human immune system and therefore are not inactivated by mounting an anti-xenoimmunoglobulin response.

Initial screening of human non-immune libraries usually yields only low-affinity antibodies. However, the affinity can be improved by random mutagenesis of the antibody coding sequence in *E. coli*, or by shuffling variable segments of the selected low affinity antibody with a library of variants in order to find one with an increased affinity (Schier *et al.*, 1996; Thompson *et al.*, 1996). The other limitation of the non-immune phage display library is its primary size. This limitation was overcome by combining two libraries using *in vivo* recombination (Sblattero and Bradbury, 2000; Tsurushita *et al.*, 1996; Waterhouse *et al.*, 1993).

Once an antibody with a sufficiently high affinity for the antigen of interest is isolated, it can be expressed independently of phage in high amounts for clinical use. Moreover, filamentous phage (nano) particles displaying the antibody variable chain can be used as detection devices or tracers. The scFv–pIII fusions are displayed on the tip of the phage filament, the body of which consists over 2000 copies of pVIII, serving as natural amplifier of the signal: anti pVIII antibodies or pVIII conjugated to fluorophores can be used to 'flag' the binding sites of the antibodies displayed on the phage, therefore allowing immunohistological detection of the antigen (Sobry *et al.*, 2005; Willats *et al.*, 1999; Zhou *et al.*, 2002).

Several companies such as Cambridge Antibody Technologies and Morphosys (see the Web resources) have established production of recombinant antibodies from pre-immune libraries using the phage display method and can deliver the product in a time as short as 10 weeks. Furthermore, a high-throughput antibody generation set-up, by screening of Ff-displayed naïve antibody libraries, has been developed within the 'Antibody Factory' of the German National Genome Research Network (NGFN). In a proof-of-concept publication, this group has obtained 347 different recombinant antibodies against a large family of cell-signalling domains (SH2 domains) used as baits (Mersmann *et al.*, 2010).

Besides the use in diagnostics and therapy, antibodies generated by phage display are also used as catalysts. This application required immunizing the experimental animal with an organic molecule that mimics the transition state intermediate in a chemical reaction. The antibodies are then displayed on the phage to select for the clones that bind to the transition state intermediate and subsequently to test whether they catalyse the reaction of interest (Baca *et al.*, 1997; Fujii *et al.*, 1998; Janda *et al.*, 1997).

Protein evolution by phage display

Phage display has been a powerful tool for evolution of proteins in the laboratory, to create proteins with novel properties. This is achieved by creating a library of variants of a protein of interest in which the residues required for binding to an interacting partner are randomized. From that library, a variant binding to a new ligand can be selected. One of the most striking examples of protein evolution is that of zinc finger-containing transcription factors carried out by Rebar and Pabo (1994). They randomized the critical DNA contact residues of the zinc finger motif which confer specificity for binding to a certain DNA sequence and then screened the library for binding to another sequence, to obtain a new zinc finger with changed specificity. Another example is selection of the human growth hormone variant with an increased affinity for the receptor (Lowman *et al.*, 1991). More recently, co-evolution of a protein domain and its ligand were monitored using a combined phage display and next-generation 'deep' sequencing approach (Ernst *et al.*, 2010). Numerous groups have focused on protease inhibitors, randomizing the protease binding sites in order to achieve increased affinity or changed specificity (Koiwa *et al.*, 2001; Roberts *et al.*, 1992; Szabo *et al.*, 2011).

Sophisticated schemes have been developed to isolate, from libraries of enzymes with randomized substrate-binding sites, variants that catalyse a reaction of interest. One such scheme is to covalently link a substrate to one copy of pIII and the enzyme to the other copy of pIII in the same particle. Because of proximity of the substrate, the variant carries out a single turnover reaction to generate a product. The selection scheme is designed so that the product is specifically captured by the solid phase, thereby capturing the enzyme-encoding phage (Forrer *et al.*, 1999; Gao *et al.*, 1997; Hansson *et al.*, 1997; Soumillion *et al.*, 1994). Protein trans-splicing on the surface of Ff phage is another innovation that increases the capacity of phage display for functional screens (Garbe *et al.*, 2010).

cDNA phage display libraries

Determination of interactions between cellular proteins is the key to understanding the essential cellular processes. Yeast and bacterial two-hybrid systems are most frequently used for this purpose. However, the interactions in those systems are identified *in vivo*. The advantage of phage display is in that the interactions are selected for *in vitro*, and that the interacting partner (or bait) need not be a protein, but any molecule or molecular complex that can be immobilized.

In the peptide and antibody libraries, as well as libraries of protein variants, the insert length is pre-set to be in frame with the downstream virion. However, for construction of phage display cDNA libraries, the lengths, termini and direction of the inserts cannot be precisely controlled, effectively decreasing the size of the library by a factor of 18. This was partially overcome by designing an indirect display system (Fig. 5.5E): the library was fused downstream of a dimerizing leucine zipper domain of the transcription factor Jun, and its interaction partner, Fos, was fused to pIII. The leucine zipper was formed in the periplasm, allowing display of cDNA fusion proteins on the surface of the phage (Crameri *et al.*, 1994). In this set-up, fusion of the library inserts to the C-terminus of Fos eliminated the requirement for the in-frame joint downstream of the insert and increased the effective library size from 1/18 to 1/6. This Ff indirect cDNA display has been used in allergy research to identify numerous allergen proteins by screening the phage display cDNA libraries of the biological material causing the allergy and thus identifying the protein(s) that induce the allergic reaction. The baits in these screens were IgE antibodies of the patients, whose binding to the allergen proteins triggers the allergic reaction (Crameri and Blaser, 1996). In addition to indirect display, a method for selection of in-frame fusions has been developed which eliminates out-of-frame clones from the library before panning (Zacchi *et al.*, 2003).

The cDNA libraries of cytosolic proteins displayed on the phages have been used to identify proteins that bind to DNA or RNA sequences of interest, or to chromatin. The λ and T7 phages, which assemble in the cytosol and therefore eliminate the problem of cytosolic protein misfolding associated with Ff display, were used as vectors for the libraries from which these proteins were identified, e.g. Cicchini *et al.* (2002), Danner and Belasc (2001) and Wang *et al.* (2004).

Display of folded cytoplasmic proteins on Ff phage has recently been achieved by utilizing a secretion pathway that is different from the classical SecYEG. This pathway, called twin arginine translocon (Tat), translocates folded proteins across the inner membrane, provided that they have a specific signal sequence, called the Tat signal sequence (Berks *et al.*, 2000). Cytosolic proteins such as green fluorescent protein, when fused to a Tat signal sequence are correctly folded, secreted into *E. coli* periplasm and displayed by indirect display (described above) through association with pIII (Velappan *et al.*, 2010).

Phage display of bacterial proteins

The Ff phage display shotgun genomic libraries have also been used to identify bacterial proteins that bind to e.g. extracellular matrix protein fibronectin or to the host cell tissues or antisera (Antonara *et al.*, 2007; Jacobsson and Frykberg, 1995; Mullen *et al.*, 2006). Recently, two methods for the selective display of bacterial surface and secreted proteins have been designed and used to systematically characterize this group of bacterial proteins. In both, the phagemid vector encodes the C-terminal domain of pIII, but not the signal sequence. Thus, only the inserts that encode signal sequence-carrying proteins in frame with pIII are incorporated into the virion. Furthermore, in the absence of pIII from the helper phage (if a *Δgene III* helper phage is used), termination of phage assembly and virion release depends on the presence of a signal sequence-containing insert fused in frame to pIII of the cloning vector. Affinity screening for a vector-encoded pIII-fused tag (Wall *et al.*, 2003) or direct selection for virion release (Jankovic *et al.*, 2007) were successfully used to obtain clone banks of surface and secreted proteins from lactobacilli. These phage display clone banks allow functional analyses of individual surface and secreted proteins without a need for systematic cloning, expression and purification. The Ff virion is an excellent hapten for immunization (van Houten *et al.*, 2010); hence clone banks of bacterial surface proteins can be used to greatly facilitate screening of vaccine candidates, as has been done for *Mycobacterium tuberculosis* (Liu *et al.*, 2011).

Phage display of peptides

Finding peptide leads as templates for design of organic compounds that bind to a target of interest has been revolutionized by phage display. In pioneering work of Scott and Smith (1990), short peptide libraries were created and screened for binding to a ligand of interest. The peptides selected from these libraries were of low affinity, partly because the peptides were short, and also because there were no constraints, resulting in poor folding of the peptide. To overcome the problem of peptide length, libraries of longer peptides were introduced. Furthermore, libraries of peptides constrained by formation of a disulphide bond between two cysteine residues engineered at the ends of the random peptide sequence were used (Felici *et al.*, 1993). A beautiful example of the successful library screen is isolation of the random peptide agonist analogous to erythropoietin, a commercially used polypeptide hormone that stimulates erythrocyte proliferation (Wrighton *et al.*, 1996). A multivalent display linear peptide library was first used to isolate low-affinity binding peptides. Building on the consensus derived from this library, a new library of constrained peptides randomized at some of the residues was constructed, displayed at one copy per virion, and used to select a high affinity peptide.

The peptide libraries are also used for basic research. For example, a major advance in the understanding of antigen presentation by the major histocompatibility complex has been achieved by isolating, from phage display libraries, MHC-binding peptides. This allowed assessment of the variability of peptides presented by different allelic variants of the major histocompatibility complex (Hammer *et al.*, 1992, 1993). Another application determined binding specificity of the eukaryotic chaperone BiP which assists in folding and membrane translocation of proteins destined for the lumen of the endoplasmic reticulum (Blond-Elguindi *et al.*, 1993). As random peptide libraries are not derived from cellular proteins and are not saturating, a new approach has recently been applied to identify epitopes of autoantibodies in patients with autoimmune diseases. Instead of random peptides, the authors have synthetized nearly a million oligonucleotides that encode the whole proteome as a collection of 36-residue-long peptides overlapping by seven amino acid residues. Cloning of these oligonucleotides into a T7 display vector resulted in a library displaying the complete proteome, which was used very effectively to identify a large number of autoantigenic epitopes (Larman *et al.*, 2011).

Phage display of peptides has been taken to a new level by their use in cell and tissue targeting as well as receptor-mediated gene delivery. This was achieved by *in situ* screening of phage display peptide libraries against cells and tissues to isolate specific cell or tissue binding peptides (Pasqualini and Ruoslahti, 1996). The tissue-binding peptides were isolated by injecting mice with the libraries and recovering the phage from specific tissues of interest, or from human breast cancer tissue xenotransplanted into mice. Peptides isolated that way were able to target the transplanted tumour for destruction by fusion of the peptide to the cancer drug doxorubicin. In addition to tumour tissue binders, peptide ligands that specifically bind to the vasculature of particular organs such as the prostate or kidney have been isolated (Arap *et al.*, 1998). These peptides can be used for therapeutic purpose, for example by fusing to cytokines or other drugs to concentrate them only to the target tissue. This strategy aims to relieve systemic toxicity and/or increase the efficiency of delivery to the affected tissue. Recently, phage display of two different peptides on the same particle has been adapted for organ-specific delivery: a tissue-specific peptide was displayed by fusion to pIII (at the tip of the phage) while an avidin-binding peptide was displayed by fusion to pVIII, in 100–200 copies along the phage filament. Thus equipped, phage can be turned into a tracking nanodevice: the avidin-binding body can be tagged with commercially available avidin-conjugated fluorescent dyes or even quantum dot particles, allowing visualization of the phage that binds to the specific ligand via its terminally displayed peptide (Chen *et al.*, 2004). Not only Ff, but also T7 phage particles have been used for tissue targeting; T7, due to the globular shape of its head, is considered a better physical mimic of clinical nanoparticles (Teesalu *et al.*, 2012).

To adapt Ff phage for use in receptor-mediated gene delivery, libraries were screened to identify cell-specific internalization-mediating peptides (Burg *et al.*, 2004; Larocca *et al.*, 2002). The selected peptides were not only used for gene delivery, but also for marking or tracking the target cells by expression in the host cell nucleus of a reporter protein encoded by the phagemid vector packaged into the internalized particles. Finally, high-density phage display can be used in simple portable diagnostic devices (Nanduri *et al.*, 2007; Petrenko, 2008; Williams *et al.*, 2003).

Bacteriophages as templates for inorganic nanostructures

In the past decade phage display was added to the arsenal of nanotechnology tools. It was used effectively to solve the long standing problem of how to control assembly of nanometer-scale inorganic structures in order to obtain useful functional structures and devices (Flynn *et al.*, 2003; Sarikaya *et al.*, 2003).

Material scientists have long looked into biomaterials for templates to assemble inorganic structures. Indeed, before phage display, several biological structures, e.g. diatoms, magnetic bacteria and viruses have been used, unmodified or chemically modified, as templates for assembly of nanostructures. However, their use was limited to only some of the materials of interest. Recently, a few research groups resorted to combinatorial libraries of peptides displayed on Ff phage to vastly extend the types of the inorganic materials that can be templated by viruses, including bacteriophages. By screening phage display libraries, peptides that bind to metals (Au, Pt, Pd, Ag), semiconductor materials (GaAs, ZnS, CdS), paramagnetic materials (FePt, CoPt) and materials for building lithium nanobatteries (CO_3O_4, $FePO_4$ and single-wall carbon nanotubes (Dang *et al.*, 2011; Lee *et al.*, 2009; Nam *et al.*, 2006)) have been identified. Peptides that specifically bind to other useful materials, such as SiO_2, zeolites, ZnO, $CaCO_3$ and Cr_3O_3 have also been isolated (Sarikaya *et al.*, 2003). Some of these peptides were also selected using cell surface display on bacteria. However, filamentous phages not only allowed the library screening to isolate inorganics-binding peptides, but also served as a template for assembly of the nanometer scale structures (Mao *et al.*, 2004). Work by the Belcher laboratory showed that peptides displayed on the surface of Ff phage nucleate formation of nanocrystals when exposed to the dissolved inorganic ions in an aqueous solution. The phases and dimensions of the nanocrystals were found to be peptide-dependent. For example, two different peptides that nucleate the semiconductor ZnS crystals, named Z8 and A7, nucleated two types of the nanocrystals, one 4nm in size, blende cubic phase, and the other 4 nm by 2 nm, hexagonal wutzite crystal phase (Mao *et al.*, 2003). These nano-scale crystals are often called 'quantum dots'. In contrast, if the ZnS solution is incubated with non-specific phage or in their absence under the same conditions, non-crystalline particles 100–500 nm in size were formed. Therefore, peptides displayed on the phage were essential to nucleate the nanocrystal or quantum dot formation. Significantly for assembly of these quantum dots into structures, the nanocrystals are physically associated to the nucleation sites (phage displayed peptides). Because of the highly ordered structure of the Ff filament, this immediately provided a template for the assembly of the nanocrystals into organized structures. As described in previous sections, peptides can be displayed along the filament, as fusions to pVIII major coat protein and at both tips of the filament. Belcher's group exploited all display positions and combinations of different peptides along the filaments and at the ends to create several types of nanostructures.

Organic–inorganic liquid crystal nanostructures

The tendency of the Ff phage to align and form liquid crystalline sheets has been exploited to utilize them as building blocks in the assembly of three-dimensional lattices and films (Chung *et al.*, 2011; Lee and Belcher, 2004; Lee *et al.*, 2002). In this set-up, the Ff particles displaying inorganic-binding peptides at the pIII end of the virion are highly concentrated, resulting in formation of liquid crystalline layers of aligned viral filaments interspersed with the inorganic layers composed of the nanocrystals nucleated at the peptides displayed at the tips. Drying of these liquid crystal structures results in formation of a very stable film (Lee *et al.*, 2002).

The liquid crystalline nature of Ff was exploited for forming fibres by wet spinning or electrospining (Lee and Belcher, 2004). In a more sophisticated assembly system, a single-step process was developed that produces long-range-ordered, supramolecular films showing multiple levels of hierarchical organization and helical twist. Using this approach, materials with distinctive optical and photonic properties, functioning as chiral reflector/filters and structural colour matrices were produced. When peptide ligands to both cell-adhesion molecules and bone minerals were displayed on the phages, these bacteriophage films were shown to direct both soft and hard tissue growth in a hierarchically organized manner (Chung *et al.*, 2011).

Semiconductor nanowires and nanorings

The high density display of nucleating peptides along the filament of Ff has been used to create entirely inorganic nanostructures, single-crystal nanotubes. In this application, Ff virions have been used both as nucleators and templates for nanotube assembly. The nucleating peptides were fused to pVIII expressed from a phagemid and assembled into the filament in combination with the wild-type pVIII, providing multiple nucleation sites along the filament. Exposure to the ZnS solution at 0°C resulted in formation of a tube of nanocrystals wrapping the filament completely. High resolution transmission electron microscopy showed a double layered coat of the highly oriented crystals in close contact surrounded the phage filament. Fast Fourier transforms showed that most of the nanocrystals assumed the same orientation relative to the filament. Combination of two peptide specificities, for ZnS and CdS, nucleated the crystals of the same wutzite phase along the filament and formed nanowires composed of both semiconductors (heterostructures).

These nanowires formed of virus and inorganic crystal can be transformed into inorganic nanowires by annealing, essentially heating to 350°C to eliminate the organic component, most importantly the nucleating peptides which prevent fusion of the nanocrystals into a single-crystal nanowire. The resulting nanowires were 475–650 nm in length and about 20 nm in diameters, retaining the same crystal orientation as prior to annealing. The same method was applied to the assembly of ferromagnetic CoPt and FePt monocrystalline nanowires, with potential application in ultrahigh-density recording media (Mao *et al.*, 2004).

Not only wires but more complex shapes can be produced by adding different functionalities to the ends of the Ff filaments. A recent report by the Belcher group described a set-up in which an adapter molecule with two binding functions, Ni-NTA and avidin, was used in conjunction with a phage that carried ligands for the two binding specificities. Poly-histidine peptide that binds nickel ions was displayed at the pVII/pIX end of the phage and the streptavidin-binding peptide was displayed at the opposite, pIII end of the filament. When adapter was mixed with these double-displaying particles, ring structures were formed. Heterofunctional phage can also be used for assembling different materials in spatially distinct areas or assembling different viruses associated with different inorganic ligands in specific order, leading to complex materials (Nam *et al.*, 2004).

Filamentous phages that displayed $FePO_4$-binding peptides along the filament and single-wall carbon nanotube (SWNT) binding peptide at one of the ends was used to assemble nanowires for cathodes, resulting in a high-power nanobattery (Lee *et al.*, 2009). Ff phage-templated nanomaterials were also combined to construct a nano structure that catalyses sustained light-driven water oxidation (Nam *et al.*, 2010) and for an increased-efficiency solar cell (Dang *et al.*, 2011).

Conclusions

Bacteriophages are highly organized self-assembled biological structures which carry encapsulated coding sequences. They are also fast-evolving biological entities, due to their short generation time and large number of progeny from a single reproductive cycle. These essential characteristics allow a wide range of applications. The ability to modify the structure by fusion of foreign peptides, coupled with rapid amplification and physical link to the encoding sequence, has led to phages being used in display technologies where large libraries of variants are mined for a peptide of interest.

Phage display technology is particularly important in the pharmaceutical industry, where it is instrumental in designing therapeutic antibodies and peptide leads. However, the nanotechnology applications of phage particles are of rapidly growing importance, from use in localization of tumours to construction of semi-conductor nanowires. Phages will undoubtedly provide resources for the nanotechnology in the future.

Web resources

Science sites

George Smith Laboratory web site containing phage display protocols: http://www.biosci.missouri.edu/smithgp/PhageDisplayWebsite/PhageDisplayWebsiteIndex.html

Company sites

Cambridge Antibody Technology http://www.cambridgeantibody.com/html/
Morphosys (Antibodies) http://www.morphosys.com/
Dyax corporation (Therapeutic protease inhibitors from phage display libraries) http://www.dyax.com/

References

Ageno, M., Donelli, G., and Guglielmi, F. (1973). Structure and physico-chemical properties of bacteriophage G. II, The shape and symmetry of the capsid. Micron *4*, 376–403.

Akita, F., Chong, K.T., Tanaka, H., Yamashita, E., Miyazaki, N., Nakaishi, Y., Suzuki, M., Namba, K., Ono, Y., Tsukihara, T., *et al.* (2007). The crystal structure of a virus-like particle from the hyperthermophilic archaeon *Pyrococcus furiosus* provides insight into the evolution of viruses. J. Mol. Biol. *368*, 1469–1483.

Aksyuk, A.A., Leiman, P.G., Shneider, M.M., Mesyanzhinov, V.V., and Rossmann, M.G. (2009). The structure of gene product 6 of bacteriophage T4, the hinge-pin of the baseplate. Structure *17*, 800–808.

Anany, H., Chen, W., Pelton, R., and Griffiths, M.W. (2011). Biocontrol of *Listeria monocytogenes* and *Escherichia coli* O157:H7 in meat by using phages immobilized on modified cellulose membranes. Appl. Environ. Microbiol. *77*, 6379–6387.

Antonara, S., Chafel, R.M., LaFrance, M., and Coburn, J. (2007). *Borrelia burgdorferi* adhesins identified using in vivo phage display. Mol. Microbiol. *66*, 262–276.

Arap, W., Pasqualini, R., and Ruoslahti, E. (1998). Cancer treatment by targeted drug delivery to tumor vasculature in a mouse model. Science *279*, 377–380.

Arslan, D., Legendre, M., Seltzer, V., Abergel, C., and Claverie, J.M. (2011). Distant Mimivirus relative with a larger genome highlights the fundamental features of Megaviridae. Proc. Natl. Acad. Sci. U.S.A. *108*, 17486–17491.

Asano, S., Higashitani, A., and Horiuchi, K. (1999). Filamentous phage replication initiator protein gpII forms a covalent complex with the 5′ end of the nick it introduced. Nucleic Acids Res. *27*, 1882–1889.

Baca, M., Scanlan, T., Stephenson, R., and Wells, J. (1997). Phage display of a catalytic antibody to optimize affinity for transition-state analogue binding. Proc. Natl. Acad. Sci. U.S.A. *94*, 10063–10068.

Barbas III, C.F., Burton, D.R., Scott, J.K., and Silverman, G.J. (2001). Phage Display: A Laboratory Manual (Cold Spring Harbor, NY: Cold Spring Harbor Laboratory Press).

Barbas III, C.F., Kang, A.S., Lerner, R.A., and Benkovic, S.J. (1991). Assembly of combinatorial antibody libraries on phage surfaces: the gene III site. Proc. Natl. Acad. Sci. U.S.A. *88*, 7978–7982.

Belnap, D.M., and Steven, A.C. (2000). 'Deja vu all over again': the similar structures of bacteriophage PRD1 and adenovirus. Trends Microbiol. *8*, 91–93.

Bennett, N.J., Gagic, D., Sutherland-Smith, A.J., and Rakonjac, J. (2011). Characterization of a dual-function domain that mediates membrane insertion and excision of Ff filamentous bacteriophage. J. Mol. Biol. *411*, 972–985.

Benson, S.D., Bamford, J.K., Bamford, D.H., and Burnett, R.M. (2002). The X-ray crystal structure of P3, the major coat protein of the lipid-containing bacteriophage PRD1, at 1.65 A resolution. Acta Crystallogr. D Biol. Crystallogr. *58*, 39–59.

Berks, B.C., Sargent, F., and Palmer, T. (2000). The Tat protein export pathway. Mol. Microbiol. *35*, 260–274.

Beuron, F., Maurizi, M.R., Belnap, D.M., Kocsis, E., Booy, F.P., Kessel, M., and Steven, A.C. (1998). At sixes and sevens: characterization of the symmetry mismatch of the ClpAP chaperone-assisted protease. J. Struct Biol. *123*, 248–259.

Blond-Elguindi, S., Cwirla, S.E., Dower, W.J., Lipshutz, R.J., Sprang, S.R., Sambrook, J.F., and Gething, M.J. (1993). Affinity panning of a library of peptides displayed on bacteriophages reveals the binding specificity of BiP. Cell *75*, 717–728.

Booy, F.P., Newcomb, W.W., Trus, B.L., Brown, J.C., Baker, T.S., and Steven, A.C. (1991). Liquid-crystalline, phage-like packing of encapsidated DNA in herpes simplex virus. Cell *64*, 1007–1015.

Booy, F.P., Trus, B.L., Newcomb, W.W., Brown, J.C., Conway, J.F., and Steven, A.C. (1994). Finding a needle in a haystack: detection of a small protein (the 12-kDa VP26) in a large complex (the 200-MDa capsid of herpes simplex virus). Proc. Natl. Acad. Sci. U.S.A. *91*, 5652–5656.

Boyer, M., Yutin, N., Pagnier, I., Barrassi, L., Fournous, G., Espinosa, L., Robert, C., Azza, S., Sun, S., Rossmann, M.G., *et al.* (2009). Giant Marseillevirus highlights the role of amoebae as a melting pot in emergence of chimeric microorganisms. Proc. Natl. Acad. Sci. U.S.A. *106*, 21848–21853.

Branston, S., Stanley, E., Ward, J., and Keshavarz-Moore, E. (2011). Study of robustness of filamentous bacteriophages for industrial applications. Biotechnol. Bioeng. *108*, 1468–1472.

Brissette, J.L., and Russel, M. (1990). Secretion and membrane integration of a filamentous phage-encoded morphogenetic protein. J. Mol. Biol. *211*, 565–580.

Burg, M., Ravey, E.P., Gonzales, M., Amburn, E., Faix, P.H., Baird, A., and Larocca, D. (2004). Selection of internalizing ligand-display phage using rolling circle amplification for phage recovery. DNA Cell Biol. *23*, 457–462.

Cardone, G., Winkler, D.C., Trus, B.L., Cheng, N., Heuser, J.E., Newcomb, W.W., Brown, J.C., and Steven, A.C. (2007). Visualization of the herpes simplex virus portal in situ by cryo-electron tomography. Virology *361*, 426–434.

Cerritelli, M.E., Conway, J.F., Cheng, N., Trus, B.L., and Steven, A.C. (2003a). Molecular mechanisms in bacteriophage T7 procapsid assembly, maturation, and DNA containment. Adv. Protein Chem. *64*, 301–323.

Cerritelli, M.E., Trus, B.L., Smith, C.S., Cheng, N., Conway, J.F., and Steven, A.C. (2003b). A second symmetry mismatch at the portal vertex of bacteriophage T7: 8-fold symmetry in the procapsid core. J. Mol. Biol. *327*, 1–6.

Chames, P., Willemsen, R.A., Rojas, G., Dieckmann, D., Rem, L., Schuler, G., Bolhuis, R.L., and Hoogenboom, H.R. (2002). TCR-like human antibodies expressed on human CTLs mediate antibody affinity-dependent cytolytic activity. J. Immunol. *169*, 1110–1118.

Chang, C.Y., Kemp, P., and Molineux, I.J. (2010). Gp15 and gp16 cooperate in translocating bacteriophage T7 DNA into the infected cell. Virology *398*, 176–186.

Chen, C.L., Strop, P., Lebl, M., and Lam, K.S. (1996). One bead-one compound combinatorial peptide library: different types of screening. Methods Enzymol. *267*, 211–219.

Chen, L., Zurita, A.J., Ardelt, P.U., Giordano, R.J., Arap, W., and Pasqualini, R. (2004). Design and validation of a bifunctional ligand display system for receptor targeting. Chem Biol *11*, 1081–1091.

Chibani-Chennoufi, S., Dillmann, M.L., Marvin-Guy, L., Rami-Shojaei, S., and Brussow, H. (2004). Lactobacillus plantarum bacteriophage LP65: a new member of the SPO1-like genus of the family Myoviridae. J. Bacteriol. *186*, 7069–7083.

Chung, W.J., Oh, J.W., Kwak, K., Lee, B.Y., Meyer, J., Wang, E., Hexemer, A., and Lee, S.W. (2011). Biomimetic self-templating supramolecular structures. Nature *478*, 364–368.

Cicchini, C., Ansuini, H., Amicone, L., Alonzi, T., Nicosia, A., Cortese, R., Tripodi, M., and Luzzago, A. (2002). Searching for DNA–protein interactions by lambda phage display. J. Mol. Biol. *322*, 697–706.

Clackson, T., Hoogenboom, H.R., Griffiths, A.D., and Winter, G. (1991). Making antibody fragments using phage display libraries. Nature *352*, 624–628.

Colson, P., Yutin, N., Shabalina, S.A., Robert, C., Fournous, G., La Scola, B., Raoult, D., and Koonin, E.V. (2011). Viruses with more than 1,000 genes: Mamavirus, a new Acanthamoeba polyphaga mimivirus strain, and reannotation of Mimivirus genes. Genome Biol. Evol. *3*, 737–742.

Condron, B.G., Atkins, J.F., and Gesteland, R.F. (1991). Frameshifting in gene 10 of bacteriophage T7. J. Bacteriol. *173*, 6998–7003.

Conway, J.F., Duda, R.L., Cheng, N., Hendrix, R.W., and Steven, A.C. (1995). Proteolytic and conformational control of virus capsid maturation: the bacteriophage HK97 system. J. Mol. Biol. *253*, 86–99.

Conway, J.F., Wikoff, W.R., Cheng, N., Duda, R.L., Hendrix, R.W., Johnson, J.E., and Steven, A.C. (2001). Virus maturation involving large subunit rotations and local refolding. Science *292*, 744–748.

Crameri, R., and Blaser, K. (1996). Cloning Aspergillus fumigatus allergens by the pJuFo filamentous phage display system. Int. Arc. Allergy Immunol. *110*, 41–45.

Crameri, R., Jaussi, R., Menz, G., and Blaser, K. (1994). Display of expression products of cDNA libraries on phage surfaces. A versatile screening system for selective isolation of genes by specific gene–product/ligand interaction. Eur. J. Biochem. *226*, 53–58.

Crissman, J.W., and Smith, G.P. (1984). Gene-III protein of filamentous phages: evidence for a carboxyl-terminal domain with a role in morphogenesis. Virology *132*, 445–455.

Cull, M.G., Miller, J.F., and Schatz, P.J. (1992). Screening for receptor ligands using large libraries of peptides linked to the C terminus of the lac repressor. Proc. Natl. Acad. Sci. U.S.A. *89*, 1865–1869.

Cwirla, S.E., Peters, E.A., Barrett, R.W., and Dower, W.J. (1990). Peptides on phage: a vast library of peptides for identifying ligands. Proc. Natl. Acad. Sci. U.S.A. *87*, 6378–6382.

Daefler, S., Russel, M., and Model, P. (1997). Module swaps betwen related translocator proteins pIVf1, pIVIKe, and PulD: Identification of a specificity domain. J. Mol. Biol. *266*, 978–992.

Dang, X., Yi, H., Ham, M.H., Qi, J., Yun, D.S., Ladewski, R., Strano, M.S., Hammond, P.T., and Belcher, A.M. (2011). Virus-templated self-assembled single-walled carbon nanotubes for highly efficient electron collection in photovoltaic devices. Nat. Nanotechnol. *6*, 377–384.

Danner, S., and Belasco, J.G. (2001). T7 phage display: a novel genetic selection system for cloning RNA-binding proteins from cDNA libraries. Proc. Natl. Acad. Sci. U.S.A. *98*, 12954–12959.

Day, L.A. (2011). Family Inoviridae. In Virus taxonomy: classification and nomenclature of viruses: Ninth Report of the International Committee on Taxonomy of Viruses, A.M.Q. King, M.J. Adams, E.B. Carstens, and E.J. Lefkowitz, eds. (San Diego, Elsevier Academic Press), pp. 375–384.

Day, L.A., Marzec, C.J., Reisberg, S.A., and Casadevall, A. (1988). DNA packing in filamentous bacteriophages Ann. Rev. Biophys. Biophys. Chem. *17*, 509–539.

de Haard, H.J., van Neer, N., Reurs, A., Hufton, S.E., Roovers, R.C., Henderikx, P., de Bruine, A.P., Arends, J.W., and Hoogenboom, H.R. (1999). A large non-immunized human Fab fragment phage library that permits rapid isolation and kinetic analysis of high affinity antibodies. J. Biol. Chem. *274*, 18218–18230.

de Wildt, R.M., Tomlinson, I.M., Ong, J.L., and Holliger, P. (2002). Isolation of receptor–ligand pairs by capture of long-lived multivalent interaction complexes. Proc. Natl. Acad. Sci. U.S.A. *99*, 8530–8535.

Dente, L., Cesareni, G., and Cortese, R. (1983). pEMBL: a new family of single stranded plasmids. Nucleic Acids Res. *11*, 1645–1655.

Devlin, J.J., Panganiban, L.C., and Devlin, P.E. (1990). Random peptide libraries: a source of specific protein binding molecules. Science *249*, 404–406.

Dokland, T., McKenna, R., Ilag, L.L., Bowman, B.R., Incardona, N.L., Fane, B.A., and Rossmann, M.G. (1997). Structure of a viral procapsid with molecular scaffolding. Nature *389*, 308–313.

Duda, R.L. (1998). Protein chainmail: catenated protein in viral capsids. Cell *94*, 55–60.

Duda, R.L., Hendrix, R.W., Huang, W.M., and Conway, J.F. (2006). Shared architecture of bacteriophage SPO1 and herpesvirus capsids. Curr. Biol. *16*, R11–13.

Duda, R.L., Martincic, K., Xie, Z., and Hendrix, R.W. (1995). Bacteriophage HK97 head assembly. FEMS Microbiol. Rev *17*, 41–46.

Duenas, M., and Borrebaeck, C.A.K. (1994). Clonal selection and amplification of phage displayed antibodies by linking antigen recognition and phage replication. Biotechnology *12*, 999–1002.

Dunn, I.S. (1995). Assembly of functional bacteriophage lambda virions incorporating C-terminal peptide or protein fusions with the major tail protein. J. Mol. Biol. *248*, 497–506.

Dunn, I.S. (1996a). In vitro alpha-complementation of beta-galactosidase on a bacteriophage surface. Eur. J. Biochem. *242*, 720–726.

Dunn, I.S. (1996b). Total modification of the bacteriophage lambda tail tube major subunit protein with foreign peptides. Gene *183*, 15–21.

Effantin, G., Boulanger, P., Neumann, E., Letellier, L., and Conway, J.F. (2006). Bacteriophage T5 structure reveals similarities with HK97 and T4 suggesting evolutionary relationships. J. Mol. Biol. *361*, 993–1002.

Endemann, H., and Model, P. (1995). Location of filamentous phage minor coat proteins in phage and in infected cells. J. Mol. Biol. *250*, 496–506.

Ernst, A., Gfeller, D., Kan, Z., Seshagiri, S., Kim, P.M., Bader, G.D., and Sidhu, S.S. (2010). Coevolution of PDZ domain–ligand interactions analysed by high-throughput phage display and deep sequencing. Mol. Biosyst. *6*, 1782–1790.

Felici, F., Luzzago, A., Folgori, A., and Cortese, R. (1993). Mimicking of discontinuous epitopes by phage-displayed peptides, II. Selection of clones recognized by a protective monoclonal antibody against the Bordetella pertussis toxin from phage peptide libraries. Gene *128*, 21–27.

Feng, J.-n. (2000). Studies of the assembly/export proteins and energy requirements for filamentous phage assembly (The Rockefeller University).

Feng, J.-n., Model, P., and Russel, M. (1999). A trans-envelope protein complex needed for filamentous phage assembly and export. Mol. Microbiol. *34*, 745–755.

Feng, J.N., Russel, M., and Model, P. (1997). A permeabilized cell system that assembles filamentous bacteriophage. Proc. Natl. Acad. Sci. U.S.A. *94*, 4068–4073.

Flynn, C., Lee, S., Peelle, B., and Belcher, A. (2003). Viruses as vehicles for growth, organization and assembly of materials. ACTA MATER *51*, 5867–5880.

Fokine, A., Leiman, P.G., Shneider, M.M., Ahvazi, B., Boeshans, K.M., Steven, A.C., Black, L.W., Mesyanzhinov, V.V., and Rossmann, M.G. (2005). Structural and functional similarities between the capsid proteins of bacteriophages T4 and HK97 point to a common ancestry. Proc. Natl. Acad. Sci. U.S.A. *102*, 7163–7168.

Forrer, P., Jung, S., and Pluckthun, A. (1999). Beyond binding: using phage display to select for structure, folding and enzymatic activity in proteins. Curr Opin Struct Biol *9*, 514–520.

Fujii, I., Fukuyama, S., Iwabuchi, Y., and Tanimura, R. (1998). Evolving catalytic antibodies in a phage-displayed combinatorial library. Nat Biotechnol *16*, 463–467.

Gailus, V., and Rasched, I. (1994). The adsorption protein of bacteriophage fd and its neighbour minor coat protein build a structural entity. Eur. J. Biochem. *222*, 927–931.

Gan, L., Conway, J.F., Firek, B.A., Cheng, N., Hendrix, R.W., Steven, A.C., Johnson, J.E., and Duda, R.L. (2004). Control of cross-linking by quaternary structure changes during bacteriophage HK97 maturation. Mol. Cell *14*, 559–569.

Gan, L., Speir, J.A., Conway, J.F., Lander, G., Cheng, N., Firek, B.A., Hendrix, R.W., Duda, R.L., Liljas, L., and Johnson, J.E. (2006). Capsid conformational sampling in HK97 maturation visualized by X-ray crystallography and cryo-EM. Structure *14*, 1655–1665.

Gao, C., Lin, C.H., Lo, C.H., Mao, S., Wirsching, P., Lerner, R.A., and Janda, K.D. (1997). Making chemistry selectable by linking it to infectivity. Proc. Natl. Acad. Sci. U.S.A. *94*, 11777–11782.

Gao, C., Mao, S., Lo, C.H., Wirsching, P., Lerner, R.A., and Janda, K.D. (1999). Making artificial antibodies: a format for phage display of combinatorial heterodimeric arrays. Proc. Natl. Acad. Sci. U.S.A. *96*, 6025–6030.

Garbe, D., Thiel, I.V., and Mootz, H.D. (2010). Protein trans-splicing on an M13 bacteriophage: towards directed evolution of a semisynthetic split intein by phage display. J. Pept. Sci. *16*, 575–581.

Gearhart, D.A., Toole, P.F., and Warren Beach, J. (2002). Identification of brain proteins that interact with 2-methylnorharman. An analogue of the parkinsonian-inducing toxin, MPP+. Neurosci. Res. *44*, 255–265.

Goldbourt, A., Gross, B.J., Day, L.A., and McDermott, A.E. (2007). Filamentous phage studied by magic-angle spinning NMR: resonance assignment and secondary structure of the coat protein in Pf1. J. Am. Chem. Soc. *129*, 2338–2344.

Grey, C.W., Brown, R.S., and Marvin, D.A. (1979). Direct visualization of adsorption protein of fd phage. J. Supramol. Str. *1979*, 91–91.

Greene, B., and King, J. (1994). Binding of scaffolding subunits within the P22 procapsid lattice. Virology *205*, 188–197.

Griffiths, A.D., Malmqvist, M., Marks, J.D., Bye, J.M., Embleton, M.J., McCafferty, J., Baier, M., Holliger, K.P., Gorick, B.D., Hughes-Jones, N.C., *et al.* (1993). Human anti-self antibodies with high specificity from phage display libraries. Embo J. *12*, 725–734.

Guenther, S., Huwyler, D., Richard, S., and Loessner, M.J. (2009). Virulent bacteriophage for efficient biocontrol of Listeria monocytogenes in ready-to-eat foods. Appl. Environ. Microbiol. *75*, 93–100.

Guenther, S., and Loessner, M.J. (2011). Bacteriophage biocontrol of Listeria monocytogenes on soft ripened white mould and red-smear cheeses. Bacteriophage *1*, 94–100.

Gupta, A., Onda, M., Pastan, I., Adhya, S., and Chaudhary, V.K. (2003). High-density functional display of proteins on bacteriophage lambda. J. Mol. Biol. *334*, 241–254.

Haigh, N.G., and Webster, R.E. (1998). The major coat protein of filamentous bacteriophage f1 specifically pairs in the bacterial cytoplasmic membrane. J. Mol. Biol. *279*, 19–29.

Haigh, N.G., and Webster, R.E. (1999). The pI and pXI assembly proteins serve separate and essential roles in filamentous phage assembly. J. Mol. Biol. *293*, 1017–1027.

Hammer, J., Takacs, B., and Sinigaglia, F. (1992). Identification of a motif for HLA-DR1 binding peptides using M13 display libraries. J. Exp. Med. *176*, 1007–1013.

Hammer, J., Valsasnini, P., Tolba, K., Bolin, D., Higelin, J., Takacs, B., and Sinigaglia, F. (1993). Promiscuous and allele-specific anchors in HLA-DR-binding peptides. Cell *74*, 197–203.

Hansson, L.O., Widersten, M., and Mannervik, B. (1997). Mechanism-based phage display selection of active-site mutants of human glutathione transferase A1–1 catalysing SNAr reactions. Biochemistry *36*, 11252–11260.

Hatfull, G.F. (2008). Bacteriophage genomics. Curr. Opin. Microbiol. *11*, 447–453.

Helgstrand, C., Wikoff, W.R., Duda, R.L., Hendrix, R.W., Johnson, J.E., and Liljas, L. (2003). The refined structure of a protein catenane: the HK97 bacteriophage capsid at 3.44 A resolution. J. Mol. Biol. *334*, 885–899.

Hendrix, R.W. (2003). Bacteriophage genomics. Curr. Opin. Microbiol. *6*, 506–511.

Heymann, J.B., Cheng, N., Newcomb, W.W., Trus, B.L., Brown, J.C., and Steven, A.C. (2003). Dynamics of herpes simplex virus capsid maturation visualized by time-lapse cryo-electron microscopy. Nat. Struct. Biol.*10*, 334–341.

Higashitani, A., Greenstein, D., Hirokawa, H., Asano, S., and Horiuchi, K. (1994). Multiple DNA conformational changes induced by an initiator protein precede the nicking reaction in a rolling circle replication origin. J. Mol. Biol. *237*, 388–400.

Higashitani, N., Higashitani, A., Guan, Z.W., and Horiuchi, K. (1996). Recognition mechanisms of the minus-strand origin of phage f1 by *Escherichia coli* RNA polymerase. Genes Cells *1*, 829–841.

Hoess, R.H. (2002). Bacteriophage lambda as a vehicle for peptide and protein display. Curr. Pharm. Biotechnol. *3*, 23–28.

Holliger, P., Riechmann, L., and Williams, R.L. (1999). Crystal structure of the two N-terminal domains of g3p from filamentous phage fd at 1.9 A: evidence for conformational lability. J. Mol. Biol. *288*, 649–657.

Horabin, J.I., and Webster, R.E. (1988). An amino acid sequence which directs membrane insertion causes loss of membrane potential. J. Biol. Chem. *263*, 11575–11583.

Huet, A., Conway, J.F., Letellier, L., and Boulanger, P. (2010). In vitro assembly of the T=13 procapsid of bacteriophage T5 with its scaffolding domain. J. Virol. *84*, 9350–9358.

Iannolo, G., Minenkova, O., Petruzzelli, R., and Cesareni, G. (1995). Modifying filamentous phage capsid: limits in the size of the major capsid protein. J. Mol. Biol. *248*, 835–844.

Jacobsson, K., and Frykberg, L. (1995). Cloning of ligand-binding domains of bacterial receptors by phage display. Biotechniques *18*, 878–885.

Janda, K.D., Lo, L.C., Lo, C.H., Sim, M.M., Wang, R., Wong, C.H., and Lerner, R.A. (1997). Chemical selection for catalysis in combinatorial antibody libraries. Science *275*, 945–948.

Jankovic, D., Collett, M.A., Lubbers, M.W., and Rakonjac, J. (2007). Direct selection and phage display of a Gram-positive secretome. Genome Biol. *8*, R266.

Jespers, L.S., De Keyser, A., and Stanssens, P.E. (1996). LambdaZLG6: a phage lambda vector for high-efficiency cloning and surface expression of cDNA libraries on filamentous phage. Gene *173*, 179–181.

Jiang, W., Baker, M.L., Jakana, J., Weigele, P.R., King, J., and Chiu, W. (2008). Backbone structure of the infectious epsilon15 virus capsid revealed by electron cryomicroscopy. Nature *451*, 1130–1134.

Jiang, W., Li, Z., Zhang, Z., Baker, M.L., Prevelige, P.E., Jr., and Chiu, W. (2003). Coat protein fold and maturation transition of bacteriophage P22 seen at subnanometer resolutions. Nat. Struct. Biol. *10*, 131–135.

Johnson, J.E., and Speir, J.A. (1997). Quasi-equivalent viruses: a paradigm for protein assemblies. J. Mol. Biol. *269*, 665–675.

Junge, W., Panke, O., Cherepanov, D.A., Gumbiowski, K., Muller, M., and Engelbrecht, S. (2001). Inter-subunit rotation and elastic power transmission in F0F1-ATPase. FEBS Lett. *504*, 152–160.

Kanamaru, S., Leiman, P.G., Kostyuchenko, V.A., Chipman, P.R., Mesyanzhinov, V.V., Arisaka, F., and Rossmann, M.G. (2002). Structure of the cell-puncturing device of bacteriophage T4. Nature *415*, 553–557.

Kay, B.K., Winter, J., and McCafferty, J. (2002). Phage Display of Peptides and Proteins (London: Academic Press).

Khayat, R., Tang, L., Larson, E.T., Lawrence, C.M., Young, M., and Johnson, J.E. (2005). Structure of an archaeal virus capsid protein reveals a common ancestry to eukaryotic and bacterial viruses. Proc. Natl. Acad. Sci. U.S.A. *102*, 18944–18949.

Kohler, G., and Milstein, C. (1975). Continuous cultures of fused cells secreting antibody of predefined specificity. Nature *256*, 495–497.

Koiwa, H., D'Urzo, M.P., Assfalg-Machleidt, I., Zhu-Salzman, K., Shade, R.E., An, H., Murdock, L.L., Machleidt, W., Bressan, R.A., and Hasegawa, P.M. (2001). Phage display selection of hairpin loop soyacystatin variants that mediate high affinity inhibition of a cysteine proteinase. Plant J. *27*, 383–391.

Kostyuchenko, V.A., Leiman, P.G., Chipman, P.R., Kanamaru, S., van Raaij, M.J., Arisaka, F., Mesyanzhinov, V.V., and Rossmann, M.G. (2003). Three-dimensional structure of bacteriophage T4 baseplate. Nat. Struct. Biol. *10*, 688–693.

Krebber, C., Spada, S., Desplancq, D., and Pluckthun, A. (1995). Co-selection of cognate antibody–antigen pairs by selectively-infective phages. FEBS Lett. *377*, 227–231.

Krumpe, L.R., Atkinson, A.J., Smythers, G.W., Kandel, A., Schumacher, K.M., McMahon, J.B., Makowski, L., and Mori, T. (2006). T7 lytic phage-displayed peptide libraries exhibit less sequence bias than M13 filamentous phage-displayed peptide libraries. Proteomics *6*, 4210–4222.

Kurakin, A., Wu, S., and Bredesen, D.E. (2004). Target-assisted iterative screening of phage surface display cDNA libraries. Methods Mol. Biol. *264*, 47–60.

La Scola, B., Audic, S., Robert, C., Jungang, L., de Lamballerie, X., Drancourt, M., Birtles, R., Claverie, J.M., and Raoult, D. (2003). A giant virus in amoebae. Science *299*, 2033.

Larman, H.B., Zhao, Z., Laserson, U., Li, M.Z., Ciccia, A., Gakidis, M.A., Church, G.M., Kesari, S., Leproust, E.M., Solimini, N.L., *et al.* (2011). Autoantigen discovery with a synthetic human peptidome. Nat. Biotechnol. *29*, 535–541.

Larocca, D., Jensen-Pergakes, K., Burg, M.A., and Baird, A. (2002). Gene transfer using targeted filamentous bacteriophage. Methods Mol. Biol. *185*, 393–401.

Lata, R., Conway, J.F., Cheng, N., Duda, R.L., Hendrix, R.W., Wikoff, W.R., Johnson, J.E., Tsuruta, H., and Steven, A.C. (2000). Maturation dynamics of a viral capsid: visualization of transitional intermediate states. Cell *100*, 253–263.

Lee, K.K., Gan, L., Tsuruta, H., Moyer, C., Conway, J.F., Duda, R.L., Hendrix, R.W., Steven, A.C., and Johnson, J.E. (2008). Virus capsid expansion driven by the capture of mobile surface loops. Structure *1*, 1491–1502.

Lee, S., and Belcher, A. (2004). Virus-based fabrication of micro- and nanofibers using electrospinning. NANO LETT *4*, 387–390.

Lee, S., Mao, C., Flynn, C., and Belcher, A. (2002). Ordering of quantum dots using genetically engineered viruses. Science *296*, 892–895.

Lee, Y.J., Yi, H., Kim, W.J., Kang, K., Yun, D.S., Strano, M.S., Ceder, G., and Belcher, A.M. (2009). Fabricating genetically engineered high-power lithium-ion batteries using multiple virus genes. Science *324*, 1051–1055.

Leiman, P.G., Arisaka, F., van Raaij, M.J., Kostyuchenko, V.A., Aksyuk, A.A., Kanamaru, S., and Rossmann, M.G. (2010). Morphogenesis of the T4 tail and tail fibers. Virol. J. *7*, 355.

Leiman, P.G., Chipman, P.R., Kostyuchenko, V.A., Mesyanzhinov, V.V., and Rossmann, M.G. (2004). Three-dimensional rearrangement of proteins in the tail of bacteriophage T4 on infection of its host. Cell *118*, 419–429.

Leverentz, B., Conway, W.S., Janisiewicz, W., Abadias, M., Kurtzman, C.P., and Camp, M.J. (2006). Biocontrol of the food-borne pathogens *Listeria monocytogenes* and *Salmonella enterica* serovar Poona on fresh-cut apples with naturally occurring bacterial and yeast antagonists. Appl. Environ. Microbiol. *72*, 1135–1140.

Linderoth, N.A., Simon, M.N., and Russel, M. (1997). The filamentous phage pIV multimer visualized by scanning transmission electron microscopy. Science *278*, 1635–1638.

Liu, D.J., and Day, L.A. (1994). Pf1 virus structure: helical coat protein and DNA with paraxial phosphates. Science *265*, 671–674.

Liu, S., Han, W., Sun, C., Lei, L., Feng, X., Yan, S., Diao, Y., Gao, Y., Zhao, H., Liu, Q. *et al.* (2011). Subtractive screening with the *Mycobacterium tuberculosis* surface protein phage display library. Tuberculosis (Edinb) *91*, 579–586.

Lowman, H.B., Bass, S.H., Simpson, N., and Wells, J.A. (1991). Selecting high-affinity binding proteins by monovalent phage display. Biochemistry *30*, 10832–10838.

Lubkowski, J., Hennecke, F., Pluckthun, A., and Wlodawer, A. (1998). The structural basis of phage display elucidated by the crystal structure of the N-terminal domains of g3p. Nat. Struct. Biol.*5*, 140–147.

Lubkowski, J., Hennecke, F., Pluckthun, A., and Wlodawer, A. (1999). Filamentous phage infection: crystal structure of g3p in complex with its coreceptor, the C-terminal domain of TolA. Structure Fold Des *15*, 711–272.

Mao, C., Flynn, C., Hayhurst, A., Sweeney, R., Qi, J., Georgiou, G., Iverson, B., and Belcher, A. (2003). Viral assembly of oriented quantum dot nanowires. Proc. Natl. Acad. Sci. U.S.A. *100*, 6946–6951.

Mao, C., Solis, D., Reiss, B., Kottmann, S., Sweeney, R., Hayhurst, A., Georgiou, G., Iverson, B., and Belcher, A. (2004). Virus-based toolkit for the directed synthesis of magnetic and semiconducting nanowires. Science *303*, 213–217.

Marciano, D., Russel, M., and Simon, S.M. (1999). An aqueous channel for filamentous phage export. Science *284*, 1516–1519.

Marciano, D.K., Russel, M., and Simon, S.M. (2001). Assembling filamentous phage occlude pIV channels. Proc. Natl. Acad. Sci. U.S.A. *98*, 9359–9364.

Marks, J.D., Hoogenboom, H.R., Bonnert, T.P., McCafferty, J., Griffiths, A.D., and Winter, G. (1991). By-passing immunization. Human antibodies from V-gene libraries displayed on phage. J. Mol. Biol. *222*, 581–597.

Marlovits, T.C., Kubori, T., Sukhan, A., Thomas, D.R., Galan, J.E., and Unger, V.M. (2004). Structural insights into the assembly of the type III secretion needle complex. Science *306*, 1040–1042.

Martin, A., and Schmid, F.X. (2003). The folding mechanism of a two-domain protein: folding kinetics and domain docking of the gene-3 protein of phage fd. J. Mol. Biol. *329*, 599–610.

Maruyama, I.N., Maruyama, H.I., and Brenner, S. (1994). Lambda foo: a lambda phage vector for the expression of foreign proteins. Proc. Natl. Acad. Sci. U.S.A. *91*, 8273–8277.

Marvin, D.A. (1998). Filamentous phage structure, infection and assembly. Curr. Opin. Struct. Biol. *8*, 150–158.

Marvin, D.A., Welsh, L.C., Symmons, M.F., Scott, W.R., and Straus, S.K. (2006). Molecular structure of fd (f1, M13) filamentous bacteriophage refined with respect to X-ray fibre diffraction and solid-state NMR data supports specific models of phage assembly at the bacterial membrane. J. Mol. Biol. *355*, 294–309.

Mattheakis, L.C., Bhatt, R.R., and Dower, W.J. (1994). An in vitro polysome display system for identifying ligands from very large peptide libraries. Proc. Natl. Acad. Sci. U.S.A. *91*, 9022–9026.

McCafferty, J., Griffiths, A.D., Winter, G., and Chiswell, D.J. (1990). Phage antibodies: filamentous phage displaying antibody variable domains. Nature *348*, 552–554.

McDonnell, P.A., Shon, K., Kim, Y., and Opella, S.J. (1993). fd coat protein structure in membrane environments. J. Mol. Biol. *233*, 447–463.

McLeod, S.M., Kimsey, H.H., Davis, B.M., and Waldor, M.K. (2005). CTXφ and *Vibrio choleree*: exploring a newly recognized type of phage-host cell relationship. Mol. Microbiol. *57*, 347–356.

Mersmann, M., Meier, D., Mersmann, J., Helmsing, S., Nilsson, P., Graslund, S., Colwill, K., Hust, M., and Dubel, S. (2010). Towards proteome scale antibody selections using phage display. Nat. Biotechnol *27*, 118–128.

Michel, B., and Zinder, N.D. (1989). Translational repression in bacteriophage f1: characterization of the gene V protein target on the gene II mRNA. Proc. Natl. Acad. Sci. U.S.A. *86*, 4002–4006.

Mikawa, Y.G., Maruyama, I.N., and Brenner, S. (1996). Surface display of proteins on bacteriophage lambda heads. J. Mol. Biol. *262*, 21–30.

Model, P., and Russel, M. (1988). Filamentous Bacteriophage. In The Bacteriophages, R. Calendar, ed. (New York, Plenum Publishing), pp. 375–456.

Morais, M.C., Choi, K.H., Koti, J.S., Chipman, P.R., Anderson, D.L., and Rossmann, M.G. (2005). Conservation of the capsid structure in tailed dsDNA bacteriophages: the pseudoatomic structure of φ29. Mol. Cell *18*, 149–159.

Morais, M.C., Tao, Y., Olson, N.H., Grimes, S., Jardine, P.J., Anderson, D.L., Baker, T.S., and Rossmann, M.G. (2001). Cryoelectron-microscopy image reconstruction of symmetry mismatches in bacteriophage φ29. J. Struct. Biol. *135*, 38–46.

Mullen, L.M., Nair, S.P., Ward, J.M., Rycroft, A.N., and Henderson, B. (2006). Phage display in the study of infectious diseases. Trends Microbiol. *14*, 141–147.

Murphy, J.S., and Philipson, L. (1962). Purification of *B. megatherium* Phage G and evidence for a muralytic enzyme as an integral part of the phage. J.Gen. Physiol. *45*, 155–168.

Nakayama, K., Takashima, K., Ishihara, H., Shinomiya, T., Kageyama, M., Kanaya, S., Ohnishi, M., Murata, T., Mori, H., and Hayashi, T. (2000). The R-type pyocin of *Pseudomonas aeruginosa* is related to P2 phage, and the F-type is related to lambda phage. Mol. Microbiol. *38*, 213–231.

Nam, K., Peelle, B., Lee, S., and Belcher, A. (2004). Genetically driven assembly of nanorings based on the M13 virus. Nano. Lett. *4*, 23–27.

Nam, K.T., Kim, D.W., Yoo, P.J., Chiang, C.Y., Meethong, N., Hammond, P.T., Chiang, Y.M., and Belcher, A.M. (2006). Virus-enabled synthesis and assembly of nanowires for lithium ion battery electrodes. Science *312*, 885–888.

Nam, Y.S., Magyar, A.P., Lee, D., Kim, J.W., Yun, D.S., Park, H., Pollom, T.S., Jr., Weitz, D.A., and Belcher, A.M. (2010). Biologically templated photocatalytic nanostructures for sustained light-driven water oxidation. Nat. Nanotechnol. *5*, 340–344.

Nandhagopal, N., Simpson, A.A., Gurnon, J.R., Yan, X., Baker, T.S., Graves, M.V., Van Etten, J.L., and Rossmann, M.G. (2002). The structure and evolution of the major capsid protein of a large, lipid-containing DNA virus. Proc. Natl. Acad. Sci. U.S.A. *99*, 14758–14763.

Nanduri, V., Sorokulova, I.B., Samoylov, A.M., Simonian, A.L., Petrenko, V.A., and Vodyanoy, V. (2007). Phage as a molecular recognition element in biosensors immobilized by physical adsorption. Biosens. Bioelectron. *22*, 986–992.

Newcomb, W.W., Juhas, R.M., Thomsen, D.R., Homa, F.L., Burch, A.D., Weller, S.K., and Brown, J.C. (2001). The UL6 gene product forms the portal for entry of DNA into the herpes simplex virus capsid. J. Virol. *75*, 10923–10932.

Newcomb, W.W., Trus, B.L., Booy, F.P., Steven, A.C., Wall, J.S., and Brown, J.C. (1993). Structure of the herpes simplex virus capsid. Molecular composition of the pentons and the triplexes. J. Mol. Biol. *232*, 499–511.

Newman, J., Swinney, H.L., and Day, L.A. (1977). Hydrodynamic properties and structure of fd virus. J. Mol. Biol. *116*, 593–603.

Opalka, N., Beckmann, R., Boisset, N., Simon, M.N., Russel, M., and Darst, S.A. (2003). Structure of the filamentous phage pIV multimer by cryo-electron microscopy. J. Mol. Biol. *325*, 461–470.

Papavoine, C.H., Christiaans, B.E., Folmer, R.H., Konings, R.N., and Hilbers, C.W. (1998). Solution structure of the M13 major coat protein in detergent micelles: a basis for a model of phage assembly involving specific residues. J. Mol. Biol. *282*, 401–419.

Parmley, S.F., and Smith, G.P. (1988). Antibody-selectable filamentous fd phage vectors – affinity purification of target genes Gene *73*, 305–318.

Pasqualini, R., and Ruoslahti, E. (1996). Organ targeting in vivo using phage display peptide libraries. Nature *380*, 364–366.

Petrenko, V. (2008). Evolution of phage display: from bioactive peptides to bioselective nanomaterials. Expert Opin. Drug Deliv. *5*, 825–836.

Poul, M.A., Becerril, B., Nielsen, U.B., Morisson, P., and Marks, J.D. (2000). Selection of tumour-specific internalizing human antibodies from phage libraries. J. Mol. Biol. *301*, 1149–1161.

Rakonjac, J., Bennett, N.J., Spagnuolo, J., Gagic, D., and Russel, M. (2011). Filamentous bacteriophage: biology, phage display and nanotechnology applications. Curr. Issues Mol. Biol. *13*, 51–76.

Rakonjac, J., Feng, J.-n., and Model, P. (1999). Filamentous phage are released from the bacterial mambrane by a two-step mechanism involving a short C-terminal fragment of pIII. J. Mol. Biol. *289*, 1253–1265.

Rakonjac, J., Jovanovic, G., and Model, P. (1997). Filamentous phage infection-mediated gene expression: construction and propagation of the *gIII* deletion mutant helper phage R408d3. Gene *198*, 99–103.

Rakonjac, J., and Model, P. (1998). The roles of pIII in filamentous phage assembly. J. Mol. Biol. *282*, 25–41.

Raoult, D., Audic, S., Robert, C., Abergel, C., Renesto, P., Ogata, H., La Scola, B., Suzan, M., and Claverie, J.M. (2004). The 1.2-megabase genome sequence of Mimivirus. Science *306*, 1344–1350.

Rapoza, M.P., and Webster, R.E. (1993). The filamentous bacteriophage assembly proteins require the bacterial SecA protein for correct localization to the membrane. J. Bacteriol. *175*, 1856–1859.

Rapoza, M.P., and Webster, R.E. (1995). The products of gene I and the overlapping in-frame gene XI are required for filamentous phage assembly. J. Mol. Biol. *248*, 627–638.

Rebar, E.J., and Pabo, C.O. (1994). Zinc finger phage: affinity selection of fingers with new DNA-binding specificities. Science *263*, 671–673.

Reichmann, L., and Holliger, P. (1997). The C-terminal domain of TolA is the coreceptor for filamentous phage infection of *E. coli*. Cell *90*, 351–360.

Rice, G., Tang, L., Stedman, K., Roberto, F., Spuhler, J., Gillitzer, E., Johnson, J.E., Douglas, T., and Young, M. (2004). The structure of a thermophilic archaeal virus shows a double-stranded DNA viral capsid type that spans all domains of life. Proc. Natl. Acad. Sci. U.S.A. *101*, 7716–7720.

Roberts, B.L., Markland, W., Ley, A.C., Kent, R.B., White, D.W., Guterman, S.K., and Ladner, R.C. (1992). Directed evolution of a protein: selection of potent neutrophil elastase inhibitors displayed on M13 fusion phage. Proc. Natl. Acad. Sci. U.S.A. *89*, 2429–2433.

Rosenberg, A., Griffin, K., Studier, F.W., McCormick, M., Berg, J., Novy, R., and Mierendorf, R. (1996). T7Select phage display system: a powerful new protein display system based on bacteriophage T7. inNovations *6*, 1–6.

Ross, P.D., Cheng, N., Conway, J.F., Firek, B.A., Hendrix, R.W., Duda, R.L., and Steven, A.C. (2005). Cross-linking renders bacteriophage HK97 capsid maturation irreversible and effects an essential stabilization. EMBO J. *24*, 1352–1363.

Ross, P.D., Conway, J.F., Cheng, N., Dierkes, L., Firek, B.A., Hendrix, R.W., Steven, A.C., and Duda, R.L. (2006). A free energy cascade with locks drives assembly and maturation of bacteriophage HK97 capsid. J. Mol. Biol. *364*, 512–525.

Ruigrok, R.W.H., Schoehn, G., and Conway, J.F. (2005). The morphlogy and structure of viruses. In Topley and Wilson's Microbiology and Microbial Infections, 10th Edition: Virology, B.W.J. Mahy, and V. ter Meulen, eds. (London, Hodder Arnold), pp. 57–85.

Russel, M. (1993). Protein–protein interactions during filamentous phage assembly. J. Mol. Biol. *231*, 689–697.

Russel, M. (1995). Moving through the membrane with filamentous phages. Trends Microbiol. *3*, 223–228.

Russel, M., and Model, P. (1989). Genetic analysis of the filamentous bacteriophage packaging signal and of the proteins that interact with it. J. Virol. *63*, 3284–3295.

Sambrook, J., and Russell, D.W. (2001). Molecular cloning: a laboratory manual (3rd. Edn.), 3rd edn (Cold Spring Harbor, Cold Spring Harbor).

Samuelson, J.C., Jiang, F., Yi, L., Chen, M., de Gier, J.W., Kuhn, A., and Dalbey, R.E. (2001). Function of YidC for the insertion of M13 procoat protein in *Escherichia coli*: translocation of mutants that show differences in their membrane potential dependence and Sec requirement. J. Biol. Chem. *276*, 34847–34852.

Santi, E., Capone, S., Mennuni, C., Lahm, A., Tramontano, A., Luzzago, A., and Nicosia, A. (2000). Bacteriophage lambda display of complex cDNA libraries: a new approach to functional genomics. J. Mol. Biol. *296*, 497–508.

Sarikaya, M., Tamerler, C., Jen, A., Schulten, K., and Baneyx, F. (2003). Molecular biomimetics: nanotechnology through biology. Nat. Mater. *2*, 577–585.

Sblattero, D., and Bradbury, A. (2000). Exploiting recombination in single bacteria to make large phage antibody libraries. Nat. Biotechnol. *18*, 75–80.

Schier, R., Bye, J., Apell, G., McCall, A., Adams, G.P., Malmqvist, M., Weiner, L.M., and Marks, J.D. (1996). Isolation of high-affinity monomeric human anti-c-erbB-2 single chain Fv using affinity-driven selection. J. Mol. Biol. *255*, 28–43.

Scholl, D., Gebhart, D., Williams, S.R., Bates, A., and Mandrell, R. (2012). Genome sequence of *E. coli* O104:H4 leads to rapid development of a targeted antimicrobial agent against this emerging pathogen. PLoS One *7*, e33637.

Scott, J.K., and Smith, G.P. (1990). Searching for peptide ligands with an epitope library. Science *249*, 386–390.

Simpson, A.A., Leiman, P.G., Tao, Y., He, Y., Badasso, M.O., Jardine, P.J., Anderson, D.L., and Rossmann, M.G. (2001). Structure determination of the head-tail connector of bacteriophage φ29. Acta Crystallogr. D Biol. Crystallogr. *57*, 1260–1269.

Smith, G.P. (1985). Filamentous fusion phage: novel expression vectors that display cloned antigens on the virion surface. Science *228*, 1315–1317.

Sobry, S., Havelange, A., and Van Cutsem, P. (2005). Immunocytochemistry of pectins in shoot apical meristems: consequences for intercellular adhesion. Protoplasma *225*, 15–22.

Soumillion, P., Jespers, L., Bouchet, M., Marchand-Brynaert, J., Winter, G., and Fastrez, J. (1994). Selection of beta-lactamase on filamentous bacteriophage by catalytic activity. J. Mol. Biol. *237*, 415–422.

Souza, G.R., Christianson, D.R., Staquicini, F.I., Ozawa, M.G., Snyder, E.Y., Sidman, R.L., Miller, J.H., Arap, W., and Pasqualini, R. (2006). Networks of gold nanoparticles and bacteriophage as biological sensors and cell-targeting agents. Proc. Natl. Acad. Sci. U.S.A. *103*, 1215–1220.

Spencer, J.V., Newcomb, W.W., Thomsen, D.R., Homa, F.L., and Brown, J.C. (1998). Assembly of the herpes simplex virus capsid: Preformed triplexes bind to the nascent capsid. J. Virol. *72*, 3944–3951.

Stengele, I., Bross, P., Garces, X., Giray, J., and Rasched, I. (1990). Dissection of functional domains in phage fd adsorption protein – discrimination between attachment and penetration sites. J. Mol. Biol. *212*, 143–149.

Sternberg, N., and Hoess, R.H. (1995). Display of peptides and proteins on the surface of bacteriophage lambda. Proc. Natl. Acad. Sci. U.S.A. *92*, 1609–1613.

Steven, A.C., Greenstone, H.L., Booy, F.P., Black, L.W., and Ross, P.D. (1992). Conformational changes of a viral capsid protein. Thermodynamic rationale for proteolytic regulation of bacteriophage T4 capsid expansion, co-operativity, and super-stabilization by soc binding. J. Mol. Biol. *228*, 870–884.

Steven, A.C., Heymann, J.B., Cheng, N., Trus, B.L., and Conway, J.F. (2005). Virus maturation: dynamics and mechanism of a stabilizing structural transition that leads to infectivity. Curr. Opin. Struct. Biol. *15*, 227–236.

Stopar, D., Trcek, T., and Strancar, J. (2002). Electron paramagnetic resonance characterization of the exopolysaccharide layer produced by bacteria. Cell Mol. Biol. Lett. *7*, 167–169.

Sutter, M., Boehringer, D., Gutmann, S., Gunther, S., Prangishvili, D., Loessner, M.J., Stetter, K.O., Weber-Ban, E., and Ban, N. (2008). Structural basis of enzyme encapsulation into a bacterial nanocompartment. Nat. Struct. Mol. Biol. *15*, 939–947.

Szabo, A., Heja, D., Szakacs, D., Zboray, K., Kekesi, K.A., Radisky, E.S., Sahin-Toth, M., and Pal, G. (2011). High affinity small protein inhibitors of human chymotrypsin C (CTRC) selected by phage display reveal unusual preference for P4' acidic residues. J. Biol. Chem. *286*, 22535–22545.

Teesalu, T., Sugahara, K.N., and Ruoslahti, E. (2012). Mapping of vascular ZIP codes by phage display. Methods Enzymol. *503*, 35–56.

Thompson, J., Pope, T., Tung, J.S., Chan, C., Hollis, G., Mark, G., and Johnson, K.S. (1996). Affinity maturation of a high-affinity human monoclonal antibody against the third hypervariable loop of human immunodeficiency virus: use of phage display to improve affinity and broaden strain reactivity. J. Mol. Biol. *256*, 77–88.

Trus, B.L., Booy, F.P., Newcomb, W.W., Brown, J.C., Homa, F.L., Thomsen, D.R., and Steven, A.C. (1996). The herpes simplex virus procapsid: structure, conformational changes upon maturation, and roles of the triplex proteins VP19c and VP23 in assembly. J. Mol. Biol. *263*, 447–462.

Trus, B.L., Cheng, N., Newcomb, W.W., Homa, F.L., Brown, J.C., and Steven, A.C. (2004). Structure and polymorphism of the UL6 portal protein of herpes simplex virus type 1. J. Virol. *78*, 12668–12671.

Tsuboi, M., Tsunoda, M., Overman, S.A., Benevides, J.M., and Thomas, G.J., Jr. (2010). A structural model for the single-stranded DNA genome of filamentous bacteriophage Pf1. Biochemistry *49*, 1737–1743.

Tsurushita, N., Fu, H., and Warren, C. (1996). Phage display vectors for in vivo recombination of immunoglobulin heavy and light chain genes to make large combinatorial libraries. Gene *172*, 59–63.

Uchiyama, A., and Fane, B.A. (2005). Identification of an interacting coat-external scaffolding protein domain required for both the initiation of φX174 procapsid morphogenesis and the completion of DNA packaging. J. Virol. *79*, 6751–6756.

van Houten, N.E., Henry, K.A., Smith, G.P., and Scott, J.K. (2010). Engineering filamentous phage carriers to improve focusing of antibody responses against peptides. Vaccine *28*, 2174–2185.

Velappan, N., Fisher, H.E., Pesavento, E., Chasteen, L., D'Angelo, S., Kiss, C., Longmire, M., Pavlik, P., and Bradbury, A.R. (2010). A comprehensive analysis of filamentous phage display vectors for cytoplasmic proteins: an analysis with different fluorescent proteins. Nucleic Acids Res. *38*, e22.

Waldor, M.K., and Mekalanos, J.J. (1996). Lysogenic conversion by a filamentous phage endocing cholera toxin. Science *272*, 1910–1914.

Wall, T., Roos, S., Jacobsson, K., Rosander, A., and Jonsson, H. (2003). Phage display reveals 52 novel extracellular and transmembrane proteins from *Lactobacillus reuteri* DSM 20016(T). Microbiology *149*, 3493–3505.

Wang, L., Zheng, A., Yi, L., Xu, C., Ding, M., and Deng, H. (2004). Identification of potential nuclear reprogramming and differentiation factors by a novel selection method for cloning chromatin-binding proteins. Biochem. Biophys. Res. Commun. *325*, 302–307.

Waterhouse, P., Griffiths, A.D., Johnson, K.S., and Winter, G. (1993). Combinatorial infection and in vivo recombination: a strategy for making large phage antibody repertoires. Nucleic Acids Res. *21*, 2265–2266.

Webster, R.E. (1996). Biology of the filamenous bacteriophage. In Phage display of peptides and proteins, B.K. Kay, J. Winter, and J. McCafferty, eds. (London: Academic Press), pp. 1–20.

Webster, R.E. (2001). Filamentous phage biology. In Phage display: a laboratory manual, C.F. Barbas III, D.R. Burton, J.K. Scott, and G.J. Silverman, eds. (Cold Spring Harbor, NY: Cold Spring Harbor Laboratory Press), pp. 1.1–1.37.

Wikoff, W.R., Conway, J.F., Tang, J., Lee, K.K., Gan, L., Cheng, N., Duda, R.L., Hendrix, R.W., Steven, A.C., and Johnson, J.E. (2006). Time-resolved molecular dynamics of bacteriophage HK97 capsid maturation interpreted by electron cryo-microscopy and X-ray crystallography. J. Struct. Biol. *153*, 300–306.

Wikoff, W.R., Liljas, L., Duda, R.L., Tsuruta, H., Hendrix, R.W., and Johnson, J.E. (2000). Topologically linked protein rings in the bacteriophage HK97 capsid. Science *289*, 2129–2133.

Willats, W.G., Gilmartin, P.M., Mikkelsen, J.D., and Knox, J.P. (1999). Cell wall antibodies without immunization: generation and use of de-esterified homogalacturonan block-specific antibodies from a naive phage display library. Plant J *18*, 57–65.

Williams, D.D., Benedek, O., and Turnbough, C.L., Jr. (2003). Species-specific peptide ligands for the detection of *Bacillus anthracis* spores. Appl. Environ. Microbiol. *69*, 6288–6293.

Wrighton, N.C., Farrell, F.X., Chang, R., Kashyap, A.K., Barbone, F.P., Mulcahy, L.S., Johnson, D.L., Barrett, R.W., Jolliffe, L.K., and Dower, W.J. (1996). Small peptides as potent mimetics of the protein hormone erythropoietin. Science *273*, 458–464.

Yan, X., Olson, N.H., Van Etten, J.L., Bergoin, M., Rossmann, M.G., and Baker, T.S. (2000). Structure and assembly of large lipid-containing dsDNA viruses. Nat. Struct. Biol. *7*, 101–103.

Yang, F., Forrer, P., Dauter, Z., Conway, J.F., Cheng, N., Cerritelli, M.E., Steven, A.C., Pluckthun, A., and Wlodawer, A. (2000). Novel fold and capsid-binding properties of the lambda-phage display platform protein gpD. Nat. Struct. Biol.*7*, 230–237.

Zacchi, P., Sblattero, D., Florian, F., Marzari, R., and Bradbury, A.R.M. (2003). Selecting open reading frames from DNA. Genome Research *13*, 980–990.

Zhou, B., Wirsching, P., and Janda, K.D. (2002). Human antibodies against spores of the genus Bacillus: a model study for detection of and protection against anthrax and the bioterrorist threat. Proc. Natl. Acad. Sci. U.S.A. *99*, 5241–5246.

Zwick, M.B., Shen, J., and Scott, J.K. (2000). Homodimeric peptides displayed by the major coat protein of filamentous phage. J. Mol. Biol. *300*, 307–320.

Bio-inspired Biomolecular Supramolecular Self-assemblies and their Applications

6

Dong Li and Chuanbin Mao

Abstract

A variety of naturally occurring biological materials exhibit supramolecular self-assembly properties. By incorporation of signaling motifs, biological information and functional units, these biological materials can find extensive applications in developing nanotechnology, material science, tissue engineering and nanomedicine. In this chapter, some naturally occurring materials, which can be genetically engineered to display foreign peptide or chemically modified to incorporate signaling motifs, are summarized. The self-assembly behaviours of these biological materials generates hierarchically organized structures from the bottom up. The presentation of functional peptides on these biological materials enables the production of biomaterials for different applications. More and more naturally occurring biological materials are to be studied with the development of biotechnology and nanotechnology.

Introduction

The assembly of small molecules into higher ordered structures from the nano-scale to functional macroscopic objects is an essential theme in bionanotechnology. During this 'bottom-up' process, molecular components spontaneously aggregate via non–covalent interactions termed supramolecular recognition and self-assembly to form one-dimensional (1-D), two-dimensional (2-D), and even three-dimensional (3-D) structures, which are of interest for a variety of practical applications. Almost all biological materials of natural origin, including DNA, RNA, peptides, proteins, lipids and saccharides, exhibit supramolecular self-assembly properties. Because the biological materials have such advanced structural complexity, it is very difficult to fabricate and copy their natural structures and functions at molecular level *in vitro*. However, after decades of efforts, a wide range of biological materials have been designed and fabricated for particular uses, especially in biology and medicine. In recent years, assembling biological nanocomposites within an aqueous environment has received much attention (Zayed *et al.*, 2010).

Supramolecular chemistry regarding molecular self-assembly refers to a well-defined complex of molecules combined by non–covalent interactions that has emerged as a leading discipline. In nature, it is crucial that biological systems form complex and well-defined microstructures to achieve their cellular functions. For example, the cell membrane is self-assembled from lipids to protect cells; double helical DNA is bonded from individual strands to form stable genetic material; and many enzymes with tertiary or quaternary structures are self-assembled from their monomers. A variety of naturally occurring materials including collagen, fibrin, silk, spider silk, and biological surfactants exhibit characteristics of supramolecular assemblies (Barthelat, 2007; Lin and Mao, 2011; Numata and Kaplan, 2010a; Sanchez *et al.*, 2005). With development of biotechnology and nanotechnology, more and more biological materials, their derivatives and synthetic materials have been developed for the fabrication of functional nanomaterials. In material science and nanotechnology, due to their own complexity, DNA, RNA and saccharide are preferably used as a structural material rather than as a carrier of signal information (Bandy *et al.*, 2011; Lin and Mao, 2011; Sadhasivam and Yun, 2010). Peptide and protein are defined by their amino acid sequences, however, can carry much more signal information due to their functional side chains, sequences and complex secondary and tertiary structural variations including α-helices, β-sheets, random coil and coiled-coils. They serve as the structural matrix as well as the functional constructs in biological system (Hauser and Zhang, 2010). Because most naturally occurring biological materials are not ideal or specifically designed for medical applications, tuning the surface chemistry of biological materials is an effective approach (Dickerson *et al.*, 2008; Gomes *et al.*, 2012; Tamerler *et al.*, 2007).

A variety of supramolecular, self-assembled complexes of peptides and proteins that have a widespread impact on biology, regenerative medicine and nanotechnology have been designed (Dickerson *et al.*, 2008; Gomes *et al.*, 2012; Tamerler *et al.*, 2007; Yanlian *et al.*, 2009). These complexes have led to the development of 'bottom-up' assembly to generate hierarchically nanostructured entities. Integrating signal peptides into these peptides and proteins based on bioengineering or surface display technique has enabled the development of 'smart nanomaterials' with potentially far-reaching applications as biomaterials. In this chapter, we highlight recent advances in supramolecular assembly of naturally occurring peptide and protein-based biological materials whose surface chemistry can be genetically or chemically modified to suit their varied applications.

Many excellent reviews have been dedicated to the supramolecular self-assembly of DNA, RNA, lipid and saccharide (Aldaye *et al.*, 2008; Bandy *et al.*, 2011; Sadhasivam and Yun, 2010; Zhou and Yan, 2010). Moreover, the far-reaching effects in many different fields of synthetic peptides with supramolecular self-assembly are also summarized in many reviews (Cui *et al.*, 2010; Hauser and Zhang, 2010; Yanlian *et al.*, 2009; Zelzer and Ulijn, 2010). The reader is referred to those. Many naturally occurring nanomaterials can be easily purified in large scale using bioengineering and biotechnology, providing an economic approach to their synthesis. More importantly, ideal biological materials should demonstrate supramolecular chemistry as well as enough biofunctionality (George and Ravindran, 2010). For example, in tissue engineering, hierarchically self-assembly of the nanobiomaterials into higher ordered 2-D and 3-D composites is essential to support and guide cell attachment and growth. In biological systems, tremendous supramolecular self-assembly systems can be observed and most biological materials exhibit excellent self-assembly properties *in vitro*. Moreover, these biological materials must incorporate signal information in order to interact with surrounding tissues, combine with gene or drugs as carriers, direct cell attachment, proliferation and differentiation. For synthetic molecules, normally only short signal peptides can be integrated onto the material's surface (Cui *et al.*, 2010). However, longer signal peptide (in the range of 20–30 amino acids residues) is usually required (Frank and Sippl, 2008; Nielson and Kroh, 1998). Some naturally occurring nanomaterials with self-assembly combined with surface display technique, however, can integrate longer signal peptide. Therefore, their utility is highly extended in many areas including genetically engineered peptides for inorganics (GEPI), gene or drug delivery, tissue engineering (TE) and nanomedicine (Abbineni and Mao, 2007; Cao and Mao, 2011). All of these research areas have taken advantage of multivalent properties of the biological nanomaterials with supramolecular self-assembly. Among various nanomaterials studied for the biomedical applications and nanomedicine, naturally occurring nanomaterials show some advantages:

1 Most biological materials represent very stable and delicate self-assembled architectures at the nanometer level from comparatively simple building blocks.
2 The surface chemistry and composition can be controlled by chemistry or bioengineering to bear desired functionality.
3 Most wild-type or recombinant biological materials can be prepared cost-efficiently at the large-scale; and
4 Under genetic control, most biological materials exhibit mono-dispersed and identical individuals.

Naturally occurring biological materials with genetic modification

Virus

The protein shells of viruses, also referred to as capsids, serve to protect enclosed genetic nucleic acids from degradation and often participate in the cell-specific transfer of genetic material to new host cells (Harrison, 1983). The self-assembling protein shells of viruses are highly organized nanoscale materials with robust chemical and physical properties which can be considered supramolecular entities (Dedeo *et al.*, 2011; Mao *et al.*, 2003, 2004; Merzlyak and Lee, 2006; Nam *et al.*, 2006; Souza *et al.*, 2006). They are assembled from many copies of distinct protein monomers with delicate hierarchical architectures which can be used for the construction of new nanomaterials. Under the right pH and ionic strength conditions, even *in vitro*, the monomers of phage proteins can form complete capsids by self-assembly. In nature, various shapes, from tubular to spherical-like viruses of different sizes and morphologies are available (Fig. 6.1) (Dedeo *et al.*, 2011; Lee *et al.*, 2009).

There are mainly three kinds of viruses in nature: bacteriophages, plant viruses and animal viruses (Goicochea *et al.*, 2007; Petrenko, 2008). Owing to their low toxicity to humans, bacteriophages and plant viruses have gained much attention. Plant viruses include the isosahedral, sphere-like cowpea chlorotic mottle virus (CCMV), brome mosaic virus (BMV), cowpea mosaic virus (CPMV), and the rod-shaped tobacco mosaic virus (TMV). Their application in bionanotechnology as biological templates has been nicely reviewed recently (Young *et al.*, 2008). Filamentous bacteriophages belong to the Ff class, which includes the M13, f1, fd, and ft strains (Pande *et al.*, 2006). The M13 bacteriophage is a filamentous virus that is the most well studied class of virus in nanomedicine, molecular biology, gene and drug delivery, and bionanotechnology (Abbineni and Mao, 2007; Dedo *et al.*, 2011; Smith, 2011). It is composed of circular single-stranded DNA (ssDNA) and is approximately 950 nm in length and 6.5 nm in diameter. Owing to the high degree of interparticle attraction along the rod axis, liquid crystalline domains

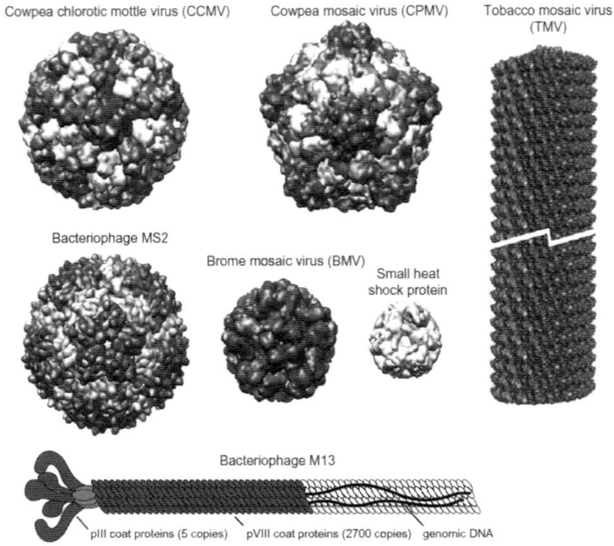

Figure 6.1 Structure of several viral capsids used for materials applications. (Reproduced with permission from Dedeo *et al.*, 2011.) A colour version of this figure is located in the plate section at the back of the book.

are formed on their surfaces. The ssDNA genome is encapsulated by ~2700 copies of a pVIII major coat protein. The virus displays approximately five copies each of the pIII and pVI minor coat proteins at one end and approximately five copies of pVII and pIX proteins at the other end (Smith, 2011).

The surface-exposed N-terminus of any protein on the M13 phage can be genetically modified with a target foreign peptide by phage display technique (Pande *et al.*, 2011). By construction of type pIII or type VIII phage display libraries, a variety of target-recognizing peptides or proteins can be selected against target substrates. Briefly, this process includes repeated cycles of selection of phage with desired binding capabilities to their target substrates in the libraries. Another approach for the genetic modification of phage surface chemistry is using a phagemid/phage as a vector. A designed oligonucleotide encoding a desired signal peptide is inserted into the vector. Then, the expressed foreign peptide can be displayed on the exposed surface of the bacteriophage. Moreover, double or triple display of different peptides can be obtained on a single phage virion by displaying different peptides on different coat proteins, such pIII and pVI, simultaneously. The resulting double or triple displayed phages can be used as multi-functional novel materials under site-specific control (Pande *et al.*, 2011). Typically, wild-type phages serve as scaffolds for the nucleation and growth of inorganic materials, while genetically modified phages are often used in biomedical and catalytic applications (Dedeo *et al.*, 2011; Merzlyak and Lee, 2006).

In addition to the controllable surface chemistry of the phage virions, the other attractive property is their hierarchical self-assembly to higher ordered liquid crystal structures under certain concentrations. With increasing concentration, filamentous phages exhibit the following phase transitions: isotropic, nematic, cholesteric, smectic A and smectic C, which has been used to fabricate highly ordered functional nanocomposites (Dogic and Fraden, 2006).

Silk proteins

Silk proteins are naturally occurring materials that are highly malleable and lightweight. They are biodegradable and biocompatible and have supramolecular self-assembly properties (Numata and Kaplan, 2010a). Some insects and most spiders produce silk proteins in nature. Silkworm silk proteins, also referred to as fibroin, obtained from the cocoon of silkworm *Bombyx mori* consist of one light (~25 kDa) and one heavy chain (~325 kDa). In the heavy chain of fibroin, there are many alanine-glycine repeats (Numata and Kaplan, 2010a). The spider silk proteins consist of block copolymers containing a large hydrophobic polyalanine-rich segment and a glycine-rich hydrophilic block with molecular weights ranging from 70 to 700 kDa (Jin and Kaplan, 2003; Vollrath and Knight, 2001). The hydrophobic blocks are cross-linked and form crystalline β-sheets domains in silk proteins stabilized by hydrogen bonds (Fig. 6.2) (Keten *et al.*, 2010). The multiple repeats of hydrophobic blocks contribute to the high tensile strength of silk fibres. The crystalline β-sheets domains are separated by less organized hydrophilic blocks which provide silk with elasticity (Heim *et al.*, 2009). The production of spider silk is a dynamic process involving the self-organization of its peptide components. Within the spider, silk exists in a native unfolded state with little defined local structure. After production, it self-assembles into either nanofibrils or microspheres in a kinetic pathway that is determined by pH and salts which act as chaotropes and kosmotropes (Rammensee *et al.*, 2008; Russo *et al.*, 2008).

Generating silk proteins in large scale in a cost-effective way is essential for their actual applications as high-performance biological materials. During the last decade, with the development of biotechnology and molecular biology, native and bioengineered silk proteins have been obtained in many different host systems using synthetic oligonucleotide versions of consensus repeats and their derivatives (Heim *et al.*, 2009). With the incorporation of desired signal peptides or domains, recombinant silk proteins lead to many new applications.

Bacterial flagella

Flagella are the linear appendage on surface of bacteria. These proteinaceous organelles are responsible for swimming as well as swarming when rotated by the motor at their base as a propeller (Van Gerven *et al.*, 2011; Fig. 6.3 left). Bacterial flagellum is a helix-shaped long filament with uniform thickness of approximately 12–25 nm depending on its species. It is composed of several to thousand copies of over 30 different proteins. The filament is a hollow nanotube composed of up to 20,000 copies of major protein (FliC) with roughly 11 subunits per two turns of the 1-start helix. The inner diameter of flagellum is about 2 nm. The FliC subunit of *Salmonella typhimurium* has an overall shape of the Greek letter 'Γ' and composed of four different domains, named D0, D1, D2, and D3 (Yonekura *et al.*, 2003). The D0 and D1 domains, which interact with neighbouring flagellins through coiled-coil interactions and are required for filament assembly, are highly conserved, whereas the surface-exposed D2 and D3 domains of the flagella are highly variable. Based on this characteristic, the surface of flagella can be genetically modified with desired foreign peptides (Lu *et al.*, 1995; Westerlund-Wikström *et al.*, 2000). Previous studies have shown that up to 302 amino acid residues can be displayed in the functional form on the variable region of FliC (Westerlund-Wikström *et al.*, 2000).

Flagella also experience higher ordered supramolecular self-assembly at a certain concentration (Barry *et al.*, 2006). When the concentration of flagella is less than 17 mg/ml,

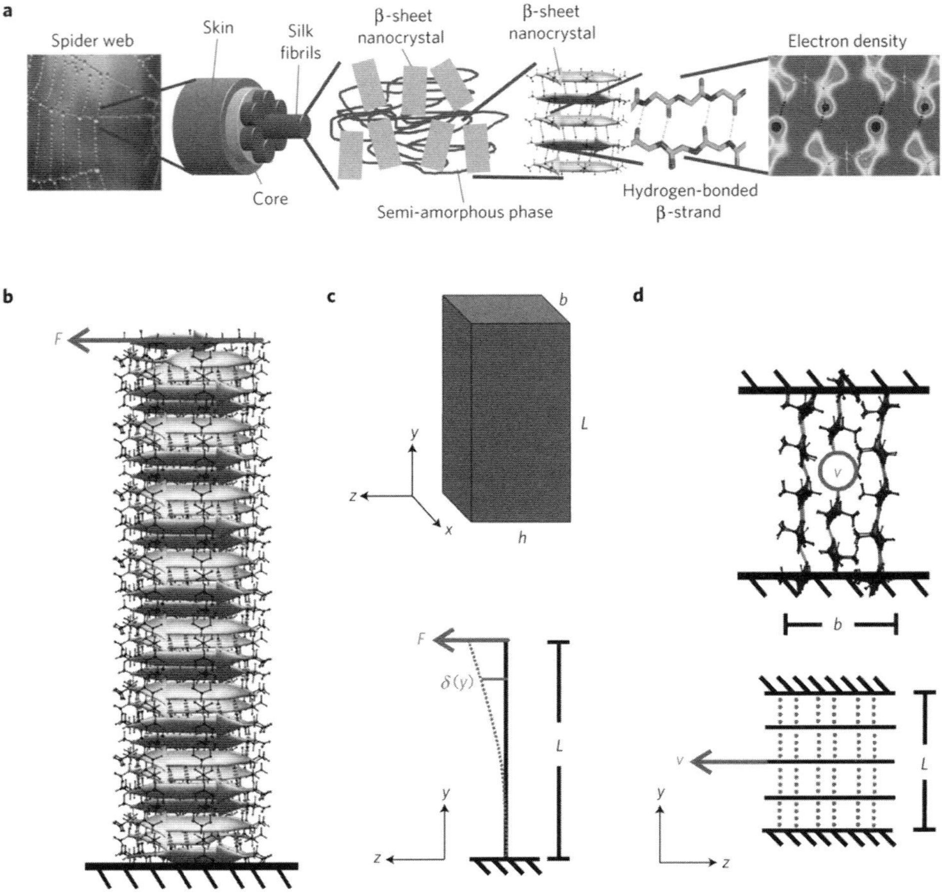

Figure 6.2 Hierarchical structure of spider silk, simulation set-up and theoretical considerations. (a) Schematic that highlights the key features of the hierarchical structure of spider silk from nano to macro, including the electron density at the Angstrom scale, hydrogen bonded β-strands, β-sheet nanocrystals, a hetero-nanocomposite of stiff nanocrystals embedded in a softer semi-amorphous phase and silk fibrils, which assembles into macroscopic silk fibres. (b) The atomistic structure of the silk β-sheet nanocrystal obtained from the Protein Data Bank (identification code 2slk) and replicated to build β-sheet nanocrystals of different sizes. In the first set of simulations, the β-sheet nanocrystal is subject to loading conditions similar to a cantilever beam with a constant tip loading, used to identify the bending rigidity and other structural properties. This loading mimics the characteristic lateral loading relevant to silk mechanics. c) Schematic representation of the β-sheet nanocrystal and definition of coordinates used here (upper part, where parameters b and h describe geometric parameters related to the number of sheets and the length of strands in the nanocrystal, and L is the size of the nanocrystal in the y direction). The lower part shows the geometry of the bending study and defines the displacement variable. d) Set-up for pull-out simulations, where to characterize fracture resistance of the nanocrystals, the central strand of the middle sheet is pulled out with constant velocity, while the top and bottom strands are restrained. (Reproduced with permission from Keten *et al.*, 2010.) A colour version of this figure is located in the plate section at the back of the book.

wild-type flagella SJW1103 form isotropic phase. As the concentration of flagella increases, bright birefringent droplets appeared from a dark isotropic background with a well-defined

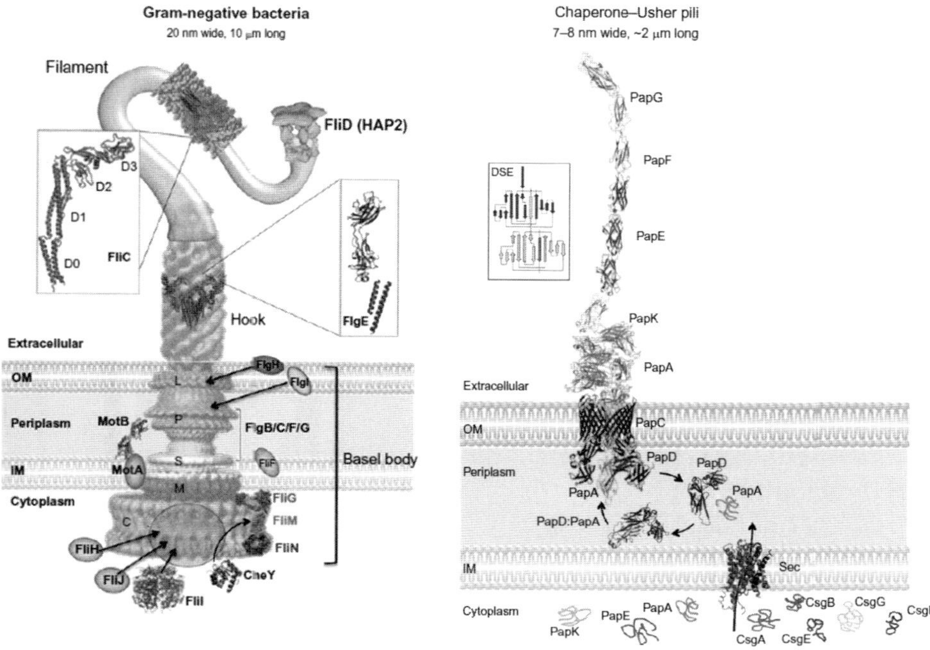

Figure 6.3 Bacterial flagella (left). The bacterial flagellum is a large macromolecular complex composed of about 30 different proteins and consisting of three major substructures: the filament, the hook, and the basal body. The flagellar basal body is formed by oligomeric ring complexes joined together by a rod and represented here by the cryo-electron microscopy 3-D structure (light blue). The filament itself consists of a single flagellin protein. The flagellin subunit of S. typhimurium FliC has an overall shape of the Greek letter 'Γ' and consists of four linearly connected domains, D0 to D3. Chaperone–Usher pili (right). Schematic representation of the biosynthetic pathway of chaperone–usher (CU) pili, here exemplified by P pili. P pili are composed of two major subassemblies: a thick and rigid rod made up of subunits of PapA (light blue) arranged in a right-handed helical cylinder; and a thin, flexible tip fibre in the distal end of the rod. In P pili, the tip fibre is composed of 5–10 PapE subunits (dark green) flanked by two minor subunits, PapF (orange) and PapK (yellow). The adhesin subunit PapG is responsible for interaction with the host tissues and present at the tip of the fibre. (Reproduced with permission from Van Gerven *et al*., 2011.) A colour version of this figure is located in the plate section at the back of the book.

striped pattern of 2.4 μm periodicity. After the concentration of flagella reached to 29 mg/ml, uniform birefringence can be observed. Fluorescence-labelled flagella indicate that the filaments are always in phase with respect to each other. With flagella display technique, 6 or 12 cysteine residues can be displayed on the surface of flagella (Kumara *et al.*, 2006). These purified recombinant flagella can form parallel bundles through disulphide bonds between neighbouring flagellar filaments. These bundles can reach up to 10 μm in length and 200 nm in diameter.

Pili

Pili, or fimbriae, are the hair-like appendages on Gram-negative bacteria (Fig. 6.3 right). They play an important role in cell recognition, colonization, biofilm formation, shielding and host subversion, motility and signalling events (Klemm and Schembri, 2000a). There are five classes of pili on bacteria: chaperone/usher (CU) pili, curli, type IV pili, type III secretion needle, and type IV secretion pili. The most widely distributed group is the pili synthesized by the CU pathway (Sauer *et al.*, 2004; Waksman and Hultgren, 2009). CU-assembled pilus is assembled by linear, unbranched polymers of several hundreds to thousands of monomers-pilins. The pilin subunits polymerize through non-covalent interactions in a head-to-tail fashion. Several pili can also be employed for surface display (Klemm and Schembri, 2000b; Samuelson *et al.*, 2002). However, unlike flagella, which can tolerate up to 302 displayed amino acids, the major and minor tip subunits of pili can only tolerate about 34 or 52 amino acids, respectively.

Bionanomaterials synthesis, assembly with their applications

Synthesis and assembly of bionanomaterials and organic–inorganic hybrid nanomaterials

Nanomaterials such as nanoparticles, nanorods, nanotubes and nanowires have gained widespread applications in nanotechnology due to their unique physical and chemical characteristics. In nature, there are a great variety of biological nanomaterials with different morphologies available. Recently, more and more of the biological materials have been employed for generating novel organic–inorganic hybrid nanomaterials, fabricating 2-D and 3-D hierarchal scaffolds in tissue engineering and combining with signal molecules in nanosensor, gene and drug delivery. Moreover, peptides or proteins play critical roles in controlling the biomineralization of inorganic nanostructures in natural systems. Combined with surface display technique, the bioengineered nanomaterials have been utilized for the fabrication and assembly of novel inorganic nanomaterials.

Assembling nanomaterials into well-defined structures and integrating them into large-scale devices are essential for their actual application. Most biomolecules exhibit sophisticated hierarchical nanostructures through precise intermolecular recognition between building blocks in the self-assembly of nanomaterials or between ordered scaffold and inorganic molecules in the synthesis of inorganic nanomaterials. In this way, 2-D or 3-D nanostructured materials with controlled size, shape, alignment and orientation can be synthesized.

Previous studies have shown that certain peptides with strong binding affinities to their targets can be selected by phage display screening (Bassindale *et al.*, 2007; Gaskin *et al.*, 2000; Rajesh *et al.*, 2002; Sarikaya *et al.*, 2003; Willats, 2002). These selected phages can be used as biotemplates for the synthesis and assembly of inorganic nanomaterials such as SiO_2, TiO_2, GeO_2, Cu_2O, Cr_2O_3, Fe_2O_3, PbO_2, $CaCO_3$, CdS, Ag, Au, Pt and Pd (Segvich and Kohn, 2009). Mao *et al.* (2003, 2004) identified several substrate-binding peptides: CNNPMHQNC to ZnS, SLTPLTTSIILRS to CdS, HNKHLPSTQPLA to FePt and CNAGDHANC to CoPt using pIII phage library. After these peptides were inserted to the N-terminus of pVIII proteins, an ordered arrangement of desired peptides was displayed on the surfaces of the M13 virion. Then, semiconductor and magnetic nanowires were successfully synthesized on these bioengineered phage templates. The biological template inside the inorganic nanomaterials can be further removed by annealing at 350–400°C forming single-crystalline nanowires (Fig. 6.4). By incubation of pre-synthesized inorganic nanoparticles with their corresponding bioengineered phage templates, the inorganic nanoparticles

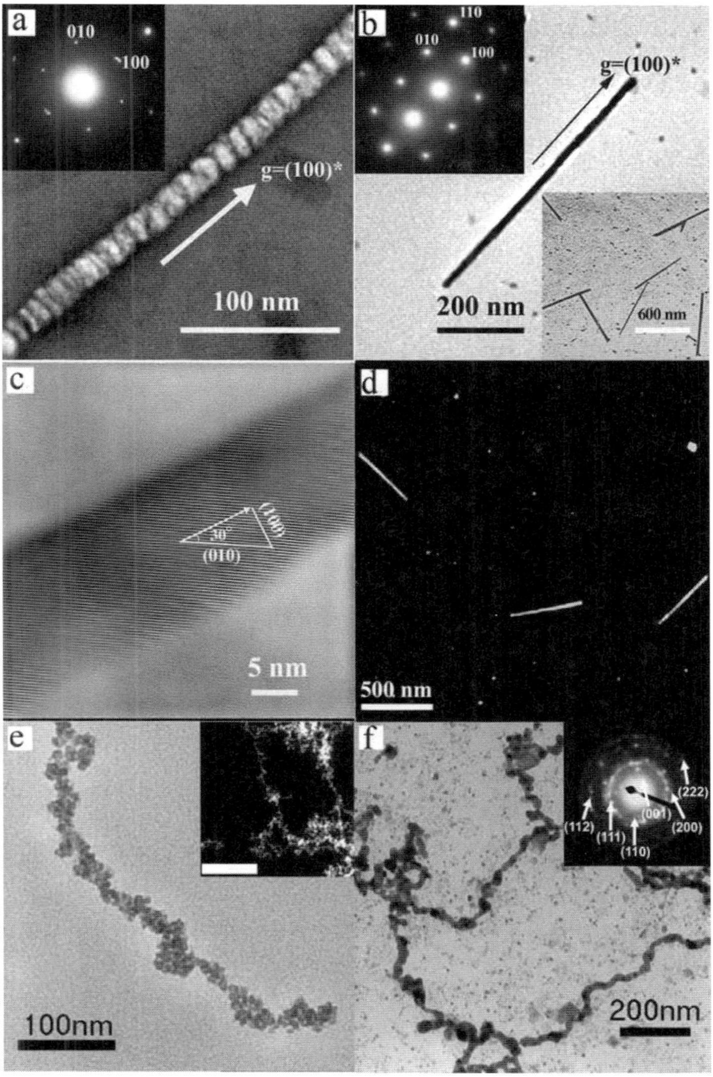

Figure 6.4 Electron microscopy images of ZnS, CdS, CoPt and FePt nanowires synthesized by phage templates. a) Dark-field diffraction contrast image of pre-annealed ZnS crystals nucleated and oriented along the phage template. b) An individual ZnS single-crystal nanowire formed after annealing. c) A lattice image of a single ZnS nanowire; 0.33nm corresponds to the (010) planes in wurtzite ZnS crystals. d) Image of CdS nanowires. e) CoPt nanowires. f) FePt nanowires. (Reproduced with permission from Mao *et al*., 2004.)

can be assembled on the existing phage templates. For example, peptides VSGSSPDS and $[\text{MHGKTQATSGTIQS}]_{3,}$ which bind specifically to the surface of gold due to a preferential binding of the rich hydroxyl groups, were selected using a type pVIII random phage library (Sarikaya *et al.*, 2003; So *et al.*, 2009). By incubation of the selected phage virions with pre-synthesized 5-nm gold nanoparticles (AuNP), AuNPs were aligned into an ordered 1-D array along the phage surfaces. Huang *et al.* (2005) demonstrated double displayed phage virion on the pVIII and pIII proteins with gold- and streptavidin-binding peptides and assembled Au and CdSe nanocrystals into ordered 1-D arrays. Lee *et al.* (2005) used bioengineered bacteriophage as a framework for the assembly of inorganic (Co–Pt) hybrid materials. A cobalt ion-binding peptide was identified by screening type pVIII phage library. Then, Co–Pt alloys were synthesized on the bioengineered phage scaffold. Owing to the presence of nanophase Co–Pt alloys, these hybrid nanocomposites exhibited extraordinary magnetic properties.

Wild-type phage can also be used to synthesize and assemble inorganic nanomaterials. Rh, Ru and Pd nanowires were synthesized on M13 phage by electrostatic interactions between cationic aqueous metal complexes and the exposed anionic carboxylate groups on phage pVIII subunit (Avery *et al.*, 2009). The organic–inorganic hybrid nanocomposites, which use genetically engineered bacteriophage as a template for the assembly of single and multiple material nanocrystals, have great potential application in the fabrication of nanodevices.

Phage can be further assembled into higher ordered structures. Liquid crystalline phase can be obtained at high concentration due to their monodispersity and rod-like shape. Moreover, by changing the virion concentration, the phases of liquid crystalline structures can be modulated [isotropic (<5 mg/ml), nematic (10–20 mg/ml), cholesteric (20–80 mg/ml), smectic A and C (>100 mg/ml)] (Dogic and Fraden, 2006; Merzylak and Lee, 2006). A high ordered ZnS nanocrystal film was fabricated on pIII engineered M13 virions through a liquid crystal self-assembly (Lee *et al.*, 2002). In this work, phage displaying a ZnS binding peptide (CNNPMHQNC) on pIII protein was incubated with $ZnCl_2$ and Na_2S solutions resulting in a virion–ZnS crystal liquid crystalline suspension. After slow evaporation of the suspension, a highly organized phage–ZnS film was obtained.

Yoo *et al.* (2006, 2008a) prepared a 2-D highly ordered phage film through the interdiffusion of two oppositely charged weak polyelectrolytes. First, negatively charged phages adsorbed on positively charged layer of linear polyethylenimine (LPEI) through electrostatic interaction. Next, LPEI and poly(acrylic acid) (PAA). Next, LPEI and PAA were repeatedly and alternately adsorbed on the top phage layer forming the thin LPEI–PAA–phage composite. Due to competitive electrostatic interactions and mutually repulsive interactions between neighbouring phage, an ordered monolayer of phage floated atop the cohesive polyelectrolyte multilayer induced by spontaneous repositioning of the phages. This kind of ordered nanostructure can be used for the assembly of inorganic nanoparticles as an ordered biological scaffold.

Bacterial flagella with surface display are also employed for the synthesis of nanomaterials. Muralidharan and co-workers did pioneer work for the synthesis of functional nanomaterials on genetically engineered flagella (Fig. 6.5) (Kumara *et al.*, 2007a,b). Several functional groups such as anionic, cationic, thiol and imidazole groups were displayed on surface of flagella. Various metal, inorganic and organic materials were then nucleated on surface of flagella resulting the formation of inorganic or organic nanotubes. Recently, bioengineered flagella were used as a template for the synthesis of pearl-necklace-like 1D silica nanotubes in aqueous solution and may be particularly useful for drug and gene delivery (Fig. 6.6) (Wang *et al.*, 2008).

Pili are also used for the synthesis of new nanomaterials or nanodevices. Yu *et al.* (2007)

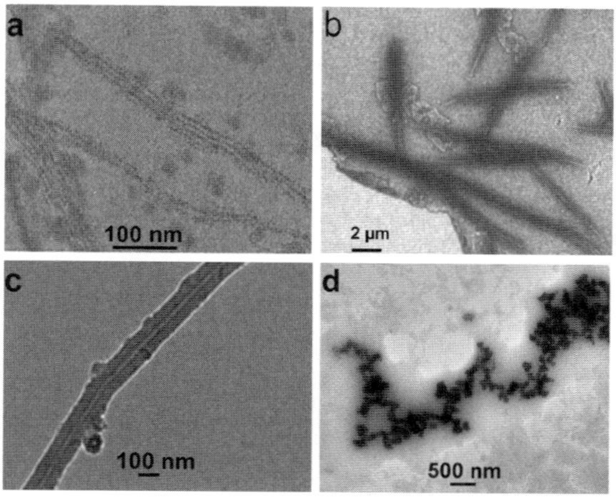

Figure 6.5 Self-Assembly of metal nanoparticles and nanotubes on bioengineered flagella scaffolds. (a) Gold nanoparticle–flagella nanotube composite materials obtained by reduction of Au(I). (b) Gold nanoparticle–flagellin nanotube composite materials prepared by covalent attachment of pre-made gold nanoparticles. (c) Copper–flagellin composite nanomaterials. Copper nanotubes synthesized by reduction of Cu(II) on histidine loop flagella. (d) Cobalt–flagella composite nanomaterials. Cobalt nanoparticles synthesized by reduction of Co(II) on histidine loop flagella. (Reproduced with permission from Kumara *et al.*, 2007).

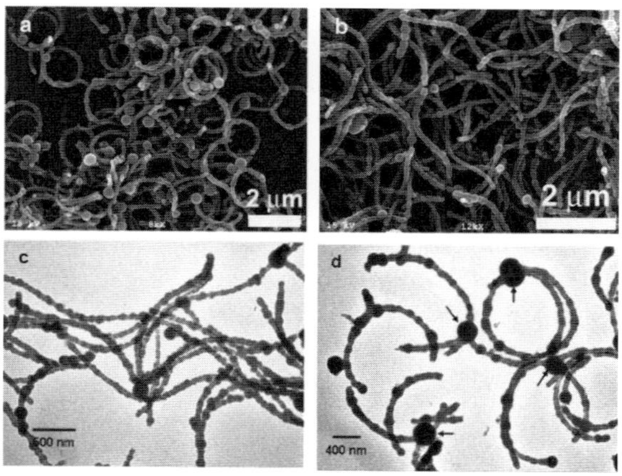

Figure 6.6 SEM and TEM images of the obtained hybrid flagella/silica nanotubes by using different flagella as template. (a, b) wild-type flagella. (b, c) Fla-KKCC flagella. (Reproduced with permission from Wang *et al.*, 2008.)

demonstrated a pilin-derived protein nanotube. Removal of the N-terminal region of the pilin's α-helix changes it into a highly soluble monomeric pilin protein. These monomeric proteins can be repolymerized into nanotube structures upon incubation with undecanethiol. The biofilms formed by pili have electrical conductivities that are comparable to those measured in conducting polymers (Qian and Li, 2011). Recently, the type 1 bacterial pili were self-assembled into highly ordered 1-D bundle, 2-D double-layer lattices, and 3-D multilayer lattices through molecular recognition in the presence of suitable inducers (Fig. 6.7). The self-assembled pili scaffold can mediate synthesis of inorganic nanomaterials as ordered biological template (Cao *et al.*, 2011).

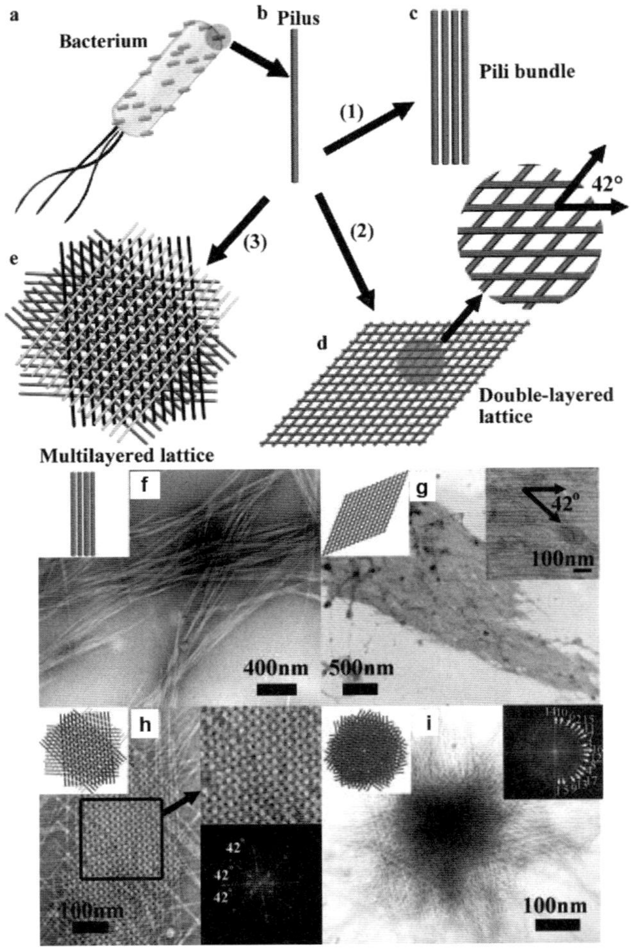

Figure 6.7 Self-assembly of pili into 1-D bundles, 2-D lattices, and 3-D lattices. (a) Illustration of a bacterial cell with many pili protruding from its surface. (b) An individual pilus. (c) Pili bundles. (d) Double-layer pili lattices. The two neighbouring layers of pili have a fixed angle of 428°. (e) Four-layer pili lattices. A red layer of parallel pili was formed, with subsequent assembly of a blue layer of pili on top of the red layer with a twist angle of 428°. After that, a black and a yellow layer are deposited in turn on top of the blue layer and black layer, with the neighbouring layer twisted at the same angle of 428°, resulting in the formation of a four-layer pili lattice. (f) TEM images of pili bundles. (g) A double-layer lattice with a fixed twist angle of 428° between two layers. (h) A four-layer pili lattice. FFT (inset, lower right) of selected area confirms the highly ordered structure and the twist angle between neighbouring layers. (i) A 17-layer pili lattice. FFT image (inset, upper right) confirms there are 17 layers of pili in the lattice with a fixed twist angle of 428° between adjacent layers. The spot corresponding to each layer is marked in the 2D-FFT image. The inset in the upper left corner of each TEM image illustrates the corresponding pili assemblies. (Reproduced with permission from Cao *et al.*, 2011.) A colour version of this figure is located in the plate section at the back of the book.

Biomaterials and tissue engineering

In material science, fabrication of complex structures self-assembled from nano-levelled components has attracted much attention due to their unique properties. Another field in which the self-assembly of biological materials may have very intriguing applications is in regenerative medicine or, more specifically, in tissue engineering (TE). It is well known that the purpose of TE is the development of biological substitutes to restore, maintain or improve tissue function or a whole organ. Designing biological scaffolds similar to their natural host is of interest to support, guide and stimulate the developing tissue. At the same time, degradation of the implanted biomaterial, formation of new extracellular matrix (ECM) components and replacing of defective tissues are also key areas of focus in TE.

Most of the time, naturally occurring biological materials can be purified or synthesized in large scale with low cost. These bionanomaterials can form higher ordered 2-D or 3-D structures under further supramolecular self-assembly. The amino acid composition of these nanomaterials makes them easy to be degraded. With surface display technique, both bioactive and cell behaviour directing signal peptides can be integrated within biological scaffold, which make them adaptable to their host environments.

In recent years, wild-type and genetically engineered bacteriophage displaying cell-recognizing peptides have been exploited in biomedical materials for biomaterials and TE (Kaur *et al.*, 2008, 2010a; Merzylak and Lee, 2006; Merzylak *et al.* 2009; Sanghvi *et al.*, 2005). As one of the most versatile assembly methods, layer-by-layer (LBL) assembly is often used for the construction of 3-D scaffolds. It involves the alternate deposition of complementarily charged species to fabricate multilayered nanomaterials. A series of phage-based thin films were prepared by LBL deposition. Recently, we reported the osteogenic differentiation of BMSCs on grooved films assembled from bioengineered bacteriophages (Zhu *et al.*, 2011). The films exhibited nearly parallel nanostructures separated by grooves, which can guide the elongation and alignment of BMSCs. After an osteogenic peptide (OGP) was displayed on phage, the proliferation and differentiation of BMSCs into osteoblasts was significantly accelerated (Fig. 6.8). Some other methods such as wet-spinning (Mao *et al.*, 2010), phage-based hydrogel (Souza *et al.*, 2006, 2008) solvent-assisted capillary moulding (Yoo *et al.*, 2008b) and electrospinning (Lee and Belcher, 2004) were also reported for the assembly of phage virions.

Merzlyak *et al.* (2009) reported that filamentous bacteriophage genetically engineered with arginine–glycine–aspartic acid (RGD), a tripeptide that binds to the integrin receptors present on all mammalian cells and IKVAV peptides, can support and guide the differentiation of neural progenitor cells. Wang and co-worker reported (Bruckmann *et al.*, 2007; Kaur *et al.*, 2008, 2010a,b; Yin *et al.*, 2008) osteogenic differentiation of bone marrow stem cells (BMSCs) based on wild-type, genetically or chemically modified TMV scaffold. Chung *et al.* (2011) recently reported biomimetic self-templating supramolecular structures based on M13 phages. Long-range-ordered, supramolecular films using M13 phage can be fabricated in a single-step, which show multiple levels of hierarchical organization and helical twist. The distinct supramolecular structures exhibit three kinds of structures: nematic orthogonal twists, cholesteric helical ribbons and smectic helicolidal nanofilaments. Moreover, using bioengineered M13 phages incorporated with bioactive peptide ligands can direct both soft and hard tissue growth in a hierarchically organized manner (Fig. 6.9).

Owing to their excellent mechanical toughness, flexibility, biodegradability and biocompatibility, silk proteins exhibit tremendous applications for TE. Both *in vivo* and *in vitro*

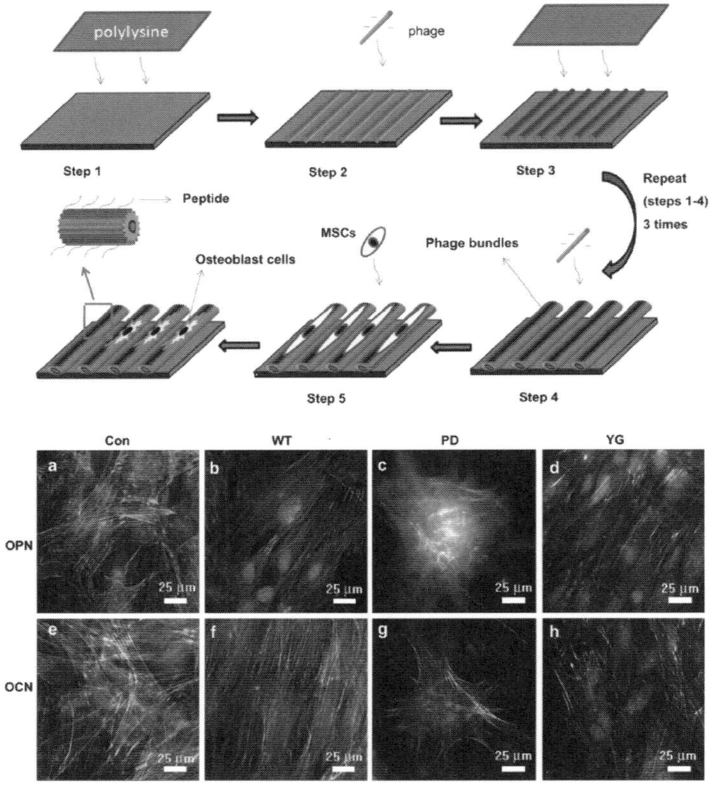

Figure 6.8 Schematic illustration (top) of the formation of the phage film by the layer-by-layer method and the MSCs growth on the phage film. After several layers were alternately deposited, the phage nanofibres assembled into parallel bundles separated by grooves. The BMSCs were oriented and elongated along the phage bundles. Fluorescent images (a–h) show the effect of phages with different surface chemistries on the differentiation of MSCs into osteoblasts. The phage film (b–d, f–h) show strong osteocalcin and osteopontin expression compared to those in the control groups (i.e. cells on glass slides coated with polylysine) (a, e). The phage film derived from phage displaying the OGP peptide (PDPLEPRREVCE) can stimulate MSCs aggregation to form a calcified nodule-like substance. Cell nuclei are stained by DAPI (blue) and F-actin is stained by FITC-labelled phalloidin (green). Osteopontin and osteocalcin are stained by rhodamine-labelled antibody (red). (Reproduced with permission from Zhu *et al.*, 2011.) A colour version of this figure is located in the plate section at the back of the book.

studies have confirmed that silk proteins are biocompatible and less inflammatory than other common biodegradable polymers such as poly(lactide) and collagen (Meinel *et al.*, 2004; Numata *et al.*, 2010). After the 3-D silk fibroin scaffold was loaded with bone morphogenetic protein-2 (BMP-2), it supported and induced osteogenic differentiation of human bone marrow stromal cells (hBMSCs) (Karageorgiou *et al.*, 2006). Silk-based 3-D scaffolds were also loaded with other small molecules such as Horseradish peroxidase (HRP) enzyme gradients to control cell and tissue outcomes, adenosine for refractory epilepsy treatments, nerve growth factor (NGF) to guide the sprouting of axons and to physically protect the axonal cone for peripheral nerve repair, and insulin-like growth factor I (IGF-I) for cartilage repair (Li *et al.*, 2009; Uebersax *et al.*, 2007; Vepari and Kaplan, 2006; Wilz *et al.*, 2008).

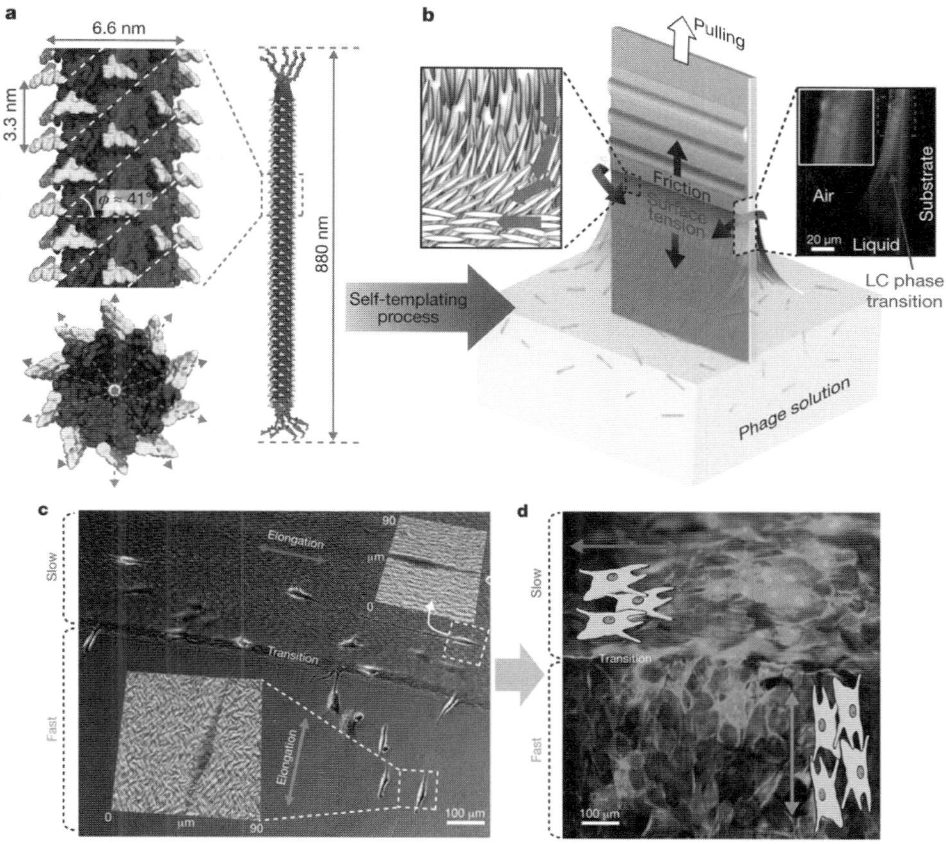

Figure 6.9 (a) Schematic illustration of the phage structure covered by about 2700 α-helical major coat (pVIII) protein subunits having five-fold helical symmetry and a two-fold screw rotation axis (red and green arrows). The phage has a helical angle (ϕ) of 41 with periodic subunit spacing of 3.3 nm. (b) Schematic illustration of the helical self-templating assembly of phage particles controlled by competing interfacial forces at the meniscus where liquid crystal phase transitions occur. The polarized optical microscopy image shows iridescent colours originating from liquid crystal phase formation at the air–liquid–solid interface. (c) Growth of biomimetic soft and hard tissue on the self-templated structures. Phase-contrast optical microscopy image of preosteoblast cells (MC3T3-E1) cultured on RGD phage films with SHN (top, slow pulling) and nematically ordered (bottom, fast pulling) structures. Insets: AFM images showing the directional guidance of cellular growth in response to the orientation of the underlying film microstructure. (d) Composite fluorescence image showing that the pre-osteoblast cells (MC-3T3 E1) grew to confluence (day 3) in the desired directions on the RGD phage films. This image was merged from two fluorescence images taken at two different focal planes. (Reproduced with permission from Chung *et al.*, 2011.) A colour version of this figure is located in the plate section at the back of the book.

Various studies also demonstrated TE applications of genetically engineered silk proteins (Bini *et al.*, 2006; Greish *et al.*, 2009; Huang *et al.*, 2007; Megeed *et al.*, 2004; Ritter *et al.*, 2002; Szela *et al.*, 2000; Wang *et al.*, 2006; Wong Po Foo *et al.*, 2006). After the *Nephila clavipes* consensus sequence from the major protein in the spider dragline silk was incorporated with other protein motifs with specific bioactivities such as dentin matrix protein (DMP) and bone sialoprotein (BSP), the chimeric proteins were shown to help accumulate calcium phosphate deposition on their surface and to be useful as novel scaffolds for hard tissue reparation (Gomes *et al.*, 2011a; Huang *et al.*, 2007). In both cases, the chimeric proteins keep their self-assembly properties. The fusion of the *N. clavipes* consensus sequence from the major protein in the spider dragline silk with antimicrobial peptides can reduce the risk of infection when biomaterials are implanted (Gomes *et al.*, 2011b).

Gene and drug delivery

The efficiency of gene and drug delivery can be highly improved using nanocarriers. Bacteriophage has been exploited in drug delivery systems as a powerful and efficient nanocarrier (Domingo-Espin *et al.*, 2011; Grasso and Santi, 2010; Petrenko and Jayanna, 2011; Mutuberria *et al.*, 2001). With phage display, a peptide that specifically binds almost any target can be selected. Moreover, the development of *in vivo* phage display has made it a more powerful tool for the identification of target-specific molecules in gene and drug delivery (Kelly *et al.* ,2006; Molek *et al.*, 2011; Newton and Deutscher, 2009). Viral capsids serve as excellent protective shells for the transport of encapsulated genes or drugs. After the ZnPc was entrapped into liposomes and anionic genetically engineered M13 phage, the self-assembly complex can be used as a novel carrier for targeted drug delivery (Ngweniform *et al.*, 2009).

Bioengineered silk proteins are also excellent candidates for gene and drug delivery due to their self-assembly properties, high mechanical toughness, flexibility, biodegradability and biocompatibility (Heim *et al.*, 2009; Numata and Kaplan, 2010b; Numata *et al.*, 2009). Moreover, silks proteins can be reconstructed into different formats depending on specific applications such as films, hydrogels, nanofibres and nanoparticles (Frandsen and Ghandehari, 2012). By adjusting the crystalline state of the silk proteins, the degradation rate and, thus, the release profile of carried molecules can be regulated (Numata and Kaplan, 2010). The microcapsules prepared by bioengineered dragline silk proteins from the spider *Araneus diadematus* were used for drug delivery at an emulsion interface (Hermanson *et al.*, 2007). The spider silk protein and poly(L-lysine) block copolymer are mixed with DNA to form polyelectrolyte complexes through electrostatic interactions, which are used for gene delivery (Numata *et al.*, 2009). The bioengineered spider silk proteins containing cell-penetrating peptides (CPPs, ppTG1) and their pDNA complexes were prepared for gene delivery to human embryonic kidney (HEK) cells (Numata and Kaplan, 2010). CPPs are short peptides that can efficiently penetrate cellular lipid bilayers or destabilize cellular membrane.

Future perspectives

In this chapter, we summarized some naturally occurring biological materials that undergo self-assembly and can be genetically modified. With genetic modification or surface display technique, the biological properties of these materials can be highly changed according to their specific applications. Nowadays, integrating multicomponent materials with nanoscale architectures and supramolecular self-assembly are adopted for various applications. The important applications including biomaterials synthesis, tissue engineering and gene/drug delivery are highlighted. In material science, preparing nanomaterial with self-assembly properties in a well-ordered fashion using traditional organic synthesis and lithographic patterning techniques is challenging. Despite the many recent advances in the construction of self-assembling materials *in vitro*, nature, with billions of years of evolution, has still created the most complex and functional examples.

While repairing damaged tissue with permanent implants or devices is an established approach to tissue engineering, regeneration of natural tissue is a more desirable approach. With the development of self-assembling biological materials, mimicking the natural nanostructures and composition becomes feasible. The recombinant or bioengineered nanostructures with the self-assembling features could be designed for highly specific functions that promote regenerative processes. Moreover, the designed functional biological materials might be able to self-assemble into nanostructures at the damaged sites through controlling the pH value, temperature or ion concentration when the molecular monomers are transported to the target sites, thus, providing a non-invasive pathway. Most biological materials also exhibit excellent biocompatibility and biodegradability.

For drug and gene delivery, a stimuli-responsive system can be fabricated using supramolecular self-assembled nanomaterials as carriers in which their structure and function can be manipulated by environmental stimulus. This property makes the biological nanomaterials effectively 'smart' and able to release the gene or drug specifically at the target sites.

With the development of biology, molecular biology and nanotechnology, more and more biological materials that can self-assemble are developed and exploited. With the incorporation of biofunctional domains into the emerging nanomaterials, their biocompatibility, biofunctionality and bioactivity can be greatly improved, leading to various applications in material science, supramolecular chemistry, nanomedicine and nanobiotechology.

Acknowledgements

We would like to thank the financial support from National Science Foundation (DMR-0847758, CBET-0854414, CBET-0854465, CMMI-1234957 and CBET-1229309), National Institutes of Health (5R01HL092526-02, 5R21EB009909-02, 4R03AR056848-03, 5R01DE015633-09, and 1R21EB015190-01A1), Oklahoma Center for the Advancement of Science and Technology (HR11-006), Oklahoma Center for Adult Stem Cell Research (434003) and Department of Defense Peer Reviewed Medical Research Program Discovery Award (W81XWH-12-1-0384). We also thank Christina Jenson for editing the manuscript.

References

Abbineni, G., and Mao, C. (2007). Design and applications of genetically engineered nanocomposites. In Nanotechnologies for the Life Sciences, Challa S.S.R. Kumar, ed. (Weinheim, Germany, Wiley), pp. 241–272.

Aldaye, F.A., Palmer, A.L., and Sleiman, H.F. (2003). Assembling materials with DNA as the guide. Science *321*, 1795–1799.

Altman, G.H., Diaz, F., Jakuba, C., Calabro, T., Horan, R.L., Chen, J., Lu, H., Richmond, J., and Kaplan, D.L. (2003). Silk-based biomaterials. Biomaterials *24*, 401–416.

Avery, K.N., Schaak, J.E., and Schaak, R.E. (2009). M13 Bacteriophage as a biological scaffold for magnetically-recoverable metal nanowire catalysts: combining specific and nonspecific interactions to design multifunctional nanocomposites. Chem. Mater. *21*, 2176–2178.

Bandy, T.J., Brewer, A., Burns, J.R., Marth, G., Nguyen, T., and Stulz, E. (2011). DNA as supramolecular scaffold for functional molecules: progress in DNA nanotechnology. Chem. Soc. Rev. *40*, 138–148.

Barry, E., Hensel, Z., Dogic, Z., Shribak, M., and Oldenbourg, R. (2006). Entropy-driven formation of a chiral liquid-crystalline phase of helical filaments. Phys. Rev. Lett. *96*, 018305.

Barthelat, F. (2007). Biomimetics for next generation materials. Phil. Trans. Roy. Soc. A *365*, 2907–2919.

Bassindale, A.R., Codina-Barrios, A., Frascione, N., and Taylor, P.G. (2007). An improved phage display methodology for inorganic nanoparticle fabrication. Chem. Commun. *28*, 2956–2958.

Bini, E., Foo, C.W.P., Huang, J., Karageorgiou, V., Kitchel, B., and Kaplan, D.L. (2006). RGD-functionalized bioengineered spider dragline silk biomaterial. Biomacromolecules *7*, 3139–3145.

Bruckman, M., Niu, Z., Li, S., Lee, L., Nelson, T., Lavigne, J., Wang, Q., and Varazo, K. (2007). Development of nanobiocomposite fibers by controlled assembly of rod-like tobacco mosaic virus. NanoBioTechnology *3*, 31–39.

Cao, B., and Mao, C. (2011). Filamentous phage-templated synthesis and assembly of inorganic nanomaterials. In Phage Nanobiotechnology, George P. Smith, ed. (London: The Royal Society of Chemistry), pp. 220–244.

Cao, B., Xu, H., and Mao, C. (2011). Controlled self-assembly of rodlike bacterial pili particles into ordered lattices. Angew. Chem. Int. Ed. *50*, 6264–6268.

Chung, W.-J., Oh, J.-W., Kwak, K., Lee, B.Y., Meyer, J., Wang, E., Hexemer, A., and Lee, S.-W. (2011). Biomimetic self-templating supramolecular structures. Nature *478*, 364–368.

Cui, H., Webber, M.J., and Stupp, S.I. (2010). Self-assembly of peptide amphiphiles: from molecules to nanostructures to biomaterials. Peptide Sci. *94*, 1–18.

Dedeo, M.T., Finley, D.T., and . Francis, M.B. (2011). Viral capsids as self-assembling templates for new materials. In H. Stefan (ed.) Progress in Molecular Biology and Translational Science (New York: Academic Press), pp. 353–392.

Dickerson, M.B., Sandhage, K.H., and Naik, R.R. (2008). Protein- and Peptide-directed syntheses of inorganic materials. Chem. Rev. *108*, 4935–4978.

Dogic, Z., and Fraden, S. (2006). Ordered phases of filamentous viruses. Curr. Opin. Colloid Interface Sci. *11*, 47–55.

Domingo-Espin, J., Unzueta, U., Saccardo, P., Rodriguez-Carmona, E., Corchero, J.L., Vazquez, E., and Ferrer-Miralles, N. (2011). Engineered biological entities for drug delivery and gene therapy protein nanoparticles. Prog. Mol. Biol. Transl. Sci. *104*, 247–298.

Frandsen, J.L., and Ghandehari, H. (2012). Recombinant protein-based polymers for advanced drug delivery. Chem. Soc. Rev. *41*, 2696–2706.

Frank, K., and Sippl, M.J. (2008). High-performance signal peptide prediction based on sequence alignment techniques. Bioinformatics *24*, 2172–2176.

Gaskin, D.J.H., Starck, K., and Vulfson, E.N. (2000). Identification of inorganic crystal-specific sequences using phage display combinatorial library of short peptides: a feasibility study. Biotechnol. Lett. *22*, 1211–1216.

George, A., and Ravindran, S. (2010). Protein templates in hard tissue engineering. Nano Today *5*, 254–266.

Goicochea, N.L., De, M., Rotello, V.M., Mukhopadhyay, S., and Dragnea, B. (2007). Core-like particles of an enveloped animal virus can self-assemble efficiently on artificial templates. Nano Lett. *7*, 2281–2290.

Gomes, S., Leonor, I.B., Mano, J.F., Reis, R.L., and Kaplan, D.L. (2011a) Spider silk-bone sialoprotein fusion proteins for bone tissue engineering. Soft Matter *7*, 4964–4973.

Gomes, S.C., Leonor, I.B., Mano, J.F., Reis, R.L., and Kaplan, D.L. (2011b) Antimicrobial functionalized genetically engineered spider silk. Biomaterials *32*, 4255–4266.

Gomes, S., Leonor, I.B., Mano, J.F., Reis, R.L., and Kaplan, D.L. (2012). Natural and genetically engineered proteins for tissue engineering. Prog. Polymer Sci. *37*, 1–17.

Grasso, S., and . Santi, L. (2010). Viral nanoparticles as macromolecular devices for new therapeutic and pharmaceutical approaches. Int. J. Physiol. Pathophysiol. Pharmacol. *2*,161–178.

Greish, K., Araki, K., Li, D., O'Malley, B.W., Dandu, R., Frandsen, J., Cappello, J., and Ghandehari, H. (2009). Silk-elastinlike protein polymer hydrogels for localized adenoviral gene therapy of head and neck tumors. Biomacromolecules *10*, 2183–2188.

Harrison, S.C. (1983). Virus structure: high-resolution perspectives. In A.L. Max and M. Karl (eds.) Advances in Virus Research (New York: Academic Press), pp. 175–240.

Hauser, C.A.E., and Zhang, S. (2010). Designer self-assembling peptide nanofiber biological materials. Chem. Soc. Rev. *39*, 2780–2790.

Heim, M., Keerl, D., and Scheibel, T. (2009). Spider silk: from soluble protein to extraordinary fiber. Angew. Chem. Int. Ed. *48*, 3584–3596.

Hermanson, K.D., Huemmerich, D., Scheibel, T., and Bausch, A.R. (2007). Engineered microcapsules fabricated from reconstituted spider silk. Advanc. Mater. *19*, 1810–1815.

Huang, Y., Chiang, C.-Y., Lee, S.K., Gao, Y., Hu, E.L., Yoreo, J.D., and Belcher, A.M. (2005). Programmable assembly of nanoarchitectures using genetically engineered viruses. Nano Lett. *5*, 1429–1434.

Huang, J., Wong, C., George, A., and Kaplan, D.L. (2007). The effect of genetically engineered spider silk-dentin matrix protein 1 chimeric protein on hydroxyapatite nucleation. Biomaterials *28*, 2358–2367.

Jin, H.-J., and Kaplan, D.L. (2003). Mechanism of silk processing in insects and spiders. Nature *424*, 1057–1061.

Karageorgiou, V., Tomkins, M., Fajardo, R., Meinel, L., Snyder, B., Wade, K., Chen, J., Vunjak-Novakovic, G., and Kaplan, D.L. (2006). Porous silk fibroin 3-D scaffolds for delivery of bone morphogenetic protein-2 in vitro and in vivo. J. Biomed. Mater. Res. A *78*, 324–334.

Kaur, G., Valarmathi, M.T., Potts, J.D., and Wang, Q. (2008). The promotion of osteoblastic differentiation of rat bone marrow stromal cells by a polyvalent plant mosaic virus. Biomaterials *29*, 4074–4081.

Kaur, G., Valarmathi, M.T., Potts, J.D., Jabbari, E., Sabo-Attwood, T., and Wang, Q. (2010a). The synergistic effects of multivalent ligand display and nanotopography on osteogenic differentiation of rat bone marrow stem cells. Biomaterials *31*, 5813–5824.

Kaur, G., Wang, C., Sun, J., and Wang, Q. (2010b). Regulation of osteogenic differentiation of rat bone marrow stromal cells on 2D nanorod substrates. Biomaterials *31*, 1732–1741.

Kelly, K.A., Nahrendorf, M., Yu, A.M., Reynolds, F., and Weissleder, R. (2006). In vivo phage display selection yields atherosclerotic plaque targeted peptides for imaging. Mol. Imaging Biol. *8*, 201–207.

Keten, S., Xu, Z., Ihle, B., and Buehler, M.J. (2010). Nanoconfinement controls stiffness, strength and mechanical toughness of [beta]-sheet crystals in silk. Nat. Mater. *9*, 359–367.

Klemm, P., and Schembri, M.A. (2000a) Bacterial adhesins: function and structure. Int. J. Med. Microbiol. *290*, 27–35.

Klemm, P., and Schembri, M.A. (2000b) Fimbrial surface display systems in bacteria: from vaccines to random libraries. Microbiology *146*, 3025–3032.

Kumara, M.T., Srividya, N., Muralidharan, S., and Tripp, B.C. (2006). Bioengineered flagella protein nanotubes with cysteine loops: self-assembly and manipulation in an optical trap. Nano Lett. *6*, 2121–2129.

Kumara, M.T., Muralidharan, S., and Tripp, B.C. (2007a). Generation and characterization of inorganic and organic nanotubes on bioengineered flagella of mesophilic bacteria. J. Nanosci. Nanotechnol. *7*, 2260–2272.

Kumara, M.T., Tripp, B.C., and Muralidharan, S. (2007b). Self-assembly of metal nanoparticles and nanotubes on bioengineered flagella scaffolds. Chem. Mater. *19*, 2056–2064.

Lee, L., Niu, Z., and Wang, Q. (2009). Viruses and virus-like protein assemblies – chemically programmable nanoscale building blocks. Nano Res. *2*, 349–364.

Lee, S.-W., and Belcher, A.M. (2004). Virus-based fabrication of micro- and nanofibers using electrospinning. Nano Lett. *4*, 387–390.

Lee, S.-W., Mao, C., Flynn, C.E., and Belcher, A.M. (2002). Ordering of quantum dots using genetically engineered viruses. Science *296*, 892–895.

Lee, S.-K., Yun, D.S., and Belcher, A.M. (2005). Cobalt ion mediated self-assembly of genetically engineered bacteriophage for biomimetic Co–Pt hybrid material. Biomacromolecules *7*, 14–17.

Li, T., Ren, G., Kaplan, D.L., and Boison, D. (2009). Human mesenchymal stem cell grafts engineered to release adenosine reduce chronic seizures in a mouse model of CA3-selective epileptogenesis. Epilepsy Res. *84*, 238–241.

Lin, Y., and Mao, C. (2011). Bio-inspired supramolecular self-assembly towards soft nanomaterials. Front. Mater. Sci. *5*, 247–265.

Lin, Y., Su, Z., Niu, Z., Li, S, Kaur, G., Lee, L.A., and Wang, Q. (2008). Layer-by-layer assembly of viral capsid for cell adhesion. Acta Biomater. *4*, 838–843.

Lu, Z., Murray, K.S., Cleave, V.V., LaVallie, E.R., Stahl, M.L., and McCoy, J.M. (1995). Expression of thioredoxin random peptide libraries on the *Escherichia coli* cell surface as functional fusions to flagellin: a system designed for exploring protein–protein interactions. Nat. Biotechnol. *13*, 366–372.

Mao, C., Flynn, C.E., Hayhurst, A., Sweeney, R., Qi, J., Georgiou, G., Iverson, B., and Belcher, A.M. (2003). Viral assembly of oriented quantum dot nanowires. Proc. Natl. Acad. Sci. U.S.A. *100*, 6946–6951.

Mao, C., Solis, D.J., Reiss, B.D., Kottmann, S.T., Sweeney, R.Y., Hayhurst, A., Georgiou, G., Iverson, B., and Belcher, A.M. (2004). Virus-based toolkit for the directed synthesis of magnetic and semiconducting nanowires. Science *303*, 213–217.

Mao, J.Y., Belcher, A.M., and Van Vliet, K.J. (2010). Antibacterial coatings: genetically engineered phage fibers and coatings for antibacterial applications. Advanc. Funct. Mater. 20(2).

Megeed, Z., Haider, M., Li, D., O'Malley Jr, B.W., Cappello, J., and Ghandehari, H. (2004). In vitro and in vivo evaluation of recombinant silk-elastinlike hydrogels for cancer gene therapy. J. Controlled Release *94*, 433–445.

Meinel, L., Karageorgiou, V., Hofmann, S., Fajardo, R., Snyder, B., Li, C., Zichner, L., Langer, R., Vunjak-Novakovic, G., and Kaplan, D.L. (2004). Engineering bone-like tissue in vitro using human bone marrow stem cells and silk scaffolds. J. Biomed. Mater. Res. A *71*, 25–34.

Merzlyak, A., and Lee, S.-W. (2006). Phage as templates for hybrid materials and mediators for nanomaterial synthesis. Curr. Opin. Chem. Biol. *10*, 246–252.

Merzlyak, A., Indrakanti, S., and Lee, S.-W. (2009). Genetically engineered nanofiber-like viruses for tissue regenerating materials. Nano Lett. *9*, 846–852.

Molek, P., Strukelj, B., and Bratkovic, T. (2011). Peptide phage display as a tool for drug discovery: targeting membrane receptors. Molecules *16*, 857–887.

Mutuberria, R., Arends, J.-W., Griffioen, A.W., and Hoogenboom, H.R. (2001). Phage display technology for target discovery in drug delivery research. In Drug Targeting (Weinheim: Wiley-VCH), pp. 255–273.

Nam, K.T., Kim, D.-W., Yoo, P.J., Chiang, C.-Y., Meethong, N., Hammond, P.T., Chiang, Y.-M., and Belcher, A.M. (2006). Virus-enabled synthesis and assembly of nanowires for lithium ion battery electrodes. Science 312, pp. 885–888.

Newton, J.R., and Deutscher, S.L. (2009). In vivo bacteriophage display for the discovery of novel peptide-based tumor-targeting agents. Methods Mol Biol. *504*, 275–290.

Ngweniform, P., Abbineni, G., Cao, B., and Mao, C. (2009). Self-assembly of drug-loaded liposomes on genetically engineered target-recognizing m13 phage: a novel nanocarrier for targeted drug delivery. Small *5*, 1963–1969.

Nielsen, H., and Krogh, A. (1998). Prediction of signal peptides and signal anchors by a hidden Markov model. Proc. Int. Conf. Intell. Syst. Mol. Biol. *6*, 122–130.

Numata, K., and Kaplan, D.L. (2010a). Silk-based delivery systems of bioactive molecules. Advanc. Drug Delivery Rev. *62*, 1497–1508.

Numata, K., and Kaplan, D.L. (2010b). Silk-based gene carriers with cell membrane destabilizing peptides. Biomacromolecules *11*, 3189–3195.

Numata, K., Subramanian, B., Currie, H.A., and Kaplan, D.L. (2009). Bioengineered silk protein-based gene delivery systems. Biomaterials *30*, 5775–5784.

Numata, K., Hamasaki, J., Subramanian, B., and Kaplan, D.L. (2010). Gene delivery mediated by recombinant silk proteins containing cationic and cell binding motifs. Journal of Controlled Release *146*, 136–143.

Pande, J., Szewczyk, M.M., and Grover, A.K. (2010). Phage display: concept, innovations, applications and future. Biotechnol. Advances, 2010. *28*, 849–858.

Petrenko, V. (2008). Evolution of phage display: from bioactive peptides to bioselective nanomaterials. Exp. Opin. Drug Delivery *5*, 825–836.

Petrenko, V.A., and Jayanna, P.K. (2011). Phage-mediated drug delivery. In Phage Nanobiotechnology, George P. Smith, ed. (London: The Royal Society of Chemistry), pp. 55–82.

Qian, F., and Li, Y. (2011). Biomaterials: a natural source of nanowires. Nat. Nanotechnol. *6*, 538–539.

Rajesh, R.N., Lawrence, L.B., Stephen, J.C., and Morley, O.S. (2002). Silica-precipitating peptides isolated from a combinatorial phage display peptide library. J. Nanosci. Nanotechnol. i, 95–100.

Rammensee, S., Slotta, U., Scheibel, T., and Bausch, A.R. (2008). Assembly mechanism of recombinant spider silk proteins. Proc. Natl. Acad. Sci. U.S.A. *105*, 6590–6595.

Rittner, K., Benavente, A., Bompard-Sorlet, A., Heitz, F., Divita, G., Brasseur, R., and Jacobs, E. (2002). New basic membrane-destabilizing peptides for plasmid-based gene delivery in vitro and in vivo. Mol. Ther. *5*, 104–114.

Russo, D., Ollivier, J., and Teixeira, J. (2008). Water hydrogen bond analysis on hydrophilic and hydrophobic biomolecule sites. Phys. Chem. Chem. Phys. *10*, 4968–4974.

Sadhasivam, S., and Yun, K. (2010). DNA self-assembly: prospectus and its future application. J. Mater. Sci. *45*, 2543–2552.

Samuelson, P., Gunneriusson, E., Nygren, P.-Å., and Ståhl, S. (2002). Display of proteins on bacteria. J. Biotechnol. *96*, 129–154.

Sanchez, C., Arribart, H., and Giraud Guille, M.M. (2005). Biomimetism and bioinspiration as tools for the design of innovative materials and systems. Nat. Mater. *4*, 277–288.

Sanghvi, A.B., Miller, K.P.H., Belcher, A.M., and Schmidt, C.E. (2005). Biomaterials functionalization using a novel peptide that selectively binds to a conducting polymer. Nat. Mater. *4*, 496–502.

Sarikaya, M., Tamerler, C., Jen, A.K.Y., Schulten, K., and Baneyx, F. (2003). Molecular biomimetics: nanotechnology through biology. Nat Mater. *2*, 577–585.

Sauer, F.G., Remaut, H., Hultgren, S.J., and Waksman, G. (2004). Fiber assembly by the chaperone – usher pathway. Biochim. Biophys. Acta Mol.Cell Res. *1694*, 259–267.

Segvich, S., and Kohn, D.H. (2009). Phage display as a strategy for designing organic/inorganic biomaterials. In D.A. Puleo and R. Bizios (eds.) Biological Interactions on Materials Surfaces (New York, Springer), pp. 115–132.

Smith, G.P. (2011). The phage nanoparticle toolkit. In Phage Nanobiotechnology (London: The Royal Society of Chemistry), pp. 1–11.

So, C.R., Kulp, J.L., Oren, E.E., Zareie, H., Tamerler, C., Evans, J.S., and Sarikaya, M. (2009). Molecular recognition and supramolecular self-assembly of a genetically engineered gold binding peptide on Au{111}. ACS Nano. *3*, 1525–1531.

Souza, G.R., Christianson, D.R., Staquicini, F.I., Ozawa, M.G., Snyder, E.Y., Sidman, R.L., Miller, J.H., Arap, W., and Pasqualini, R. (2006). Networks of gold nanoparticles and bacteriophage as biological sensors and cell-targeting agents. Proc. Natl. Acad. Sci. U.S.A. *103*, 1215–1220.

Souza, G.R., Yonel-Gumruk, E., Fan, D., Easley, J., Rangel, R., Guzman-Rojas, L., Miller, J.H., Arap, W., and Pasqualini, R. (2008). Bottom-up assembly of hydrogels from bacteriophage and au nanoparticles: the effect of cis- and trans-acting factors. PLoS ONE *3*, e2242.

Szela, S., Avtges, P., Valluzzi, R., Winkler, S., Wilson, D., Kirschner, D., and Kaplan, D.L. (2000). Reduction–oxidation control of beta-sheet assembly in genetically engineered silk. Biomacromolecules *1*, 534–542.

Tamerler, C., Kacar, T., Sahin, D., Fong, H., and Sarikaya, M. (2007). Genetically engineered polypeptides for inorganics: a utility in biological materials science and engineering. Materi. Sci. Eng. C *27*, 558–564.

Uebersax, L., Mattotti, M., Papaloïzos, M., Merkle, H.P., Gander, B., and Meinel, L. (2007). Silk fibroin matrices for the controlled release of nerve growth factor (NGF). Biomaterials *28*, 4449–4460.

Van Gerven, N., Waksman, G., and Remaut, H. (2011). Pili and flagella biology, structure, and biotechnological applications. Prog. Mol. Biol. Transl. Sci. *103*, 21–72.

Vepari, C.P., and Kaplan, D.L. (2006). Covalently immobilized enzyme gradients within three-dimensional porous scaffolds. Biotechnol. Bioeng. *93*, 1130–1137.

Vollrath, F., and Knight, D.P. (2001). Liquid crystalline spinning of spider silk. Nature *410*, 541–548.

Waksman, G., and Hultgren, S.J. (2009). Structural biology of the chaperone-usher pathway of pilus biogenesis. Nat. Rev. Microbiol. *7*, 765–774.

Wang, Y., Kim, H.-J., Vunjak-Novakovic, G., and Kaplan, D.L. (2006). Stem cell-based tissue engineering with silk biomaterials. Biomaterials *27*, 6064–6082.

Wang, F., Li, D., and Mao, C. (2008). Genetically modifiable flagella as templates for silica fibres: from hybrid nanotubes to 1D periodic nanohole arrays. Advanc. Funct. Mater. *18*, 4007–4013.

Westerlund-Wikström, B. (2000). Peptide display on bacterial flagella: Principles and applications. Int. J. Med. Microbiol. *290*, 223–230.

Westerlund-Wikström, B., Tanskanen, J., Virkola, R., Hacker, J., Lindberg, M., Skurnik, M., and Korhonen, T.K. (1997). Functional expression of adhesive peptides as fusions to *Escherichia coli* flagellin. Protein Eng. *10*, 1319–1326.

Willats, W.G.T. (2002). Phage display: practicalities and prospects. Plant Mol. Biol. *50*, 837–854.

Wilz, A., Pritchard, E.M., Li, T., Lan, J.-Q., Kaplan, D.L., and Boison, D. (2008). Silk polymer-based adenosine release: therapeutic potential for epilepsy. Biomaterials *29*, 3609–3616.

Wong Po Foo, C., Patwardhan, S.V., Belton, D.J., Kitchel, B., Anastasiades, D., Huang, J., Naik, R.R., Perry, C.C., and Kaplan, D.L. (2006). Novel nanocomposites from spider silk–silica fusion (chimeric) proteins. Proc. Natl. Acad. Sci. U.S.A. *103*, 9428–9433.

Yanlian, Y., Ulung, K., Xiumei, W., Horii, A., Yokoi, H., and Shuguang, Z. (2009). Designer self-assembling peptide nanomaterials. Nanotechnol. Today *4*, 193–210.

Yonekura, K., Maki-Yonekura, S., and Namba, K. (2003). Complete atomic model of the bacterial flagellar filament by electron cryomicroscopy. Nature *424*, 643–650.

Yoo, P.J., Nam, K.T., Qi, J., Lee, S.-K., Park, J., Belcher, A.M., and Hammond, P.T. (2006). Spontaneous assembly of viruses on multilayered polymer surfaces. Nat. Mater. *5*, 234–240.

Yoo, P.J., Zacharia, N.S., Doh, J., Nam, K.T., Belcher, A.M., and Hammond, P.T. (2008a). Controlling surface mobility in interdiffusing polyelectrolyte multilayers. ACS Nanotechnol. *2*, 561–571.

Yoo, P.J., Nam, K.T., Belcher, A.M., and Hammond, P.T. (2008b). Solvent-assisted patterning of polyelectrolyte multilayers and selective deposition of virus assemblies. Nano Lett. *8*, 1081–1089.

Young, M., Debbie, W., Uchida, M., and Douglas, T. (2008). Plant viruses as biotemplates for materials and their use in nanotechnology. Annu. Rev. Phytopathol. *46*, 361–384.

Yu, B., Giltner, C.L., Van Schaik, E.J., Bautista, D.L., Hodges, R.S., Audette, G.F., Li, D.Y., and Irvin, R.T. (2007). A novel biometallic interface: high affinity tip-associatedbinding by pilin-derived protein nanotubes. J. Biosci. *1*, 73–83.

Zayed, J.M., Nouvel, N., Rauwald, U., and Scherman, O.A. (2010). Chemical complexity-supramolecular self-assembly of synthetic and biological building blocks in water. Chem. Soc. Rev. *39*, 2806–2816.

Zelzer, M., and Ulijn, R.V. (2010). Next-generation peptide nanomaterials: molecular networks, interfaces and supramolecular functionality. Chem. Soc. Rev. *39*, 3351–3357.

Zhou, Y., and Yan, D. (2009). Supramolecular self-assembly of amphiphilic hyperbranched polymers at all scales and dimensions: progress, characteristics and perspectives. Chem. Commun. *10*, 1172–1188.

Zhu, H., Cao, B., Zhen, Z., Laxmi, A.A., Li, D., Liu, S., and Mao, C. (2011). Controlled growth and differentiation of MSCs on grooved films assembled from monodisperse biological nanofibers with genetically tunable surface chemistries. Biomaterials *32*, 4744–4752.

Virus-like Particles 7

Rob Noad and Polly Roy

Abstract

Virus-like particles (VLPs) are self-assembling nanoparticles that mimic viruses. They have been used as highly immunogenic, safe, vaccines; as carriers for antigen and epitope display; and for the delivery of small molecules to cells. Formed from the proteins that normally make up the structure of the virus particle, VLPs often have characteristics that are indistinguishable from the parent virus. In addition, the ability to form VLPs appears to be an intrinsic property of many, if not most, viral structural proteins. Given that viruses are ubiquitous in the environment, the VLP approach offers the potential to match particle characteristics to biotechnological application. This chapter reviews the current state of VLP research with a focus on the diverse applications for which VLPs have been used and the potential for further development of the technology.

Introduction

On viruses

Viruses are autonomously replicating nucleic acids that rely on the machinery of the host cell for the production of protein, and sometimes other functions, including transcription and replication of the virus genome. They are distinguished from other 'parasitic' nucleic acids by the presence of a particle that protects the viral genome from degradation and facilitates its movement from cell-to-cell and host-to-host. In this respect, virus particles are already fully functioning nanomachines with highly specific targeting to particular host cells, they are adapted to survival in the environment, and in many cases can be extremely stable. Although viruses are arguably the most widely distributed pathogens on the planet, infecting bacteria, archea, plants, animals and fungi (and other viruses) the structure of all viruses can be simplified to a few essential components. All virus particles contain the viral genome (which can be either RNA or DNA) and a protein coat (which can be present in one or multiple layers and encloses the genome). Some viruses also additionally have enzymatic proteins that are minor structural components but are essential to processing of the protein coat or replication and transcription of the viral genome and some also have a lipid bilayer in which virally encoded proteins are present (Fig. 7.1).

Viral genomes vary greatly in size, from less than 2000 bases to over 1 Mbp (Hamel *et al.*, 1998; Colson *et al.*, 2011), however most viruses that are pathogenic in plants and animals have genomes of less than 300 kbp of DNA or fewer than 30,000 nucleotides of RNA (Fields *et al.*, 2007). Crick and Watson pointed out in 1956 that, particularly for viruses with small

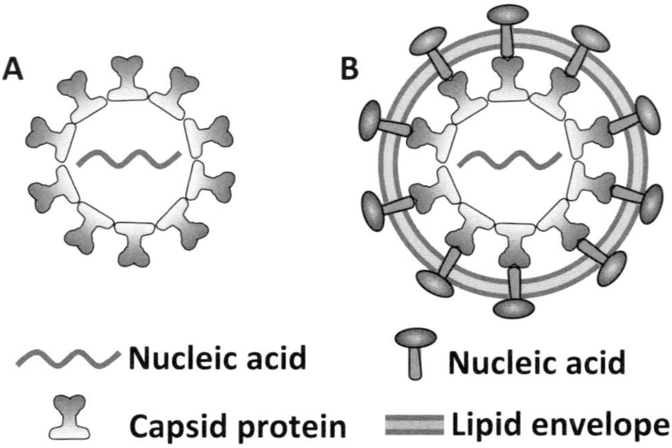

Figure 7.1 Generic structure of icosahedral non-enveloped (A) and enveloped (B) virus. The viral nucleic acid is enclosed by a highly repetitive icosahedral capsid formed by one or more virally encoded proteins. This capsid serves to protect the nucleic acid from degradation and to transport it from cell-to-cell and host-to-host. In the case of enveloped viruses the nucleocapsid (nucleic acid + capsid) is enclosed by a lipid bilayer in which one or more virally coded proteins is embedded. These proteins mediate specificity of interaction between the virus and the host cell and allow fusion of the viral and cellular membranes to facilitate virus entry.

RNA genomes, the only way for the nucleic acid to encode sufficient protein to enclose the genome is for the virus particle to be built of repeating subunits of a small number of proteins (Crick and Watson, 1956). Since then, this suggestion has been supported by hundreds of studies on the structure of virus particles (Fields *et al.*, 2007). In some cases a single protein is sufficient to serve as the major structural component of the capsid of the virus (McAleer *et al.*, 1984; Jiang *et al.*, 1992; Kirnbauer *et al.*, 1992), indeed for most RNA viruses four proteins or less are sufficient (Noad and Roy, 2003). This is also true for viruses with small DNA genomes (Montross *et al.*, 1991; Kirnbauer *et al.*, 1992; Forstova *et al.*, 1993; Nawagitgul *et al.*, 2000), although it should be noted that viruses with large (> 30 kb) DNA genomes can have more complex capsid architectures (Rohrmann, 1992; Fields *et al.*, 2007; Sathish *et al.*, 2012).

Ultrastructural studies of virus particles from cell culture and field isolates had revealed that the presence of 'empty particles' was a common feature of infections with some viruses. These particles were indistinguishable from authentic virus particles apart from the fact that they were apparently lacking the genetic material that is normally present. In the late 1980s the use of recombinant DNA technologies to express the viral proteins that form the majority the particle for some simple viruses demonstrated that these proteins have the intrinsic ability to self-assemble into 'empty' virus-like particles (VLPs) (McAleer *et al.*, 1984; French *et al.*, 1990; French and Roy, 1990; Kirnbauer *et al.*, 1992). Although the initial observation was for particles formed from a single capsid protein (McAleer *et al.*, 1984), this was quickly followed by the demonstration that more complex particles formed from two, three or four separate proteins would self-assembly in the absence of any virus scaffold proteins (French *et al.*, 1990; French and Roy, 1990; Crawford *et al.*, 1994). Indeed, formation of VLPs is not only a property of viruses which lack a lipid envelope, as VLPs have now been demonstrated for a broad range of viruses from many different virus families (Noad and Roy, 2003; Roy and Noad, 2008). We would argue that there is now sufficient evidence to suggest that the potential to form VLPs is an inherent property of viruses, and although there are some viruses that form VLPs relatively inefficiently because the nucleic acid is itself a structural component of the particle, or the particle requires cell specific post-translational processing, most viruses will form VLPs under the right conditions.

For the purpose of this chapter, VLPs will be used to refer to non-replicating structures derived by the recombinant expression of viral proteins in a heterologous system. We will not discuss, replicating or partially replicating virus-derived genomes, which we would regard as replicons, or disabled virus, rather than VLPs.

Virus-like particles as immunogens

Direct application of particles as vaccines

Problems with the use of recombinant protein as a basis for vaccination often stem from poor dose response relative to other types of vaccine. One of the reasons for this is that not all the recombinant protein is necessarily in the authentic conformation to elicit an immune response. In addition, the immune system is less sensitive to single proteins and peptides than it is to larger invading pathogens, such as viruses and bacteria. Subunit vaccines based on VLPs overcome these problems because only correctly folded protein is assembled into particles. Furthermore, the size range of VLPs mimics the size of the virus from which the structural genes are derived, and is recognized by the immune system as potentially an invading pathogen, and readily taken up into dendritic cells (Bachmann *et al.*, 1996; Rudolf *et al.*, 2001; Moron *et al.*, 2002, 2003; Ruedl *et al.*, 2002; Fausch *et al.*, 2003; Tsunetsugu-Yokota *et al.*, 2003; Bosio *et al.*, 2004; Fifis *et al.*, 2004; Freyschmidt *et al.*, 2004; Yang *et al.*, 2004; Zhang *et al.*, 2004; Arico *et al.*, 2005; Barth *et al.*, 2005; Lenz *et al.*, 2005; Buonaguro *et al.*, 2006; Garcia-Pineres *et al.*, 2006; Gedvilaite *et al.*, 2006; Ye *et al.*, 2006; Istrate *et al.*, 2007; Tegerstedt *et al.*, 2007; Ding *et al.*, 2010; Song *et al.*, 2010; Chang *et al.*, 2011; Win *et al.*, 2011; Link *et al.*, 2012).

An example where incorporation of a protein into a VLP enhances its immunogenicity is the case of Bluetongue virus VP2. This protein is the serotype determinant of a midge-transmitted ruminant pathogen that has recently resulted in major economic losses in Europe (Roy *et al.*, 2009; Velthuis *et al.*, 2009; Eschbaumer *et al.*, 2010). The structure of the BTV particle is complex, with 4 major capsid proteins (VP2, VP3, VP5, VP7) arranged in layers, three minor structural proteins (VP1, VP4, VP6), and a segmented double-stranded RNA genome (Fig. 7.2). VP2 is the most exposed capsid protein on the mature virion, although significantly neither VP2 alone, nor VP2 plus the other outer capsid protein, VP5, is able to form VLPs without the scaffold provided by VP3 and VP7. Sheep vaccinated with 100 μg recombinant VP2 produced in insect cells using the baculovirus expression system were partially protected from challenge with virulent BTV (Roy *et al.*, 1990). Combining 50 μg of VP2 with 25 μg of VP5 enhanced its immunogenicity to the point that all vaccinated animals were protected. In contrast, when VP2 was incorporated into a VLP consisting of all four major structural proteins of BTV, the equivalent of 1–2 μg of VP2 was sufficient to completely protect sheep against virulent virus challenge (Roy *et al.*, 1990). Subsequent studies have confirmed that BTV VLPs containing VP2 afford high level long term protection

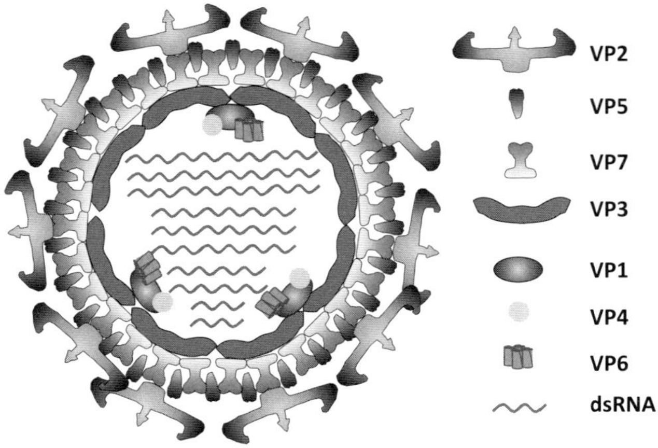

Figure 7.2 Cartoon of the structure of bluetongue virus. The mature virus particle consists of concentric layers of VP3, VP7, VP5 and VP2. In the centre of the particle there are 10 different segments of double stranded RNA (dsRNA) which make up the virus genome, and for each of these segments a transcriptase complex consisting of the RNA-dependent-RNA polymerase, VP1, a dimer of the capping enzyme, VP4, and a hexamer of the helicase, VP6. The transcriptase complex and the RNA genome are dispensable for the formation of virus-like particles which can be formed by co-expression of VP3 and VP7 or by expression of VP3, VP7, VP5 and VP2.

against virulent virus challenge in sheep (Roy *et al.*, 1992, 1994; Stewart *et al.*, 2009, 2012; Perez de Diego *et al.*, 2011).

The effect of particle assembly on immunogenicity has been explored for other VLP-based immunogens and is broadly in line with what was described for BTV VP2. For human papillomavirus, 20–40 times more L1 capsomer is required to give an equivalent immune response to the same protein assembled as VLP (Thones *et al.*, 2008). For influenza A, 600 ng VLPs containing the viral HA, NA and M1 proteins elicited serum anti-HA antibodies that were one $\log_{10}$ higher than 600 ng of recombinant HA alone (Bright *et al.*, 2007).

We and others have previously reviewed the wide range of different virus particles that have been mimicked with the VLP approach (Noad and Roy, 2003; Grgacic and Anderson, 2006; Roy and Noad, 2009). However it is worth noting that many of the major viral pathogens of humans have been produced as VLPs, including influenza A (Latham and Galarza, 2001; Galarza *et al.*, 2005), rotavirus (Labbe *et al.*, 1991; Crawford *et al.*, 1994), caliciviruses (Jiang *et al.*, 1992), HIV (Yao *et al.*, 2003; Young *et al.*, 2006), filoviruses (Noda *et al.*, 2002; Warfield *et al.*, 2003; Bosio *et al.*, 2004; Swenson *et al.*, 2004), hepatitis B, C and E viruses (McAleer *et al.*, 1984; Li *et al.*, 1997; Wu *et al.*, 1997; Baumert *et al.*, 1999; Wang *et al.*, 2009), bunyaviruses (Liu *et al.*, 2008), hantaviruses (Betenbaugh *et al.*, 1995; Li *et al.*, 2010) flaviviruses (Konishi *et al.*, 1992; Allison *et al.*, 1995; Ferlenghi *et al.*, 2001), paramyxoviruses (Patch *et al.*, 2007; Walpita *et al.*, 2011), alphaviruses (Akahata *et al.*, 2010) and papillomavirus (Kirnbauer *et al.*, 1992; Kirnbauer *et al.*, 1993). Although VLPs for some of these examples do not include all of the structural proteins found in native virions, they do all self-assemble into highly repetitive non-replicating particulate structures in the 20–150 nm size range. Most VLP vaccines are currently at a preclinical phase or undergoing clinical trials, however VLP vaccines to human papillomavirus (Cervarix, GSK and

Gardasil, Merck), hepatitis B virus (Recombivax HB, Merck; Engerix B, GSK; Bio-HepB, Biotechnology General; Sci-B-Vac, SciGen) and porcine circovirus 2 (Ingelvax, Boehringer Ingelheim Vetmedica) are licensed and commercially available.

VLPs as carriers for other epitopes

In addition to acting as highly immunogenic vaccines for the cognate virus, VLPs have been demonstrated to be effective ways to display antigen and epitopes for other pathogens and diseases to the immune system. The underlying rationale to this is that VLP are highly repetitive structures (each VLP would normally have a minimum of 60 copies of capsid protein, depending on the virus) that are readily taken up by dendritic cells, and any cell taking up VLP receives a concentrated dose of any antigen linked to the capsid protein. Therefore, linking otherwise poorly immunogenic peptides or other antigens to the capsid will enhance its immunogenicity. In these instances the VLP serves as a scaffold for the presentation of the foreign antigen. There is good evidence that many VLPs have an inherent adjuvant function (Harris *et al.*, 1992; Roy *et al.*, 1992; Layton *et al.*, 1993; Lo-Man *et al.*, 1998; Niikura *et al.*, 2002; Bright *et al.*, 2007; Denis *et al.*, 2008; Savard *et al.*, 2012) that makes them ideal for this purpose. An advantage of this approach is that it uncouples the VLP and the antigenic target, without losing all the excellent immunogenic properties elicited by the particle. Using this approach VLPs have been engineered that target other viruses (Adams *et al.*, 1987; Harris *et al.*, 1992; von Brunn *et al.*, 1993; Sedlik *et al.*, 1997; Adler *et al.*, 1998; Peng *et al.*, 1998; Crisci *et al.*, 2012), non-viral pathogens (Schodel *et al.*, 1994; Oliveira-Ferreira *et al.*, 2000) and mammalian antigens including those linked to cancer and Alzheimer's disease (Tegerstedt *et al.*, 2005; Chackerian, 2010; Pejawar-Gaddy *et al.*, 2010; Cubas *et al.*, 2011).

The precise strategy for insertion of foreign epitopes is dependent on the type of VLPs, however they can be divided into three broad classes (Fig. 7.3). By far the most common approach to the display of epitopes on the surface of viruses and VLPs is the incorporation of the epitope into an exposed loop on the surface of the capsid protein. This is usually based on an understanding of the structure of the capsid protein in the mature particle informed by structural models. Success of the approach is dependent on the VLP retaining the ability to successfully self-assemble in the presence of the modified loop and the ability of the antigen/epitope to retain any structure necessary for immunogenicity when inserted into the capsid protein. In general, this approach is most successful when short, linear epitopes in target proteins have already been identified but are insufficiently immunogenic on their own to stimulate a strong immune response.

Some capsid proteins retain the ability to self-assemble even when relatively large protein sequences are inserted. For example, the particle formed by hepatitis B core antigen (183 amino acids) will form even with green fluorescent protein (GFP, 238 amino acids), flanked by flexible glycine-serine linkers, inserted into a major immunodominant loop (Kratz *et al.*, 1999). However, GFP has a structure that might be expected to facilitate surface display on particles, both the amino (N) and the carboxyl (C) termini of GFP are solvent exposed and are on the same side of the correctly folded protein (Ormo *et al.*, 1996). Other large antigens have been successfully expressed on the Hepatitis B core antigen particle, but in some cases there is clear disruption of the efficiency of core assembly (Vogel *et al.*, 2005b; Nassal *et al.*, 2008). Detailed investigation of this particular particle has suggested that the sequence and solubility of the inserted polypeptide is an important determinant of potential particle

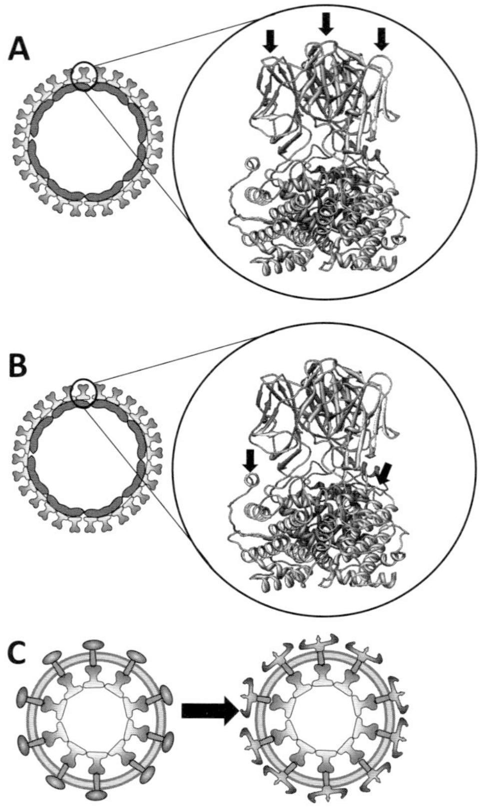

Figure 7.3 Strategies for foreign epitope display from VLPs without chemical modification. These can be summarized as three basic approaches: the insertion of foreign epitopes into an exposed, immunogenic loop on the outer surface of the capsid (A); the insertion of epitopes onto the end (usually the carboxy terminus) of the capsid protein (B); the display of foreign antigen on the surface of enveloped virus particles by generation of chimeric particles in which there is a mismatch between the capsid and the envelope protein (C).

formation (Karpenko *et al.*, 2000). Given the need to preserve the ability of the capsid protein to self-assemble these characteristics are likely to be general features of other VLP based loop display approaches. Although different particles may have more or less flexibility in the amount of foreign peptide sequence that will be accommodated the need for correct folding and subunit–subunit interaction of the capsid protein multimers are the key determinants in VLP assembly.

The second widely explored strategy for the display of epitopes on the surface of VLPs involves the fusion of the foreign sequence to the termini, more often the carboxyl terminus, of the capsid protein. The rationale here is that since the foreign epitope is anchored only at one end it will have better ability to fold into a native configuration. This approach has been used with human papillomavirus, which appears to have some limited capacity for insertion of epitopes into internal loops (Slupetzky *et al.*, 2001; Varsani *et al.*, 2003; Slupetzky *et al.*, 2007; Kondo *et al.*, 2008) but a larger capacity for insertion of sequences (up to 60 amino acids) at the carboxyl-terminus of the L1 capsid protein (Liu *et al.*, 1998;

Peng *et al.*, 1998; Dale *et al.*, 2002). VLPs with similar extensions that have potential for epitope display have been constructed for a variety of other viruses including bluetongue virus core (Adler *et al.*, 1998), hepatitis B core (von Brunn *et al.*, 1993), hepatitis E (Niikura *et al.*, 2002) and Tomato bushy stunt virus (Kumar *et al.*, 2009). However, extension of the C-terminus of other VLPs abrogates the formation of particles (Rueda *et al.*, 1999). Although in principle this extension approach should result in the possibility of addition of larger and more complex fusion proteins, in practice the C-terminus of the capsid protein is rarely displayed at a highly exposed position on the surface of the VLP. For example, in the case of hepatitis B core particles comparison of immune response for particles displaying the same epitope from an exposed surface loop and the C-terminus of the protein suggest that the loop insertion results in a stronger immune response in mice (von Brunn *et al.*, 1993). The tomato bushy stunt virus VLP appears to be unusual in that proteins fused to the C-terminus of the capsid protein are highly exposed on the surface of the particle (Kumar *et al.*, 2009).

Attempts have been made to improve the carrying capacity of some VLPs by modifying the expression conditions, or by modifying the capsid protein itself to improve its potential for antigen display. One idea is that one of the reasons that capsid protein displaying large antigen fails to assemble is that there is a steric hindrance caused by the close packing of the foreign antigen that prevents normal assembly of the VLP. This has been addressed by co-expressing tagged and untagged capsid protein in the same cell to assemble chimeric particles where not all the capsid subunits display the foreign antigen (Koletzki *et al.*, 1997; Ulrich *et al.*, 1999; Beterams *et al.*, 2000; Vogel *et al.*, 2005a). An alternative approach has been the engineering of the capsid protein itself to effectively move the C-terminus of the capsid protein to a more immunogenic position by splitting the protein into two separate domains. The first demonstration of this for the hepatitis B core antigen was the insertion of a protease cleavage site in the major antigenic loop on the surface of the capsid protein. Resulting capsid protein retained the ability to assemble into VLP even with large and highly structured inserts when co-expressed with the corresponding protease (Walker *et al.*, 2008). This idea was subsequently taken to its logical conclusion by the separate expression of the N- and C-terminal domains of the capsid protein without need for protease co-expression (Walker *et al.*, 2012). This modification effectively displays the C-terminus of the N-terminal domain at the same position as the major immunodominant loop on the unmodified capsid protein.

The third approach used to display foreign antigens on VLPs is based on those particles that are based on viruses that have a lipid envelope. These particles usually comprise a capsid protein that acquires a lipid envelope by budding from the cellular plasma membrane, nuclear membrane or Golgi, depending on the VLP, and in which viral surface proteins are embedded. There are two challenges with such enveloped VLPs, one is the production of sufficient, correctly folded, surface protein to decorate the particle and the other is the efficiency with which VLPs bud in heterologous expression systems. For some VLPs the budding is a characteristic of the proteinaceous capsid rather than a property of the glycoproteins. For example, expression of the nucleocapsid protein of Rift Valley fever virus in insect cells leads to the accumulation of large amounts of nucleocapsid like particle in the culture supernatant (Liu *et al.*, 2008). This feature is common to other enveloped VLPs and has been used to efficiently produce enveloped particles in which there is a mismatch of the surface protein of the VLP with the capsid protein that drives budding (Liu *et al.*, 2011).

Another way in which the ability to engineer non-native VLPs has been used is to assemble a rationally designed, synthetic surface protein in the lipid layer onto an authentic viral core. For example, consensus influenza A haemagglutinin, designed to give broadly cross-reactive immune responses has been incorporated into influenza VLPs (Giles *et al.*, 2011; Giles and Ross, 2011).

Overall there is strong evidence that VLPs are highly effective at the display of foreign epitopes and antigens and have an immunostimulatory effect *in vivo* even in the absence of other adjuvants.

Virus-derived self-assembling non-VLP structures

In addition to VLPs, which are authentic mimics of the particle of the cognate virus, overexpression of viral proteins in heterologous systems has led to the observation that other higher order multimers can be formed which have some similar properties. Some of these structures are oligomers of capsid protein and are formed when fusions displaying foreign epitopes disrupt the ability of the capsid to form VLPs. These can either be capsid assembly intermediates consisting of assemblies 2–24 monomers depending on the virus (Delos *et al.*, 1995; Rose *et al.*, 1998; Ohlschlager *et al.*, 2003; Campos and Barry, 2006; Dell *et al.*, 2006; Bian *et al.*, 2008; Thones *et al.*, 2008; Jagu *et al.*, 2009; Murata *et al.*, 2009; Schadlich *et al.*, 2009; Tamminen *et al.*, 2011) or they can be higher order aggregates of capsid protein (Rutkowska *et al.*, 2011). In the case of the capsid assembly intermediates, there is evidence that they can stimulate immune response; however, when compared to VLP they are rarely as immunogenic (Rose *et al.*, 1998; Thones *et al.*, 2008; Schadlich *et al.*, 2009; Tamminen *et al.*, 2011), possibly because they do not produce particles of the size readily taken up into dendritic cells (Bachmann *et al.*, 1996; Rudolf *et al.*, 2001; Moron *et al.*, 2002; Ruedl *et al.*, 2002; Fausch *et al.*, 2003; Moron *et al.*, 2003; Tsunetsugu-Yokota *et al.*, 2003; Bosio *et al.*, 2004; Fifis *et al.*, 2004; Freyschmidt *et al.*, 2004; Yang *et al.*, 2004; Zhang *et al.*, 2004; Arico *et al.*, 2005; Barth *et al.*, 2005; Lenz *et al.*, 2005; Buonaguro *et al.*, 2006; Garcia-Pineres *et al.*, 2006; Gedvilaite *et al.*, 2006; Ye *et al.*, 2006; Istrate *et al.*, 2007; Tegerstedt *et al.*, 2007; Ding *et al.*, 2010; Song *et al.*, 2010; Chang *et al.*, 2011; Win *et al.*, 2011; Link *et al.*, 2012).

Another class of non-VLP structures do self-assemble into highly repetitive structures with a similar size to virus particles. An example of this would be the tubule structures formed by the Bluetongue virus NS1 protein (Fig. 7.4). When overexpressed, these tubules are assembled to form a coiled ribbon of dimers forming a structure which is 52.3 nm in diameter and up to 1000 nm in length with approximately 22 dimers per turn (Marshall *et al.*, 1990; Hewat *et al.*, 1992). With a pitch of 9.1 nm, each 1000-nm tubule contains 4835 copies of NS1 monomer. Characterization of antibodies raised to native tubules revealed that the C-terminus of monomers was exposed and immunogenic (Monastyrskaya *et al.*, 1995). Further studies linking B-cell and T-cell epitopes for a variety of pathogens to the C-terminus of NS1 revealed that tubules were an excellent display platform (Mikhailov *et al.*, 1996; Ghosh *et al.*, 2002; Larke *et al.*, 2005). Indeed, it was demonstrated that multivalent tubules could be assembled by incorporating subunits linked to different antigens (Mikhailov *et al.*, 1996) and that tubules would self-assemble even in the presence of a 527 amino acid C-terminal extension (Larke *et al.*, 2005). Although currently at a less advanced stage than VLP studies, the potential for non-VLP particulate structures based on self-assembling repeating protein structures is clear.

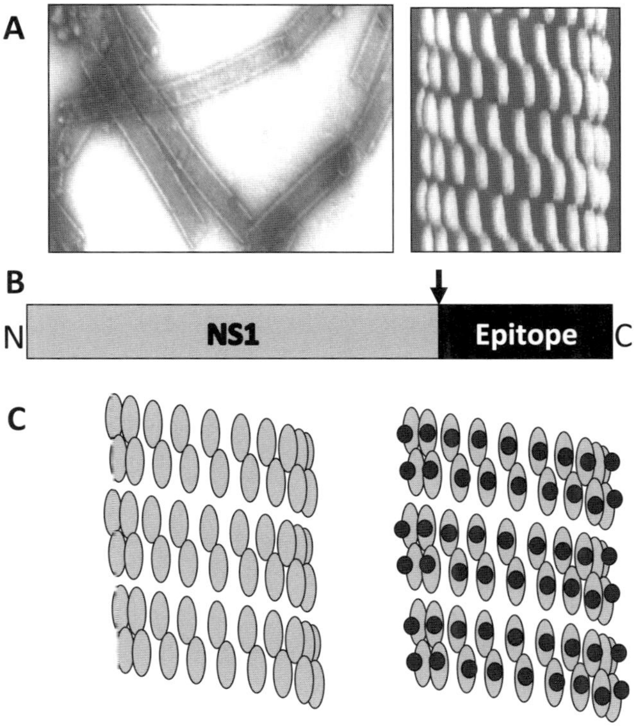

Figure 7.4 Particulate, self-assembling particles produced by over expression of NS1. Electron micrograph (A, left) and reconstruction (A, right) of tubules produces by the self-assembly of BTV NS1 protein. This particle has similar physical characteristics to VLPs, in that it forms a self-assuming particle <˙ mm that does not contain viral RNA. The NS1 monomer can be extended at the C terminus (B) with up to 512 amino acids of foreign protein and displays one copy of this extension on each monomer of the assembled tubule (C, right). Unmodified tubules (C, left). Modified tubules can be purified as particle and are highly immunogenic.

Virus-like particles as nanomachines

Vaccine studies with VLPs focus on their structural and immunological similarity to virus particles. However, another view of these particles is that they are functional nanomachines that are adapted for the highly target intracellular delivery of viral nucleic acids to specific cells. Since VLPs do not need to package replicative viral nucleic acid this payload function is available for exploitation for other purposes.

Delivery of nucleic acids

One way in which VLPs have been exploited as vehicles is for the delivery of foreign DNA and RNA to cells. This is similar to viable virus where one of the functions of the virion is the delivery of virus genome to cells. VLPs that have demonstrated capacity for delivery of non-replicating nucleic acids to cells include those derived from tobacco mosaic virus (RNA packaging) (Gallie *et al.*, 1987; Hwang *et al.*, 1994), satellite tobacco necrosis virus (RNA packaging) (Bunka *et al.*, 2011), JC virus (DNA packaging) (Forstova *et al.*, 1995; Ou *et al.*, 1999, 2001; Chen *et al.*, 2010; Chou *et al.*, 2010), bacteriophage MS2 (RNA packaging) (Wu *et al.*, 2005; Sun *et al.*, 2011), feline immunodeficiency virus (RNA delivery, reverse transcribing) (Poeschla *et al.*, 1998) and HIV (RNA delivery, reverse transcribing) (Naldini *et al.*, 1996). However, since one of the main roles of virus particles is nucleic acid delivery it is likely that any VLP has the potential to be used for gene delivery in this way provided the basis for nucleic acid packaging is understood. Intriguingly VLPs for some viruses appear to be particularly non-specific in their packaging activity and will readily package host nucleic acid if no viral nucleic acid is available (Krishna *et al.*, 2003; Kler *et al.*, 2012; Routh *et al.*, 2012). This is not true of all VLPs, as some require specific packaging signals to be present on the nucleic acid before it is packaged (Hwang *et al.*, 1994; Naldini *et al.*, 1996; Poeschla *et al.*, 1998; Bunka *et al.*, 2011). Apart from lentiviral vector systems the VLP that is most developed with regards to non-replicating nucleic acid delivery is the polyomavirus system. Foreign DNA can be introduced into JC virus VLPs by osmotic shock *in vitro* and the particles are capable of delivering the DNA to eukaryotic cells (Forstova *et al.*, 1995; Ou *et al.*, 1999). Later it was observed that it was also possible to capture DNA into VLPs *in vivo* (Chen *et al.*, 2010).

Targeted drug delivery

An extension of the idea that VLPs, like virus particles, can carry nucleic acid is the idea that VLPs can contain other molecules such as other proteins and drugs. The packaging of foreign proteins in the space on the inside of particles by fusion to the capsid protein has been described for a variety of VLPs including bluetongue (Kar *et al.*, 2005), rotavirus (Charpilienne *et al.*, 2001), hepatitis B core (Beterams *et al.*, 2000) and murine polyomavirus (Lipin *et al.*, 2008). One of the critical factors in the assembly of particles modified in this way is the available space inside the particle. Although in some cases this is sufficient to accommodate fusion proteins (Charpilienne *et al.*, 2001; Kar *et al.*, 2005), in other cases a mixture of modified and unmodified capsid protein was required to allow proper VLP assembly (Beterams *et al.*, 2000). An alternative approach has been the packaging of the peptide drug molecule by covalent linkage to a nucleic acid packaging signal. This has been used for packaging the ricin A toxin into the bacteriophage MS2 (Wu *et al.*, 1995).

Incorporation of drug and other non-protein non-nucleic acid molecules into VLPs presents more of a problem but has been achieved for some VLPs. Brome mosaic virus particles have been assembled *in vitro* around carboxylated polyethylene glycol functionalized gold and iron nanoparticles by mixing capsid protein and nanoparticles (Dragnea *et al.*, 2003; Sun *et al.*, 2007; Huang *et al.*, 2011). These particles offer the opportunity to use magnetic resonance imaging to follow the migration and localization of particles in live hosts (Huang *et al.*, 2011).

The potential for packaging drug molecules in the internal space of virus particles by the reversible disassembly and reassembly of virus particles has also been raised for several viruses, including papillomaviruses (McCarthy *et al.*, 1998; Mach *et al.*, 2006; Zhao *et al.*, 2012a,b), cowpea chlorotic mottle virus (CCMV) (Lavelle *et al.*, 2007; Comellas-Aragones *et al.*, 2009; Cadena-Nava *et al.*, 2011) and parvovirus B19 (Sanchez-Rodriguez *et al.*, 2011). Intriguingly, this process for human papillomavirus appears to result in more uniform, and more immunologically active VLPs (Zhao *et al.*, 2012a,b). The potential for developing these observations as useful tools in nanomedicine will depend on the efficiency of the reassembly and packaging reactions. However, the CCMV system, which is the most widely reassembly based packaging system has demonstrated that this approach is viable for a number of laboratories including the packaging of active enzyme molecules into VLPs (Comellas-Aragones *et al.*, 2007, 2009; Lavelle *et al.*, 2007; Cadena-Nava *et al.*, 2011).

In contrast to the packaging of drugs to the inside of VLPs, some studies have modified exposed sites on the outside of particles as potential sites for drug linkage. This approach has been used both for drug linkage and the prevention of steric effects observed with fusion protein insertion of foreign epitopes and antigen into capsid protein. The general approach is the insertion of specific amino acids at key immunogenic sites on the surface of the capsid followed by chemical linking of other molecules at these positions. The highly repetitive structure of the VLPs results in high-level incorporation of molecules onto the VLP surface (Banerjee *et al.*, 2010; Pejawar-Gaddy *et al.*, 2010; Patel and Swartz, 2011). In some cases this approach has been highly successful but in others linkage has resulted in a reduction in the bioactivity of the attached molecule (Patel and Swartz, 2011). A twist on this approach has been the use of the intrinsic sialic acid binding activity of JC virus capsid to conjugate sialic acid coated gold nanoparticles to the surface of VLPs (Niikura *et al.*, 2009). Chemical linkage to the surface of VLPs has the advantage of providing a large surface area for the incorporation of foreign antigen but the drawback of this approach is that some of the intrinsic targeting of the VLP is lost. This can be overcome by the modification of VLP by linking specific cell targeting peptides and proteins to the surface of the VLP (Banerjee *et al.*, 2010).

Future outlook

VLP technology as a vaccine platform can now be regarded as a mature field; VLP-based immunogens are now available commercially targeting human papillomavirus, hepatitis B virus and porcine circovirus, and there are many more VLP based immunogens in the development pipeline. It is quite likely that the immediate pay-off in the VLP field will be related to the development of these VLPs as direct immunogens for cognate viruses. As always, commercial and production factors will determine which VLP successfully reaches the market. However, the biggest general challenges facing the field are more related to producing, particularly enveloped, VLPs in sufficient quantity and quality in a cost-effective manner than any concern that VLPs are not effective immunogens.

The use of VLPs as display platforms for other epitopes and antigens has a strong laboratory success rate but so far has not had a large commercial impact. This is likely to be a matter of time, as the effectiveness of VLPs as immunogens becomes more accepted, the willingness of companies to risk the second step becomes more likely. In our opinion, the first immunogens in mass use from this strategy are likely to be based on one of the currently available or newly licensed unmodified VLP platforms rather than VLPs developed solely for antigen display. This is because technical challenges in the large scale production and purification of the basic VLP will already be established and investors and regulators will already be familiar with the core technology. The only further issue that display-platform type VLPs would need to overcome would be the possibility that individuals already vaccinated with the unmodified VLP might respond preferentially to this rather than the displayed epitope.

The potential of VLPs as nanocarriers of drugs and nucleic acid, platforms for enzyme reactions and nanoreactors are extremely exciting. In principle, these self-assembling particles are ideal for this purpose, and the discovery that some VLPs can be reversibly disassembled and reassembled offers exciting opportunities for drug delivery and targeting. One major challenge regarding their use *in vivo* relates to their similarity to viruses and the ready immune response that is developed. Would a VLP based drug delivery system be able to be used more than once on the same patient? Or would immune responses to the VLP result in uptake of the drug into phagocytic cells and off target effects? Not all applications of these technologies rely on *in vivo* use and the use of VLPs as surfaces for the deposition of metal ions and other molecules has clear potential for further development. The challenge here is to integrate the VLP technology with the chemistry and the target application. In effect, viruses are already adapted as environmentally and biologically stable nanomachines and the real challenge is the commercialization of the concept rather than its technical feasibility.

Lentivirus-based gene delivery is already well established although most would not recognize it as a VLP technology, even though it is one. As the line between VLPs and defective, engineered virus replicons becomes increasingly blurred (Roy *et al.*, 2009; Qing *et al.*, 2010; Matsuo *et al.*, 2011), it is likely that many virus based non-infectious RNA and DNA delivery systems will become used in laboratories and *in vivo*

Conclusions

Recent years have seen a rapid expansion in the number of VLPs reported and the different applications that these particles can be put to. Having proven their worth as vaccines for cognate diseases, VLPs are likely to become a conventional tool in the fight against disease. However the potential of these self-assembling nanoparticles extends far beyond conventional vaccines and has implications for control of other diseases and nanotechnology.

References

Adams, S.E., Dawson, K.M., Gull, K., Kingsman, S.M., and Kingsman, A.J. (1987). The expression of hybrid HIV:Ty virus-like particles in yeast. Nature *329*, 68–70.

Adler, S., Reay, P., Roy, P., and Klenk, H.D. (1998). Induction of T cell response by bluetongue virus core-like particles expressing a T cell epitope of the M1 protein of influenza A virus. Med. Microbiol. Immunol. *187*, 91–96.

Akahata, W., Yang, Z.Y., Andersen, H., Sun, S., Holdaway, H.A., Kong, W.P., Lewis, M.G., Higgs, S., Rossmann, M.G., Rao, S., *et al.* (2010). A virus-like particle vaccine for epidemic Chikungunya virus protects nonhuman primates against infection. Nat. Med. *16*, 334–338.

Allison, S.L., Stadler, K., Mandl, C.W., Kunz, C., and Heinz, F.X. (1995). Synthesis and secretion of recombinant tick-borne encephalitis virus protein E in soluble and particulate form. J. Virol. *69*, 5816–5820.

Arico, E., Wang, E., Tornesello, M.L., Tagliamonte, M., Lewis, G.K., Marincola, F.M., Buonaguro, F.M., and Buonaguro, L. (2005). Immature monocyte derived dendritic cells gene expression profile in response to Virus-Like Particles stimulation. J. Transl. Med. *3*, 45.

Bachmann, M.F., Lutz, M.B., Layton, G.T., Harris, S.J., Fehr, T., Rescigno, M., and Ricciardi-Castagnoli, P. (1996). Dendritic cells process exogenous viral proteins and virus-like particles for class I presentation to CD8+ cytotoxic T lymphocytes. Eur J. Immunol. *26*, 2595–2600.

Banerjee, D., Liu, A.P., Voss, N.R., Schmid, S.L., and Finn, M.G. (2010). Multivalent display and receptor-mediated endocytosis of transferrin on virus-like particles. ChemBioChem *11*, 1273–1279.

Barth, H., Ulsenheimer, A., Pape, G.R., Diepolder, H.M., Hoffmann, M., Neumann-Haefelin, C., Thimme, R., Henneke, P., Klein, R., Paranhos-Baccala, G., *et al.* (2005). Uptake and presentation of hepatitis C virus-like particles by human dendritic cells. Blood *105*, 3605–3614.

Baumert, T.F., Vergalla, J., Satoi, J., Thomson, M., Lechmann, M., Herion, D., Greenberg, H.B., Ito, S., and Liang, T.J. (1999). Hepatitis C virus-like particles synthesized in insect cells as a potential vaccine candidate. Gastroenterology *117*, 1397–1407.

Betenbaugh, M., Yu, M., Kuehl, K., White, J., Pennock, D., Spik, K., and Schmaljohn, C. (1995). Nucleocapsid- and virus-like particles assemble in cells infected with recombinant baculoviruses or vaccinia viruses expressing the M and the S segments of Hantaan virus. Virus Res. *38*, 111–124.

Beterams, G., Bottcher, B., and Nassal, M. (2000). Packaging of up to 240 subunits of a 17 kDa nuclease into the interior of recombinant hepatitis B virus capsids. FEBS Lett. *481*, 169–176.

Bian, T., Wang, Y., Lu, Z., Ye, Z., Zhao, L., Ren, J., Zhang, H., Ruan, L., and Tian, H. (2008). Human papillomavirus type 16 L1E7 chimeric capsomeres have prophylactic and therapeutic efficacy against papillomavirus in mice. Mol. Cancer Ther. *7*, 1329–1335.

Bosio, C.M., Moore, B.D., Warfield, K.L., Ruthel, G., Mohamadzadeh, M., Aman, M.J., and Bavari, S. (2004). Ebola and Marburg virus-like particles activate human myeloid dendritic cells. Virology *326*, 280–287.

Bright, R.A., Carter, D.M., Daniluk, S., Toapanta, F.R., Ahmad, A., Gavrilov, V., Massare, M., Pushko, P., Mytle, N., Rowe, T., *et al.* (2007). Influenza virus-like particles elicit broader immune responses than whole virion inactivated influenza virus or recombinant hemagglutinin. Vaccine *25*, 3871–3878.

Bunka, D.H., Lane, S.W., Lane, C.L., Dykeman, E.C., Ford, R.J., Barker, A.M., Twarock, R., Phillips, S.E., and Stockley, P.G. (2011). Degenerate RNA packaging signals in the genome of Satellite Tobacco Necrosis Virus: implications for the assembly of a T=1 capsid. J. Mol. Biol. *413*, 51–65.

Buonaguro, L., Tornesello, M.L., Tagliamonte, M., Gallo, R.C., Wang, L.X., Kamin-Lewis, R., Abdelwahab, S., Lewis, G.K., and Buonaguro, F.M. (2006). Baculovirus-derived human immunodeficiency virus type 1 virus-like particles activate dendritic cells and induce ex vivo T-cell responses. J. Virol. *80*, 9134–9143.

Cadena-Nava, R.D., Hu, Y., Garmann, R.F., Ng, B., Zelikin, A.N., Knobler, C.M., and Gelbart, W.M. (2011). Exploiting fluorescent polymers to probe the self-assembly of virus-like particles. J. Phys. Chem. B *115*, 2386–2391.

Campos, S.K., and Barry, M.A. (2006). Comparison of adenovirus fiber, protein IX, and hexon capsomeres as scaffolds for vector purification and cell targeting. Virology *349*, 453–462.

Chackerian, B. (2010). Virus-like particle based vaccines for Alzheimer disease. Hum. Vaccin. *6*, 926–930.

Chang, M.O., Suzuki, T., Suzuki, H., and Takaku, H. (2011). HIV-1 Gag-virus-like particles induce natural killer cell immune responses via activation and maturation of dendritic cells. J. Innate Immun. *4*, 187–200.

Charpilienne, A., Nejmeddine, M., Berois, M., Parez, N., Neumann, E., Hewat, E., Trugnan, G., and Cohen, J. (2001). Individual rotavirus-like particles containing 120 molecules of fluorescent protein are visible in living cells. J. Biol. Chem. *276*, 29361–29367.

Chen, L.S., Wang, M., Ou, W.C., Fung, C.Y., Chen, P.L., Chang, C.F., Huang, W.S., Wang, J.Y., Lin, P.Y., and Chang, D. (2010). Efficient gene transfer using the human JC virus-like particle that inhibits human colon adenocarcinoma growth in a nude mouse model. Gene Ther. *17*, 1033–1041.

Chou, M.I., Hsieh, Y.F., Wang, M., Chang, J.T., Chang, D., Zouali, M., and Tsay, G.J. (2010). In vitro and in vivo targeted delivery of IL-10 interfering RNA by JC virus-like particles. J. Biomed. Sci. *17*, 51.

Colson, P., Yutin, N., Shabalina, S.A., Robert, C., Fournous, G., La Scola, B., Raoult, D., and Koonin, E.V. (2011). Viruses with more than 1,000 genes: Mamavirus, a new Acanthamoeba polyphaga mimivirus strain, and reannotation of Mimivirus genes. Genome Biol. Evol. *3*, 737–742.

Comellas-Aragones, M., de la Escosura, A., Dirks, A.T., van der Ham, A., Fuste-Cune, A., Cornelissen, J.J., and Nolte, R.J. (2009). Controlled integration of polymers into viral capsids. Biomacromolecules *10*, 3141–3147.

Comellas-Aragones, M., Engelkamp, H., Claessen, V.I., Sommerdijk, N.A., Rowan, A.E., Christianen, P.C., Maan, J.C., Verduin, B.J., Cornelissen, J.J., and Nolte, R.J. (2007). A virus-based single-enzyme nanoreactor. Nat. Nanotechnol. *2*, 635–639.

Crawford, S.E., Labbe, M., Cohen, J., Burroughs, M.H., Zhou, Y.J., and Estes, M.K. (1994). Characterization of virus-like particles produced by the expression of rotavirus capsid proteins in insect cells. J. Virol. *68*, 5945–5952.

Crick, F.H., and Watson, J.D. (1956). Structure of small viruses. Nature *177*, 473–475.

Crisci, E., Fraile, L., Moreno, N., Blanco, E., Cabezon, R., Costa, C., Mussa, T., Baratelli, M., Martinez-Orellana, P., Ganges, L., *et al.* (2012). Chimeric calicivirus-like particles elicit specific immune responses in pigs. Vaccine *30*, 2427–2439.

Cubas, R., Zhang, S., Li, M., Chen, C., and Yao, Q. (2011). Chimeric Trop2 virus-like particles: a potential immunotherapeutic approach against pancreatic cancer. J Immunother *34*, 251–263.

Dale, C.J., Liu, X.S., De Rose, R., Purcell, D.F., Anderson, J., Xu, Y., Leggatt, G.R., Frazer, I.H., and Kent, S.J. (2002). Chimeric human papilloma virus-simian/human immunodeficiency virus virus-like-particle vaccines: immunogenicity and protective efficacy in macaques. Virology *301*, 176–187.

Dell, K., Koesters, R., Linnebacher, M., Klein, C., and Gissmann, L. (2006). Intranasal immunization with human papillomavirus type 16 capsomeres in the presence of non-toxic cholera toxin-based adjuvants elicits increased vaginal immunoglobulin levels. Vaccine *24*, 2238–2247.

Delos, S.E., Cripe, T.P., Leavitt, A.D., Greisman, H., and Garcea, R.L. (1995). Expression of the polyomavirus minor capsid proteins VP2 and VP3 in *Escherichia coli*: in vitro interactions with recombinant VP1 capsomeres. J. Virol. *69*, 7734–7742.

Denis, J., Acosta-Ramirez, E., Zhao, Y., Hamelin, M.E., Koukavica, I., Baz, M., Abed, Y., Savard, C., Pare, C., Lopez Macias, C., *et al.* (2008). Development of a universal influenza A vaccine based on the M2e peptide fused to the papaya mosaic virus (PapMV) vaccine platform. Vaccine *26*, 3395–3403.

Ding, F.X., Xian, X., Guo, Y.J., Liu, Y., Wang, Y., Yang, F., Wang, Y.Z., Song, S.X., Wang, F., and Sun, S.H. (2010). A preliminary study on the activation and antigen presentation of hepatitis B virus core protein virus-like particle-pulsed bone marrow-derived dendritic cells. Mol Biosyst *6*, 2192–2199.

Dragnea, B., Chen, C., Kwak, E.S., Stein, B., and Kao, C.C. (2003). Gold nanoparticles as spectroscopic enhancers for in vitro studies on single viruses. J. Am. Chem. Soc. *125*, 6374–6375.

Eschbaumer, M., Hoffmann, B., Moss, A., Savini, G., Leone, A., Konig, P., Zemke, J., Conraths, F., and Beer, M. (2010). Emergence of bluetongue virus serotype 6 in Europe – German field data and experimental infection of cattle. Vet. Microbiol. *143*, 189–195.

Fausch, S.C., Da Silva, D.M., and Kast, W.M. (2003). Differential uptake and cross-presentation of human papillomavirus virus-like particles by dendritic cells and Langerhans cells. Cancer Res *63*, 3478–3482.

Ferlenghi, I., Clarke, M., Ruttan, T., Allison, S.L., Schalich, J., Heinz, F.X., Harrison, S.C., Rey, F.A., and Fuller, S.D. (2001). Molecular organization of a recombinant subviral particle from tick-borne encephalitis virus. Mol Cell *7*, 593–602.

Fields, B.N., Knipe, D.M., and Howley, P.M. (2007). Fields' Virology, 5th edn/editors-in-chief, David M. Knipe, Peter M. Howley (Philadelphia: Wolters Kluwer/Lippincott Williams & Wilkins).

Fifis, T., Gamvrellis, A., Crimeen-Irwin, B., Pietersz, G.A., Li, J., Mottram, P.L., McKenzie, I.F., and Plebanski, M. (2004). Size-dependent immunogenicity: therapeutic and protective properties of nano-vaccines against tumors. J. Immunol. *173*, 3148–3154.

Forstova, J., Krauzewicz, N., Sandig, V., Elliott, J., Palkova, Z., Strauss, M., and Griffin, B.E. (1995). Polyoma virus pseudocapsids as efficient carriers of heterologous DNA into mammalian cells. Hum. Gene Ther. *6*, 297–306.

Forstova, J., Krauzewicz, N., Wallace, S., Street, A.J., Dilworth, S.M., Beard, S., and Griffin, B.E. (1993). Cooperation of structural proteins during late events in the life cycle of polyomavirus. J. Virol. *67*, 1405–1413.

French, T.J., Marshall, J.J., and Roy, P. (1990). Assembly of double-shelled, viruslike particles of bluetongue virus by the simultaneous expression of four structural proteins. J. Virol. *64*, 5695–5700.

French, T.J., and Roy, P. (1990). Synthesis of bluetongue virus (BTV) corelike particles by a recombinant baculovirus expressing the two major structural core proteins of BTV. J. Virol. *64*, 1530–1536.

Freyschmidt, E.J., Alonso, A., Hartmann, G., and Gissmann, L. (2004). Activation of dendritic cells and induction of T cell responses by HPV 16 L1/E7 chimeric virus-like particles are enhanced by CpG ODN or sorbitol. Antivir. Ther. *9*, 479–489.

Galarza, J.M., Latham, T., and Cupo, A. (2005). Virus-like particle vaccine conferred complete protection against a lethal influenza virus challenge. Viral Immunol. *18*, 365–372.

Gallie, D.R., Sleat, D.E., Watts, J.W., Turner, P.C., and Wilson, T.M. (1987). In vivo uncoating and efficient expression of foreign mRNAs packaged in TMV-like particles. Science *236*, 1122–1124.

Garcia-Pineres, A.J., Hildesheim, A., Trivett, M., Williams, M., Wu, L., Kewalramani, V.N., and Pinto, L.A. (2006). Role of DC-SIGN in the activation of dendritic cells by HPV-16 L1 virus-like particle vaccine. Eur J. Immunol. *36*, 437–445.

Gedvilaite, A., Dorn, D.C., Sasnauskas, K., Pecher, G., Bulavaite, A., Lawatscheck, R., Staniulis, J., Dalianis, T., Ramqvist, T., Schonrich, G., *et al.* (2006). Virus-like particles derived from major capsid protein VP1 of different polyomaviruses differ in their ability to induce maturation in human dendritic cells. Virology *354*, 252–260.

Ghosh, M.K., Borca, M.V., and Roy, P. (2002). Virus-derived tubular structure displaying foreign sequences on the surface elicit CD4+ Th cell and protective humoral responses. Virology *302*, 383–392.

Giles, B.M., Bissel, S.J., Dealmeida, D.R., Wiley, C.A., and Ross, T.M. (2011). Antibody breadth and protective efficacy are increased by vaccination with computationally optimized haemagglutinin but not with polyvalent haemagglutinin-based H5N1 virus-like particle vaccines. Clin. Vaccine Immunol. *19*, 128–139.

Giles, B.M., and Ross, T.M. (2011). A computationally optimized broadly reactive antigen (COBRA) based H5N1 VLP vaccine elicits broadly reactive antibodies in mice and ferrets. Vaccine *29*, 3043–3054.

Grgacic, E.V., and Anderson, D.A. (2006). Virus-like particles: passport to immune recognition. Methods *40*, 60–65.

Hamel, A.L., Lin, L.L., and Nayar, G.P. (1998). Nucleotide sequence of porcine circovirus associated with postweaning multisystemic wasting syndrome in pigs. J. Virol. *72*, 5262–5267.

Harris, S.J., Gearing, A.J., Layton, G.T., Adams, S.E., and Kingsman, A.J. (1992). Enhanced proliferative cellular responses to HIV-1 V3 peptide and gp120 following immunization with V3:Ty virus-like particles. Immunology *77*, 315–321.

Hewat, E.A., Booth, T.F., Wade, R.H., and Roy, P. (1992). 3-D reconstruction of bluetongue virus tubules using cryoelectron microscopy. J. Struct. Biol. *108*, 35–48.

Huang, X., Stein, B.D., Cheng, H., Malyutin, A., Tsvetkova, I.B., Baxter, D.V., Remmes, N.B., Verchot, J., Kao, C., Bronstein, L.M., *et al.* (2011). Magnetic virus-like nanoparticles in N. benthamiana plants: a new paradigm for environmental and agronomic biotechnological research. ACS Nano. *5*, 4037–4045.

Hwang, D.J., Roberts, I.M., and Wilson, T.M. (1994). Expression of tobacco mosaic virus coat protein and assembly of pseudovirus particles in *Escherichia coli*. Proc. Natl. Acad. Sci. U.S.A. *91*, 9067–9071.

Istrate, C., Douagi, I., Charpilienne, A., McInerney, G.M., Hidmark, A., Johansen, K., Larsson, M., Magnusson, K.E., Poncet, D., Svensson, L., *et al.* (2007). Bone marrow dendritic cells internalize live RF-81 bovine rotavirus and rotavirus-like particles (RF 2/6-GFP-VLP and RF 8*2/6/7-VLP) but are only activated by live bovine rotavirus. Scand. J. Immunol. *65*, 494–502.

Jagu, S., Karanam, B., Gambhira, R., Chivukula, S.V., Chaganti, R.J., Lowy, D.R., Schiller, J.T., and Roden, R.B. (2009). Concatenated multitype L2 fusion proteins as candidate prophylactic pan-human papillomavirus vaccines. J. Natl. Cancer Inst. *101*, 782–792.

Jiang, X., Wang, M., Graham, D.Y., and Estes, M.K. (1992). Expression, self-assembly, and antigenicity of the Norwalk virus capsid protein. J. Virol. *66*, 6527–6532.

Kar, A.K., Iwatani, N., and Roy, P. (2005). Assembly and intracellular localization of the bluetongue virus core protein VP3. J. Virol. *79*, 11487–11495.

Karpenko, L.I., Ivanisenko, V.A., Pika, I.A., Chikaev, N.A., Eroshkin, A.M., Veremeiko, T.A., and Ilyichev, A.A. (2000). Insertion of foreign epitopes in HBcAg: how to make the chimeric particle assemble. Amino Acids *18*, 329–337.

Kirnbauer, R., Booy, F., Cheng, N., Lowy, D.R., and Schiller, J.T. (1992). Papillomavirus L1 major capsid protein self-assembles into virus-like particles that are highly immunogenic. Proc. Natl. Acad. Sci. U.S.A. *89*, 12180–12184.

Kirnbauer, R., Taub, J., Greenstone, H., Roden, R., Durst, M., Gissmann, L., Lowy, D.R., and Schiller, J.T. (1993). Efficient self-assembly of human papillomavirus type 16 L1 and L1-L2 into virus-like particles. J. Virol. *67*, 6929–6936.

Kler, S., Asor, R., Li, C., Ginsburg, A., Harries, D., Oppenheim, A., Zlotnick, A., and Raviv, U. (2012). RNA encapsidation by SV40-derived nanoparticles follows a rapid two-state mechanism. J Am Chem Soc. *134*, 8823–8830.

Koletzki, D., Zankl, A., Gelderblom, H.R., Meisel, H., Dislers, A., Borisova, G., Pumpens, P., Kruger, D.H., and Ulrich, R. (1997). Mosaic hepatitis B virus core particles allow insertion of extended foreign protein segments. J. Gen. Virol. *78*, 2049–2053.

Kondo, K., Ochi, H., Matsumoto, T., Yoshikawa, H., and Kanda, T. (2008). Modification of human papillomavirus-like particle vaccine by insertion of the cross-reactive L2-epitopes. J. Med. Virol. *80*, 841–846.

Konishi, E., Pincus, S., Paoletti, E., Shope, R.E., Burrage, T., and Mason, P.W. (1992). Mice immunized with a subviral particle containing the Japanese encephalitis virus prM/M and E proteins are protected from lethal JEV infection. Virology *188*, 714–720.

Kratz, P.A., Bottcher, B., and Nassal, M. (1999). Native display of complete foreign protein domains on the surface of hepatitis B virus capsids. Proc. Natl. Acad. Sci. U.S.A. *96*, 1915–1920.

Krishna, N.K., Marshall, D., and Schneemann, A. (2003). Analysis of RNA packaging in wild-type and mosaic protein capsids of flock house virus using recombinant baculovirus vectors. Virology *305*, 10–24.

Kumar, S., Ochoa, W., Singh, P., Hsu, C., Schneemann, A., Manchester, M., Olson, M., and Reddy, V. (2009). Tomato bushy stunt virus (TBSV), a versatile platform for polyvalent display of antigenic epitopes and vaccine design. Virology *388*, 185–190.

Labbe, M., Charpilienne, A., Crawford, S.E., Estes, M.K., and Cohen, J. (1991). Expression of rotavirus VP2 produces empty corelike particles. J. Virol. *65*, 2946–2952.

Larke, N., Murphy, A., Wirblich, C., Teoh, D., Estcourt, M.J., McMichael, A.J., Roy, P., and Hanke, T. (2005). Induction of human immunodeficiency virus type 1-specific T cells by a bluetongue virus tubule-vectored vaccine prime-recombinant modified virus Ankara boost regimen. J. Virol. *79*, 14822–14833.

Latham, T., and Galarza, J.M. (2001). Formation of wild-type and chimeric influenza virus-like particles following simultaneous expression of only four structural proteins. J. Virol. *75*, 6154–6165.

Lavelle, L., Michel, J.P., and Gingery, M. (2007). The disassembly, reassembly and stability of CCMV protein capsids. J. Virol. Methods *146*, 311–316.

Layton, G.T., Harris, S.J., Gearing, A.J., Hill-Perkins, M., Cole, J.S., Griffiths, J.C., Burns, N.R., Kingsman, A.J., and Adams, S.E. (1993). Induction of HIV-specific cytotoxic T lymphocytes in vivo with hybrid HIV-1 V3:Ty-virus-like particles. J. Immunol. *151*, 1097–1107.

Lenz, P., Lowy, D.R., and Schiller, J.T. (2005). Papillomavirus virus-like particles induce cytokines characteristic of innate immune responses in plasmacytoid dendritic cells. Eur J. Immunol. *35*, 1548–1556.

Li, C., Liu, F., Liang, M., Zhang, Q., Wang, X., Wang, T., Li, J., and Li, D. (2010). Hantavirus-like particles generated in CHO cells induce specific immune responses in C57BL/6 mice. Vaccine *28*, 4294–4300.

Li, T.C., Yamakawa, Y., Suzuki, K., Tatsumi, M., Razak, M.A., Uchida, T., Takeda, N., and Miyamura, T. (1997). Expression and self-assembly of empty virus-like particles of hepatitis E virus. J. Virol. *71*, 7207–7213.

Link, A., Zabel, F., Schnetzler, Y., Titz, A., Brombacher, F., and Bachmann, M.F. (2012). Innate Immunity Mediates Follicular Transport of Particulate but Not Soluble Protein Antigen. J Immunol.

Lipin, D.I., Chuan, Y.P., Lua, L.H., and Middelberg, A.P. (2008). Encapsulation of DNA and non-viral protein changes the structure of murine polyomavirus virus-like particles. Arch Virol *153*, 2027–2039.

Liu, L., Celma, C.C., and Roy, P. (2008). Rift Valley fever virus structural proteins: expression, characterization and assembly of recombinant proteins. Virol. J. *5*, 82.

Liu, X.S., Abdul-Jabbar, I., Qi, Y.M., Frazer, I.H., and Zhou, J. (1998). Mucosal immunisation with papillomavirus virus-like particles elicits systemic and mucosal immunity in mice. Virology *252*, 39–45.

Liu, Y.V., Massare, M.J., Barnard, D.L., Kort, T., Nathan, M., Wang, L., and Smith, G. (2011). Chimeric severe acute respiratory syndrome coronavirus (SARS-CoV) S glycoprotein and influenza matrix 1 efficiently form virus-like particles (VLPs) that protect mice against challenge with SARS-CoV. Vaccine *29*, 6606–6613.

Lo-Man, R., Rueda, P., Sedlik, C., Deriaud, E., Casal, I., and Leclerc, C. (1998). A recombinant virus-like particle system derived from parvovirus as an efficient antigen carrier to elicit a polarized Th1 immune response without adjuvant. Eur J. Immunol. *28*, 1401–1407.

Mach, H., Volkin, D.B., Troutman, R.D., Wang, B., Luo, Z., Jansen, K.U., and Shi, L. (2006). Disassembly and reassembly of yeast-derived recombinant human papillomavirus virus-like particles (HPV VLPs). J. Pharm. Sci. *95*, 2195–2206.

Marshall, J.J., Fayard, B., and Roy, P. (1990). Biophysical studies on the morphology of baculovirus-expressed bluetongue virus tubules. J. Gen. Virol. *71*, 1839–1844.

Matsuo, E., Celma, C.C., Boyce, M., Viarouge, C., Sailleau, C., Dubois, E., Breard, E., Thiery, R., Zientara, S., and Roy, P. (2011). Generation of replication-defective virus-based vaccines that confer full protection in sheep against virulent bluetongue virus challenge. J. Virol. *85*, 10213–10221.

McAleer, W.J., Buynak, E.B., Maigetter, R.Z., Wampler, D.E., Miller, W.J., and Hilleman, M.R. (1984). Human hepatitis B vaccine from recombinant yeast. Nature *307*, 178–180.

McCarthy, M.P., White, W.I., Palmer-Hill, F., Koenig, S., and Suzich, J.A. (1998). Quantitative disassembly and reassembly of human papillomavirus type 11 viruslike particles in vitro. J. Virol. *72*, 32–41.

Mikhailov, M., Monastyrskaya, K., Bakker, T., and Roy, P. (1996). A new form of particulate single and multiple immunogen delivery system based on recombinant bluetongue virus-derived tubules. Virology *217*, 323–331.

Monastyrskaya, K., Gould, E.A., and Roy, P. (1995). Characterization and modification of the carboxy-terminal sequences of bluetongue virus type 10 NS1 protein in relation to tubule formation and location of an antigenic epitope in the vicinity of the carboxy terminus of the protein. J. Virol. *69*, 2831–2841.

Montross, L., Watkins, S., Moreland, R.B., Mamon, H., Caspar, D.L., and Garcea, R.L. (1991). Nuclear assembly of polyomavirus capsids in insect cells expressing the major capsid protein VP1. J. Virol. *65*, 4991–4998.

Moron, G., Rueda, P., Casal, I., and Leclerc, C. (2002). CD8alpha- CD11b+ dendritic cells present exogenous virus-like particles to CD8+ T cells and subsequently express CD8alpha and CD205 molecules. J. Exp. Med. *195*, 1233–1245.

Moron, V.G., Rueda, P., Sedlik, C., and Leclerc, C. (2003). In vivo, dendritic cells can cross-present virus-like particles using an endosome-to-cytosol pathway. J. Immunol. *171*, 2242–2250.

Murata, Y., Lightfoote, P.M., Rose, R.C., and Walsh, E.E. (2009). Antigenic presentation of heterologous epitopes engineered into the outer surface-exposed helix 4 loop region of human papillomavirus L1 capsomeres. Virol J *6*, 81.

Naldini, L., Blomer, U., Gallay, P., Ory, D., Mulligan, R., Gage, F.H., Verma, I.M., and Trono, D. (1996). In vivo gene delivery and stable transduction of nondividing cells by a lentiviral vector. Science *272*, 263–267.

Nassal, M., Skamel, C., Vogel, M., Kratz, P.A., Stehle, T., Wallich, R., and Simon, M.M. (2008). Development of hepatitis B virus capsids into a whole-chain protein antigen display platform: new particulate Lyme disease vaccines. Int. J. Med. Microbiol. *298*, 135–142.

Nawagitgul, P., Morozov, I., Bolin, S.R., Harms, P.A., Sorden, S.D., and Paul, P.S. (2000). Open reading frame 2 of porcine circovirus type 2 encodes a major capsid protein. J. Gen. Virol. *81*, 2281–2287.

Niikura, K., Nagakawa, K., Ohtake, N., Suzuki, T., Matsuo, Y., Sawa, H., and Ijiro, K. (2009). Gold nanoparticle arrangement on viral particles through carbohydrate recognition: a non-cross-linking approach to optical virus detection. Bioconjug. Chem. *20*, 1848–1852.

Niikura, M., Takamura, S., Kim, G., Kawai, S., Saijo, M., Morikawa, S., Kurane, I., Li, T.C., Takeda, N., and Yasutomi, Y. (2002). Chimeric recombinant hepatitis E virus-like particles as an oral vaccine vehicle presenting foreign epitopes. Virology *293*, 273–280.

Noad, R., and Roy, P. (2003). Virus-like particles as immunogens. Trends Microbiol. *11*, 438–444.

Noda, T., Sagara, H., Suzuki, E., Takada, A., Kida, H., and Kawaoka, Y. (2002). Ebola virus VP40 drives the formation of virus-like filamentous particles along with GP. J. Virol. *76*, 4855–4865.

Ohlschlager, P., Osen, W., Dell, K., Faath, S., Garcea, R.L., Jochmus, I., Muller, M., Pawlita, M., Schafer, K., Sehr, P., *et al.* (2003). Human papillomavirus type 16 L1 capsomeres induce L1-specific cytotoxic T lymphocytes and tumor regression in C57BL/6 mice. J. Virol. *77*, 4635–4645.

Oliveira-Ferreira, J., Miyahira, Y., Layton, G.T., Savage, N., Esteban, M., Rodriguez, D., Rodriguez, J.R., Nussenzweig, R.S., and Zavala, F. (2000). Immunogenicity of Ty-VLP bearing a CD8(+) T cell epitope of the CS protein of P. yoelii: enhanced memory response by boosting with recombinant vaccinia virus. Vaccine *18*, 1863–1869.

Ormo, M., Cubitt, A.B., Kallio, K., Gross, L.A., Tsien, R.Y., and Remington, S.J. (1996). Crystal structure of the Aequorea victoria green fluorescent protein. Science *273*, 1392–1395.

Ou, W.C., Hseu, T.H., Wang, M., Chang, H., and Chang, D. (2001). Identification of a DNA encapsidation sequence for human polyomavirus pseudovirion formation. J. Med. Virol. *64*, 366–373.

Ou, W.C., Wang, M., Fung, C.Y., Tsai, R.T., Chao, P.C., Hseu, T.H., and Chang, D. (1999). The major capsid protein, VP1, of human JC virus expressed in *Escherichia coli* is able to self-assemble into a capsid-like particle and deliver exogenous DNA into human kidney cells. J. Gen. Virol. *80*, 39–46.

Patch, J.R., Crameri, G., Wang, L.F., Eaton, B.T., and Broder, C.C. (2007). Quantitative analysis of Nipah virus proteins released as virus-like particles reveals central role for the matrix protein. Virol J *4*, 1.

Patel, K.G., and Swartz, J.R. (2011). Surface functionalization of virus-like particles by direct conjugation using azide-alkyne click chemistry. Bioconjug. Chem. *22*, 376–387.

Pejawar-Gaddy, S., Rajawat, Y., Hilioti, Z., Xue, J., Gaddy, D.F., Finn, O.J., Viscidi, R.P., and Bossis, I. (2010). Generation of a tumor vaccine candidate based on conjugation of a MUC1 peptide to polyionic papillomavirus virus-like particles. Cancer Immunol. Immunother. *59*, 1685–1696.

Peng, S., Frazer, I.H., Fernando, G.J., and Zhou, J. (1998). Papillomavirus virus-like particles can deliver defined CTL epitopes to the MHC class I pathway. Virology *240*, 147–157.

Perez de Diego, A.C., Athmaram, T.N., Stewart, M., Rodriguez-Sanchez, B., Sanchez-Vizcaino, J.M., Noad, R., and Roy, P. (2011). Characterization of protection afforded by a bivalent virus-like particle vaccine against bluetongue virus serotypes 1 and 4 in sheep. PLoS One *6*, e26666.

Poeschla, E.M., Wong-Staal, F., and Looney, D.J. (1998). Efficient transduction of nondividing human cells by feline immunodeficiency virus lentiviral vectors. Nat Med *4*, 354–357.

Qing, M., Liu, W., Yuan, Z., Gu, F., and Shi, P.Y. (2010). A high-throughput assay using dengue-1 virus-like particles for drug discovery. Antiviral Res *86*, 163–171.

Rohrmann, G.F. (1992). Baculovirus structural proteins. J. Gen. Virol. *73*, 749–761.

Rose, R.C., White, W.I., Li, M., Suzich, J.A., Lane, C., and Garcea, R.L. (1998). Human papillomavirus type 11 recombinant L1 capsomeres induce virus-neutralizing antibodies. J. Virol. *72*, 6151–6154.

Routh, A., Domitrovic, T., and Johnson, J.E. (2012). Host RNAs, including transposons, are encapsidated by a eukaryotic single-stranded RNA virus. Proc. Natl. Acad. Sci. U.S.A. *109*, 1907–1912.

Roy, P., Bishop, D.H., LeBlois, H., and Erasmus, B.J. (1994). Long-lasting protection of sheep against bluetongue challenge after vaccination with virus-like particles: evidence for homologous and partial heterologous protection. Vaccine *12*, 805–811.

Roy, P., Boyce, M., and Noad, R. (2009). Prospects for improved bluetongue vaccines. Nat. Rev .Microbiol. *7*, 120–128.

Roy, P., French, T., and Erasmus, B.J. (1992). Protective efficacy of virus-like particles for bluetongue disease. Vaccine *10*, 28–32.

Roy, P., and Noad, R. (2008). Virus-like particles as a vaccine delivery system: myths and facts. Hum. Vaccin. *4*, 5–12.

Roy, P., and Noad, R. (2009). Virus-like particles as a vaccine delivery system: myths and facts. Adv. Exp. Med. Biol. *655*, 145–158.

Roy, P., Urakawa, T., Van Dijk, A.A., and Erasmus, B.J. (1990). Recombinant virus vaccine for bluetongue disease in sheep. J. Virol. *64*, 1998–2003.

Rudolf, M.P., Fausch, S.C., Da Silva, D.M., and Kast, W.M. (2001). Human dendritic cells are activated by chimeric human papillomavirus type-16 virus-like particles and induce epitope-specific human T cell responses in vitro. J. Immunol. *166*, 5917–5924.

Rueda, P., Hurtado, A., del Barrio, M., Martinez-Torrecuadrada, J.L., Kamstrup, S., Leclerc, C., and Casal, J.I. (1999). Minor displacements in the insertion site provoke major differences in the induction of antibody responses by chimeric parvovirus-like particles. Virology *263*, 89–99.

Ruedl, C., Storni, T., Lechner, F., Bachi, T., and Bachmann, M.F. (2002). Cross-presentation of virus-like particles by skin-derived CD8(–) dendritic cells: a dispensable role for TAP. Eur J. Immunol. *32*, 818–825.

Rutkowska, D.A., Meyer, Q.C., Maree, F., Vosloo, W., Fick, W., and Huismans, H. (2011). The use of soluble African horse sickness viral protein 7 as an antigen delivery and presentation system. Virus Res. *156*, 35–48.

Sanchez-Rodriguez, S.P., Munch-Anguiano, L., Echeverria, O., Vazquez-Nin, G., Mora-Pale, M., Dordick, J.S., and Bustos-Jaimes, I. (2011). Human parvovirus B19 virus-like particles: In vitro assembly and stability. Biochimie *94*, 870–878.

Sathish, N., Wang, X., and Yuan, Y. (2012). Tegument Proteins of Kaposi's sarcoma-associated herpesvirus and related gamma-herpesviruses. Front. Microbiol. *3*, 98.

Savard, C., Laliberte-Gagne, M.E., Babin, C., Bolduc, M., Guerin, A., Drouin, K., Forget, M.A., Majeau, N., Lapointe, R., and Leclerc, D. (2012). Improvement of the PapMV nanoparticle adjuvant property through an increased of its avidity for the antigen [influenza NP]. Vaccine *30*, 2535–2542.

Schadlich, L., Senger, T., Gerlach, B., Mucke, N., Klein, C., Bravo, I.G., Muller, M., and Gissmann, L. (2009). Analysis of modified human papillomavirus type 16 L1 capsomeres: the ability to assemble into larger particles correlates with higher immunogenicity. J. Virol. *83*, 7690–7705.

Schodel, F., Wirtz, R., Peterson, D., Hughes, J., Warren, R., Sadoff, J., and Milich, D. (1994). Immunity to malaria elicited by hybrid hepatitis B virus core particles carrying circumsporozoite protein epitopes. J. Exp. Med. *180*, 1037–1046.

Sedlik, C., Saron, M., Sarraseca, J., Casal, I., and Leclerc, C. (1997). Recombinant parvovirus-like particles as an antigen carrier: a novel nonreplicative exogenous antigen to elicit protective antiviral cytotoxic T cells. Proc. Natl. Acad. Sci. U.S.A. *94*, 7503–7508.

Slupetzky, K., Gambhira, R., Culp, T.D., Shafti-Keramat, S., Schellenbacher, C., Christensen, N.D., Roden, R.B., and Kirnbauer, R. (2007). A papillomavirus-like particle (VLP) vaccine displaying HPV16 L2 epitopes induces cross-neutralizing antibodies to HPV11. Vaccine *25*, 2001–2010.

Slupetzky, K., Shafti-Keramat, S., Lenz, P., Brandt, S., Grassauer, A., Sara, M., and Kirnbauer, R. (2001). Chimeric papillomavirus-like particles expressing a foreign epitope on capsid surface loops. J. Gen. Virol. *82*, 2799–2804.

Song, H., Wittman, V., Byers, A., Tapia, T., Zhou, B., Warren, W., Heaton, P., and Connolly, K. (2010). In vitro stimulation of human influenza-specific CD8+ T cells by dendritic cells pulsed with an influenza virus-like particle (VLP) vaccine. Vaccine *28*, 5524–5532.

Stewart, M., Bhatia, Y., Athmaran, T.N., Noad, R., Gastaldi, C., Dubois, E., Russo, P., Thiery, R., Sailleau, C., Breard, E., *et al.* (2009). Validation of a novel approach for the rapid production of immunogenic virus-like particles for bluetongue virus. Vaccine *28*, 3047–3054.

Stewart, M., Dovas, C.I., Chatzinasiou, E., Athmaram, T.N., Papanastassopoulou, M., Papadopoulos, O., and Roy, P. (2012). Protective efficacy of Bluetongue virus-like and subvirus-like particles in sheep: presence of the serotype-specific VP2, independent of its geographic lineage, is essential for protection. Vaccine *30*, 2131–2139.

Sun, J., DuFort, C., Daniel, M.C., Murali, A., Chen, C., Gopinath, K., Stein, B., De, M., Rotello, V.M., Holzenburg, A., *et al.* (2007). Core-controlled polymorphism in virus-like particles. Proc. Natl. Acad. Sci. U.S.A. *104*, 1354–1359.

Sun, S., Li, W., Sun, Y., Pan, Y., and Li, J. (2011). A new RNA vaccine platform based on MS2 virus-like particles produced in *Saccharomyces cerevisiae*. Biochem. Biophys. Res. Commun. *407*, 124–128.

Swenson, D.L., Warfield, K.L., Kuehl, K., Larsen, T., Hevey, M.C., Schmaljohn, A., Bavari, S., and Aman, M.J. (2004). Generation of Marburg virus-like particles by co-expression of glycoprotein and matrix protein. FEMS Immunol. Med. Microbiol. *40*, 27–31.

Tamminen, K., Huhti, L., Koho, T., Lappalainen, S., Hytonen, V.P., Vesikari, T., and Blazevic, V. (2011). A comparison of immunogenicity of norovirus GII-4 virus-like particles and P-particles. Immunology *135*, 89–99.

Tegerstedt, K., Franzen, A., Ramqvist, T., and Dalianis, T. (2007). Dendritic cells loaded with polyomavirus VP1/VP2Her2 virus-like particles efficiently prevent outgrowth of a Her2/neu expressing tumor. Cancer Immunol. Immunother. *56*, 1335–1344.

Tegerstedt, K., Franzen, A.V., Andreasson, K., Joneberg, J., Heidari, S., Ramqvist, T., and Dalianis, T. (2005). Murine polyomavirus virus-like particles (VLPs) as vectors for gene and immune therapy and vaccines against viral infections and cancer. Anticancer Res. *25*, 2601–2608.

Thones, N., Herreiner, A., Schadlich, L., Piuko, K., and Muller, M. (2008). A direct comparison of human papillomavirus type 16 L1 particles reveals a lower immunogenicity of capsomeres than viruslike particles with respect to the induced antibody response. J. Virol. *82*, 5472–5485.

Tsunetsugu-Yokota, Y., Morikawa, Y., Isogai, M., Kawana-Tachikawa, A., Odawara, T., Nakamura, T., Grassi, F., Autran, B., and Iwamoto, A. (2003). Yeast-derived human immunodeficiency virus type 1 p55(gag) virus-like particles activate dendritic cells (DCs) and induce perforin expression in Gag-specific CD8(+) T cells by cross-presentation of DCs. J. Virol. *77*, 10250–10259.

Ulrich, R., Koletzki, D., Lachmann, S., Lundkvist, A., Zankl, A., Kazaks, A., Kurth, A., Gelderblom, H.R., Borisova, G., Meisel, H., *et al.* (1999). New chimaeric hepatitis B virus core particles carrying hantavirus (serotype Puumala) epitopes: immunogenicity and protection against virus challenge. J. Biotechnol. *73*, 141–153.

Varsani, A., Williamson, A.L., de Villiers, D., Becker, I., Christensen, N.D., and Rybicki, E.P. (2003). Chimeric human papillomavirus type 16 (HPV-16) L1 particles presenting the common neutralizing epitope for the L2 minor capsid protein of HPV-6 and HPV-16. J. Virol. *77*, 8386–8393.

Velthuis, A.G., Saatkamp, H.W., Mourits, M.C., de Koeijer, A.A., and Elbers, A.R. (2009). Financial consequences of the Dutch bluetongue serotype 8 epidemics of 2006 and 2007. Prev. Vet. Med. *93*, 294–304.

Vogel, M., Diez, M., Eisfeld, J., and Nassal, M. (2005a). In vitro assembly of mosaic hepatitis B virus capsid-like particles (CLPs): rescue into CLPs of assembly-deficient core protein fusions and FRET-suited CLPs. FEBS Lett. *579*, 5211–5216.

Vogel, M., Vorreiter, J., and Nassal, M. (2005b). Quaternary structure is critical for protein display on capsid-like particles (CLPs): efficient generation of hepatitis B virus CLPs presenting monomeric but not dimeric and tetrameric fluorescent proteins. Proteins *58*, 478–488.

von Brunn, A., Brand, M., Reichhuber, C., Morys-Wortmann, C., Deinhardt, F., and Schodel, F. (1993). Principal neutralizing domain of HIV-1 is highly immunogenic when expressed on the surface of hepatitis B core particles. Vaccine *11*, 817–824.

Walker, A., Skamel, C., and Nassal, M. (2012). SplitCore: An exceptionally versatile viral nanoparticle for native whole protein display regardless of 3D structure. Sci. Rep. *1*, 5.

Walker, A., Skamel, C., Vorreiter, J., and Nassal, M. (2008). Internal core protein cleavage leaves the hepatitis B virus capsid intact and enhances its capacity for surface display of heterologous whole chain proteins. J. Biol. Chem. *283*, 33508–33515.

Walpita, P., Barr, J., Sherman, M., Basler, C.F., and Wang, L. (2011). Vaccine potential of Nipah virus-like particles. PLoS One *6*, e18437.

Wang, X.J., Gu, K., Xiong, Q.Y., Shen, L., Cao, R.Y., Li, M.H., Li, T.M., Wu, J., and Liu, J.J. (2009). A novel virus-like particle based on hepatitis B core antigen and substrate-binding domain of bacterial molecular chaperone DnaK. Vaccine *27*, 7377–7384.

Warfield, K.L., Bosio, C.M., Welcher, B.C., Deal, E.M., Mohamadzadeh, M., Schmaljohn, A., Aman, M.J., and Bavari, S. (2003). Ebola virus-like particles protect from lethal Ebola virus infection. Proc. Natl. Acad. Sci. U.S.A. *100*, 15889–15894.

Win, S.J., McMillan, D.G., Errington-Mais, F., Ward, V.K., Young, S.L., Baird, M.A., and Melcher, A.A. (2011). Enhancing the immunogenicity of tumour lysate-loaded dendritic cell vaccines by conjugation to virus-like particles. Br. J. Cancer *106*, 92–98.

Wu, H.L., Chen, P.J., Mu, J.J., Chi, W.K., Kao, T.L., Hwang, L.H., and Chen, D.S. (1997). Assembly of hepatitis delta virus-like empty particles in yeast. Virology *236*, 374–381.

Wu, M., Brown, W.L., and Stockley, P.G. (1995). Cell-specific delivery of bacteriophage-encapsidated ricin A chain. Bioconjug Chem *6*, 587–595.

Wu, M., Sherwin, T., Brown, W.L., and Stockley, P.G. (2005). Delivery of antisense oligonucleotides to leukaemia cells by RNA bacteriophage capsids. Nanomedicine *1*, 67–76.

Yang, R., Murillo, F.M., Lin, K.Y., Yutzy, W.H.t., Uematsu, S., Takeda, K., Akira, S., Viscidi, R.P., and Roden, R.B. (2004). Human papillomavirus type-16 virus-like particles activate complementary defence responses in key dendritic cell subpopulations. J. Immunol. *173*, 2624–2631.

Yao, Q., Bu, Z., Vzorov, A., Yang, C., and Compans, R.W. (2003). Virus-like particle and DNA-based candidate AIDS vaccines. Vaccine *21*, 638–643.

Ye, L., Lin, J., Sun, Y., Bennouna, S., Lo, M., Wu, Q., Bu, Z., Pulendran, B., Compans, R.W., and Yang, C. (2006). Ebola virus-like particles produced in insect cells exhibit dendritic cell stimulating activity and induce neutralizing antibodies. Virology *351*, 260–270.

Young, K.R., McBurney, S.P., Karkhanis, L.U., and Ross, T.M. (2006). Virus-like particles: designing an effective AIDS vaccine. Methods *40*, 98–117.

Zhang, R., Li, M., Chen, C., and Yao, Q. (2004). SHIV virus-like particles bind and activate human dendritic cells. Vaccine *23*, 139–147.

Zhao, Q., Allen, M.J., Wang, Y., Wang, B., Wang, N., Shi, L., and Sitrin, R.D. (2012a). Disassembly and reassembly improves morphology and thermal stability of human papillomavirus type 16 virus-like particles. Nanomedicine [Epub ahead of print].

Zhao, Q., Modis, Y., High, K., Towne, V., Meng, Y., Wang, Y., Alexandroff, J., Brown, M., Carragher, B., Potter, C.S., *et al.* (2012b). Disassembly and reassembly of human papillomavirus virus-like particles produces more virion-like antibody reactivity. Virol J. *9*, 52.

Plant Oil Bodies and Oleosins: Structure, Function and Biotechnological Applications

8

Anisha David, Sunita Yadav and Satish Chander Bhatla

Abstract

Although oil bodies are present in a wide variety of tissues in plants, it is their abundance in the oilseed cotyledons that has been most extensively investigated for their biogenesis, structure, physiological roles, isolation and biotechnological applications. The phospholipid monolayer membrane of the oil bodies encasing the triacylglycerol (TAG) matrix not only possesses a set of structural and functional proteins (oleosins, steroleosins and caleosins), they also exhibit quite a few enzymatic and non-enzymatic proteins on their surface (lipoxygenase, protease and phospholipase) whose expression is transient and depends on the stage of oil body mobilization during seed germination. These transiently expressed signalling molecules are under the influence of various environmental and consequent physiological factors for their roles in oil body mobilization during seed germination. Based on these features of oil bodies to attract and bind a variety of biomolecules on their surface, oil body preparations have been put to extensive biotechnological applications, which are also being discussed in this review.

Distribution of lipid bodies in oleaginous plants

Oilseeds store energy in the form of lipids to support periods of active metabolism during seed germination and seedling growth. Lipids are stored in specialized organelles called as lipid bodies, oil bodies (OBs) or oleosomes (Fig. 8.1). Most of our knowledge pertaining to OBs has been attained using oilseeds as model systems. OBs have been reported in the cotyledons of exalbuminous dicot seeds (rapeseed and sunflower), endosperm of albuminous dicot seeds (Shewry and Casey, 1999), scutellum (Qu and Huang, 1990) and aleurone layer (Aalen *et al.*, 1994) of monocot seeds (Murphy and Vance, 1999). Plants have also been reported to accumulate OBs in postembryonic tissues, such as fruit mesocarp (Ross *et al.*, 1993; Ozdemir and Topuz, 2004), pollen grains, anther tapetum, leaves, apical buds of temperate perennials, like *Betula, Populus* (Rinne *et al.*, 2011), carnation petals, stem (Mongolian oil wood) and underground organs (such as the roots and tubers of cotton and nutsedge). Lipid bodies have also been reported in root tips or root caps of rice (Sargent and Osborne, 1980), pea and maize (Craig and Staehelin, 1988) and garden cress (Hensel, 1986; Busch *et al.*, 1993). The root cap cells act as statocytes (gravity sensors) and their OBs, apart from the storage function, appear to determine polar distribution of ER, possibly enabling the positive orthogravitropic root growth (Murphy *et al.*, 2001). A large and conspicuous oil droplet is also present in the cytoplasm of mesophyll cells (Lersten *et al.*,

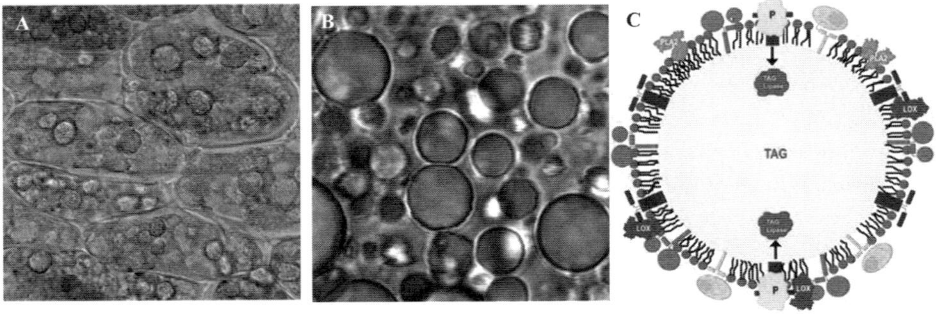

Figure. 8.1 (A) Distribution of oil bodies in intact cells of sunflower seedling (2-day-old) cotyledon cross section (400×). (B) Isolated oil bodies from 2-day-old sunflower seedling cotyledons (1000×). (C) Representation of the detailed structure of the OB from seedling cotyledon, associated with various structural and functional proteins (modified from Bhatla *et al.*, 2009), namely oleosin, caleosin, steroleosin, protease, phospholipase, TAG lipase and lipoxygenase. A colour version of this figure is located in the plate section at the back of the book.

2006). Starchy seed plants, like rice, have abundant storage lipids (up to 34%) accumulated in OBs present in the embryo and aleurone layer.

Structure of oil bodies

Oil bodies range from 0.5–2.5 Mm in diameter, depending on the seed type. Orthodox seeds, that undergo desiccation, contain large number of small OBs of 0.5–1 Mm diameter whereas recalcitrant or non-desiccating seeds house larger OBs (Frandsen *et al.*, 2001; Purkrtova *et al.*, 2008; Roberts *et al.*, 2008)). Preparations from seeds contain 94–98% neutral lipids, 0.6–2% phospholipids and 0.6–3% proteins (Tzen *et al.*, 1993). Triacylglycerols (TAGs) are the principal constituents of OB matrix. The central core of lipids is enveloped by a phospholipid (PL) monolayer (Fig. 8.1). Its acyl moieties interact with the TAG matrix and the PL head is exposed to the cytoplasm (Frandsen *et al.*, 2001; Purkrtova *et al.*, 2008; Tzen *et al.*, 1993; Beisson *et al.*, 1996). The principal phospholipid constituents of OB membrane include phosphatidylcholine (50–60%), phosphatidylserine (20–30%) and phosphatidylethanolamine, phosphatidylinositol and phosphatidic acid as minor constituents. Three major groups of proteins are specifically associated with plant OBs, out of which two constitute the structural component (oleosins and caleosins) and the third (steroleosin) is an enzyme. Apart from them, some minor proteins, like aquaporins, are also present in the OB membrane (Turesson *et al.*, 2010; Capuano *et al.*, 2007).

Oil bodies from tissues other than seeds may exhibit certain structural variations. Thus, OBs from the mesocarp cells of olive (*Olea europaea*), palm (*Elaeis guineensis*) and avocado (*Persea americana*) are very large, ranging from 10 to 20 μm in diameter (Capuano *et al.*, 2007; Shimada and Nishimura, 2010). These OBs are apparently deficient in oleosins and their lipidic constituents serve to attract animals for seed dispersal. Lack of oleosins does not restrain coalescence of OBs and the oil droplets become large during fruit maturation (Salas *et al.*, 2000). Tapetal cells also accumulate high amount of lipids in two distinct types of cell organelles, namely elioplasts and tapetosomes (Platt *et al.*, 1998). Elioplasts are specialized chloroplasts, 3–4 μm in diameter. They are devoid of thylakoids and contain OBs (plastoglobuli) rich in steryl esters enclosed by plastid lipid-associated proteins or fibrillin-like proteins (Wu *et al.*, 1997, 1999; Hernandez-Pinzon *et al.*, 1999). Tapetosomes are multivesicular bodies of 2–3 μm diameter, containing TAGs enclosed in a PL monolayer, stabilized by oleosins, as in seeds (Wu *et al.*, 1997; Murphy and Ross, 1998; Platt *et al.*, 1998; Ting *et al.*, 1998; Hernandez-Pinzon *et al.*, 1999). Tapetosomes have primarily been reported in insect/self-pollinating Brassicaceae family, especially in *Brassica* and *Arabidopsis* (de Oliveria *et al.*, 1993; Roberts *et al.*, 1994). A tapetosome consists of a cluster of small OBs (0.1–0.5 μm in diameter) sequestered among numerous membranous vesicles. OBs in tapetum are similar to the seed OBs, consisting of a neutral lipid matrix, rich in TAGs, and enveloped by a PL monolayer studded with amphipathic structural proteins, oleosins. The membranous vesicles possess ER-derived calreticulin, luminal binding proteins and other unknown molecules of the ER lumen. In the early stages of tapetum cell development, calreticulin exists as a network and is devoid of oleosins. Subsequently, oleosins appear with calreticulin in the ER network and gradually outnumber calreticulin. Numerous maturing tapetosomes are interconnected with ER cisternae but at maturity this cisternal network disappears and solitary tapetosomes containing numerous OBs and membranous vesicles held together by ionic interactions remain. These vesicles are proposed to aid transfer of oleosins onto the pollen exine. In a recent survey of 302 angiosperm species, Lersten *et al.* (2006) reported that 24% species have OBs in leaves, with usually a large and conspicuous OB per mesophyll cell. OBs from leaves are reported to range from 1 to 18 μm in diameter. Analysis of OBs from wheat leaves showed oleosin to have 50–60% triglycerides and 15–40% esters of wax (Murphy and Parker, 1984). Though a specific role has not been assigned to leaf TAGs but they are envisaged to be participating in carbon storage and membrane lipid remodelling (Murphy and Parker, 1984; Murphy, 2001; Kaup *et al.*, 2002; Lin and Oliver, 2008).

Oil body biogenesis

Oil bodies are formed at distinct sub-domains of the ER by the accumulation of TAGs between the PL bilayer. The constituents of OBs, which includes PL, TAG and oleosins, are also synthesized on the ER. Diacylglycerol acyl-transferase (DAG-AT), an enzyme facilitating the last step of TAG synthesis, is exclusively localized on ER within discrete regions. TAGs thus synthesized get simultaneously deposited between the phospholipid bilayer where OBs emerge, with an envelope of PL monolayer and eventually pinched off into the cytoplasm (Sarmiento *et al.*, 1997; Hsieh and Huang, 2004; Purktrova *et al.*, 2008). In the process, the PL monolayer is embedded with oleosins and it is the oleosin/TAG ratio which determines the size of OBs. Oleosins are synthesized on the polyribosomes at the specialized ER domains. Oleosins probably contain a functional ER targeting signal sequence, either in the N-terminus or the central hydrophobic region, as their deletion impaired the oleosin targeting to lipid bodies (Abell *et al.*, 1997). In tissues where oleosin is less abundant during the early phase of seed development, the OBs formed are comparatively deficient in oleosins than in mature seeds. Later, during seed development, when oleosin is more abundant, OBs either coalesce or fuse again with the ER to attain the appropriate oleosin/TAG ratio (Sarmiento *et al.*, 1997).

Integral protein constituents of oil bodies

Three plant specific oil body membrane protein families have been identified through proteomic analysis of isolated OBs from seeds. These are oleosins, caleosins and steroleosins.

Oleosins

These are the most abundant oil body-associated structural proteins, which are ubiquitous in oil storing seeds and their primary function is to stabilize OBs. They represent up to 0.4–1.6% of seed weight (Huang, 1992) and approximately 10% of the total seed proteins in sunflower (Murphy *et al.*, 2001). Oleosins were first isolated from *Arachis hypogaea* seeds (Jacks *et al.*, 1967). Although only one oleosin isoform has been detected in gymnosperms, more than two immunologically distinct oleosin isoforms are known to occur in angiosperms (Murphy *et al.*, 2001; Lee *et al.*, 1994). Oleosins are the unique constituents of plant OBs only. Seed oleosins can be categorized into high- and low-molecular-weight proteins (Tzen *et al.*, 1990; Kalinski *et al.*, 1991; Radetzky and Langheinrich, 1994). Oleosin genes are expressed during seed development and maturation, and are also induced via abscisic acid and desiccation (Hatzopoulos *et al.*, 1990; Plant *et al.*, 1994). In *Arabidopsis*, oleosins are encoded by a multigenic family composed of around 16 genes-5 in seed type, eight in tapetum and three in seed and anther-type oleosins (Shimada and Nishimura, 2010). OLEOSIN 1 and OLEOSIN 2 proteins are found to be abundant in *Arabidopsis* seeds (Shimada and Nishimura, 2010). However, all the oleosin proteins are still not biochemically identified.

The primary structure of oleosins remains highly conserved within all plant species. It consists of three domains – a highly hydrophobic central domain flanked by relatively polar C and N terminus (Vance and Huang, 1987). Oleosins are a class of alkaline proteins of 15–30 kDa (in seeds) and 10–50 kDa (in anther). This variation is predominantly due to different lengths of the N- and C-terminal domains. The central region is composed of 72 non-polar amino acids (Huang, 1996; Purkrtova *et al.*, 2008), which are aliphatic in nature (alanine, glycine, leucine, isoleucine and valine). Hydrophobic region penetrates into the PL monolayer while the amphipathic N and C terminal domains are exposed to the cytoplasm, thus interacting with the aqueous surroundings and PL head groups (Huang, 1992; Murphy, 1993) (Fig. 8.2). The negatively charged residues of the N- and C-terminal amphipathic domains facing outward are responsible for maintaining OBs as distinct entities via steric hindrance and electrostatic repulsion (Tzen and Huang, 1992). In orthodox seeds, high amounts of oleosins are present especially at the end of seed maturation and rehydration during germination. This protects the OBs from aggregation and coalescence during desiccation in seeds and also in pollen grains. Wild type seeds with high oleosin content have small OBs while double mutant seeds with low oleosin content have larger OBs. This property is crucial during the compaction process accompanying cellular dehydration in maturing orthodox seeds.

The N-terminal region can be long or short (6–68 residues) while the C-terminal can be short or very long (28–1000 residues) (Hsieh and Huang, 2004). These terminal domains lie at or on the OB surface and they also act as receptors for the binding of lipases or glyoxysomes, thereby facilitating rapid release of energy during seed germination and post-germinative growth of seedlings. The hydrophobic domain is highly conserved in plants and thought to be the longest hydrophobic domain known to occur in nature (Hsieh and Huang, 2004). The central region of this hydrophobic stretch comprises three proline residues and one Ser residue (-PX_5SPX_3P-, where X represents hydrophobic amino acids) (Fig. 8.2). These residues being relatively less hydrophobic, interact to form a Pro-knot, thus acting as

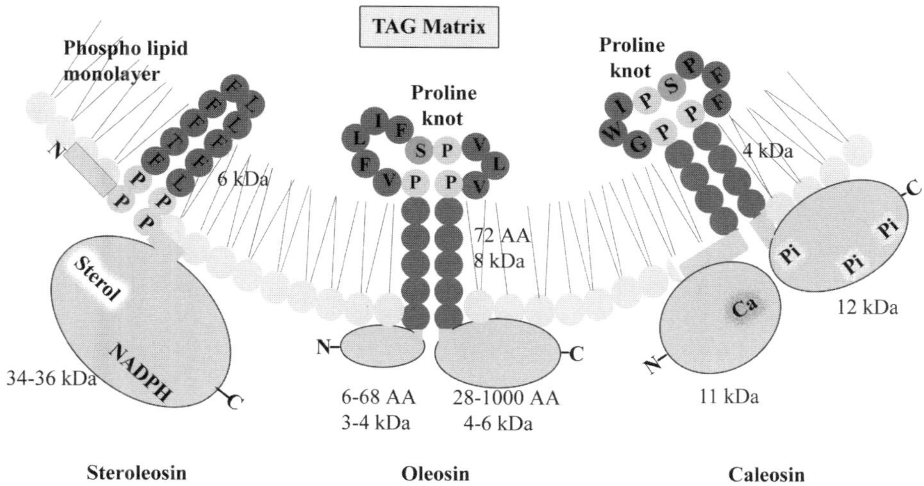

Figure 8.2 Detailed structure of steroleosin (A), oleosin (B) and caleosin (C) anchored to the phospholipid monolayer of oil body with their conserved hydrophobic stretch and their hydrophilic N and C terminals facing the cytosol. The amino acid F represents phenylalanine, L (leucine), T (threonine), P (proline), S (serine), V (valine), I (isoleucine), W (tryptophan), and G (glycine). The phosphorylation sites in the hydrophilic C-terminus of caleosin are depicted as 'Pi' and the calcium binding EF hand as 'Ca'. A colour version of this figure is located in the plate section at the back of the book.

breakers or turners and provide a hairpin structure to the whole hydrophobic stretch. This motif is known to be indispensable for the stable anchoring of protein in the TAG core of the lipid body (Abell *et al.*, 1997). Owing to low solubility of oleosins in aqueous media, the secondary, tertiary and quaternary structures of the central domain are still not clear. The two arms of the hydrophobic stretch are proposed to be present in an antiparallel α-helical structure (Lacey *et al.*, 1998) or in two antiparallel β strands (Huang, 1996) with the proline knot deeply embedded in the TAG core (Napier *et al.*, 1996; Hsieh and Huang, 2004; Capuano *et al.*, 2007). The β strands were further suggested to be folded into a 6-stranded antiparallel β-barrel. An intermolecular β-sheet structure extending around the surface of oil body is believed to be formed by the parallel arrangement of separate oleosin strands (Li *et al.*, 1992, 2002). Natural formation of hydrogen bonds between the adjacent oleosin molecules could be supported by the formation of natural homoligomers when purifying OBs even with strong denaturants. Two additional tertiary and quartenary arrangements can be extrapolated from the intermolecular β-sheet structure (antiparallel β-sheets) and a mixture of parallel and antiparallel β-sheets (Roberts *et al.*, 2008). Furthermore, the hydrophobic domain has been proposed to consist of two α helices with the proline knot motif forming a 180° turn and the whole structure is thought to be monomeric and stabilized by internal H bonds with no β-sheet (Lacey *et al.*, 1998; Alexander *et al.*, 2002). Thus, a single oil body can simultaneously contain more than one of these forms based on the nature and length of hydrophobic domain and also the plasticity of the surrounding environment (Roberts *et al.*, 2008). The C-terminal region comprises of several repeats of short peptides, which are not conserved among oleosins and they may not be functionally significant. The N- terminal domain is highly variable and no specific secondary structure is described for it. In comparison, the C-terminal domain contains α helical domain of 30–40 residues.

Presence of oleosins maintains small OB size, rendering high surface to volume ratio in OBs, thus facilitating access by lipases and enabling accelerated TAG mobilization (Dmitriev *et al.*, 1999). Lipases are known to be membrane – associated enzymes that cleave mono-, di- and triacylglycerol into free fatty acids and glycerol at the oil/water interphase during seed germination and seedling growth (Graham, 2008). It has been suggested that oleosin may contain a lipase binding site or may act as a lipase activator (Huang, 1992; Murphy and Vance, 1999; Chapman *et al.*, 2012), thus enabling lipase-mediated TAG hydrolysis. Thus, oleosins play a pivotal role during seed germination and seedling growth. By using single and multiple mutants for oleosin, Shimada *et al.* (2008) observed that seed germination rates were positively co-related with oleosin levels in seeds.

Oleosin is reported to confer freezing tolerance in *Arabidopsis* seeds (Shimada *et al.*, 2008). It has been demonstrated that mutants for oleosin were sensitive to freezing stress whereas, the wild type seeds exposed to freezing germinated normally. Secondly, in oleosin mutants freezing caused OBs to fuse and OB size enlargement is suggested to cause germination defects. Thus, oleosins increase the viability of overwintering seeds by preventing the abnormal fusion of OBs during seed germination. This function of oleosins could be exploited by their overexpression to produce seeds tolerant to freezing. Some of the oleosins reported in OBs from dormant apices of temperate perennials like *Betula* and *Populus* appear similar to *Arabidopsis* seed and pollen oleosin, suggesting that these OBs also play a similar role during overwintering (Van Der Schoot *et al.*, 2011). It is speculated that OBs in mesophyll cells act as intermediate storage for photosynthates and could have other functions as well (Lersten *et al.*, 2006). A possible role of lipid bodies in leaves is adaptation to cold as *Diapensia lapponica* (an evergreen plant of the Arctic and sub-Arctic regions) has a single large OB in summers but many smaller OB per mesophyll cell during winters (Pihakaski *et al.*, 1987). The smaller OB size suggests increase in the amount of oleosin that helps to withstand the extreme winter temperatures. It is speculated that OBs in mesophyll cells act as intermediate storage for photosynthates and could have other functions as well (Lersten *et al.*, 2006). Though the function of oleosins released from tapetosomes is not clear, but it appears that due to its amphipathic property it acts first as an emulsifying agent to uniformly coat the pollen with steryl esters for waterproofing. Secondly, it can also act as a wick and aid in the uptake of water from the interior of stigma, required for pollen germination, especially in Brassicaceae members which have dry stigma. The pollen oleosin has numerous repeats of short peptides which are rich in glycine, serine and lysine and because of this nature they are speculated to be involved in the interaction with cell walls of stigma cells (Huang, 2006).

Caleosins

Apart from oleosins, OBs contain a minor group of 30-kDa proteins, named as caleosins. These proteins appear to have both structural and functional roles in the life cycle of lipid body. Specific subdomains of ER and lipid bodies are found to have different caleosin isoforms. Structurally they are similar to oleosins with a conserved hydrophobic domain containing a proline knot motif, flanked by hydrophilic N and C terminal domains. However, caleosins differ from oleosins in having significantly larger N and C terminal domains (Fig. 8.2). Furthermore, the N terminal region in caleosins has a single calcium binding site, called as a helix–loop–helix EF hand motif, and has an enzymatic activity as peroxygenase (Chen *et al.*, 1999; Hanano *et al.*, 2006). The C terminal hydrophilic region in most of the plant proteins has four conserved protein kinase phosphorylation sites, three for casein kinase II and one for a protein tyrosine kinase (Frandsen *et al.*, 2001). Some of these sites can also be observed in algal and fungal proteins. Although phosphorylation by casein kinase II has been documented in plants, tyrosine phosphorylation seems less likely. Both N- and C-terminal domains have conserved histidine residues which together coordinate the binding of a haem prosthetic group. They may also be associated with lipid trafficking, membrane expansion and docking of peroxisomes to lipid bodies. Certain plant species, such as cycads, contain caleosins rather than oleosins as their major lipid body-associated protein, thus providing an evidence for the role of caleosin in structural maintenance of lipid bodies. Although caleosin potentially contributes to lipid body stability, its primary role is in cytoplasmic signalling through calcium binding motif and a few phosphorylation sites at the hydrophilic C terminal domain. Some caleosin homologues are found to be induced in leaves and stem in response to drought and osmotic stress. Caleosin has a unique position among the calcium binding proteins as it has only one EF hand. Thus, caleosins may form cis- or trans-dimeric associations with other caleosins, like in promoting fusion of lipid bodies with vacuoles for mobilization of storage lipids. Seven caleosin genes have been identified in *Arabidopsis*, one of which is strongly expressed in seeds. They are also associated with seed lipid bodies, root tips, pollen and shoots. Binding of calcium is crucial for their peroxygenase activity to catalyse the hydroperoxide-dependent oxygenation of unsaturated fatty acids to produce epoxy fatty acids. Such lipid transformations are commonly related with oxylipin metabolism in plants, which involves the production of various hormone-like molecules connected with plant stress response and innate immunity. They are also proposed to be associated in the stress signalling pathway in shoots (Van Der Schoot *et al.*, 2011). Caleosins are also involved in the degradation of polyunsaturated fatty acids. Caleosin proteins are N-acetylated, phosphorylated and ubiquitinated (like oleosins) under some conditions, thus indicating their regulation in plant cells in a highly coordinated manner. Caleosins have important signalling roles in both seed and non-seed tissues. Like in *Arabidopsis*, out of the five caleosin genes present in its genome, only one is highly expressed in developing seeds while the remaining four are expressed throughout the plant during normal growth and development, although at lower abundance. However, the manner in which caleosins interact with lipid bodies to influence the signalling components or in generating peroxygenase-derived lipid metabolites still remains to be elucidated.

Steroleosins

A second minor 40 kDa protein belonging to SDR (short chain dehydrogenase reductase) (Purkrtova *et al.*, 2008) superfamily is also found in lipid bodies. It is called sterol dehydrogenase or steroleosin. Steroleosin comprises of two structural domains- a N terminal, oil body anchoring domain and a sterol binding dehydrogenase domain, consisting of a conserved NADPH binding subdomain and a diversified sterol binding sub domain (Bhatla *et al.*, 2009; Van Der Schoot *et al.*, 2011). Unlike oleosins and caleosins, the pro-knot motif (Pro-Pro/Phe/Ile/Leu-10 hydrophobic amino acids-Pro-Pro/Phe) is displaced to the hydrophobic N-terminal domain and perhaps because of this shift, steroleosins cannot stabilize artificial oil bodies (Shimada and Nishimura, 2010). This N-terminal domain anchors steroleosin in the TAG matrix (Fig. 8.2). Steroleosin is proposed to specifically facilitate mobilization of lipid bodies during seed germination and is also involved in signal transduction in various plant tissues. Similar to caleosin, steroleosin is not seed specific and is also present in post-embryonic tissues. Eight steroleosin genes having variable sterol binding domains are reported in *Arabidopsis,* suggesting that they have diversified to bind different regulatory sterols during the initiation of diverse signalling events. Steroleosin, like oleosin and caleosin, does not have a cleavable signal sequence and post-translational modification. Although it is supposed to be related to sterol biosynthesis, its physiological functions still remain unknown.

Oil body-associated signalling molecules

Although oleosins have long been known to serve as structural proteins in the OB membranes, reports in the past decade have also implicated crucial roles for caleosins and steroleosins in the OBs. In addition, lipid body fractions might reveal the presence of small amounts of other lipid- or protein-associated polypeptides (Fig. 8.1C). Peripherally associated proteins may participate in the biosynthesis or degradation of oil. They include ER-associated enzymes, glyoxysome receptors, lipase and putative lipase receptors and also unanticipated enzymes acting as signalling molecules. Example of a OB-associated enzyme in the endosperm of castor bean is RcOBL1, which has acid lipase activity and shares homology with lipases in taxonomically diverse species including plants, fungi and mammals. It is relatively amphipathic but it lacks a prominent hydrophobic region like the one present in oleosins. It was shown to be immunologically localized on the OB surface confirming that it is not a cytoplasmic contaminant and it could hydrolyse a range of TAGs.

In oilseeds, TAGs stored in OBs provide energy and membrane building blocks upon lipolysis during early stages of seedling growth. In *Arabidopsis* seeds five lipase homologues are present on the OB surface. Similarly, developing seeds of *Arabidopsis* possess a protein which is localized on the LB surface and possesses a patatin-like acyl hydrolase domain which hydrolyses TAGs and DAGs *in vitro*. Lipoxygenase (LOX) activity leads to the oxygenation of polyunsaturated fatty acids containing a (Z,Z)-1,4-pentadiene system to produce an unsaturated fatty acid hydroperoxide in some oil seeds. These derivatives are consequently cleaved by lipases. A controlled proteolytic degradation of oleosin isoforms and lipolytic degradation of phospholipids constituting the oil body membrane is quite expected before lipase or LOX could access oil body membrane for their action on the TAG matrix. Tzen and Huang (1992) reported that trypsin treatment renders OBs susceptible to the action of phospholipase A_2 (PLA_2) and C and Matsui *et al.* (1999) also observed that trypsin-assisted digestion of oil bodies isolated from the cotyledons of cucumber seedlings rendered them more susceptible to lipoxygenase action. These two observations lead to the hypothesis that proteolysis of oil body membrane proteins is a precondition for oil body mobilization. Phospholipases play a pivotal role in releasing a diverse array of lipid-derived products by degrading membrane phospholipids. PLA_2 is an acyl hydrolase that catalyses phospholipid hydrolases at the sn-2 position, thereby generating a fatty acid and a lysophospholipid. A transient, temporal and tissue specific expression of PLA_2 is associated with OBs during seed germination in cucumber. A patatin-like protein with PLA_2 activity is believed to function in lipid mobilization and signalling. *In-situ* localization of development-associated changes in the expression of PLA_2 activity on the surface of OBs in protoplasts from sunflower seedling cotyledons has been observed in the author's laboratory (Gupta and Bhatla, 2007). The principal intracellular sites exhibiting PLA_2 activity in oilseed cotyledon protoplasts are the OBs, and the enzyme is apparently expressed or activated on the surface of OBs during early phase of seedling growth prior to lipase action. Fluorometric estimation of phospholipase in the author's laboratory indicates the possibility of preferential accumulation of PLA_2 on/around the OBs for the degradation of the OB membranes at a specific stage of seed germination. A PLA_2 might possibly be causing alteration or hydrolysis of the oil body membrane phospholipids, rendering OBs susceptible to lipase action, leading to TAG mobilization from the oil body matrix during seed germination. Association of the activities of PLA_2 and a cytosolic protease with OBs during early stages of seed germination in sunflower has been shown in the author's laboratory (Gupta and Bhatla, 2007).

Lipase is known to be synthesized in the cytosol and to be associated with oil body membrane during seed germination (Huanh, 1992; Beisson *et al.*, 2001). Lipase activity has been shown to be localized at its storage sites in the protein storage vacuoles (PSVs) prior to seed germination and subsequently it gets transferred to OB surface during early phase of seed germination in sunflower (Gupta and Bhatla, 2005). Spatial and temporal changes in lipase activity sites during OB mobilization in protoplasts from sunflower seedling cotyledons during seed germination has also been investigated in the author's laboratory (Gupta and Bhatla, 2005). The probable association of a thiol protease with OBs has also been demonstrated in the author's laboratory (Vandana and Bhatla, 2006). The associated thiol protease is suggested to gradually degrade oleosins during sunflower seedling growth. Maximum protease activity coincides with the phase of active TAG hydrolysis. As a result of the interaction of various enzymes on the OB membrane subsequently leads to TAG degradation and the remaining OBs tend to fuse with the vacuolar membranes. A model depicting the expression of various probable signalling molecules on oil body membranes accompanying lipolysis during germination has also been recently presented (Bhatla *et al.*, 2009) based on the current knowledge and the author's recent findings on sunflower as a model system for localization of various oil body associated molecules. Further work is, however, required for detecting the pattern of expression of LOX, caleosins and steroleosins. Work in the author's laboratory and similar future investigations are likely to decipher the various signalling molecules associated with oil body biogenesis and germination. Protein–protein interaction also needs to be investigated on the oil body membranes using techniques like fluorescence resonance energy transfer (FRET) coupled with multiphoton fluorescence lifetime imaging to provide crucial information on the oil body membrane protein interaction with various signalling proteins such as proteases.

Polyoleosins

Polyoleosins are homo or hetero multimeric oleosin tandem repeats, formed by their head to tail associations. The homoligomeric configuration probably occurs via hydrogen bonding between the adjacent parallel beta strands, thus forming a tightly knit beta sheet structure which probably extends around the entire OB surface (Banilas *et al.*, 2011). This stabilizes the OB even in presence of strong denaturants, like SDS. Use of oleosin homo-dimer or -trimer fusion strategy does not retard lipid mobilization since the germination rates of these seedlings has been observed to be comparable to the wild type *Arabidopsis* plants (Banilas *et al.*, 2011). Formation of polyoleosins alters the emulsification properties of oleosins as well as the stability of OBs and artificial OBs (Scott *et al.*, 2007). Individual oleosin molecules in polyoleosins could be interlinked with small peptides of specific amino acid residues to allow sufficient flex between the repeats and also provide subcloning sites for insertion of other recombinant peptide sequences (Roberts *et al.*, 2008). Up to six tandem repeats of oleosin have been successfully produced in plants. Polyoleosins alter the properties of OBs rendering them more beneficial in many ways as follows:

- enhanced thermostability, thus broadening their application as emulsifiers in reconstituted feeds, pharmaceuticals, beauty products, topical applications etc.;
- slow release of lipid soluble bioactive compounds or recombinant bioactive peptides from OBs used in dermatological and pharmaceutical applications (Roberts *et al.*, 2008);
- less susceptibility to exopeptidase action as the number of N-termini available for degradation is reduced, thus increasing the longevity of OBs present in cattle feed; these OBs protect the encapsulated TAGs from biohydrogenation in the rumen, allowing TAGs to be later absorbed in the intestine as unsaturated fats.
- tighter anchoring of fused peptides to the OBs, thereby allowing the purification of individual and multiple fused peptides;
- higher expression of the recombinant polypeptide resulting in their stable association to the OBs. Thus, a combination of oligomeric repeats of oleosins fused to desired peptide would provide advantages over conventional single-oleosin C-terminal fusion.

Production and biotechnological applications of artificial oil bodies

The process of generating artificial oil bodies (AOBs) involves mixing of TAGs, phospholipids and membrane proteins. Among the three types of membrane proteins, oleosins and caleosins, but not steroleosins have been found to stabilize AOBs. Thus, the stabilizing capability of these proteins is more likely to be due to their structural features, as both share a similarity in their structural domains (Chen *et al.*, 2004). The size and structural/thermal stability of AOBs can be modulated by varying the proportions of the membrane proteins (Peng *et al.*, 2003; Chen *et al.*, 2004). Increase in the amount of proteins leads to the formation of smaller and more stable AOBs. Moreover, AOBs constituted with caleosins are reported to be 10 times smaller and more stable than those constituted with oleosins (Chen *et al.*, 2004). They are maintained as discrete particles at lower pH surroundings (pH 6.5) in contrast to AOBs constituted with oleosins and native OBs (Chen *et al.*, 2004). AOBs constituted with 10 times more oleosin possess higher thermo stability (up to 60°C) while those constituted with caleosin are thermally even more stable up to 70°C (Chen *et al.*, 2004). Even though caleosins can also be used to constitute AOBs, still oleosins make a preferred choice. Furthermore, the most suitable isoform of oleosin needs to be selected. Different oleosin isoforms have been found to provide differential stability to AOBs. As in case of sesame, L-oleosin isoform provides better structural stability than H-isoform (Tai *et al.*, 2002). The pH at which AOBs are most stable should also be considered and all the procedures should be carried out at a pH when the negative charges borne by oleosins on the surface of AOBs are not neutralized, thus preventing their aggregation. AOBs can be put to various important applications:

1 Modified oleosins can be incorporated into AOBs and the resulting effects on oil body stability can be investigated without waiting for the expression of modified oleosin in a transgenic crop (Peng *et al.*, 2007).

2 AOBs can also be used for encapsulation of probiotics in food products to increase their nutritional value. In order to reach the colon in viable and active state, these bacteria have to surpass the strongly acidic conditions of the stomach and the bile salts in the small intestine. Encapsulation of these bacteria into AOBs is one approach adopted to increase their viability. Sesame oil has been used in such an encapsulation of *Lactobacillus delbrueckii* subsp. *bulgaricus* where, addition of commercial oil containing *Lactobacillus* to disintegrated native OBs and vortexing formed AOBs. The survival rate of these encapsulated *Lactobacilli* increased 10^4 times at pH 2 and in the presence of bile salts (Hou *et al.*, 2003).

3 Another application of AOBs is pertaining to purification, immobilization and refolding of recombinant proteins expressed in microbial systems as insoluble aggregates (Chiang *et al.*, 2006; Hudebine, 2009; Peng *et al.*, 2004a). A recombinant oleosin fused with the desired protein, interlinked by a cleavable peptide (in case if the protein is intended to be cleaved from oleosin fusion), is overexpressed in *E. coli.* where it accumulates as insoluble aggregates and is isolated as a pellet. This pellet when mixed with TAGs and phospholipids regenerate AOBs wherein recombinant oleosin gets incorporated into the PL monolayer, thus exposing the protein of interest on the AOB surface where it refolds spontaneously. The protein of interest can be used while still immobilized on the AOBs (Chiang *et al.*, 2006) or it can be cleaved from the oleosin fusion for further applications (Chiang *et al.*, 2005, 2006, 2007; Peng *et al.*, 2004a). Various economically and scientifically important proteins like hydantoinase, nattokinase and cystatin have been obtained at much lower cost of production using this strategy (Chiang *et al.*, 2005, 2007; Peng *et al.*, 2004b). Enzyme hydantoinase is used in the production of D or L amino acids that are in turn used in the synthesis of antibodies and peptide hormones. This enzyme is linked to intein–oleosin gene fusion, and later incorporated into AOBs. This hydantoinase immobilized onto AOBs is more tolerant to heat and can be used for seven cycles. Otherwise it can be cleaved from the AOBs by self-splicing of intein linker in response to dithioerythritol (DTT) treatment.

Biotechnological applications of oil bodies and oleosins

Hirudin production

Hirudin is the most potent and specific thrombin inhibitor and used in the treatment of blood clotting disorders. Earlier it was being obtained from the salivary glands of *Hirudo medicinalis*, an endangered annelid. Expression of hirudin as oleosin fusion protein on OBs offers an effective substitute for its production. Fusion of hirudin gene with oleosin gene from *Arabidopsis*, interspersed with an endoprotease recognition site, has been successfully introduced into *Brassica napus* and *B. carinata*. This recombinant hirudin extracted from OBs of recombinant seeds exhibits correct confirmation and anti-thrombin activity after its release from the fusion (Parmenter *et al.*, 1995; Chaudhary *et al.*, 1998) (Fig. 8.3).

Insulin production

Recombinant human precursor insulin (Des-B30) fused with oleosin and interlinked with trypsin cleavable propeptide, has been expressed in *Arabidopsis* (Nykiforuk *et al.*, 2005).

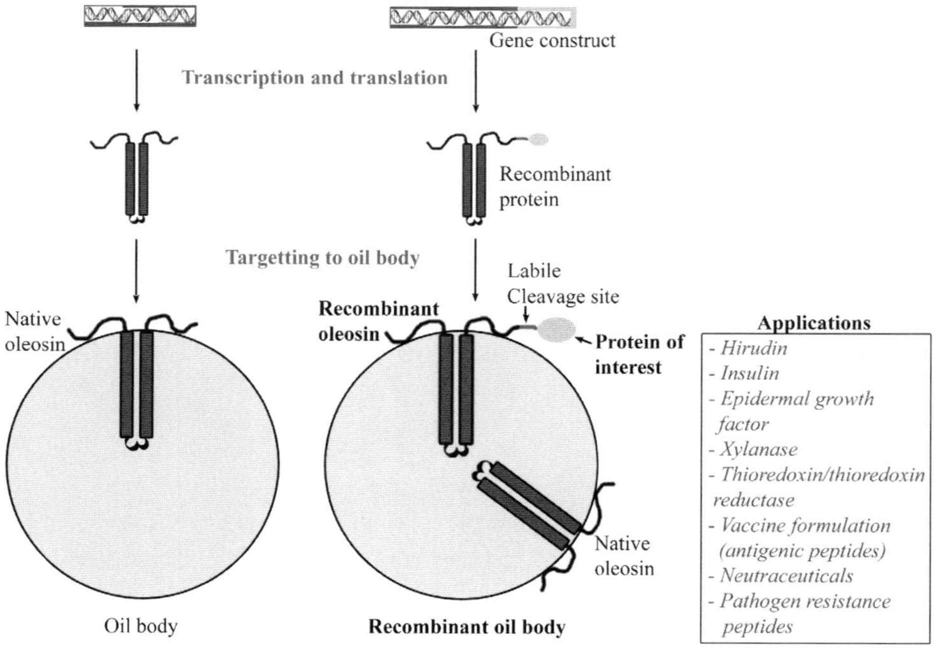

Figure. 8.3 Synthesis of OB with recombinant protein and their biotechnological applications. A colour version of this figure is located in the plate section at the back of the book.

This fusion protein (34.2 kDa) was detected on the oil body surface. Upon treatment of purified oil body with trypsin, recombinant insulin was cleaved from the fusion and matured *in vitro*. This plant derived Des-B30 insulin (0.13% of total seed protein) was purified further and was found to be equally effective as Humulin and Roche insulin without any side effects. It is also recognized by insulin receptors and can elicit signal transduction.

Expression of epidermal growth factor

Human epidermal growth factor (hEGF) has been expressed in *Arabidopsis* fused with oleosin, linked by chymosin cleavage factor with an accumulation level of 0.12%. EGFs of other organisms can also be produced in a similar way (Moloney and Van-Rooijen, 2006).

Vaccine formulations

In addition to being as an emulsifier in vaccine formulations, OBs can also serve as an adjuvant, physically carrying the antigenic peptides. Such vaccines, comprising recombinant OBs containing antigenic peptide, can be used to elicit an immune response against an antigen through any route of administration, including transdermal or through mucosa. TbpB N-lobe gene of M982 strain of *Neisseria meningitidis* was recently introduced into *Arabidopsis* as oleosin fusion, resulting in the expression of the desired antigenic protein on the oil body surface (Deckers *et al.*, 2004).

Xylanase expression and immobilization

Xylanases, often very expensive to produce and immobilize, have been successfully expressed and localized on OBs in *B. napus* as oleosin fusion. These OBs themselves serve as immobilization matrix for xylanase molecules, thereby cutting the cost of production and immobilization. Liu *et al.* (1997) have successfully fused the Xyn C gene (encoding a xylanase from *Neocallimastix patriciarum*) with oleosin gene. The fusion protein expressed on OBs exhibits xylanase activity and can be exploited through several cycles.

Production of thioredoxin/thioredoxin reductase

Thioredoxin is involved in the regulation of photosynthesis in chloroplasts and in seeds it acts as an important cellular redox buffer. Thioredoxin has been used in personal care products, to reduce protein allergens in food items and pharmaceutical products. Thioredoxin/thioredoxin reductase has also been produced as a fusion protein with oleosin (Szarka *et al.*, 2006; Moloney and Dalmia, 2004).

Production of nutraceuticals

Methionine levels have been enhanced in canola meal by expressing seed storage protein zein from maize in fusion with oleosin (Moloney, 1997). Oleosin–phytase fusion linked with collagenase cleavage site, has been expressed in *B. napus*. Phytase (that enable hydrolytic splitting of phosphate groups of phytic acid) expressed on OBs may be purified following collagenase treatment or the transgenic seed may be used as a feed additive as such (Moloney and Van-Rooijen, 2004). Utility of canola meal has also been increased by expressing somatotropin (fish growth hormone) peptide as oleosin fusion protein on OBs of canola (Moloney and Habibi, 2001). Similarly, somatotropin from other organisms can also be expressed. Moloney and Van-Rooijen (2004) have achieved a decrease in seed glucosinolate levels by expressing tryptophan decarboxylase as oleosin fusion protein.

Pathogen resistance peptides

Peptides involved in pathogen resistance have been expressed as oleosin fusion proteins and the linker protein in these fusions can be cleaved in a stage specific manner during seed germination. These oleosin fusion proteins are found to accumulate in the developing seeds. The desired peptides are then released from the fusion product (during germination) when the seedlings require pathogen resistance. Two examples of such peptides, which are linked to oleosin via collagen, include trypsin inhibitor and chitinase (Moloney and Van-Rooijen, 2004).

Multimeric proteins

Different subunits of multimeric proteins can be expressed in plant cells in order to target them onto OBs. Also, their polypeptides could be fused directly to oleosin proteins or some appropriate linker could be used to attach the polypeptide to oleosin/oil body. The subunits of these proteins can then interact to form the desired protein and the OBs having such multimeric proteins could be extracted. This methodology has been successfully employed in producing immunoglobulin comprising of heavy and light chains (Szarka *et al.*, 2006).

Other applications

Oil bodies can even be used for purification of non-proteinaceous molecules by varying the ligand (Moloney *et al.*, 2008). Oil bodies having oleosins fused with single-stranded nucleic acid binding proteins can be employed to isolate single stranded nucleic acids. This method could be successfully used in diagnostics and research applications, such as in hybridization reactions for differential screening of expressed genes, where non-reannealing single stranded DNA or RNA are to be removed. Oil bodies having protein– metalloprotein fusion can be used to recover or eliminate metals. Metalloproteins may include haemoglobin (iron binding), parvalbumin (having affinity for calcium), metallothionein (protein having affinity for zinc and other metal ions). Oleosin–metallothionein fusion protein expressed in *B. carinata* has been used for removal of Cd^{2+} from solution. Such removal of metal ions can help in bioremediation. By using appropriate ligands, OBs can be exploited for the separation of whole cells and cell organelles. This method could have various industrial and medical uses. In recombinant DNA technology, transformed cells expressing surface protein/proteins which can bind to an affinity ligand (attached to oil bodies) can be separated from non-transformed cells. Likewise haematopoietic stem cells (required for bone marrow transplantation and stem cell gene therapy) can then also be segregated from other blood cells.

Conclusion

Oleosins, the principal constituent structural proteins in the oil body membrane, are responsible for maintaining the structural stability and systematic mobilization of oil bodies. Caleosins are also capable of performing similar functions in some plant tissues. Preparation of multimeric repeats (polyoleosins) has also been exploited along with oleosins for the synthesis of various pharmaceutically important fusion products, such as growth factors, vaccine formulations, and enzyme immobilization. Oil bodies have further been exploited for purification of non-proteinaceous molecules through their fusion with certain metalloproteins, as well for separation of whole cells and cell organelles. The unique features of oil bodies and oleosins discussed in this chapter shall continue to offer many more biotechnological applications in future.

References

Aalen, R.B., Opsahl-Fferstad, H.G., Linnestad, C., and Olsen, O.A. (1994). Transcripts encoding an oleosin and a dormancy-related protein are present in both the aleurone layer and the embryo of developing barley (*Hordeum vulgare*) seeds. Plant J. *5*, 385–396.

Abell, B.M., Holbrook, L.A., Albenes, M., Murphy, D.J., Hills, M.J., and Moloney, M.M. (1997). Role of the proline knot motif in oleosin endoplasmic reticulum topology and oil body targeting. Plant Cell *9*, 1481–1493.

Alexander, L.G., Sessions, R.B., and Clarke, A.R. (2002). Characterization and modelling of the hydrophobic domain of a sunflower oleosin. Planta *214*, 546–512.

Banilas, G., Daras, G., Rigas, S., and Moloney, M.M. (2011). Oleosin di-or tri- meric fusions with GFP undergo correct targeting and provide advantages for recombinant protein production. Plant Physiol. Biochem. *49*, 216–222.

Beisson, F., Ferte, N., and Noat, G. (1996). Oil bodies from sunflower (*Helianthus annuus* L.) seeds. Biochem. J. *317*, 955–958.

Beisson, F., Ferte, N., Bruley, S., Voultoury, R., Verger, R., and Arondel, V. (2001). Oil bodies as substrate for lipolytic enzymes. Biochim. Biophys. Acta. *1531*, 47–58.

Busch, M.B., Koertje, K.H., Rahmann, H., and Sievers, A. (1993). Characteristic and differential calcium signals from cell structures of the root cap detected by energy-filtering electron microscopy (EELS/ESI). Eur. J. Cell Biol. *60*, 88–100.

Bhatla, S.C., Vandana, S., and Kaushik, V. (2009). Recent developments in the localization of oil-body associated signalling molecules during lipolysis in oil seeds. Plant Signal. Behav. *4*, 176–182.

Capuano, F., Beaudoin, F., Napier, J.A., and Shewry, P.R. (2007). Properties and exploitation of oleosins. Biotechnol. Advanc. *25*, 203–206.

Chapman, K.D., Dyer, J.M., and Mullen, R.T. (2011). Biogenesis and functions of lipid droplets in plants. J. Lipid Res. *53*, 215–226.

Chaudhary, S., Parmenter, D.L., and Moloney, M.M. (1998). Transgenic *Brassica carinata* as a vehicle for the production of recombinant proteins in seeds. Plant Cell Rep. *17*, 195–200.

Chiang, C.J., Chen, H.C., Chao, Y.P., and Tzen, J.T.C. (2005). Efficient system of artificial oil bodies for functional expression and purification of recombinant nattokinase in *Escherichia coli*. J. Agric. Food Chem. *53*, 4799–4804.

Chiang, C.J., Chen, H.C., Kuo, H.F., Chao, Y.P., and Tzen, J.T.C. (2006). A simple and effective method to prepare immobilized enzymes using artificial oil bodies. Enzyme Microb. Tech. *39*, 1152–1158.

Chiang, C.J., Chen, H.C., Chao, Y.P., and Tzen, J.T.C. (2007). One-step purification of insoluble hydantoinase overproduced in *Escherichia coli*. Protein Expr. Purif. *52*, 14–18.

Chen, J.C., Tsai, C.C., and Tzen, J.T. (1999). Cloning and secondary structure analysis of caleosin, a unique calcium-binding protein in oil bodies of plant seeds. Plant Cell Physiol. *40*, 1079–1086.

Chen, M.C.M., Chyan, C.L., Lee, T.T.T., Huang, S.H., and Tzen, J.T.C. (2004). Constitution of stable artificial oil bodies with triacylglycerol, phospholipid, and caleosin. J. Agric. Food Chem. *52*, 3982–3987.

Craig, S., and Staehelin, L.A. (1988). High pressure freezing of intact plant tissues. Evaluation and characterization of novel features of the endoplasmic reticulum and associated membrane systems. Eur. J. Cell Biol. *46*, 80–93.

de Oliveira, D.E., Franco, L.O., Simoens, C., Seurinck, J., Coppieters, J., Botterman, J., van Montagu, M. (1993). Inflorescence-specific genes from *Arabidopsis* thaliana encoding glycine-rich proteins. Plant J. *3*, 495–507.

Deckers, H.M., Van-Rooijen, G., Boothe, J., Goll, J., Moloney, M.M., and Schryvers, A.B. (2004 July) Sembiosys Genetics Inc. Immunogenic formulations comprising oil bodies. United States patent US 6761914.

Dmitriev, O., Jones, P.C., Jiang, W., and Fillingame, R.H. (1999). Structure of the membrane domain of subunit b of the *Escherichia coli* F0F1 ATP synthase. J. Biol. Chem. *274*, 15598–15604.

Frandsen, G.I., Mundy, J., and Tzen, J.T.C. (2001). Oil bodies and their associated proteins, oleosin and caleosin. Physiol. Plant. *112*, 301–307.

Graham, I.A. (2008). Seed storage oil mobilization. Annu. Rev. Plant Biol. *59*, 115–142.

Gupta, A., and Bhatla, S.C. (2005). Spatial and temporal changes in the lipase activity sites during oil body mobilization in protoplasts from sunflower seedling cotyledons. Plant Growth Regul. *46*, 11–17.

Gupta, A., and Bhatla, S.C. (2007). Preferential phospholipase A_2 activity on the oil bodies in cotyledons during seed germination in *Helianthus annuus* L. cv. Morden. Plant Sci. *172*, 535–543.

Hanano, A., Burcklen, M., Flenet, M., Ivancich, A., Louwagie, M., Garin, J., and Blee, E. (2006). Plant seed peroxygenase is an original heme-oxygenase with an EF-hand calcium binding motif. J. Biol. Chem. *281*, 33140—33151.

Hatzopoulos, P., Franz, G., Choy, L., and Sung, R.Z. (1990). Interaction of nuclear factors with upstream sequences of a lipid body membrane protein gene from carrot. Plant Cell *2*, 457–467.

Hensel, W. (1986). Cytodifferentiation of polar plant cells use of antimicrotubular agents during the differentiation of statocytes from cress roots (*Lepidium sativum* L.). Planta *169*, 293–303.

Hernandez-Pinzon, I., Ross, J.H., Barnes, K.A., Damant, A.P., and Murphy, D.J. (1999). Composition and role of tapetal lipid bodies in the biogenesis of the pollen coat of *Brassica napus*. Planta *208*, 588–598.

Hou, R.C.W., Lin, M.Y., Wang, M.M.C., and Tzen, J.T.C. (2003). Increase of viability of entrapped cells of *Lactobacillus delbrueckii ssp. bulgaricus* in artificial sesame oil emulsions. J. Dairy Sci. *86*, 424–428.

Hsieh, K., and Huang, H.C. (2004). Endoplasmic reticulum, oleosins and oil in seeds and tapetum cells. Plant Physiol. *136*, 3427–3434.

Huang, A.H.C. (1992). Oil bodies and oleosins in seeds. Ann. Rev. Plant Physiol. Plant Mol. Biol. *43*, 177–200.

Huang, A.H.C. (1996). Oleosins and oil bodies in seeds and other organs. Plant Physiol. *110*, 1055–1061.

Huang, A.H.C. (2006). Oleosins and endoplasmic reticulum in seeds and anthers. In The Plant Endoplasmic Reticulum, D.G. Robinson, ed. (Berlin: Springer-Verlag), pp. 187–204.

Hudebine, L.M. (2009). Production of pharmaceutical proteins by transgenic animals. Comp. Immun. Microbiol. Infect Dis. *32*, 107–121.

Jacks, T., Yatsu, L., and Altschul, A. (1967). Isolation and characterization of peanuts spherosomes. Plant Physiol. *42*, 585–597.

Kaup, M.T., Froese, C.D., and Thompson, J.E. (2002). A role for diacylglycerol acyltransferase during leaf senescence. Plant Physiol. *129*, 1616–1626.

Kalinski, A., Loer, D., Weisemann, J., Matthews, B., and Herman, E. (1991). Isoforms of soybean seed oil body membrane protein 24 kDa oleosin are encoded by closely related cDNAs. Plant Mol. Biol. *17*, 1095–1098.

Lacey, D.J., Wellner, N., Beaudoin, F., Napier, J.A., and Shewry, P.R. (1998). Secondary structure of oleosins in oil bodies isolated from seeds of safflower (*Carthamus tinctorius* L.) and sunflower (*Helianthus annuus* L.). Biochem J. *334*, 469–477.

Lee, K., Bih, F.Y., Learn, G.H., Ting, J.T., Sellers, C., and Huang, A.H.C. (1994). Oleosins in the gametophytes of *Pinus* and *Brassica* and their phylogenetic relationship with those in the sporophytes of various species. Planta *193*, 461–469.

Lersten, N.R., Czlapinski, A.R., Curtis, J.D., Freckmann, R., and Horner, H.T. (2006). Oil bodies in leaf mesophyll cells of angiosperms: overview and a selected survey. Am. J. Bot. *93*, 1731–1739.

Li, M., Smith, L.J., Clark, D.C., Wilson, R., and Murphy, D.J. (1992). Secondary structures of a new class of lipid body proteins from oilseeds. J. Biol. Chem. *267*, 8245–8253.

Li, M., Murphy, D.J., and Lee, K.H.K. (2002). Purification and structural characterization of the central hydrophobic domain of oleosin. J. Biol. Chem. *277*, 37888–37895.

Lin, W., and Oliver, D.J. (2008). Role of triacylglycerols in leaves. Plant Sci. *175*, 233–237.

Liu, J.H., Selinger, L.B., Cheng, K.J., Beauchemin, K.A., and Moloney, M.M. (1997). Plant seed oil-bodies as an immobilization matrix for a recombinant xylanase from the rumen fungus *Neocallimastix patriciarum*. Mol. Breed. *3*, 463–470.

Matsui, K., Hijiya, K., Tabuchi, Y., and Kajiwara, T. (1999). Cucumber cotyledon lipoxygenase during postgerminative growth: its expression and action on lipid bodies. Plant Physiol. *119*, 1279–1287.

Moloney, M.M. (1997 Jul). Sembiosys Genetics Inc. Oil-body proteins as carriers of high-value peptides in plants. United States patent US 5650554.

Moloney, M.M., and Habibi, H.R. (2001 Sept). Sembiosys Genetics Inc. Expression of somatotropin in plant seeds. United States patent US 6288304.

Moloney, M.M., and Van-Rooijen, G. (Jun 2004). Sembiosys Genetics Inc. Preparation of heterologous proteins on oil bodies. United States patent US 6753167.

Moloney, M.M., and Dalmia, B.K. (Jun 2004). Sembiosys Genetics Inc. Preparation of thioredoxin and thioredoxin reductase proteins on oil bodies. United States patent US 6750046.

Moloney, M.M., and Van-Rooijen, G. (2006 Aug). Sembiosys Genetics Inc. Expression of epidermal growth factor in plant seeds. United States patent US 7091401.

Moloney, M.M., Boothe, J., and Van-Rooijen G. (2008 Feb) Sembiosys Genetics Inc. Oil and associated proteins as affinity matrices. United States patent US 7332587.

Murphy, D.J. (1993). Structure, function and biogenesis of storage lipid bodies and oleosins in plants. Prog. Lipid. Res. *32*, 247–280.

Murphy, D.J., Hernández-Pinzón, I., and Patel, K. (2001). Role of lipid bodies and lipid-body proteins in seeds and other tissues. J. Plant Physiol. *158*, 471–478.

Murphy, D.J. (2001). The biogenesis and functions of lipid bodies in animals, plants and microorganisms. Prog. Lipid Res. *40*, 325–438.

Murphy, D.J., and Ross, J.H. (1998). Biosynthesis, targeting and processing of oleosin-like proteins, which are major pollen coat components in *Brassica napus*. Plant J. *13*, 1–16.

Murphy, D.J., and Vance, J. (1999). Mechanisms of lipid body formation. Trends Biochem. Sci. *24*, 109–115.

Murphy, G.J.P., and Parker, M.L. (1984). Lipid composition and carbon turnover of wheat leaf oleosomes. J. Exp. Bot. *35*, 348–355.

Napier, J.A., Stobart, A.K., and Shewry, P.R. (1996). The structure and biogenesis of plant oil bodies: the role of the ER membrane and the oleosin class of proteins. Mol. Plant Biol. *31*, 945–956.

Nykiforuk, C.L., Boothe, J.G., Murray, E.W., Keon, R.G., Goren, J., and Markley, N.A. (2005). Transgenic expression and recovery of biologically active recombinant human insulin from *Arabidopsis thaliana* seeds. Plant Biotechnol. J. *4*, 77–85.

Ozdemir, F., and Topuz, A. (2004). Changes in dry matter, oil content and fatty acids composition of avocado during harvesting time and post-harvesting ripening period. Food Chemistry *86*, 79–83.

Parmenter, D.L., Boothe, J.G., Van-Rooijen, G.J.H., Yeung, E.C., and Moloney, M.M. (1995).Production of biologically active hirudin in plants seeds using oleosin partitioning. Plant Mol. Biol. *29*, 1167–1180.

Peng, C.C., Lin, I.P., Lin, C.K., and Tzen, J.T.C. (2003). Size and stability of reconstituted sesame oil bodies. Biotechnol. Prog. *19*, 1623–1626.

Peng, C.C., Chen, J.C.F., Shyu, D.J.H., Chen, M.J., and Tzen, J.T.C. (2004a). A system for purification of recombinant proteins in *Escherichia coli* via artificial oil bodies constituted with their oleosin-fused polypeptide. J. Biotechnol. *111*, 51–57.

Peng, C.C., Shyu, D.J.H., Chou, W.M., Chen, M.J., and Tzen, J.T.C. (2004b). Method for bacterial expression and purification of sesame cystatin via artificial oil bodies. J. Agric. Food Chem. *52*, 3115–3119.

Peng, C.C., Lee, V.S.Y., Lin, M.Y., Huang, H.Y., and Tzen, J.T.C. (2007). Minimizing the central hydrophobic domain in oleosin for the constitution of artificial oil bodies. J. Agric. Food Chem. *55*, 5604–5610.

Pihakaski, K., Pihakaski, S., Karunen, P., and Kallio, P. (1987). Seasonal changes in leaf lipids of *Diapensia lapponica*, with special reference to storage lipid bodies. Nordic Journal of Botany *7*, 281–292.

Plant, A.L., Van-Rooijen, G.J.H., Anderson, C.P., and Moloney, M.M. (1994). Regulation of an *Arabidopsis* oleosin gene promoter in transgenic *Brassica napus*. Plant Mol. Biol. *25*, 193–205.

Platt, K.A., Huang, A.H.C., and Thomson, W.W. (1998). Ultrastructural study of lipid accumulation in tapetal cells of *Brassica napus L. cv. Westar* during microsporogenesis. Int. J. Plant Sci. *159*, 724–737.

Purkrtova, Z., Jolivet, P., Miquel, M., and Chardot, T. (2008). Structure and function of seed lipid-body associated proteins. CR Biologies *331*, 746–754.

Qu, R.D., and Huang, A.H. (1990). Oleosin KD 18 on the surface of oil bodies in maize. Genomic and cDNA sequences and the deduced protein structure. J. Biol. Chem. *265*, 2238–2243.

Radetzky, R., and Langheinrich, U. (1994). Induction of accumulation and degradation of the 18.4-kDa oleosin in a triacylglycerol-storing cell culture of anise (*Pimpinella anisum* L.). Planta *194*, 1–8.

Rinne, P.L.H., Welling, A., Vahala, J., Ripel, L., Ruonala, R., Kangasjärvi, J., and van der Schoot, C. (2011). Chilling of dormant buds hyperinduces *FLOWERING LOCUS T* and recruits GA-inducible 1, 3-β-glucanases to reopen signal conduits and release dormancy in *Populus*. Plant Cell *23*, 130–146.

Roberts, L.S., Gerster, J., Allard, S., Cass, L., and Simmonds, J. (1994). Molecular characterization of two *Brassica napus* genes related to oleosins which are highly expressed in the tapetum. Plant J. *6*, 927–933.

Roberts, N.J., Scott, R.W., and Tzen, J.T.C. (2008). Recent biotechnological applications using oleosins. Open Biotechnol. J. *2*, 13–21.

Ross, J.H.E., Sanchez, J., Millan, F., and Murphy, D.J. (1993). Differential presence of oleosins in oleogenic seed and mesocarp tissues in olive (*Olea europaea*) and avocado (*Persea americana*). Plant Sci. *93*, 203–210.

Sadeghipour, H.R., and Bhatla, S.C. (2002). Differential sensitivity of oleosins to proteolysis during oil body mobilization in sunflower seedlings. Plant Cell Physiol. *43*, 1117–1126.

Sadeghipour, H.R., and Bhatla, S.C. (2003). Light-enhanced oil body mobilization in sunflower seedlings accompanies faster protease action on oleosins. Plant Physiol. Biochem. *41*, 309–316.

Salas, J.J., Sanchez, J., Ramli, U.S., Manaf, A.M., Williams, M., and Harwood, J.L. (2000). Biochemistry of lipid metabolism in olive and other oil fruits. Prog. Lipid Res. *39*, 151–180.

Sargent, J.A., and Osborne, D.J. (1980). A comparative study of the fine structure of coleorhiza and root cells during the early hours of germination of rye embryos. Protoplasma, *104*, 91–103.

Sarmiento, C., Ross, J.H.E., Herman, E., and Murphy, D.J. (1997). Expression and subcellular targeting of a soybean oleosin in transgenic rapeseed. Implications for the mechanism of oil-body formation in seeds. Plant J. *11*, 783–796.

Scott, R.W., Arcus, V.L., and Roberts, N.J. (2007 April) Agriculture Victoria Services Pty Ltd. Polyoleosins. World patent WO/2007/045019.

Shimada, T.L., Shimada, T., Takahashi, H., Fukao, Y., and Hara- Nishimura, I. (2008). A novel role of oleosins in freezing tolerance of oilseeds in *Arabidopsis thaliana*. Plant J. *55*, 798–809.

Shimada, T.L., and Nishimura, I.H. (2010). Oil-body-membrane proteins and their physiological functions in plants. Biol. Pharm. Bult. 33, 360—363.

Szarka, S., Van-Rooijen, G, and Moloney, M.M. (2006 Oct). Sembiosys Genetics Inc. Methods for the production of multimeric protein complexes and related compositions. United States patent US 7098383.

Tai, S.S.K., Chen, M.C.M., Peng, C.C., and Tzen, J.T.C. (2002). Gene family of oleosin isoforms in sesame seed oil bodies and their structural stabilization to reconstituted oil bodies. Biosci. Biotechnol. Biochem. *66*, 2146–2153.

Ting, J.T.L., Wu, S.S.H., Ratnayake, C., and Huang, A.H.C. (1998). Constituents of the tapetosomes and elaioplasts in *Brassica campestris* tapetum and their degradation and retention during microsporogenesis. Plant J. *16*, 541–551.

Turesson, H., Marttila, S., Gustavsson, K., Hofvander, P., Olsson, M.E., Bülow, L., Stymne, S., and Carlsson, A.S. (2010). Characterization of oil and starch accumulation in tubers of *cyperus esculentus* var. *sativus* (cyperaceae): a novel model system to study oil reserves in nonseed tissues. Am. J. Bot. *97*, 1884–1893.

Tzen, J.T.C., Lai, Y.K., Chan, K.L., and Huang, A.H.C. (1990). Oleosin isoforms of high and low molecular weights are present in the oil bodies of diverse seed species. Plant Physiol. *94*, 1282–1289.

Tzen, J.T.C., and Huang, A.H.C. (1992). Surface structure and properties of plant seed oil bodies. J. Cell Biol. *117*, 327–335.

Tzen, J.T.C., Lie, G.C., and Huang, A.H.C. (1992). Characterization of the charged components and their topology on the surface of plant seed oil bodies. J. Biol. Chem. *267*, 15626–15634.

Tzen, J.T.C., Cao, Y., Laurent, P., Ratnayake, C., and Huang, A.H.C. (1993). Lipids, proteins, and structure of seed oil bodies from diverse species. Plant Physiol. *101*, 267–276.

Vance, V.B., and Huang, A.H. (1987). The major protein from lipid bodies of maize. Characterization and structure based on cDNA cloning. J. Biol. Chem. *262*, 11275–11279.

Vandana, S., and Bhatla, S.C. (2006). Evidence for the probable oil body association of a thiol-protease, leading to oleosin degradation in sunflower seedling cotyledons. Plant Physiol. Biochem. *44*, 714–723.

Van Der Schoot, C., Paul, L.K., Paul, S.B., and Rinne, P.L.H. (2011). Plant lipid bodies and cell–cell signalling: A new role for an old organelle. Plant signalling and Behaviour *6*, 1732–1738.

Wu, S.S.H., Platt, K.A., Ratnayake, C., Wang, T.W., Ting, J.T.L., and Huang, A.H.C. (1997). Isolation and characterization of novel neutral-lipid-containing organelles and globuli-filled plastids from *Brassica napus* tapetum. Proc. Natl. Acad. Sci. *94*, 12711–12716.

Wu, S.S.H., Moreau, R.A.,Whitaker, B.D., and Huang, A.H.C. (1999). Steryl esters in the elaioplasts of the tapetum in developing *Brassica* anthers and their recovery on the pollen surface. Lipids *34*, 517–523.

Visual Restoration using Microbial Rhodopsins

9

Nicole L. Wagner, Jordan A. Greco and Robert R. Birge

Abstract

Biological systems are governed by nano- and microscale processes that can be efficiently exploited by various technological tools and research approaches. Harnessing the power of nanobiotechnology has modified the way scientists approach and treat diseases. Through manipulation of matter at the nanoscale regime, engineers now have the ability to attack many of the challenges of modern science and medicine. As the field of nanotechnology and nanobiotechnology grows, new paradigms are being developed to accentuate the development of nanometric and micrometric materials and devices. This review will focus on the use of retinylidene proteins, particularly microbial rhodopsins, in gene therapy, optogenetics, and visual prosthetics for the treatment of ophthalmic degenerative diseases.

Introduction

The rapid advances in nanoscale bioengineering and nanotechnology are redefining the frontiers of our knowledge and conception of the future (Sweeney *et al.*, 2003). In a multidisciplinary approach, interfacing biology and the physical sciences, scientists and engineers have started to gain control over the properties of matter, and are creating novel applications that have the potential for innovations in manufacturing, medicine, and energy production (Sanchez *et al.*, 2005). The unique interplay between biologically inspired technologies, which harness the intrinsic properties of naturally occurring materials and the physical sciences, has elucidated the possibility for informed solutions to complex problems in a variety of scientific disciplines (Bar-Cohen, 2006; Buehler, 2010; Langer and Tirrell, 2004).

Manipulation of matter at the nanoscale and the preparation of novel materials with specific manipulable physical properties and functions is the primary goal of developments in nanoscience (Jain, 2008; Mazzola, 2003). From the introduction of nanotechnology and miniaturization in 1959 by Richard Feynman to the most recent developments in materials science, the goal is the same: to promote the development of materials and devices with at least one dimension of less than 100 nm (Whitesides, 2003). Nanoelectronics, once the cynosure of nanotechnology, started with the continual minimization of the length scale of transistors on integrated circuits, a trend described by Moore's law (Moore, 1965). This field now encompasses the processing, transmission, storage, and retrieval of information at the molecular level. Perpetuated by ambitious technological, economic, and commercial aspirations, nanoelectronics has greatly impacted fundamental research to meet the demands of modern society. (Lu and Lieber, 2007; Thomas and Sayre, 2005). Addressing issues such as nanoscale transport (Goel and Vogel, 2008; Hsieh *et al.*, 2008; Nitta *et al.*, 2006), self-assembly techniques (Carmona-Ribeiro, 2010; Chen, 2005; Matharu *et al.*, 2012), and the development of new electronic architectures (Aono *et al.*, 1012; Markovic *et al.*, 2012; Nikolic *et al.*, 2011) is at the forefront of the field.

From a developmental viewpoint, nanotechnology paradigms can be placed into two classes: a top-down approach and a bottom-up approach (Konstas *et al.*, 2010). These strategies of information processing and knowledge ordering are most often found in software and computer development, however, they can be applied to any scientific field (Yu and Meyyappan, 2005). Using a top-down approach, nanoscale objects are the result of the processing of larger sized objects, whereas in a bottom-up approach, nanoproducts are assembled directly from smaller building blocks, such as molecules and atoms (Bisoyi and Kumar, 2011). Although the bottom-up approach in manufacturing technologies is more recent than the top-down approach, the bottom-up approach is gaining popularity (Hughes, 2000). Perhaps the best example of the use of the bottom-up approach is nature. Nature has exploited a bottom-up approach in the development of many of the complex structures needed to sustain life (Tu and Tirrell, 2003). Scientists and engineers are mimicking this approach in a dynamic way using molecular self-assembly techniques (Bensaude-Vincent, 2009; Tamerler and Sarikaya, 2009). The appeal of self-assembly systems for biological materials is that these materials have the potential to degrade into smaller molecules that can be broken down in the body, particularly in the case of pharmacology, drug delivery systems, and nanomedicine (Meyers *et al.*, 2008).

In recent years, the infiltration and influence of biology and natural sciences on nanotechnology has led to a deemphasized dimension criterion. Because nature functions on a different scale, one made up of proteins, cells, organelles, molecules, and individual atoms the definition of nanotechnology has been recast to stress the development of new materials and devices on a nano- and micro- metric scale to study biological systems, a term loosely defined as nanobiotechnology (Nguyen *et al.*, 2010; Whitesides, 2004). The closely related term, bionanotechnology, is often used interchangeably with nanobiotechnology; however when distinction is needed, bionanotechnology is generally used to refer to the study of how nanotechnology can be guided through improved understanding of biology and the integration of biological ideas and motifs into improving nanotechnologies (Nikolov *et al.*, 2008). Conversely, nanobiotechnology describes the application of nanotechnology on biological problems (Bruchez, 2007).

Target applications of nanotechnologies span a wide range, from medicine and pharmacology to nanomechanics and nanoelectronics. Some areas of active research and interest in nanotechnology are drug delivery (Caraglia *et al.*, 2012; Cario, 2012; Mazzola, 2003), gene therapy (Chaudhari *et al.*, 2010; Cronin and Bennett, 2011; Yang *et al.*, 2011), implantable devices (Tomsia *et al.*, 2001), and regenerative medicine (Arora *et al.*, 2012; Mata *et al.*, 2012). This chapter will explore the history of microbial rhodopsins and the important roles and influence that these rhodopsins play in the fields of nanotechnology, and most recently, neurobiology and optogenetics. In particular, we will emphasize the use of microbial rhodopsins in visual restoration and ophthalmology.

Retinylidene proteins

Retinylidene proteins are integral membrane proteins characterized by seven transmembrane alpha helices and a retinal chromophore that is covalently linked to a conserved lysine residue (Béjà *et al.*, 2000; Spudich *et al.*, 2000; Terakita, 2005). There are two distinct families of retinyldene proteins. Proteins classified as Type I rhodopsins function as microbial ion pumps and sensors, whereas Type II rhodopsins are photosensitive receptor proteins found in visual pigments that function as G-protein-coupled receptors (GPCRs) (Man *et al.*, 2003; Palczweski *et al.*, 2000; Ryazantsev *et al.*, 2012). Type I rhodopsins, not normally found in animals, have the ability to photocycle and are capable of undergoing this process without regeneration of the photoactive element, retinal (Hirayama *et al.*, 1992; Oesterhelt *et al.*, 1991; Sasaki and Spudich, 1999). Photobleaching, a process carried out by most of the Type II rhodopsins, often produces an inactive form of the protein (Birge, 1990a).

Although both Type I and Type II rhodopsins share a common retinal cofactor and an architecture of seven transmembrane alpha helices, research is still being carried out to determine whether these two protein families came from a single progenitor or are representative of convergent evolution. The question of the evolutionary relationships between these two classes of rhodopsins is still under investigation; however current research suggests lateral gene transfer as a driving force for the diversification and distribution of these proteins (Sharma *et al.*, 2006). The junction of research on visual rhodopsins and microbial rhodopsins has led to exploration and advances in the fields of structural biology, protein dynamics, vision and visual restoration, and neurobiology.

When evaluating the potential of these proteins in nanotechnology, it is important to note the native functions of these organisms and the influence that role has on the intended device application. Because Type I rhodopsins function as ion pumps and undergo a photocyclic mechanism, these rhodopsins tend to have greater potential in devices (Miesenbock, 2011). We note that Type II retinylidene proteins of higher eukaryotes have been investigated for potential in devices, but because of the complex function of Type II rhodopsins, application in devices is limited. The inactivation due to photobleaching and the function of these proteins as GPCRs has yet to be useful in applied technologies. A thorough knowledge of both classes of proteins is critical to understanding the potential use of microbial rhodopsins in visual restoration.

Type I rhodopsins

For decades, research on microbial rhodopsins and the organisms that contain them has elucidated several facets of biology. From the discovery of bacteriorhodopin (BR) (Oesterhelt and Stoeckenius, 1971) in the cell membrane of *Halobacterium salinarum* in 1970 to the more recent discovery of proteorhodopsin (PR) in 2000 (Béjà *et al.*, 2000; Gourdon *et al.*, 2008; Rangarajan *et al.*, 2007), research on these proteins has opened up a plethora of opportunities for biologically inspired technologies, including artificial retinas, chemical weapons sensors, and protein-based memories. Nature has optimized these robust molecular machines to efficiently harvest solar energy for survival in a variety of environments. Because light energy is the ultimate source of energy for most living things, the efficient energy capture of light by these organisms is a highly evolved process. Prokaryotes use many of the microbial opsin genes, which include BR, halorhodopsin (HR), sensory rhodopsin (SR), and channelrhodopsin (ChR), among others, to transduce light to regulate pH, alter chloride ion concentrations, and switch flagellar motor direction (Grote and O'Malley, 2011).

Although Type I rhodopsins were first found in the membranes of archaea they have recently been detected in all three domains of life (Fu *et al.*, 2010; Kato *et al.*, 2012). Despite a vast array of functions, including light transduction, ion translocation, and phototaxis, these seven transmembrane alpha helical proteins all contain an organic chromophore, which is covalently bound via a Schiff base to a conserved lysine residue. Upon the absorption of a single photon, the retinal chromophore undergoes an isomerization event, leading to a series of spectroscopically distinct photointermediate states. These photointermediates are observed as the protein goes through a series of thermally driven conformational changes as the protein relaxes back to the resting state.

Bacteriorhodopsin

Historically, the microbial rhodopsin, BR, has played an integral role in membrane protein research, as well as in the use of Type I rhodopsins in device applications. Bacteriorhodopsin, a 27 kDa protein located in the outer membrane of the halophilic archaeon *H. salinarum*, is expressed in the native organism when the concentration of oxygen drops below a level necessary to sustain oxidative phosphorylation (Lanyi, 2004a). Bacteriorhodopsin forms two-dimensional crystalline patches in the cell membrane of the native organism, often referred to as the purple membrane (Krebs and Isenbarger, 2000). Upon the absorption of a photon of light by the retinal chromophore, the light-adapted form of BR undergoes a complex photocyclic mechanism as shown in Fig. 9.1 (Gillespie *et al.*, 2002; Lanyi, 2000). Through a series of transient photochemical and conformational intermediates, known as the photocycle, a proton is transported through the molecule and across the membrane. The translocation of a proton from the intracellular to the extracellular milieux of the protein generates a proton gradient, thereby driving cellular ATPase to synthesize ATP under anaerobic conditions (Luecke *et al.*, 1998; Oesterhelt and Stoeckenius, 1973; Song *et al.*, 1993a; Spudich *et al.*, 1986). After nearly four decades of research, the photocyclic mechanism of proton pumping in the protein is well characterized. Several models have been generated to elucidate the complex nature of the BR photocycle, and include, but are not limited to, the Isomerization-Switch-Transfer (IST) model (Haupts *et al.*, 1997; Lanyi, 1998b; Tittor *et al.*, 1997), the Local-Access model (Brown *et al.*, 1998a,b; Lanyi, 1998a), and the Hydration model (Kandori, 2004).

The main photocycle of BR is made up of the **bR**, **K**, **L**, **M**, **N**, and **O** photostates (Lozier *et al.*, 1975). Translocation of a proton across the membrane of the protein is a fast event that has a duration of approximately 10–15 ms (Birge, 1994). The high quantum efficiency ($\varphi \sim 0.65$) of the primary photochemical event, which initiates the photocycle, makes proton pumping an efficient process (Stuart and Birge, 1996). Absorption of a photon by the retinal chromophore produces an excited state known as the **K** state, which is characterized by the isomerization of all-*trans*-15-*anti* retinal to a 13-*cis*-15-*anti* configuration (Birge *et al.*, 1987). Proton translocation is conventionally divided into five steps as the proton moves through the various photointermediates and the reader is encouraged to see reviews by (Balashov, 2000; Lanyi, 2000, 2004b; Luecke *et al.*, 2000; Schobert *et al.*, 2003). Transport of the first proton occurs during the L→M photoconversion, as a proton is transferred from the Schiff base nitrogen of Lys-216 to the primary counterion, Asp-85 (Dickopf and Heyn, 1997; Metz *et al.*, 1992; Zimanyi *et al.*, 1992a,b). The chromophore is in a distorted 13-*cis*-15-*anti* configuration during the **M** state (Schobert *et al.*, 2003). Release of a proton from the proton release complex occurs as the protein undergoes a conformal change from the **M→N** state (Balashov *et al.*, 1993). Reprotonation of the Schiff base nitrogen via Asp-96 makes

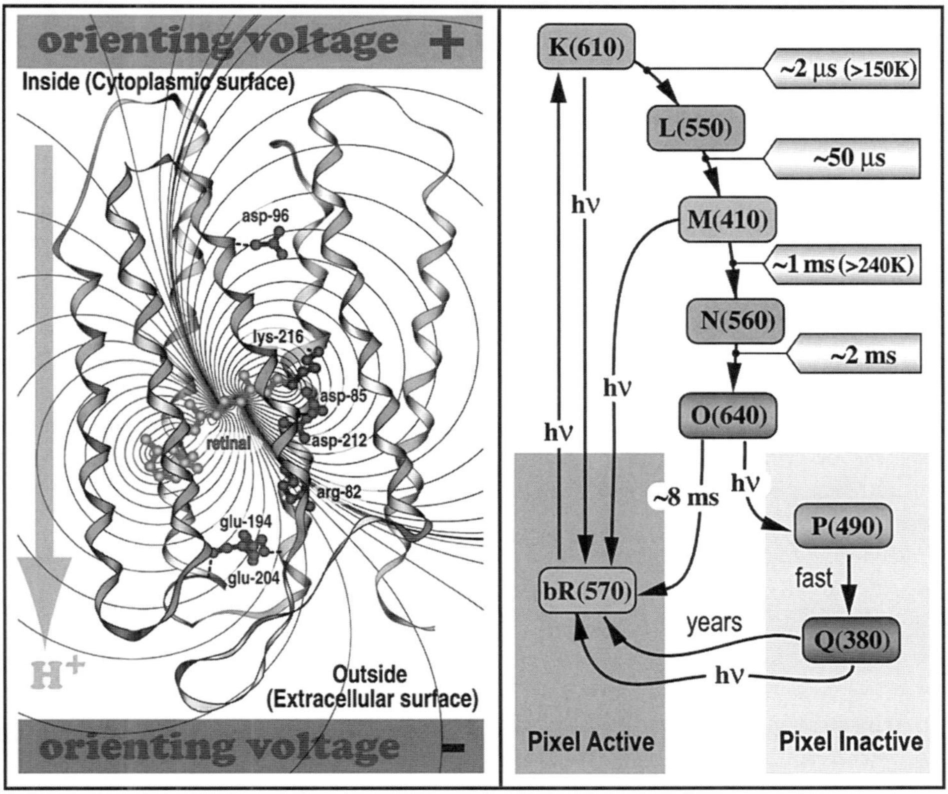

Figure 9.1 Structure (left) and photocycle (right) of bacteriorhodopsin.

up the third step (Cao *et al.*, 1991, 1993) and the subsequent reprotonation of Asp-96 from the cytoplasm and isomerization of retinal from a 13-*cis* to all-*trans* configuration comprises the fourth step, represented by the **O** state (Brown *et al.*, 1999; Holz *et al.*, 1989; Otto *et al.*, 1989). The last step in proton transfer involves the reprotonation of the proton release complex via Asp-85, which reforms the **bR** resting state (Richter *et al.*, 1996). The protein will then remain in the resting state until another photon is absorbed by the chromphore.

The branched photocycle of BR can be accessed by several different mechanisms (Birge *et al.*, 1999; Masthay *et al.*, 2002; Tallent *et al.*, 1998), however, the sequential two-photon absorbance of green and red photons is the most characterized (Popp *et al.*, 1993). The last and most red-shifted intermediate in the photocycle, known as the **O** state, contains an all-*trans* configuration of the retinal chromophore. If the chromophore absorbs a second photon of red light during the **O** state, the protonated all-*trans* chromphore will isomerize to form a 9-*cis* photoproduct (Popp *et al.*, 1993). This process produces a conformationally unstable intermediate known as the **P** state. The **P** state then thermally decays to form the **Q** state, an inactive form of the protein that contains a hydrolysed 9-*cis* retinal chromophore trapped inside the binding site (Gillespie *et al.*, 2005).

The photochemical reaction of BR induced by the absorption of a photon by the chromophore, retinal, is one of the more efficient and stable photochemical reactions found in nature. Bacteriorhodopsin has an extremely high quantum efficiency that is identical to that of native photopigments found in the photoreceptor cells of the human retina (~65%) (Birge *et al.*, 1989a,b; Tittor and Oesterhelt, 1990). The photocycle lasts roughly 15 ms in the light-adapted form of the protein, and the cyclic nature of the reaction allows the protein to reset itself after every complete cycle. Bacteriorhodopsin is also distinguished as a membrane protein with unprecedented photochemical and thermal stability. Photochemical stability is measured as the number of times a photochromic material can be converted between two states before 37% ($1/e$) of the material denatures (Birge *et al.*, 1999). This quantity, also termed cyclicity, has been found to exceed 10^6 for BR at ambient temperature (Birge, 1990b; Bräuchle *et al.*, 1991). With a melting temperature that approaches 80°C, BR is remarkably stable (Jackson and Sturtevant, 1978). Based on the extremophilic nature of the native organism, *H. salinarum*, and the ability to withstand high fluctuations of temperature, light flux, and chemical stress from a self-induced pH gradient, BR has long been investigated as an ideal material for use in photonic devices. Using BR as a medium for optical data storage and information processing has the potential to circumvent current issues and challenges of semiconductor-based computing (Henderson *et al.*, 1990; Jackson and Sturtevant, 1978).

Modern demands of miniaturization and high-speed computing necessitate that scientists and engineers continually search for materials and architectures that circumvent the inevitable limit of lithographic manipulation of bulk materials (Fig. 9.2). The complex photocycle of BR provides a time-dependent series of photointermediates that can be manipulated to read, write, and erase data. Fourier transform holographic associative processors have been developed that use the transient change of refractive indices between the **bR** resting state and the **M** state for associative recall (Hampp and Juchem, 2000; Hampp *et al.*, 1992). Furthermore, the discovery of the branched photocycle of BR has permitted the use of an isolatable photointermediate (the **Q** state) that facilitates the assignment of binary bit 0 and binary bit 1 in volumetric data storage paradigms (Birge *et al.*, 1994; Hillebrecht *et al.*, 2005). The ability of BR to undergo chemical and genetic modifications allows for further enhancement of the photophysical properties of the protein for these devices. Optical computing is only one of many applications for which BR has been investigated as a photochromic medium. These bioelectronic and photonic devices include, but are not limited to, random access thin film memories (Birge *et al.*, 1989c; Lawrence and Birge, 1984; Schick *et al.*, 1988), neural-type logic gates (Mobarry and Lewis, 1986), photon counters (Marwan *et al.*, 1988), picosecond photodetectors (Rayfield, 1989), spatial light modulators (Birge *et al.*, 1990; Song *et al.*, 1993b), holographic correlators (Thoma and Hampp, 1992), non-linear optical filters (Thoma *et al.*, 1991), dynamic time-average interferometers (Renner and Hampp, 1992), optical limiters (Song *et al.*, 1993c), pattern recognition systems (Hampp *et al.*, 1994), and multi-level logic gates (Gu *et al.*, 1996). The potential use of BR in devices now extends into novel applications in the biomedical sciences.

Halorhodopsin

Halorhodopsin, discovered in 1980 in the membrane of *H. salinarum*, functions as a light-driven anion pump involved in maintaining osmotic homeostasis (Matsung-Yagi and Mukohata, 1980). The X-ray structure of HR at 1.8 Å was published in 2000 and revealed a 31% transmembrane sequence identity with BR (Fig. 9.3) (Kolbe *et al.*, 2000). In a direction opposite of BR, HR pumps chloride ions from the extracellular milieux to the cell cytoplasm

Figure 9.2 A model of a hypothetical three-dimensional silicon computer. This artistic model of a silicon computer served as the centrepiece exhibit at the opening of the Digital Equipment Software Engineering Centre in 2002. Three-dimensional semiconductor devices are being investigated by many companies, but devices beyond four layers have not been implemented with success due to heat dissipation problems. In contrast, three-dimensional protein-based devices are easily created, are often facilitated via self-assembly, and avoid heat dissipation problems due an aqueous environment. (Image courtesy of Dr Paul Saia).

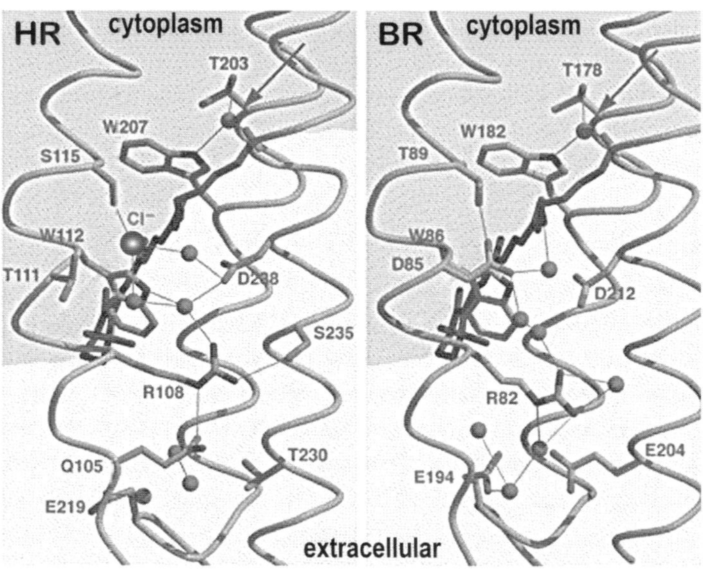

Figure 9.3 Comparison of the key residues in halorhodopsin (HR) (left) and bacteriorhodopsin (BR) (right). Halorhodopsin pumps a chloride ion from the extracellular to the cytoplasmic surface. In contract, BR pumps a proton from the cytoplasmic to the extracellular surface. Figure modified from (Essen, 2002).

to produce an electrochemical ion gradient, which contributes to the regulation of pH in *H. salinarum* (Essen, 2002; Spudich, 2000). Halorhodopsin of *H. salinarum* and *Natronomonas pharaonis* responds to illumination with yellow light (~580 nm) and pumps chloride ions in the direction of the cell cytoplasm. Devices based on HR use the *N. pharaonis* (NpHR) strain because of the enhanced stability and chloride affinity over the *Halobacterium* species (Zhang *et al.*, 2007). Despite the aforementioned advantages of the NpHR strain, one of the first devices that employed HR, an anion-sensitive biosensor, used the *H. salinarum* strain (Seki *et al.*, 1994).

Exogenous expression of NpHR in neurons and photoreceptor cells has become a popular area of research. Recent studies show that NpHR can be used to silence neuronal activity by controlling the hyperpolarization of neurons upon illumination with intense yellow light (Lynagh and Lynch, 2010). Transgenic zebrafish expressing NpHR were also investigated in the silencing of neurons (Arrenberg *et al.*, 2009). Additionally, NpHR has been used in organotypic hippocampal cultures to evaluate inhibition of hyperexcitability and epileptiform activity (Tønnesen *et al.*, 2009). Finally, NpHR has been explored as a potential optogenetic tool for visual restoration (Busskamp *et al.*, 2010) (see 'Optogenetics').

Channelrhodopsin

Channelrhodopsins (ChRs) are retinylidene proteins that function as light-gated non-specific cation channels. The proteins, channelrhodopsin 1 (ChR1) and channelrhodopsin 2 (ChR2) were uncovered during a large-scale expressed sequencing tag (EST) project in the microalgae, *Chlamydomonas reinhardtii*, and are responsible for phototaxis and photophobic responses in the organism (Nagel *et al.*, 2002; Sineshchekov *et al.*, 2002). Exposure to blue light (~460 nm) allows these proteins to conduct hydrogen, sodium, potassium, and calcium ions into the cell cytoplasm (Fig. 9.4) (Buchen, 2010). Unlike most of the Type II metabotropic rhodopsins, which open ion channels in response to second messengers, ChRs are ionotropic and directly form ion channels. The ability to form ion channels via an ionotropic response allows for rapid cellular depolarization, an important characteristic for the implementation of ChRs in bioengineering and neuroscience (Zhang *et al.*, 2006). Despite a low sequence homology to BR, several amino acids that define the retinal binding site and proton transport network are conserved in ChR (Nagel *et al.*, 2005).

Channelrhodopsin 1 is a light gated proton channel, while ChR2 is a related microbial Type I rhodopsin, which has been shown to be involved in the generation of photocurrents in *C. reinhardtii* (Nagel *et al.*, 2003). Channelrhodopsin 2, from *C. reinhardtii*, consists of 737 amino acids, although the seven transmembrane domains and photocurrent functionality are all contained within the N-terminal 300 amino acids. Experimentally, these proteins have been broadly applied to neuroscience research (Kleinlogel *et al.*, 2011) and have shown potential in optogenetics, a new method that combines the use of genetics, virology, and optics to target and regulate specific events in living tissue (Buchen, 2010). Channelrhodopsin 2, in particular, has shown fast kinetics, high conductance, and good temporal resolution of the photoresponse, making the protein a promising candidate for application in neural circuits (Nikolic *et al.*, 2006). Most importantly, ChR2 can be expressed in mammalian neurons without the need for exogenous retinal, a capability that is critical in mediating the control of action potential firing in vertebrate systems (Kramer *et al.*, 2009).

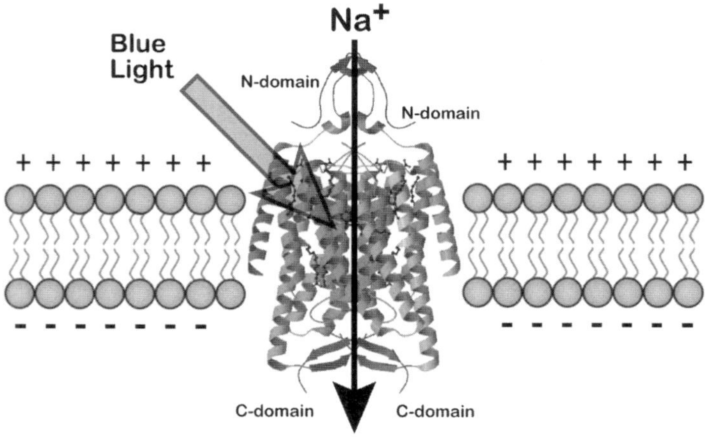

Figure 9.4 Channelrhodopsin-2 (ChR2) is a light-gated cation channel native to the green algae *C. reinhardtii*. It consists of two identical seven transmembrane proteins in a dimeric form, each containing a retinal chromophore. When either absorbs a photon of blue light (~460 nm), the conformational change induces a change in the conductivity of the channel with regard to small positively charged ions such as sodium, calcium or potassium, moving the ions into the cell cytoplasm.

A 2.3 Å resolution crystal structure of a C1C2 chimaera of the channelrhodopsin light-gated cation channel was recently published providing insight into retinal binding, ChR dimerization, and the cation conductance of these proteins. The crystal structure revealed the presence of all-*trans* retinal covalently bound to Lys-296 on TM7 and the presence of five aromatic residues located around the polyene chain and Β-ionone ring. The structural features around the retinal chromophore, Schiff base, and conduction pathway have provided insight into the blue-shifted absorption spectrum of ChR in comparison to BR (Kato *et al.*, 2012).

Type II rhodopsins

Type II rhodopsins are classically identified as the light-transducing proteins responsible for initiating a visual response in vertebrate eyes. In this review, we focus on the visual system as it specifically relates to human physiology and pathology. Human visual perception is facilitated by two general types of photopigments; rhodopsin, found in rod photoreceptor cells, and three colour pigments that reside in cone photoreceptor cells (Shichida and Imai, 1998). Rhodopsin is the primary photoreceptor of dim light, while the colour pigments facilitate colour differentiation via blue-, green-, and red-sensitive proteins (Birge, 1990a; Oprian *et al.*, 1991). The visual pigments all contain a common chromophore, 11-*cis* retinal, that is bound to the protein via a Schiff base linkage to a conserved lysine residue. The wavelength selectivity and photochemical characteristics of these photoactive proteins exist due to the molecular interactions of the chromophore with various amino acids contained within the primary sequences. Because of the relative stability of rhodopsin, the protein–chromophore interaction has been well characterized using X-ray crystallography (Palczewski *et al.*, 2000), spectroscopy (Birge, 1986; DeGrip *et al.*, 1988; Rothschild *et al.*, 1983; Siebert *et al.*, 1983), and molecular dynamics (Birge, 1982; Birge *et al.*, 1987). Conversely, the lack of structural studies of the colour pigments impede the elucidation of the colour tuning mechanism in vertebrate cone photoreceptor cells.

Rhodopsin is a 40 kDa protein that shares a common structural motif with the seven transmembrane alpha helical microbial rhodopsins that were discussed in the sections above. The absorption of a photon by rhodopsin results in a series of photochemically induced transformations of the protein, which trigger a chemical cascade response within the rod photoreceptor cells. In contrast to the BR photocycle, rhodopsin undergoes a photobleaching sequence that is initiated by the 11-*cis* to 11-*trans* photoisomerization of retinal (Fig. 9.5) (Birge, 1981). The rhodopsin to bathorhodopsin transition, which is characterized by this isomerization, represents the primary photochemical event and stores roughly 32 kcal/mol of energy (Birge and Vought, 2000; Cooper and Converse, 1976). This isomerization is followed by a series of thermally driven photointermediates, culminating in the eventual expulsion of the isomerized chromophore from the binding site of the protein. Denaturation into the apoprotein leaves rhodopsin in an inactive state, which precludes the spontaneous recovery of the photoactive protein. Retyinyl ester transferase is an enzyme that is required to reisomerize all-*trans* retinal to 11-*cis* retinal to effectively regenerate the protein into an active phototransduction element (Law *et al.*, 1989). This regeneration takes approximately 20 min in the human retina (Stuart and Birge, 1996). Further details about the role of rhodopsin and the photoreceptor cells in the phototransduction cascade can be found below.

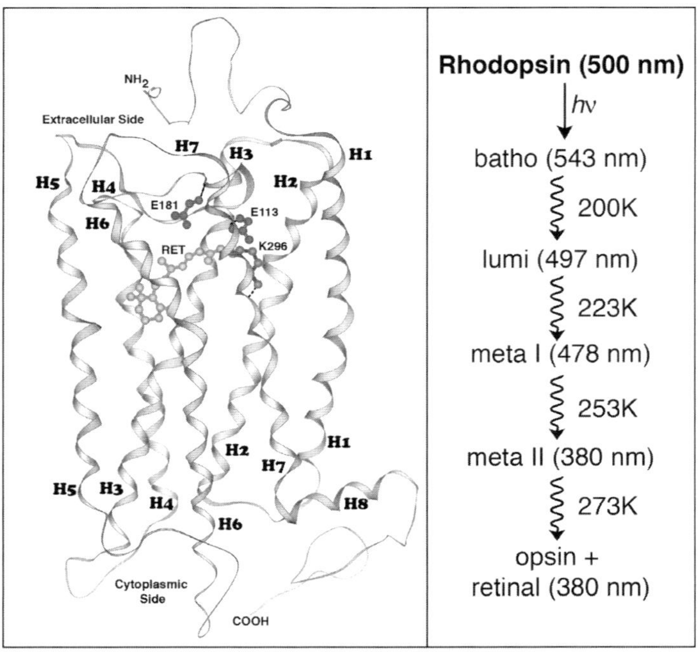

Figure 9.5 Structure (left) and photobleaching sequence (right) of rhodopsin. Unlike BR, rhodopsin has a photobleaching sequence that terminates with the expulsion of the chromophore from the binding site. An enzymatic process is required to isomerize the all-*trans* retinal that is expelled to 11-*cis* retinal, which is the photoactive component of rhodopsin (RET; retinal).

The visual process

In the vertebrate eye, light enters the through the pupil and is focused on a highly organized array of light-sensitive neurons in the retina (Nelson and Cox, 2008). The retina is approximately 0.2 mm thick and contains a complex neural circuitry responsible for converting the graded electrical activity of photoreceptor cells into a series of action potentials that travel to central targets in the brain via axons in the optic nerve (Purves *et al.*, 2008; Sung and Chuang, 2010). Five main classes of neural cells comprise the retina: photoreceptor cells, bipolar cells, amacrine cells, horizontal cells, and ganglion cells (Fig. 9.6). The cell bodies and processes of these neurons are stacked in distinct layers and act to amplify, extract, and compress signals to preserve information before transmitting it to the brain through the optic nerves (Sung and Chuang, 2010).

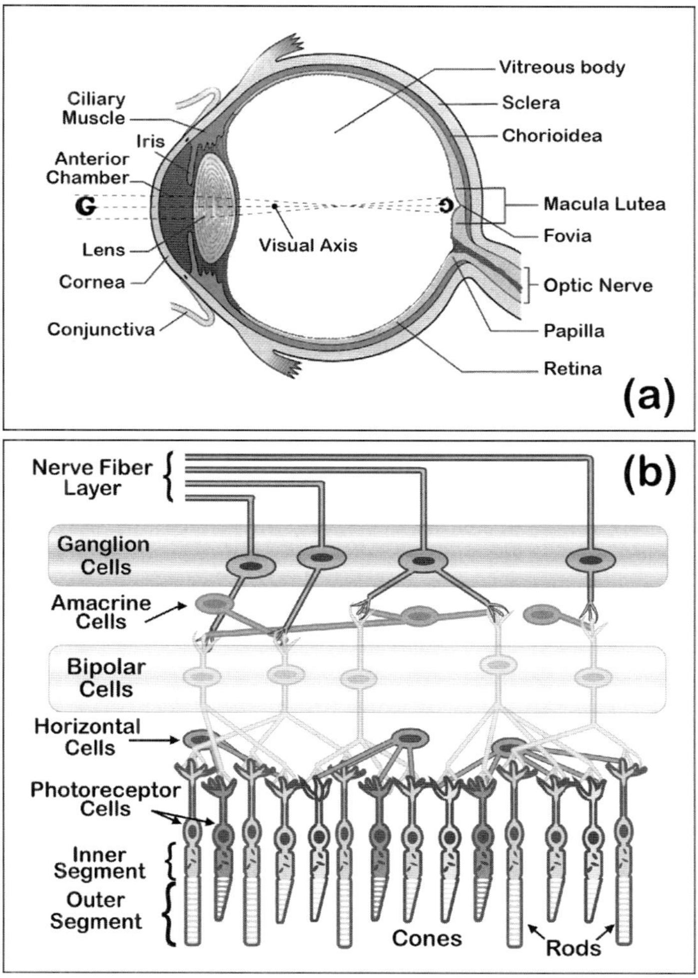

Figure 9.6 The visual system of humans showing the key components of the eye (a) and the neural machinery and photoreceptor cells of the retina (b).

The primary photosensory neurons in the retina are the photoreceptor cells. Photoreceptor cells, consisting of the rods and cones, can be distinguished by shape, type of photopigment they contain, distribution across the retina, and pattern of synaptic connections. The rod photoreceptor cells (about 10^9 rod cells/retina) are sensitive to low light conditions, and produce a response to a single photon of light. Cone photoreceptor cells (about 3×10^6 cone cells/retina) are relatively insensitive to light and require more than 100 photons of light to elicit a neural response (Nelson and Cox, 2008). Both photoreceptor types have an outer segment, composed of membranous disks containing light-sensitive photopigments, and an inner segment, which contains a cell nucleus and mitochondria and is responsible for providing the synaptic terminals to contact the bipolar or horizontal cells.

Retinal photoreceptor cells do not exhibit action potentials like most sensory systems in the body. Instead, light activation of the photoreceptors causes a graded change in membrane potential, which results in a change of the rate of neurotransmitter release onto postsynaptic neurons (Massy, 1990). The excitatory amino acid, glutamate, is used as the neurotransmitter in the photoreceptor cells to transmit signals to the next order neuron in the chain. In the presence of light, the retinal chromophore conjugated with the photoreceptor cell isomerizes, thereby initiating a signal transduction cascade that amplifies the signal and leads to the closure of cation channels on the plasma membrane. The closure of these gated channels leads to a hyperpolarization of the plasma membrane, characterized by a more negative potential inside the cell. This hyperpolarization is sensed by the synapses, which react by reducing the release of glutamate. Information is then relayed to bipolar cells and subsequently the ganglion cells.

Visual disorders

Retinitis pigmentosa (RP) refers to a heterogeneous group of heritable retinal degenerative diseases characterized by a progressive loss of visual acuity resulting from photoreceptor apoptosis (Smith, 2004). This currently incurable disease, resulting from a diverse group of genetic mutations in more 44 genes in the rod photoreceptor cells (Busskamp *et al.*, 2010) has a unique set of clinical characteristics that make it particularly difficult to treat (Shintani *et al.*, 2009). As a common pathology, patients affected by RP experience nyclalopia or night blindness, resulting from the loss of rod photoreceptor cells. This degradation is followed by the gradual loss of peripheral vision leading to tunnel vision, and subsequently to a loss of central cone-mediated vision. Current treatment modalities include nutritional drug supplements to preserve photoreceptor function, in addition to cell transplantation and gene therapy. Gene therapy has been proven to improve visual function; however, long-term safety and efficacy, immune response, and issues associated with various vectors complicate the advancement and success of gene therapy as a way to treat RP (Shintani *et al.*, 2009).

Age-related macular degeneration (AMD) is a degenerative retinal disease, which is caused by the deterioration of the central portion of the retina, leading to a loss of central vision. According to the National Eye Institute (NEI), AMD is most common in those age 70 and older (Birch and Qi Liang, 2009); however, AMD is cited as being the leading cause of irreversible vision loss and legal blindness in individuals over the age of 55 (Purves *et al.*, 2008). Although the pathogenesis of AMD remains unresolved, this multi-factorial disease has both genetic and environmental components, which play a role in its development. Current treatment options include intraocular or periocular drug administration and nanoparticle-mediated gene therapy (Farjo and Ma, 2010; Thomson and Lotery, 2009). These treatments only slow the progression of AMD and come with significant risks.

Leber's congenital amaurosis (LCA) is a rare inheritable retinal degenerative disease, which is characterized by severe vision loss in infancy. Patients with LCA can experience a variety of other eye related problems including roving eye movements, often referred to as nystagmus, sunken or deep-set eyes, and sensitivity to bright light. Symptomatic at birth, patients with LCA often display a lack of visual responsiveness and have a very low level of retinal activity. To date, there is no substantial treatment or cure for LCA and care is supportive. Patients affected by LCA often benefit from the use of low vision aids. More recent treatment for LCA has focused on gene replacement therapies, as 14 genes have been identified to be responsible for LCA (Ashtari *et al.*, 2011). In 2005, scientists were able to use gene therapy utilizing adeno-associated viral vector (AAV) mediated RPE65 to restore visual function in a Briard canine model of LCA. This LCA canine model was blinded as a result of mutations in the gene encoding retinal pigment epithelium 65-kDa protein (RPE65), the gene important for the production of vitamin A (Acland *et al.*, 2005; Herzog, 2012). These treatments have also been used in three human phase I clinical treatment trials of AAV-mediated RPE65 gene therapy, and results showed a slight improvement in vision in both bright and dim light (Bainbridge *et al.*, 2008; Cideciyan *et al.*, 2008; Hauswirth *et al.*, 2008; Maguire *et al.*, 2008).

Approaches in nanobiotechnology for visual restoration

The incidence of retinal degenerative diseases is increasing at an alarming rate due to an ever-ageing global population. The pathologies associated with diseases like AMD and RP have long been classified as irreversible and irreparable, although treatments do exist to slow the progression. Multiple genetic and environmental factors have been shown to promote the degradation of photoreceptor cells in the human retina, which has led to difficulty in the early diagnosis, prevention, and treatment of AMD and RP. Despite monumental advancements in general surgical methodology and technology, the unique complexity of the neurosensory system has precluded the same degree of innovative developments. Consequently, there is a significant need for the design of a retinal prosthesis or device that is capable of (1) mimicking the sensitivity of native photoreceptor function and (2) stimulating intact bipolar and ganglion cells within the retina. The following sections describe the use of microbial rhodopsins to restore vision and improve the quality of life for those with retinal degenerative diseases using gene therapy, optogenetics, electrical stimulation, or an ion-mediated stimulation approach.

Neurosensors

The ability to elicit excitatory and inhibitory responses is a fundamental characteristic of neural networks. As more information becomes available about the different response times for excitatory and inhibitory regions of receptor fields, the ability to detect sharp variations in image space (i.e. edge detection) and the ability to detect the motion of objects in retinal ganglion receptive fields is feasible. Researchers around the world are working to develop technologies and biomaterials which mimic the inherit function of the retinal neural network. Optoelectronic devices utilizing microbial rhodopsins, particularly BR, HR, and ChR, have shown promise at restoring functionality of the visual neural circuitry, which is the focus of this chapter.

Gene therapy

One of the most promising advances in the treatment of inherited and acquired ocular diseases is gene therapy. In the past decade gene therapy approaches have advanced significantly, and are being used for the treatment of progressive eye diseases such as AMD, RP, and LCA (Conley and Naash, 2010). To date these degenerative ophthalmic disorders lack curative treatments and regenerative treatments are a requisite to finding a cure. For many retinal diseases, including AMD, RP, and LCA, the onset of phenotype often follows the onset of degeneration, making treatment of chronic retinal diseases limited.

Treatments and therapies at the genetic level offer several advantages over pharmaceutical based treatments. In particular, gene therapies address the source of the problem, not just the symptoms, and can be locally delivered without risking systemic toxicity. Intraocularly delivered drugs are also less likely to elicit a severe immune response, in contrast to systemically delivered drugs because the eye is protected from the bloodstream by the blood–retina barrier. In gene therapy approaches, the vectors, usually AAVs, are taken into the target cells, where the genetic material is protected in the nucleus. The gene product can then be regularly expressed.

Despite the advancements and promise of genetic therapy to treat degenerative eye diseases, several challenges exist (Thomas *et al.*, 2003). Because of the pathology of many of these inherited retinal degenerative diseases, a detectable phenotype is not present until after significant cell apoptosis has occurred. Thus, gene therapies would not only need to target the remaining neural tissue and rescue it from the disease causing defect, but they would need to stimulate neuronal regeneration. Additionally, many of the gene therapy vectors do not transfect post-mitotic or neuronal cells well. For a gene expression vector to be considered a viable candidate for the treatment of these chronic ophthalmic diseases, long-term gene expression without the need for repeat dosing is critical. Finally because of the number of retinal disease causing genes identified and the complexity of some of the retinal disorders, each gene would need to be treated differently, making the cost–benefit ratio for this treatment unfavourable.

Optogenetics

With the discovery of light gated ion channels and ion pumps in green algae and bacteria, new methods for the treatment of neurological and visual disorders have emerged. New concepts, methods, and fields have arisen as a result of the investigation of microbial rhodopsins. Most recently, Type I rhodopsins have become the target of new technological applications, including optogenetics. Optogenetics is a new technology that combines genetic and optical methods to regulate the gain or loss of function of well-characterized functions in specific cells of living tissue. Through this method, distinct events in biological systems, can be targeted and controlled with precision (Deisseroth, 2011). The development of optogenetics was triggered by the functionality of microbial opsin genes to safely transfer information to neurons. The demonstration of optogenetic control of neurons with microbial opsins, including BR, HR, and ChR, is vital in the ability to control a well defined series of biological and neurological events. The method, which was developed by Karl

Deisseroth and Ed Boyden in 2005 (Buchen, 2010), began with the expression of ChR2 in neurons (Boyden *et al.*, 2005).

Recently, optogenetic therapies have been used to target and manipulate the electrical activity of specific neurons embedded within brain circuits. Through the use of recombinant DNA technologies, microbial rhodopsins have been able to function in other organisms. Using microbial rhodopsins as optogenetic tools has enabled the transfection of neurons in various organisms. These proteins are transferred into the membrane of these cells and can be used as light activated switches to turn action potentials on or off. In particular HR is often used to inhibit action potentials (Han and Boyden, 2007) and ChR2 is used to trigger a neuronal response. Because these optogenetic tools are non-invasive and do not require any pharmacological agents, they have been used for biomedical therapies of neurological diseases. Currently, the use of these optogenetics tools to restore visual function has exhibited moderate success in the inhibition and excitation of cells (Busskamp *et al.*, 2011; Busskamp and Roska, 2011).

The issues of using optogenetic treatments in visual restoration involve the use of AAVs. Halorhodopsin expression has been shown in post-mortem human photoreceptors from cultured retinas and ChR2 expression via intravitreal delivery was shown *in vivo* in marmoset retinas (Busskamp *et al.*, 2010); however, to date, there has been no AAV vectors with optogenes injected into human patients with RP or AMD. The efficacy and safety of these tools is also under investigation; however, preliminary testing in rats and mice has shown no neurotoxic effects (Ivanova and Pan, 2009; Sugano *et al.*, 2011). Additionally, cells expressing HR or ChR require high light intensity to be activated. In order for optogenetic tools to be viable, an external visual device is needed to adapt to ambient light and amplify and focus this light on the retina in an adjustable way (Degenaar *et al.*, 2009).

Electrode-based artificial retinas

The ability to convert light energy into an electrical signal is the fundamental function of the retina as a photosensory array. Based on this behaviour of the neurosensory network, scientists have considered designs for retinal prostheses that exploit the electrical nature of these signals. A number of investigations using extracellularly recorded responses from retinal ganglion cells have provided evidence that electrical stimuli are effective modes of generating neural responses (Fried *et al.*, 2006; Grumet *et al.*, 2000; Jensen, 2003; Jensen and Rizzo, 2006, 2008; Jensen *et al.*, 2005a,b; Margalit and Thoreson, 2006). These studies have resulted in the characterization of the magnitude, polarity, and duration of an electrical stimulus that is required to mediate neural activity. It was found that extremely low currents (0.5 μA) are sufficient to stimulate retinal ganglion cells using microelectrodes on the epiretinal surface (Jensen, 2003). The polarity of the applied current has also been shown to be a significant stimulus parameter (Jensen and Rizzo, 2006). Threshold currents from anodal current pulses are relatively low for both on-centre and off-centre ganglion cells, while cathodal current pulses elicit varying responses between the neural cell types. Currents at and around the magnitudes described above were capable of generating both short-latency (3–5 ms) or long-latency (>9 ms) action potentials, which were based on the duration of the electrical stimulus. A brief pulse can stimulate a single action potential, while longer pulses can produce a series of action potentials from the successive activation of ganglion cells and bipolar cells in the neurosensory network (Jensen *et al.*, 2005a). These stimulus parameters demonstrate the complexity of the spatiotemporal responsively that must be considered to develop an electrode-based implant capable of simulating typical retina function.

A retinal implant must either be placed subretinally, between the retinal pigment epithelium and the bipolar cells, or epiretinally, directly adjacent to the nerve fibre layer. Numerous research groups across the globe are currently investigating both approaches, some of which have established marginal successes in preclinical and clinical trials (Sakaguchi *et al.*, 2004; Weiland *et al.*, 2005; Yanai *et al.*, 2007). To date, most implant prototypes exploit the electrical nature of the signals produced by bipolar and ganglion cells through the implementation of microelectrode arrays or silicon-based diodes to stimulate the retina. This particular tactic has exhibited significant successes in preclinical trials, where electrodes were sufficient to generate a neural response in both the subretinal position (Jensen and Rizzo, 2006) and epiretinal position (Jensen *et al.*, 2005b). One of the more promising sets of results was produced by Optobionics, which incorporated silicon diodes as photoreceptors in a subretinal implant to generate a signal upon light activation (Chow *et al.*, 2004). Despite conclusive data that attributed responses to the presence of the implant, signals were only observed under intense light sources that are unrealistic when compared to ambient light (Chow *et al.*, 2004). An analogous implant design was investigated using an external camera coupled with a microelectrode array as an epiretinal implant. The external camera is mounted on a pair of glasses and is directly attached to the microelectrode array by a connecting cable. The epiretinal positioning of the 16 electrodes in the array leads to the stimulation of ganglion cells (Cunningham *et al.*, 2001). Using this approach in clinical trials, patients were able to detect motion and were capable of distinguishing between simple shapes (Humayun *et al.*, 2003). The protein-based artificial retinas that are in the following sections are capable of high-resolution vision with edge sensitivity. In the early 1990s optoelectronic devices using BR as the substrate demonstrated the capability of BR to stimulate excitatory and inhibitory responses and edge detection at zero crossings (Martin *et al.*, 1997; Takei *et al.*, 1992; Yao *et al.*, 1997). In 1993, a 256-pixel artificial photoreceptor generated from immobilized BR thin films at the solid–liquid interface of an electrode demonstrated the ability of BR to detect motion and edge information in real time (Miyasaka and Koyama, 1993). Differential responsivity of BR to varying light intensity allows for the ability of the protein to aid in motion detection (Hong, 1997). Furthermore, these implants do not require external power supplies or connecting wires protruding from the orbitals, which leads to less invasive implantation surgeries, lessening the chance for infection and an immune response.

Bacteriorhodopsin as a photovoltaic element

Bacteriorhodopsin exhibits a photoelectric effect that is common among pigment containing membranes. A photoelectric effect refers to the ability of a material to generate electricity upon the absorption of light, which is similar mechanistically to the electrode-based retinal implants described in the previous sections. As the array of BR molecules in the purple membrane absorb light, a significant proton motive force is generated towards the extracellular surface (Calimet and Ulmann, 2004; Oesterhelt and Stoeckenius, 1973). This force is roughly equivalent to that of 4 pH units, and can also be observed as a measurable photovoltaic response (Calimet and Ulmann, 2004). To amplify this effect, multilayered thin films of oriented BR can be engineered and utilized in biomolecular sensors. The accumulation of multiple BR layers also enhances the net capture of energy as light passes through the protein film (Fig. 9.7). There are multiple methods for creating

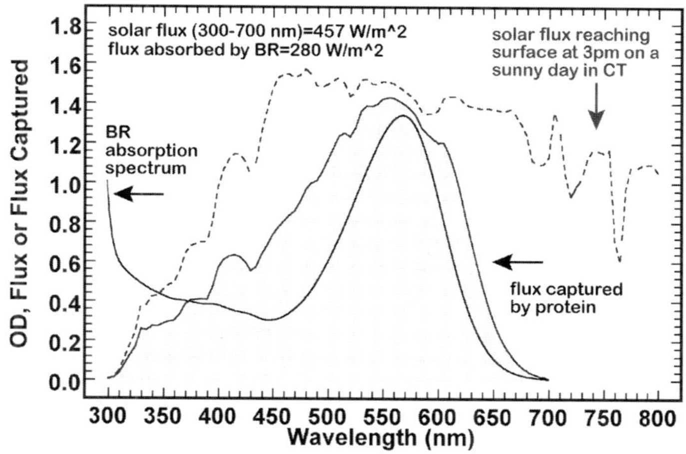

Figure 9.7 Bacteriorhodopsin (BR) has been optimized by evolution to have an absorption spectrum that couples efficiently with solar radiation. For example, on a typical day, 457 W/m^2 of solar energy hits the surface of the planet in the 300–700 nm region. A 400 layer film of BR will capture 280 W/m^2 (61%) of that energy during the first pass of the light through the protein film. If the light that exits the film is reflected back with 90% efficiency through the layers, the protein film will capture a net 83% of the energy in this region.

these oriented thin films (Chen *et al.*, 1991; Koyama *et al.*, 1994; Miyasaka and Koyama, 1992); however, the devices in the following sections specifically use electrostatic layer-by-layer adsorption (He *et al.*, 1998b).

The illumination of BR thin films produces two types of photovoltaic responses, both of which are dependent on the duration of light exposure (Hong, 1997). The first response type can be considered as a fast photoelectric pulse of positive charge and is observed as a result of the proton pumping gradient that occurs upon the absorption of light. Oriented BR thin films display three distinct components in the observed photovoltage (Fig. 9.8a). The initial response is a sharp negative voltage spike (the B1 signal) that occurs within the picosecond scale, which is followed by a large positive voltage and a subsequent exponential decay (the B2 and B3 signals, respectively) (Hong and Hong, 1995; Váró and Keszthelyi, 1983). The B1 signal is associated with an electron density shift upon the isomerization of retinal during the primary photochemical event (Groma *et al.*, 1995). Positively charged residues in the binding site of BR (e.g. arginine-82) are responsible for the transition from the primary event to a positive photovoltaic signal (Xu *et al.*, 2003). The positive B2 and B3 signals are associated with the **K**→**L**→**M** and **M**→**N**→**O**→**bR** transitions in the photocycle, respectively, and are gated by the release of H$^+$ between the **M** and **N** states.

The second type of photovoltaic response that is characteristic of BR thin films occurs when the protein is exposed to periods of long square-wave illumination (Fig. 9.8b). Under these conditions, a differential response, comprising transient positive and negative signals, is observed at the onset and termination of a light stimulus, respectively. The magnitude, duration, and polarity of this differential response is dependent on the characteristics of the BR thin film assembly. Furthermore, BR thin films are capable of modification via genetic engineering (see Genetic Optimization of Microbial Rhodopsins) to optimize

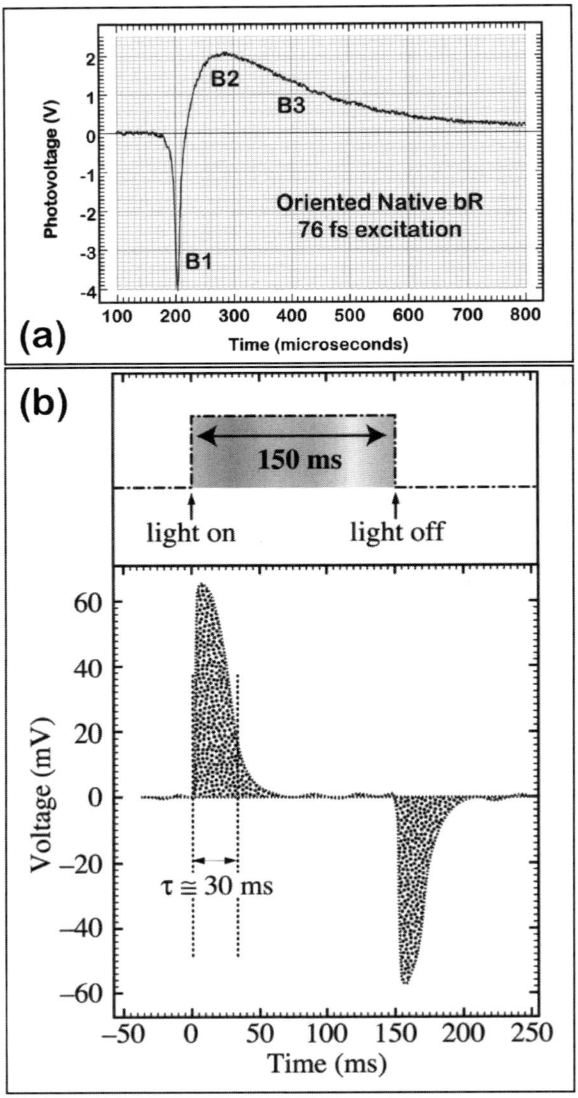

Figure 9.8 The three components of the photovoltage of light adapted bacteriorhodopsin (BR) oriented between two indium-tin-oxide transparent plates. A positive voltage is defined as the motion of positive charge in the same direction as the proton pumping or, conversely, the motion of negative charge in a direction opposite to the proton pumping (a). Differential responsivity of native BR (b). The size of the differential (reversed voltage) light-off signal (150 ms) is adjustable from 5–95% of the initial photovoltaic peak shown at 0 ms by changing the interlayer polymer permeability.

characteristics of the differential responses to match latency and threshold requirements of electrical signals in the neural network.

Artificial retinal implants based on BR have the capability of mimicking previously proposed designs that used silicon photoconvertors to generate a signal. The biological medium introduces a number of advantages over the approaches described above, the most significant of which includes an increase in voltage and responsivity. The high quantum efficiency of BR permits that each absorbed photon has a 0.65 probability of translocating a proton towards the pixel surface of the implant, thereby creating a local potential shift of many volts (Govindjee *et al.*, 1990). To generate a current that can sufficiently stimulate the neural network of the retina, many incident photons will be required to distribute this voltage over the entire pixel surface. Thin films of BR have been shown capable of reaching photocurrents of 120 μA/cm^2 and photovoltages of 3.8 V (Zhang *et al.*, 2003). These values are much higher than the experimentally derived current threshold of activation (Jensen, 2003).

Fig. 9.9 highlights two different designs that use BR as a photovoltaic element to convert light energy into an electrical stimulus. The first is an epiretinal implant that is placed directly adjacent to the nerve fibre layer. This epiretinal design places the outer surface of

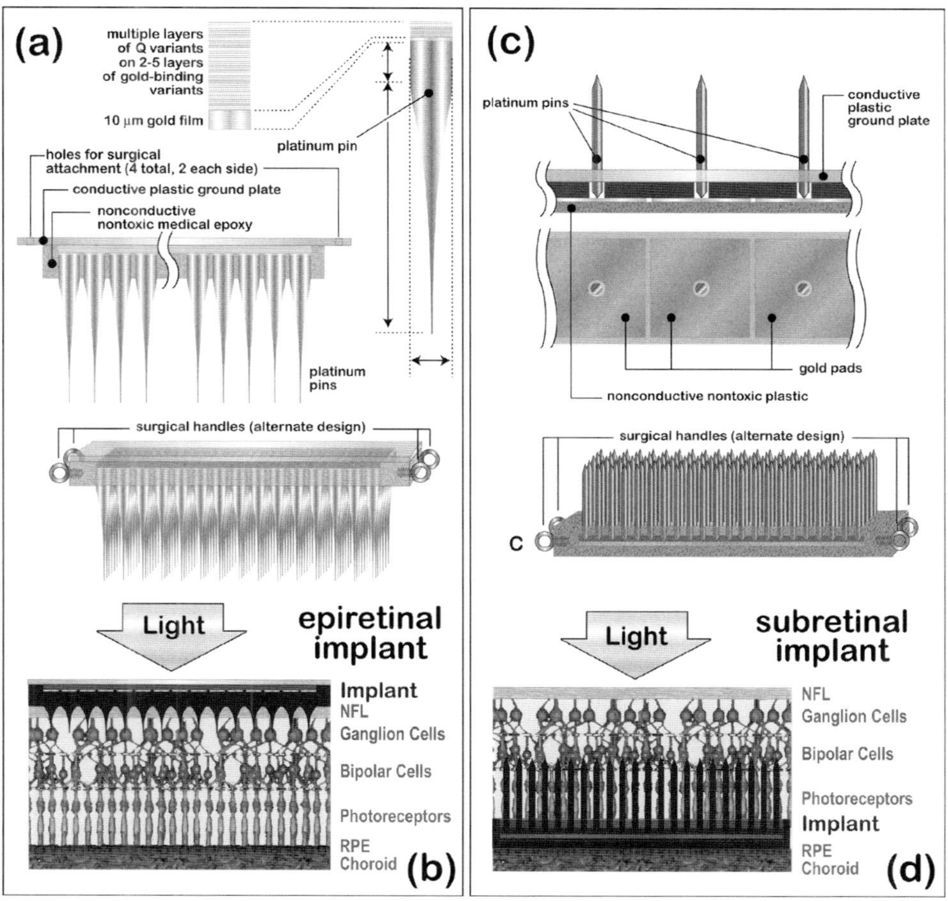

Figure 9.9 Schematic diagrams of the two protein-based artificial retinas that operate via photovoltaic excitation of the ganglion or bipolar cells. The epiretinal implant (a and b) has the probe lengths optimized to intercept the ganglion cells directly underneath the nerve fibre layer (NFL). The subretinal implant (c and d) is inserted between the retinal pigment epithelium (RPE) and the photoreceptor cells. The advantage of the subretinal implant is that a smaller current is required to stimulate the bipolar cells.

the implant in the direct path of incident sunlight that is collimated towards the back of the eye. Thin needle-like electrodes protrude from the face of the device to form a direct contact to transmit an electrical signal towards individual ganglion cells. The second design is a subretinal implant, placed between the retinal pigment epithelium and the damaged photoreceptor cells. This model also uses BR as a photovoltaic scaffold to conduct an electrical signal to individual bipolar cells via platinum needle electrodes. Both the epiretinal and subretinal protein-based implants are analogous to microelectrode-based devices, and are capable of generating an electrical stimulus to produce action potentials within ganglion or bipolar cells. However, similar problems with resolution and edge sensitivity arise due to the generation of diffuse visual perceptions from targeting individual neural cells. Implementing BR as the photovoltaic element provides greater responsivity than silicon diodes, although the protein must be harnessed in a different way to circumvent the problems of this approach.

Ion-based artificial retinas

Although there have been significant improvements towards the development of electrode-based retinal implants, much remains to be investigated for the successful implementation of these prototypes as viable sources for meaningful vision. The drawbacks for implants relying on electrical stimulation are two-fold. First, it is difficult to replicate the spatiotemporal patterns of the neurosensory network that facilitate normal visual perception. And second, these devices often require external equipment or power supplies to create a stimulus, which can be toxic to human tissue and may elicit an immune response. Theoretical simulations are being undertaken to understand the dynamics of retinal signalling using electrodes as a stimulus (Freeman *et al.*, 2011); however, coupling these advancements with materials that are biologically inert will remain a challenge for years to come. In the subsequent sections, we describe a biologically inert protein-based retinal implant that bypasses the issues observed with electrode-based counterparts by creating an ion gradient as a mechanism for retinal stimulation.

Ion-mediated stimulation of the retina by a protein-based implant has the potential to mimic the phototransduction cascade mechanism initiated by healthy photoreceptor cells. Computational models have recently been examined to predict the success of using gradients of hydrogen, potassium, or chloride ions to activate the retinal neural machinery (Theogarajan, 2007). These simulations suggest that altering the pH or ion concentration near the photoreceptor, bipolar, or ganglion cells can hyperpolarize the cell membranes sufficiently to trigger a nerve impulse. In the human retina, the absorption of a photon by the visual pigments creates a hyperpolarization of the inner segment of the photoreceptor cell, which leads to a positive potential within the extracellular milieu (Stuart and Birge, 1996). This hyperpolarization can also be generated by manipulating the extracellular environment via an induced ion gradient. This theoretical work supports previous experimental findings, which further suggest that ion gating would be an effective method to generate the impulse necessary for initiating the visual cascade (Bringmann *et al.*, 1997; Enz *et al.*, 1999; Ettaiche *et al.*, 2006; Konnerth *et al.*, 1987; Marc, 1999; Verweij *et al.*, 1996). Some of these investigations included work involving the use of ChRs as a medium for ion transport (Berthold *et al.*, 2008; Douglass *et al.*, 2008; Nagel *et al.*, 2002, 2003; Petreanu *et al.*, 2007). Retinal implants consisting of microelectrode arrays are characterized by low resolution and a weak interface to the different bipolar and ganglion cell types.

Bacteriorhodopsin as an ion source

The ion-mediated protein-based artificial retina implant exploits the unique photophysical properties of BR, which were briefly described in the above sections. The necessary functionality of the photoactive medium of these implants must be capable of efficiently absorbing incident light while simultaneously generating a net response that sufficiently interacts with the remaining neurosensory network. Based on these criteria, a BR-based implant is a viable candidate for the visual restoration of patients suffering from retinal degenerative diseases because it is capable of accomplishing both of these tasks simultaneously (Chen and Birge, 1993). The key function of this photoactive protein to the native organism is to produce metabolic energy, which is facilitated by producing a proton motive force sustained by anaerobic respiration (see Bacteriorhodopsin). The large quantity of BR trimers per unit area within the purple membrane enables a significant ion gradient to form upon the absorption of light. It is this ion gradient that can be used to generated neural responses, which will be sent to the optic nerve and then to the brain, to produce meaningful vision. Bacteriorhodopsin used as an ion gradient source circumnavigates the drawbacks of the photovoltaic protein-based artificial retinas.

The protein-based artificial retina architecture comprises a multilayered BR thin film generated via sequential electrostatic adsorption. In order to harness the ability of BR to generate a sufficient ion gradient for retinal stimulation, the protein must be uniformly oriented across a large surface area at an optimal optical density (Muneyuki *et al.*, 1999; Váró and Keszthelyi, 1983). The multilayer thin film must contain enough layers of BR to adequately absorb incident light, while also generating a substantial ion gradient (Fig. 9.10) (He *et al.*, 1998b). The inherent proton pumping action of BR is necessary for the survival of the native organism, however, one can manipulate this specific mechanism to produce an appreciable ion gradient at the macroscopic level (Uehara *et al.*, 1993; Zhang *et al.*, 2003). The native production of a proton gradient enables an implant to alter the pH around surrounding neuronal cells, which has been shown to generate nerve impulses through the indirect manipulation of secondary signal carriers (Bringmann *et al.*, 1997; Enz *et al.*, 1999; Ettaiche *et al.*, 2006; Konnerth *et al.*, 1987; Marc, 1999; Theogarajan, 2007; Verweij *et al.*, 1996). A unidirectional pH gradient is achieved through the sequential electrostatic deposition of alternating layers of purple membrane and a polycation (He *et al.*, 1998b, 1999). The use of organic cations as a platform for the layering of anionic proteins has been demonstrated in previous investigations (He *et al.*, 1998a,b; Lvov *et al.*, 1995). Bacteriorhodopsin contains a net negative charge on both the extracellular and cytoplasmic loop regions, as well as an inherent dipole moment of 570 Debye that is observed due to the relative position of charged residues in the protein (Porschke, 1996). Thus, the purple membrane fragments selectively orient onto adsorbed polymer layers, and, once repeated over several iterations, form highly ordered assemblies of polymer-protein subunits on a solid substrate. Fig. 9.11 illustrates the BR-based thin film comprising these alternating layers.

The material that serves as a platform for this layer-by-layer approach is a modified flexible synthetic bioinert microfibre, which has been shown to be medically inert when inserted into human tissue (Roll *et al.*, 2008; Scholz, 2007). In addition to being bioinert, the substrate is flexible and ion permeable membrane. These properties fulfil the paramount requisite for any retinal implant; the capability of sequestering a flexible and bioinert device that is capable of generating a neurosensory response upon light activation. This artificial retinal implant exploits the intrinsic light-transducing properties of retinylidene proteins

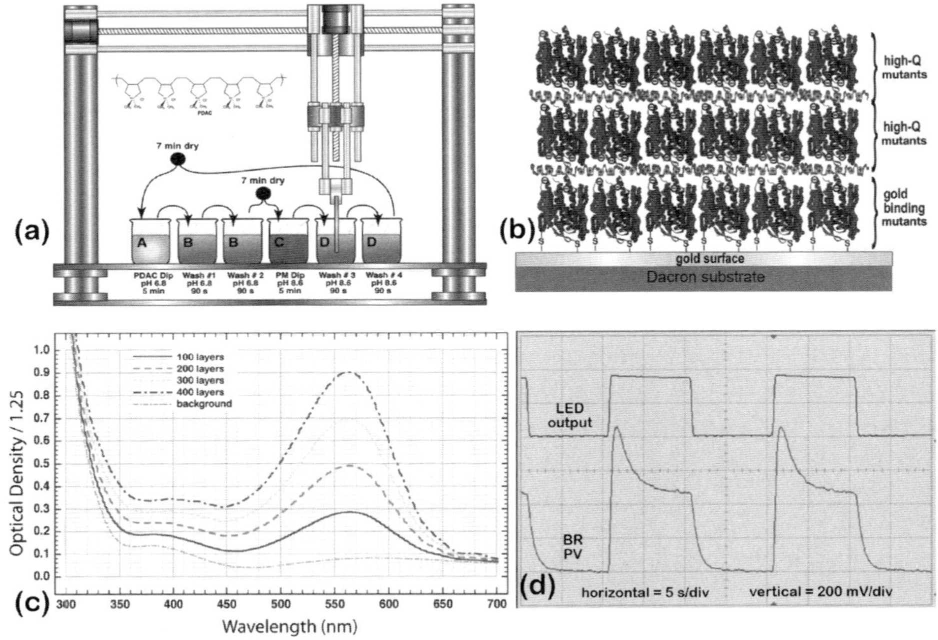

Figure 9.10 Multiple layers of bacteriorhodopsin (BR) are generated by using the layer-by-layer (LBL) method as shown in inserts (a) and (b). The first layer uses a gold binding mutant that covalently attaches to an ultrathin layer of gold placed down on a Dacron substrate. The optical density is directly proportional to the number of layers (c) and a 200 layer system will generate an 800 mV signal (d).

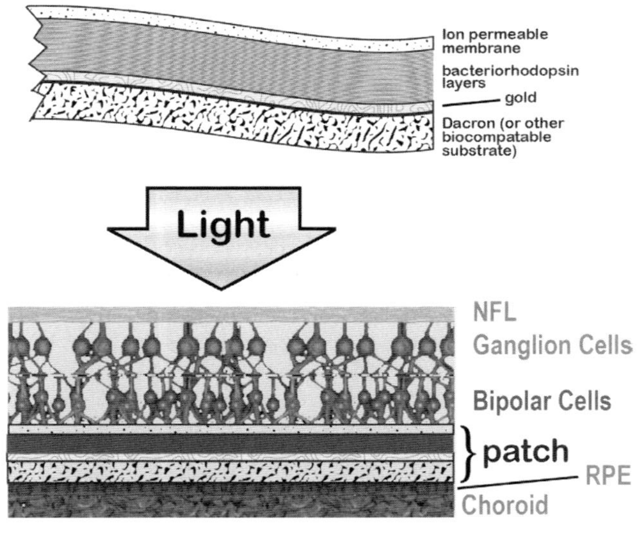

Figure 9.11 Schematic diagram of the subretinal ion patch. This implant pumps protons or chloride ions into the bipolar cell milieu. The implant can be placed in either direction so that it can increase or decrease the pH or chloride ion concentration near the bipolar cells.

to evade the potential necessity of external power supplies or camera equipment. Thus, the retinal implant will be surgically less invasive and would lessen the chance of infection due to the presence of exogenous cables or sensors. From a materials standpoint, this protein-based device is an optimal combination of biological and synthetic components that facilitate an effective design of a subretinal implant. Despite the numerous noteworthy advantages of using native BR as a photoreceptor supplement, the photophysical properties of the protein can be further enhanced using advancements in genetic engineering.

In the above sections, we have highlighted the inherent stability of BR and the unique photophysical properties that make the protein an ideal candidate for application in many bioelectronics and biophotonic devices, including an artificial retina implant. The potential of genetically modifying the protein to manipulate or enhance these properties has led to even more successes in the development of devices with BR as the photoactive medium (Wise *et al.*, 2002). More specifically, the ion-mediated approach for generating a neural response can be augmented in a number of ways. First, the proton motive force of bacteriorhodopsin has been shown capable of manipulation by reversing the surface charge asymmetry of the protein (Hsu *et al.*, 1996). The enhancement of the inherent dipole moment allows purple membrane fragments to pack more densely when using the layer-by-layer method, and also increases ion transport and photoresponse through the prevention of back diffusion of H^+ ions (He *et al.*, 1998a). Furthermore, the branched photocycle of BR directly facilitates the quantitative control of the ion gradient that is produced upon light absorption (Popp *et al.*, 1993). The unique ability to form an inactive photoproduct (Fig. 9.1), allows for the potential to regulate and control pixels, where extraneous pixels can either be turned down or turned off. Pixel mediation enables more control of the ion–gated interaction with bipolar and ganglion cells, thusly providing the option of fine-tuning the ion flux to engineer high-resolution vision. Additional modifications to BR involve enhancing the chloride pumping capabilities of the protein, which has been achieved through single site mutations (Paula *et al.*, 2001; Tittor *et al.*, 1997). Although *H. salinarum* contains HR, a native chloride ion pump in the outer membrane, the protein is comparatively fragile and does not exhibit the same potential in devices as BR (Kolbe *et al.*, 2000; Lanyi, 1986). Previous investigations have established that chloride ions are realistic candidates for activating the neural machinery within the retina (Theogarajan, 2007). The manipulation of the structure and function of BR has provided exciting possibilities in the development and design of protein-based artificial retinas. The advantages of these devices are further improved upon by the modification of the primary photophysical characteristics of the retinylidene protein. The following section provides a discussion on recent advancements in genetic engineering, including methods in site-directed mutagenesis (SDM), site-specific saturation mutagenesis (SSSM), semi-random mutagenesis (SRM), random mutagenesis (RM), and directed evolution, and how they pertain to the improvement of BR for protein-based devices.

Genetic optimization of microbial rhodopsins

Biological macromolecules and proteins have been optimized to function in native ecosystems for over three billion years. These proteins have been used by microorganisms in all three domains of life to carry out a majority of the complex biological functions required for survival. Unfortunately, nature finds no comparative advantage in optimizing proteins for non-native environments, and thus applied technologies harnessing the complex functions of native proteins have found limited success. In most cases, the native protein requires further enhancement to become commercially viable. With the recent advancements in genetic engineering, scientists are now able to tailor these complex molecular machines for function outside of the native organism or environment. Optimizing the photophysical, thermal, and spectroscopic properties of light activated proteins, such as HR, BR, and ChR, is a significant step in adapting these proteins for use as tools and materials in optogenetics, gene therapy, and prosthetics.

Protein engineers use novel mutagenesis strategies, in conjunction with high throughput screening methods, to modify biological molecules for performance in non-native environments. Through the course of evolution, mutations accumulate and manifest in structural and functional changes for the protein, which can have a positive, neutral, or negative effect on the biological macromolecule. As lethal mutations increase, cell death often occurs; however the increases in beneficial mutations may enable the host organism to overcome a select environmental pressure. For scientists, the challenge is predicting *a priori* which amino acid sequence will generate the desired properties for an intended application (see Fig. 9.12). Nature uses selective pressure to generate genetically diverse and versatile molecular machines, and it is expected that the same selective pressures can be applied in the laboratory to modify biological molecules for device applications. With respect to the optimization of BR, scientists are faced with the unique problem of optimizing a photochemical process that gains stability from a two-dimensional lattice that is generated during the expression of the protein in the native organism (Krebs and Isenbarger, 2000). Molecular modelling and theoretical platforms are often used for selecting mutations of interest, but these methods are difficult to implement when a complex photochemical reaction is to be optimized, particularly if variables such as pH require simultaneous enhancement (Boxer *et al.*, 1992; Jäckel *et al.*, 2010). In this section, we describe the use of Type I directed evolution, combining chemistry, spectroscopy, and molecular biology, for the systematic optimization of BR for devices.

Directed evolution is a method used in protein engineer to optimize the inherent properties of a molecule via iterative rounds of diversification and differential selection for a set of desired properties, not normally found in nature. Using artificial selection, proteins can be evolved to thrive outside of the biological context in which they thrive. Scientists use this capability to distinguish novel properties or combinations of properties that are biologically relevant, but may not be encountered in natural proteins (Arnold *et al.*, 2001). Current research efforts have focused on improving the thermal and chemical properties of enzymes and biopolymers for industrial and pharmaceutical applications AAVs for therapeutic gene therapy (Asuri *et al.*, 2012; Bartel *et al.*, 2012; Maheshri *et al.*, 2006). In each case, diversified genetic libraries were generated and appropriate screening methods were used to select mutants with enhanced phenotypes.

Redesigning photoactive proteins for synthetic environments has been accomplished using a variety of methods, including SDM, SSSM, SRM, RM, and now directed evolution. The most commonly used method for modifying the genetic coding sequence of a protein is SDM. Using this method, a single amino acid residue, typically a key residue in the primary sequence, is exchanged for a different amino acid in order to alter the structure and function of the molecule. Although a powerful method, the enormous number of unique substitutions that are possible in large proteins make it challenging to explore the entire genetic landscape of a protein using SDM. As an alternative, SSSM, SRM and RM, DNA

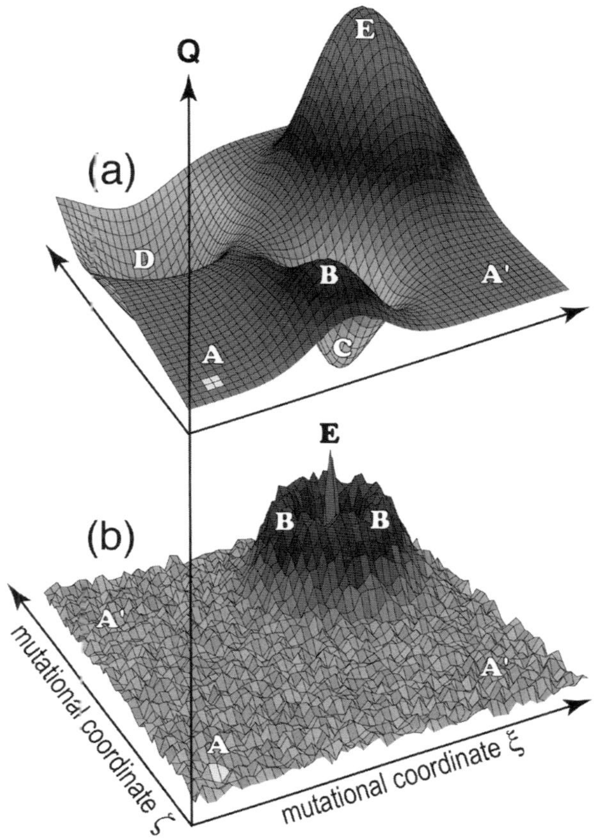

Figure 9.12 Hypothetical mutational landscapes for fluid traits (a) and spectrokinetic optimizations (b). The native protein is located at the origin of the axes, as represented by A. The mutational coordinates are arbitrary, and the vertical axis measures an arbitrary Q factor. Starting at A, the ultimate goal of our mutational strategies is to discover the optimal protein, as represented by peak E. The A and A' regions are representative of mutations that have little impact on the Q factor, B regions are local maxima in Q, and the optimal mutation is labelled at the peak, as designated by E. Modelling can be used to assist in finding E. For optimization of temperature, a fluid trait, the overall stability of the protein may decrease, as shown in C and D of (a). More complex characteristics, such as spectrokinetic properties, result in the mutational landscape shown in (b).

shuffling, and the staggered extension process, offer comparative advantages over targeted mutagenesis (Arnold *et al.*, 2001; Crameri *et al.*, 1998; Dalby, 2003; Georgescu *et al.*, 2003; Graddis *et al.*, 2002; Zhao *et al.*, 1998). Semi-random mutagenesis and RM produce a large number of indiscriminate mutants through the use of doped oligonucleotides, and SSSM, a combination of SRM and RM, allows for the exploration of multiple amino acid substitutions via saturation of a key residue at a target locus on the gene of interest. The challenge protein engineers face currently is how to efficiently couple these genetic diversification techniques with an efficient screening system. In order for SRM and RM to be cost effective and efficient new screening and selection techniques must be developed, particularly in the case of photoactive proteins, for these methods to be viable.

Photoactive proteins typically have individual residues or an isolated group of residues that contribute to the overall photochemistry and stability of the molecule. For the application of BR in devices, most of the mutagenesis has focused on the optimization of the photointermediates, **M**, **O**, and the branched photoproduct **Q**, a state rarely found in nature. Mutagenesis of BR has also been employed to enhance the innate dipole moment, gold-binding capabilities, and overall photocycle speed of the protein. In the case of gold-binding, SDM has been utilized to convert residues on the intracellular and extracellular loop regions to cysteine residues, which selectively bind to gold. The native protein contains no native cysteine residues and thus the addition of cysteine residues permits binding via a thioether linkage and by extension an ordered base structure for the orientation of subsequent protein layers. Gold as a support substrate is desirable for biological applications because it is a chemically inert and electrically conductive metal (Berthoumieu *et al.*, 2012; Crameri *et al.*, 1998; Patil *et al.*, 2012). Enhanced dipole mutants are critical for the application of BR in photovoltaic devices, including a photovoltaic artificial retina. The term dipole mutant refers generally to any mutational substitution that involves a charge substitution that alters the intrinsic dipole moment of the protein. Such mutations provide a larger photovoltaic homogeneity in thin films by improving the packing density and orientation of the protein layers. Molecular modelling programs such as CHARMM (Brooks *et al.*, 1983) and Mozyme (Stewart, 1996) have been used to theoretically determine which residues are targets for mutation.

Type I directed evolution of BR has been used to optimize the formation and reversion efficiencies of the **Q** state for application in long term data storage, including holographic memories and associative memories, in addition to a protein based artificial retina. The **Q** state is a photochemical product of the BR photocycle that is accessible via the **O** state upon the addition of a second photon of light (Gillespie *et al.*, 2002; Popp *et al.*, 1993; Wise *et al.*, 2002). As a stable photo-species, with a calculated lifetime of ~10 years, the **Q** state offers a means by which to non-invasively turn off extraneous pixels within the artificial retina. The **Q** state is minimized in BR because formation eliminates the protein from carrying out the biological function of pumping protons, and thus directed evolution offers a reliable method for tailoring BR for performance in devices. Using directed evolution, new mutants are generated via region specific SRM and then are screen and evaluated for enhanced formation and reversion of the **Q** state. The most efficient **Q** state forming proteins are then selected to serve as the parents to the next generation of genetic progeny, which are then produced via SDM, SSSM, and SRM. Only the best mutants from each round of mutagenesis are sequenced, which allows the procedures to be cost effective and efficient. After six stages of directed evolution, implementing automated screening and microgram protein characterization, over 10,000 BR mutants were generated, screened, and characterized for application in devices.

Conclusions

Considerable progress has been made in the field of nanotechnology since the introduction of the concept of nanotechnology in 1959 by Feynman. The use of nanobiotechnology and the ability to manipulate proteins via genetic engineering has allowed scientists to modify the way that they approach and treat diseases. In this review, we have summarized some of the most recent advances in visual restoration using microbial rhodopsins. The continual exploration of technology for the treatment of ocular eye diseases using gene therapy, optogenetics, and visual prosthetics has the potential to improve the quality of life for those suffering from visual loss due to retinal degenerative diseases.

References

Acland, G.M., Aguiree, G.D., Bennett, J., Aleman, T.S., Cideciyan, A.V., Bennicelli, J., Dejneka, N.S., Pearce-Kelling, S.E., Maguire, A.M., Palczewski, K., *et al.* (2005). Long-term restoration of rod and cone vision by single dose rAAV-mediated gene transfer to the retina in a canine model of childhood blindness. Mol. Ther. *12*, 1072–1082.

Aono, M., Bando, Y., and Ariga, K. (1012). Nanoarchitectonics: pioneering a new paradigm for nanotechnology in materials development. Adv. Mater. *24*, 150–151.

Arnold, F., Wintrode, P.L., Miyazaki, K., and Gershenson, A. (2001). How enzymes adapt: lessons from directed evolution. Trends Biochem. Sci. *26*, 100–106.

Arora, P., Sindhu, A., RajaKumar, G., Dilbaghi, N., Rahuman, A.A., and Chaudhury, A. (2012). Nano-regenerative medicine towards clinical outcome of stem cell and tissue engineering in humans. J. Cell. Mol. Med.

Arrenberg, A.B., Del Bene, F., and Baler, H. (2009). Optical control of zebrafish behavior with halorhodopsin. Proc. Natl. Acad. Sci. U.S.A. *106*, 17968–17973.

Ashtari, M., Cyckowski, L.L., Monroe, J.F., Marshall, K.A., Chung, D.C., Auricchio, A., Simonelli, F., Leroy, B.P., Maguire, A.M., Shindler, K.S., *et al.* (2011). The human visual cortex responds to gene therapy-mediated recovery of retinal function. J. Clin. Invest. *121*, 2140–2168.

Asuri, P., Bartel, M.A., Vazin, T., Jang, J.-H., Wong, T.B., and Schaffer, D.V. (2012). Directed evolution of adeno-associated virus for enhanced gene delivery and gene targeting in human pluripotent stem cells. Mol.Ther. *20*, 329–338.

Bainbridge, J.W., Smith, A.J., Barker, S.S., Robbie, S., Henderson, R., Balaggan, K., Viswanathan, A., Holder, G.E., Stockman, A., Tyler, N., *et al.* (2008). Effect of gene therapy on visual function in Leber's congenital amaurosis. N. Engl. J. Med. *358*, 2231–2239.

Balashov, S. (2000). Protonation reactions and their coupling in bacteriorhodopsin. Biochim. Biophys. Acta *1460*, 75–94.

Balashov, S., Govindjee, R., Kono, M., Imasheva, E., Lukashev, E., Ebrey, T., Crouch, R., Menick, D., and Feng, Y. (1993). Effect of the arginine-82 to alanine mutation in bacteriorhodopsin on dark adaptation, proton release, and the photochemical cycle. Biochemistry *32*, 10331–10343.

Bar-Cohen, Y. (2006). Introduction to Biomimetics: The wealth of inventions in nature as an inspiration for human innovation. In Biomimetics: Biologically Inspired Technologies, Y. Bar-Cohen, ed. (Boca Raton, FL: Taylor & Francis), pp. 2–40.

Bartel, M.A., Weinstein, J.R., and Schaffer, D.V. (2012). Directed evolution of novel adeno-associated viruses for therapeutic gene delivery. Gene Ther. *19*, 694–700.

Béjà, O., Aravind, L., Koonin, E.V., Suzuki, M.T., Hadd, A., Nguyen, L.P., Jovanovich, S.B., Gates, C.M., Feldman, R.A., Spudich, J.L., *et al.* (2000). Bacterial rhodopsin: evidence for a new type of phototrophy in the sea. Science *289*, 1902–1906.

Bensaude-Vincent, B. (2009). Self-assembly, Self-organization: nanotechnology and vitalism. NanoEthics *3*, 31–42.

Berthold, P., Tsunoda, S.P., Ernst, O.P., Mages, W., Gradmann, D., and Hegemann, P. (2008). Channelrhodopsin-1 initiates phototaxis and photophobic responses in *chlamydomonas* by immediate light-induced depolarization. Plant Cell *20*, 1665–1677.

Berthoumieu, O., patil, A., Wang, X., Aslimovska, L., Davis, J., and Watts, A. (2012). Molecular scale conductance photoswitching in engineered bacteriorhodopsin. Nano Lett. *12*, 899–903.

Birch, D.G., and Qi Liang, F. (2009). Age-related macular degeneration: a target for nanotechnology derived medicines. Int. J. Nanomed. *2*, 65–77.

Birge, R.R. (1981). Photophysics of light transduction in rhodopsin and bacteriorhodopsin. Annu. Rev. Biophys. Bioeng. *10*, 315–354.

Birge, R.R. (1982). Simulation of the primary event in rhodopsin photochemistry using semiempirical molecular dynamics theory. In Biological events probed by ultrafast laser spectroscopy, R.R. Alfano, ed. (New York: Academic Press), pp. 299–317.

Birge, R.R. (1986). Two-photon spectroscopy of protein bound chromophores. Accounts Chem. Res. *19*, 138–146.

Birge, R.R. (1990a). Nature of the primary photochemical events in rhodopsin and bacteriorhodopsin. Biochim. Biophys. Acta *1016*, 293–327.

Birge, R.R. (1990b). Photophysics and molecular electronic applications of the rhodopsins. Annu. Rev. Phys. Chem. *41*, 683–733.

Birge, R.R. (1994). Bacteriorhodopsin, a nonlinear proton pump. Nature *371*, 659–660.

Birge, R.R., Cooper, T.M., Lawrence, A.F., Masthay, M.B., Vasilakis, C., Zhang, C.F., and Zidovetzki, R. (1989a). A spectroscopic, photocalorimetric and theoretical investigation of the quantum efficiency of the primary event in bacteriorhodopsin. J. Am. Chem. Soc. *111*, 4063–4074.

Birge, R.R., Findsen, L.A., Lawrence, A.F., Masthay, M.B., and Zhang, C.F. (1989b). Origins of the quantum efficiency duality in the primary photochemical event of bacteriorhodopsin. In Biomolecular Spectroscopy, R.R. Birge, and H.H. Mantsch, eds. (Bellingham, Washington: The International Society for Optical Engineering), pp. 103–112.

Birge, R.R., Findsen, L.A., and Pierce, B.M. (1987). Molecular dynamics of the primary photochemical event in bacteriorhodopsin. Theoretical evidence for an excited singlet state assignment for the J intermediate. J. Am. Chem. Soc. *109*, 5041–5043.

Birge, R.R., Fleitz, P.A., Gross, R.B., Izgi, J.C., Lawrence, A.F., Stuart, J.A., and Tallent, J.R. (1990). Spatial light modulators and optical associative memories based on bacteriorhodopsin. Proc. IEEE EMBS *12*, 1788–1789.

Birge, R.R., Gillespie, N.B., Izaguirre, E.W., Kusnetzow, A., Lawrence, A.F., Singh, D., Song, Q.W., Schmidt, E., Stuart, J.A., Seetharaman, S., *et al.* (1999). Biomolecular electronics: Protein-based associative processors and volumetric memories. J. Phys. Chem. B. *103*, 10746–10766.

Birge, R.R., Govender, D.S.K., Gross, R.B., Lawrence, A.F., Stuart, J.A., Tallent, J.R., Tan, E., and Vought, B.W. (1994). Bioelectronics, three-dimensional memories and hybrid computers. IEEE IEDM Technical Digest *94*, 3–6.

Birge, R.R., and Vought, B.W. (2000). Energetics of rhodopsin photobleaching: photocalorimetric studies of energy storage in the early and later intermediates. In Methods in Enzymology, K. Palczewski, ed. (New York: Academic Press), pp. 143–163.

Bisoyi, H.K., and Kumar, S. (2011). Liquid-crystal nanoscience: an emerging avenue of soft self assembly. Chem. Soc. Rev. *40*, 306–319.

Boxer, S.G., Stocker, J., Franzen, S., and Salafsky, J. (1992). Re-engineering photosynthetic raction centers. In Molecular Electronics – Science and Technology, A. Aviram, ed. (New York: American Institute of Physics), pp. 226–241.

Boyden, E.S., Zhang, F., Bamberg, E., Nagel, G., and Deisseroth, K. (2005). Millisecond-timescale, genetically targeted optical control of neural activity. Nat. Neurosci. *8*, 1263–1268.

Bräuchle, C., Hampp, N., and Oesterhelt, D. (1991). Optical applications of bacteriorhodopsin and its mutated variants. Adv. Mater. 3, 420–428.

Bringmann, A., Faude, F., and Reichenbach, A. (1997). Mammalian retinal glial (M¸ller) cells express large-conductance Ca^{++}-activated K^{+} channels that are modulated by Mg++ and pH and activated by protein kinase A. Glia *19*, 311–323.

Brooks, B., Bruccoleri, R.E., Olafson, B.D., States, D.J., Swaminathan, S., and Karplus, M. (1983). CHARMM: A program for macromolecular energy, minimization, and dynamics calculations. J. Comput. Chem. *4*, 187–217.

Brown, L., Needleman, R., and Lanyi, J. (1999). Functional roles of aspartic acid residues at the cytoplasmic surface of bacteriorhodopsin. Biochemistry *38*, 6855–6861.

Brown, L.S., Dioumaev, A.K., Needleman, R., and Lanyi, J.K. (1998a). Connectivity of the retinal Schiff base to Asp85 and Asp96 during the bacteriorhodopsin photocycle: the local-access model. Biophys. J. *75*, 1455–1465.

Brown, L.S., Dioumaev, A.K., Needleman, R., and Lanyi, J.K. (1998b). Local-access model for proton transfer in bacteriorhodopsin. Biochemistry *37*, 3982–3993.

Bruchez, M.P. (2007). Commercialization and future developments in bionanotechnology. In Frontiers of Engineering: reports on leading-edge engineering from the 2006 symposium (Washington, DC: The National Academies Press) pp. 73–81.

Buchen, L. (2010). Illuminating the brain. Nature *465*, 26–28.

Buehler, M.J. (2010). Nanomaterials: strength in numbers. Nat. Nanotechnol. *5*, 172–174.

Busskamp, V., Duebel, J., Balya, D., Fradot, M., Viney, J.T., Siegert, S., Groner, A.C., Cabuy, E., Forster, V., Seeliger, M., *et al.* (2010). Genetic reactivation of cone photoreceptors restores visual responses in retinitis pigmentosa. Science *329*, 413–417.

Busskamp, V., Picaud, S., Sahel, J.A., and Roska, B. (2011). Optogenetic therapy for retinitis pigmentosa. Gene Ther. 1–7.

Busskamp, V., and Roska, B. (2011). Optogenetic approaches to restoring visual function in retinities pigmentosa. Curr. Opin. Neurobiol. *21*, 942–946.

Calimet, N., and Ulmann, G.M. (2004). The influence of a transmembrane pH gradient on protonation probabilities of bacteriorhodopsin: the structural basis of the back-pressure effect. J. Mol. Biol. *339*, 571–589.

Cao, Y., Varo, G., Chang, M., Ni, B., Needleman, R., and Lanyi, J.K. (1991). Water is required for proton transfer from aspartate-96 to the bacteriorhodopsin Schiff base. Biochemistry *30*, 10972–10979.

Cao, Y., Váró, G., Klinger, A.L., Czajkowsky, D.M., Braiman, M.S., Needleman, R., and Lanyi, J.K. (1993). Proton transfer from asp-96 to the bacteriorhodospin schiff base is caused by a decrease of the pKa of asp-96 which follows a protein backbone conformational change. Biochemistry *32*, 1981–1990.

Caraglia, M., Rosa, G., Salzano, G., Santini, D., Lamberti, M., Sperlongano, P., Lombardi, A., Abbruzzese, A., and Addeo, R. (2012). Nanotech Revolution for the Anti-Cancer Drug Delivery though Blood-Brain-Barrier. Curr. Cancer Drug Tar. *12*, 186–196.

Cario, E. (2012). Nanotechnology-based drug delivery in mucosal immune diseases: hype or hope? Mucosal Immunol. *5*, 1–3.

Carmona-Ribeiro, A.M. (2010). Biomimetic nanoparticles: preparation, characterization, and biomedical applicaitons. Int. J. Nanomed. *5*, 249–259.

Chaudhari, Y.S., Mody, H.R., Kokardekar, R.R., Paranjpe, P.P., and Bahuguna, Y.M. (2010). Nanobiotechnology: the emerging field in pharmaceutical sciences. J. Pharm. Res. *3*, 1297–1299.

Chen, P. (2005). Self-assembly of ionic-complementary peptidesL a physicochemical viewpoint. Interfaces *261*, 3–24.

Chen, Z., and Birge, R.R. (1993). Protein based artificial retinas. Trends Biotechnol. *11*, 292–300.

Chen, Z., Lewis, A., Takei, H., and Nabenzahl, I. (1991). Bacteriorhodopsin oriented in polyvinyl alcohol films as an erasable optical storage medium. Appl. Opt. *30*, 5188–5196.

Chow, A.Y., Chow, V.Y., Packo, K.H., Pollack, J.S., Peyman, G.A., and Schuchard, R. (2004). The Artificial Silicon Retina Microchip for the Treatment of Vision Loss From Retinitis Pigmentosa. Arch. Ophthalmol. *122*, 460–469.

Cideciyan, A.V., Aleman, T.S., Boye, S.L., Schwartz, S.B., Kaushal, S., Roman, A.J., Pang, J.J., Sumaroka, A., Windsor, E.A., Wilson, J.M., *et al.* (2008). Human gene therapy for RPE65 isomerase deficiency activates the retinoid cycle of vision but with slow rod kinetics. Proc. Natl. Acad. Sci. USA. *105*, 15112–15117.

Conley, S.M., and Naash, M.I. (2010). Nanoparticles for retinal gene therapy. Prog. Retin. Eye Res. *29*, 376–397.

Cooper, A., and Converse, C.A. (1976). Energetics of primary processes in visual excitation: Photocalorimetry of rhodopsin in rod outer segment membranes. Biochemistry *15*, 2970–2978.

Crameri, A., Raillart, S., Bermudez, E., and Stemmer, W.P.C. (1998). DNA shuffling of a family of genes from diverse species accelerates directed evolution. Nature *391*, 288–291.

Cronin, T., and Bennett, J. (2011). Switching on the Lights: The Use of Optogenetics to Advance Retinal Gene Therapy. Mol. Ther. *19*, 1190–1192.

Cunningham, W., Mathieson, K., McEwan, F.A., Blue, A., McGeachy, R., McLeod, J.A., Morris-Ellis, C., O'Shea, V., Smith, K.M., Litke, A., *et al.* (2001). Fabrication of microelectrode arrays for neural measurements from retinal tissue. J. Phys. D. Appl. Phys. *34*, 2804–2809.

Dalby, P.A. (2003). Optimizing enzyme function by directed evolution. Curr. Opin. Struct. Biol. *13*, 500–505.

Degenaar, P., Grossman, N., Memon, M.A., Burrone, J., Dawson, M., Drakakis, E., Neil, M., and Nikolic, K. (2009). Optobionic vision – a new genetically enhanced light on retinal prosthesis. J. Neural Eng. *6*, 1–10.

DeGrip, W.J., Grey, D., Gillespie, J., Bovee, P.H.M., Van den Berg, H.M.M., Lugtenburg, J., and Rothschild, K.J. (1988). Photoexcitation of rhodopsin: conformation changes in the chromophore, protein and associated lipids as determined by FTIR difference spectroscopy. Photochem. Photobiol. *48*, 497–504.

Deisseroth, K. (2011). Optogenetics. Nat. Methods *8*, 26–29.

Dickopf, S., and Heyn, M.P. (1997). Evidence for the first phase of the reprotonation switch of bacteriorhodopsin from time-resolved photovoltage and flash photolysis experiments on the photoreversal of the M-intermediate. Biophys. J. *73*, 3171–3181.

Douglass, A.D., Kraves, S., Deisseroth, K., Schier, A.F., and Engert, F. (2008). Escape Behaviour Elicited by Single, Channelrhodopsin-2-Evoked Spikes in Zebrafish Somatosensory Neurons. Curr. Biol. *18*, 1133–1137.

Enz, R., Ross, B.J., and Cutting, G.R. (1999). Expression of the voltage-gated chloride channel ClC-2 in rod bipolar cells of the rat retina. J. Neurosci. *19*, 9841–9847.

Essen, L.O. (2002). Halorhodopsin: light-driven ion pumping made simple?. Curr. Opin. Struct. Biol. *12*, 516–522.

Ettaiche, M., Deval, E., Cougnon, M., Lazdunski, M., and Voilley, N. (2006). Silencing acid-sensing ion channel 1a alters cone-mediated retinal function. J. Neurosci. *26*, 5800–5809.

Farjo, K.M., and Ma, J. (2010). The potential of nanomedicine therapies for neovascular disease in the retina. J. Angiogenes Res. *2*, 1–14.

Freeman, D.K., Rizzo, J.F., and Fried, S.I. (2011). Encoding visual information in retinal ganglion cells with prosthetic stimulation. J. Neural Eng. *8*, 1–18.

Fried, S.I., Hsueh, H.A., and Werblin, F.S. (2006). A method for generating precise temporal patterns of retinal spiking using prosthetic stimulation. J. Neurophysiol. *95*, 970–978.

Fu, H.-Y., Lin, Y.-C., Chang, Y.-N., Tseng, H., Huang, C.-C., Liu, K.-C., Huang, C.-S., Su, C.-W., Weng, R.R., Lee, Y.-Y., *et al.* (2010). A novel six-rhodopsin system in a single archaeon. J. Bacteriol. *192*, 5866–5873.

Georgescu, R., Bandara, G., and Sun, L. (2003). Saturation mutagenesis. Methods Mol. Biol. *231*, 75–83.

Gillespie, N.B., Ren, L., Ramos, L., Daniell, H., Dews, D., Utzat, K.A., Stuart, J.A., Buck, C.H., and Birge, R.R. (2005). Characterization and photochemistry of 13-desmethyl bacteriorhodopsin. J. Phys. Chem. B. *109*, 16142–16152.

Gillespie, N.B., Wise, K.J., Ren, L., Stuart, J.A., Marcy, D.L., Hillebrecht, J., Li, Q., Ramos, L., Jordan, K., Fyvie, S., *et al.* (2002). Characterization of the branched-photocycle intermediates P and Q of bacteriorhodopsin. J. Phys. Chem. B. *106*, 13352–13361.

Goel, A., and Vogel, V. (2008). Harnessing biological motors to engineer systems for nanoscale transport and assembly. Nat. Nanotechnol. *3*, 465–475.

Gourdon, P., Alfredsson, A., Pedersen, A., Malmerberg, E., Nyblom, M., Widell, M., Berntsson, R., Pinhassi, J., Braiman, M., Hansson, O., *et al.* (2008). Optimized *in vitro* and *in vivo* expression of proteorhodopsin: A seven-transmembrane proton pump. Prot. Expres. Purif. *58*, 103–113.

Govindjee, R., Balashov, S.P., and Ebrey, T.G. (1990). Quantum efficiency of the photochemical cycle of bacteriorhodopsin. Biophys. J. *58*, 597–608.

Graddis, T.J., Remmele, R.L.J., and McGrew, J.T. (2002). Designing proteins that work using recombinant technologies. Curr. Pharm. Biotechnol. *3*, 285–297.

Groma, G.I., Hebling, J., Ludwig, C., and Kuhi, J. (1995). Charge displacement in bacteriorhodopsin during forward and backward bR-K phototransition. Biophys. J. *69*, 2060–2065.

Grote, M., and O'Malley, M.A. (2011). Enlightening the life sciences: the history of halobacterial and microbial rhodopsin research. FEMS Microbiol. Rev. *35*, 1082–1099.

Grumet, A.E., Wyatt, J.L., and Rizzo, J.F. (2000). Multi-electrode stimulation and recording in the isolated retina. J. Neurosci. Meth. *101*, 31–42.

Gu, L.Q., Zhang, C.P., Niu, A.F., Li, J., Zhang, G.Y., Wang, Y.M., Tong, M.R., Pan, J.L., Song, Q.W., Parson, B., *et al.* (1996). Bacteriorhodopsin based photonic logic gate and its application to grey level image subtraction. Optics Commun. *131*, 25–30.

Hampp, N., and Juchem, T. (2000). Fringemaker – the first technical system based on bacteriorhodopsin. Paper presented at: Bioelectronic applications of photochromic pigments (Szeged, Hungary: IOS Press), pp. 44–53.

Hampp, N., Popp, A., Bräuchle, C., and Oesterhelt, D. (1992). Diffraction efficiency of bacteriorhodopsin films for holography containing bacteriorhodopsin wildtype BRwt and its variants BR_{D85E} and $BR_{D96N.}$ J. Phys. Chem. *96*, 4679–4685.

Hampp, N., Thoma, R., Zeisel, D., and Bräuchle, C. (1994). Bacteriorhodopsin variants for holographic pattern recognition. Adv. Chem. *240*, 511–526.

Han, X., and Boyden, E.S. (2007). Multiple-colour optical activation silencing, and desynchronization of neural activity, with single-spike temporal resolution. Plos One *2*, e299.

Haupts, U., Tittor, J., Bamberg, E., and Oesterhelt, D. (1997). General concept for ion translocation by halobacterial retinal proteins: The iosmerization/switch/transfer (IST) model. Biochemistry *36*, 2- 7.

Hauswirth, W.W., Aleman, T.S., Kaushal, S., Cideciyan, A.V., Schwartz, S.B., Wang, L., Conlon, T.J., Boye, S.L., Flotte, T.R., Byrne, B.J., *et al.* (2008). Treatment of leber congenital amaurosis due to RPE65 mutations by ocular subretinal injection of adeno-associated virus gene vector: short-term results of a phase I trial. Hum. Gene Ther. *19*, 979–990.

He, J., Samuelson, L., Li, L., Kumar, J., and Tripathy, S.K. (1998a). Photoelectric properties of oriented bacteriorhodopsin/polycation multilayers by electrostatic layer-by-layer assembly. J. Phys. Chem. B. *102*, 7067–7072.

He, J., Samuelson, L., Li, L., Kumar, J., and Tripathy, S.K. (1999). Bacteriorhodopsin thin film assemblies – immobilization, properties, and applications. Adv. Mater. *11*, 435–446.

He, J.-A., Samuelson, L., Li, L., Kumar, J., and Tripathy, S.K. (1998b). Oriented bacteriorhodopsin/polycation multilayers by electrostatic layer-by-layer assembly. Langmuir *14*, 1674–1679.

Henderson, R., Baldwin, J.M., Ceska, T.A., Zemlin, F., Beckmann, E., and Downing, K.H. (1990). Model for the structure of bacteriorhodopsin based on high-resolution electron cryo-microscopy. J. Mol. Biol. *213*, 899–929.

Herzog, R.W. (2012). See Eye to eye with gene therapy. Mol. Ther. *20*, 687–688.

Hillebrecht, J.R., Koscielecki, J.F., Wise, K.J., Marcy, D.L., Tetley, W., Rangarajan, R., Sullivan, J., Brideau, M., Krebs, M.P., Stuart, J.A., *et al.* (2005). Optimization of Protein-Based Volumetric Optical Memories and Associative Processors by Using Directed Evolution. NanoBiotechnol. *1*, 141–152.

Hirayama, J., Imamato, Y., Shichida, Y., Kamo, N., H., T., Yoshizawa, T., and (1992). Photocycle of phoborhodopsin from haloalkaliphilic bacterium (Natronobacterium pharaonis) studied by low-temperature spectrophotometry. Biochemistry *31*, 2093–2098.

Holz, M., Drachev, L.A., Mogi, T., Otto, H., Kaulen, A.D., Heyn, M.P., Skulachev, V.P., and Khorana, H.G. (1989). Replacement of aspartic acid-96 by asparagine in bacteriorhodopsin slows both the decay of the M intermediate and the associated proton movement. Proc. Natl. Acad. Sci. USA. *86*, 2167–2171.

Hong, F.H., and Hong, F.T. (1995). Component analysis of the fast photoelectric signal from model bacteriorhodopsin membranes Part 4. A method for isolating the B2 component and the evidence for its polarity reversal at low pH. Bioelectroch. Bioener. *37*, 91–99.

Hong, F.T. (1997). Molecular sensors based on the photoelectric effect of bacteriorhodopsin: origin of differential responsivity. Mater. Sci. Eng. C. *4*, 267–285.

Hsieh, C.-H., Chou, L.-J., Lin, G.-R., Bando, Y., and Goldberg, D. (2008). Nanophotonic Switch: Gold-in-Ga2O3 Peapod Nanowires. Nano Lett. *8*, 3081–3085.

Hsu, K.C., Rayfield, G.W., and Needleman, R. (1996). Reversal of the surface charge asymmetry in purple membrane due to single amino acid substitutions. Biophys. J. *70*, 2358–2365.

Hughes, M.P. (2000). AC electrokinetics: applications for nanotechnology. Nanotechnology *11*.

Humayun, M.S., Weiland, J.D., Fujii, G.Y., Greenberg, R., Williamson, R., Little, J., Mech, B., Cimmarusti, V., Van Boemel, G., Dagnelie, G., *et al.* (2003). Visual perception in a blind subject with a chronic microelectronic retinal prosthesis. Vision Res. *43*, 2573–2581.

Ivanova, E., and Pan, Z.H. (2009). Evaluation of the adeno-associated virus mediated long-term expression of channelrhodopsin-2 in the mouse retina. Mol. Vis. *15*, 1680–1689.

Jäckel, C., Bloom, J.D., Kast, P., Arnold, F.H., and Hilvert, D. (2010). Consensus Protein Design without Phylogenetic Bias. J. Mol. Biol. *399*, 541–546.

Jackson, M., and Sturtevant, J. (1978). Phase Transitions of the Purple Membranes of *Halobacterium halobium*. Biochemistry *17*, 911–915.

Jain, K.K. (2008). Nanomedicine: Applicaiton of Nanobiotechnology in Medical Practice. Med. Prin. Pract. *17*, 89–101.

Jensen, R.J. (2003). Thresholds for activation of rabbit retinal ganglion cells with an ultrafine, extracellular microelectrode. Invest. Ophthalmol. Vis. Sci. *44*, 3533–3543.

Jensen, R.J., and Rizzo, J.F. (2006). Thresholds for activation of rabbit retinal ganglion cells with a subretinal electrode. Exp. Eye Res. *83*, 367–373.

Jensen, R.J., and Rizzo, J.F. (2008). Activation of retinal ganglion cells in wild-type and rd1 mice through electrical stimulation of the retinal neural network. Vision Res. *48*, 1562–1568.

Jensen, R.J., Ziv, O.R., and Rizzo Iii, J.F. (2005a). Thresholds for activation of rabbit retinal ganglion cells with relatively large, extracellular microelectrodes. Invest. Ophthalmol. Vis. Sci. *46*, 1486–1496.

Jensen, R.J., Ziv, O.R., and Rizzo, J.F. (2005b). Responses of rabbit retinal ganglion cells to electrical stimulation with an epiretinal electrode. J. Neural Eng. *2*, S16–21.

Kandori, H. (2004). Hydration switch model for the protonation transfer in the Schiff base region of bacteriorhodopsin. Biochim. Biophys. Acta *1658*, 72–79.

Kato, H.E., Zhang, F., Yizhar, O., Ramakrishnan, C., Nishizawa, T., Hirata, K., Ito, J., Aita, Y., Tsukazaki, T., Hayashi, S., *et al.* (2012). Crystal structure of the channelrhodopsin light-gated cation channel. Nature,482, 369–374.

Kleinlogel, S., Feldbauer, K., Dempski, R.E., Wood, P.G., Bamann, C., and Bamberg, E. (2011). Ultra light-sensitive and fast neuronal activation with the Ca2+-permeable channelrhodopsin CatCh. Nat. Neurosci. *14*, 513–518.

Kolbe, M., Besir, H., Essen, L.-O., and Oesterhelt, D. (2000). Structure of the light-driven chloride pump halorhodopsin at 1.8 A resolution. Science *288*, 1390–1396.

Konnerth, A., Lux, H.D., and Morad, M. (1987). Proton-induced transformation of calcium channel in Chick dorsal root ganglion cells. J. Physiol. *386*, 603–633.

Konstas, K., Langford, S.J., and Latter, M.J. (2010). Advances towards synthetic machines at the molecular and nanoscale level. Int. J. Mol. Sci. *11*, 2453–2472.

Koyama, K., Yamaguchi, N., and Miyasaka, T. (1994). Antibody-mediated bacteriorhodopsin orientation for molecular device architectures. Science *265*, 762–765.

Kramer, R.H., Fortin, D.L., and Trauner, D. (2009). New photochemical tools for controlling neuronal activity. Curr. Opin. Neurobiol. *19*, 1–9.

Krebs, M., and Isenbarger, T. (2000). Structural determinants of purple membrane assembly. Biochim. Biophys. Acta *1460*, 15–26.

Langer, R., and Tirrell, D.A. (2004). Designing materials for biology and medicine. Nature *428*, 487–492.

Lanyi, J.K. (1986). Halorhodopsin: A light-driven chloride ion pump. Annu. Rev. Biophys. Biophys. Chem. *15*, 11–28.

Lanyi, J.K. (1998a). The local-access mechanism of proton transport by bacteriorhodopsin. Biochim. Biophys. Acta *1365*, 17–22.

Lanyi, J.K. (1998b). Understanding structure and function in the light-driven proton pump bacteriorhodopsin. J. Struct. Biol. *124*, 164–178.

Lanyi, J.K. (2000). Molecular mechanism of ion transport in bacteriorhodopsin: insights from crystallographic, spectroscopic, kinetic, and mutational studies. J. Phys. Chem. B. *104*, 11441–11448.

Lanyi, J.K. (2004a). Bacteriorhodopsin. Annu. Rev. Physiol. *66*, 665–688.

Lanyi, J.K. (2004b). X-ray diffraction of bacteriorhodopsin photocycle intermediates. Mol. Membr. Biol. *21*, 143–150.

Law, W.C., Kim, S., and Rando, R.R. (1989). Stereochemistry of the visual cycle. J. Am. Chem. Soc. *111*, 793–795.

Lawrence, A.F., and Birge, R.R. (1984). Communication with submicron structures. Perspectives in the application of biomolecules to computer technology. In Nonlinear electrodynamics in biological systems, H.R. Adey, and A.F. Lawrence, eds. (New York: Plenum), pp. 207–218.

Lozier, R.H., Bogomolni, R.A., and Stoeckenius, W. (1975). Bacteriorhodopsin: A light-driven proton pump in Halobacterium halobium. Biophys. J. *15*, 955–962.

Lu, W., and Lieber, C.M. (2007). Nanoelectronics from the bottom up. Nat. Mater. *6, 841–850*.

Luecke, H., Richter, H.T., and Lanyi, J.K. (1998). Proton transfer pathways in bacteriorhodopsin at 2.3 Angstrom resolution. Science *280*, 1934–1937.

Luecke, H., Schobert, B., Cartailler, J.-P., Richter, H.-T., Rosengarth, A., Needleman, R., and Lanyi, J.K. (2000). Coupling Photoisomerization of Retinal to Directional Transport in Bacteriorhodopsin. J. Mol. Biol. *300*, 1237–1255.

Lvov, Y., Ariga, K., Ichinose, I., and Kunitake, T. (1995). Assembly of multicomponent protein films by means of electrostatic layer-by-layer adsorption. J. Am. Chem. Soc. *117*, 6117–6123.

Lynagh, T., and Lynch, J.W. (2010). An improved Ivermectin-activated chloride channel receptor for inhibiting electrical activity in defined neuronal populations. J. Biol. Chem. *285*, 14890–14897.

Maguire, A.M., Simonelli, F., Pierce, E.A., Pugh, E.N., Mingozzi, F., Bennicelli, J., Banfi, S., Marshall, K.A., Testa, F., Surace, E.M., *et al.* (2008). Safety and efficacy of gene transfer for Leber's congenital amaurosis. N. Engl. J. Med. *358*, 2240–2248.

Maheshri, N., Koerber, J.T., Kasper, B.K., and Schaffer, D.V. (2006). Directed evolution of adeno-associated virus yields enhanced gene delivery vectors. Nat. Biotechnol. *24*, 198–204.

Man, D., Wang, W., Sabehi, G., Aravind, L., Post, A.F., Massana, R., Spudich, E.N., Spudich, J.L., and Béjà, O. (2003). Diversification and spectral tuning in marine proteorhodopsins. EMBO J. *22*, 1725–1731.

Marc, R.E. (1999). Kainate activation of horizontal, bipolar, amacrine, and ganglion cells in the rabbit retina. J. Comp. Neurol. *407*, 65–76.

Margalit, E., and Thoreson, W.B. (2006). Inner retinal mechanisms engaged by retinal electrical stimulation. Invest. Ophthalmol. Vis. Sci. *47*, 2606–2612.

Markovic, D.S., Zivkovic, D., Cvetkovic, D., and Popovic, R. (2012). Impact of Nanotechnology advances in ICT on sustainability and energy efficiency. Renew. Sustain. Energ. Rev. *16*, 2966–2972.

Martin, C.H., Chen, Z.P., and Birge, R.R., eds. (1997). Towards a Bacteriorhodopsin-Silicon Neuromorphic Photosensor (Maui, World Scientific).

Marwan, W., Hegemann, P., and Oesterhelt, D. (1988). Single photon detection by an archaebacterium. J. Mol. Biol. *199*, 663–664.

Massy, S.C. (1990). Cell types using glutamate as aneurotransmitter in the vertebrate retina. Prog. Retin. Res. *9*, 399–425.

Masthay, M.B., SamMeth. D.M., Helvenston, M.C., Buckman, C.B., Li, W., Cde-Baca, M.J., and Kofron, J.T. (2002). The laser-induced blue state of bacteriorhodopsin: mechanistic and color regulatory roles of protein-protein interactions, protein-lipid interactions, and metal ions. J. Am. Chem. Soc. *124*, 3418–3430.

Mata, A., Palmer, L., Tejeda-Montes, E., and Stupp, S.I. (2012). Design of biomolecules for nanoengineered biomaterials for regenerative medicine. In Nanotechnology in Regernerative Medicine: Methods and Protocols, Methods in Molecular Biology, M. Navarro, and J.A. Planell, eds. (Springer), pp. 39–49.

Matharu, Z., Bandodkar, A.J., Gupta, V., and Malhotra, B.D. (2012). Fundamentals and application of ordered molecular assemblies in affinity biosensing. Chem. Soc. Rev. *41*, 1363–1402.

Matsung-Yagi, A., and Mukohata, Y. (1980). ATP-synthesis linked to light-dependent proton-uptake in a red mutant strain of *Halobacterium* lacking bacteriorhodopsin. Arch. Biochem. Biophys. *199*, 297–303.

Mazzola, L. (2003). Commercializing nanotechnology. Nat. Biotechnol. *21*, 1137–1143.

Metz, G., Siebert, F., and Engelhard, M. (1992). ASP_{85} is the only internal aspartic acid that gets protonated in the M intermediate and the purple-to-blue transition of bacteriorhodopsin; A solid-state ^{13}C CP-MAS NMR investigation. FEBS Lett. *303*, 237–241.

Meyers, M.A., Chen, P.-Y., Lin, A.Y.-M., and Seki, Y. (2008). Biological materials: Structure and mechanical properties. Prog. Mater. Sci. *53*, 1–206.

Miesenbock, G. (2011). Annu. Rev. Cell Dev. Bi. Optogenetic Control of Cells and Circuits *27*, 731–758.

Miyasaka, T., and Koyama, K. (1992). Rectified photocurrents from purple membrane Lanmuir-Blodgett films at the electrode–electrolyte interface. Thin Solid Films *210–211*, 146–149.

Miyasaka, T., and Koyama, K. (1993). Image sensing and processing by a bacteriorhodopsin-based artificial photoreceptor. Appl. Optics *32*, 6371–6379.

Moore, G. (1965). Cramming more components onto integrated circuits. Electronics *38*, 114–117.

Muneyuki, E., Shibazaki, C., Ohtan, H., Okuno, D., Asaumi, M., and Mogi, T. (1999). Time-resolved measurements of photovoltage generation by bacteriorhodopsin and halorhodopsin adsorbed on a thin polymer film. J. Biochem. *125*, 270–276.

Nagel, G., Brauner, M., Liewald, J.F., Adeishvili, N., Bamberg, E., and Gottschalk, A. (2005). Light activation of channelrhodopsin-2 in excitable cells of caenorhabditis elegans triggers rapid behavioural responses. Curr. Biol. *15*, 2279–2284.

Nagel, G., Ollig, D., Fuhrmann, M., Kateriya, S., Musti, A., Bamberg, E., and Hegemann, P. (2002). Channelrhodopsin-1: a light-gated proton channel in green algae. Science *296*, 2395–2398.

Nagel, G., Szellas, T., Huhn, W., Kateriya, S., Adeishvili, N., Berthold, P., Ollig, D., Hegemann, P., and Bamberg, E. (2003). Channelrhodopsin-2, a directly light-gated cation-selective membrane channel. Proc. Natl. Acad. Sci. U.S.A. *100*, 13940–13945.

Nelson, D.L., and Cox, M.M. (2008). Lehninger Principles of Biochemistry, 5th edn (New York: W. H. Freeman and Company).

Nguyen, P., Meyyappan, M., and Yiu, S.C. (2010). Applications in Nanobiotechnology in Ophthalmology. Ophthalm. Res. *44*, 1–14.

Nikolic, B.K., Saha, K.K., Markussen, T., and Thygesen, K. (2011). First-principles quantum transport modelling of thermoelectricity in single-molecule nanojunctions with graphene nanoribbon electrodes. J. Comput. Electron. *11*, 78–92.

Nikolic, K., Degenaar, P., and Toumazou, C. (2006). Modelling and engineering aspects of channelrhodpsin-2 system for neural photostimulation. Proceedings of the 28th IEEE Engineering in Medicine and Biology Society.

Nikolov, L., Slavchev, S., and Stoychev, S. (2008). The convergence of biotechnology and nanobiotechnology as an accelerator of the development of biofilm technologies. Ecological Eng. Environ. Prot. *7*, 36–47.

Nitta, T., Tanahashi, A., Hirano, M., and Hess, H. (2006). Simulating molecular shuttle movements: Towards computer-aided design of nanoscale transport systems. Lab Chip *6*, 881–885.

Oesterhelt, D., Brauchle, C., and Hampp, N. (1991). Bacteriorhodopsin: a biological material for information processing. Q. Rev. Biophys. *24*, 425–278.

Oesterhelt, D., and Stoeckenius, W. (1971a). Rhodopsin-like Protein from the purple membrane of *Halobacterium halobium*. Nature 233, 149–152.

Oesterhelt, D., and Stoeckenius, W. (1971b). Rhodopsin-like protein from the purple membrane of *Halobacterium halobium*. Nat. New Biol. *233*, 149–152.

Oesterhelt, D., and Stoeckenius, W. (1973). Functions of a new photoreceptor membrane. Proc. Natl. Acad. Sci. U.S.A. *70*, 2853–2857.

Oprian, D.D., Asenjo, A.B., Lee, N., and Pelletier, S.L. (1991). Design, chemical synthesis and expression of genes for the three human colour vision pigments. Biochemistry *30*, 11367–11372.

Otto, H., Marti, T., Holz, M., Lindau, M., Khorana, H.G., and Heyn, M.P. (1989). Aspartic acid-96 is the internal proton donor in the repronation of the Schiff base of bacteriorhodopsin. Proc. Natl. Acad. Sci. U.S.A. *86*, 9228–9232.

Palczweski, K., Kumasaka, T., Hori, T., Behnke, C.A., Motoshima, H., Fox, B.A., Le Trong, I., Teller, D.C., Okada, T., Stenkamp, R.E., *et al.* (2000). Crystal structure of rhodopsin: a g protein-coupled receptor. Science *289*, 739–745.

Patil, A., Premaruban, T., Berthoumieu, O., Watts, A., and Davis, J. (2012). Enhanced photocurrent in engineered bacteriorhodopsin monolayer films. J. Phys. Chem. B. *116*, 683–689.

Paula, S., Tittor, J., and Oesterhelt, D. (2001). Roles of cytoplasmic arginine and threonine in chloride transport by the bacteriorhodopsin mutant D85T. Biophys. J. *80*, 2386–2395.

Petreanu, L., Huber, D., Sobczyk, A., and Svoboda, K. (2007). Channelrhodopsin-2-assisted circuit mapping of long-range callosal projections. Nat. Rev. Neurosci. *10*, 663–668.

Popp, A., Wolperdinger, M., Hampp, N., Bräuchle, C., and Oesterhelt, D. (1993). Photochemical conversion of the O-intermediate to 9-cis-retinal-containing products in bacteriorhodopsin films. Biophys. J. *65*, 1449–1459.

Porschke, D. (1996). Electrostatics and electrodynamics of bacteriorhodopsin. Biophys. J. *71*, 3381–3391.

Purves, D., Augustine, G.J., Fitzpatrick, D., Hall, W.C., LaMantia, A.-S., McNamara, J.O., and White, L.E., eds. (2008). Neuroscience, 4th edn (Sunderland, Sinauer Associates, Inc.).

Rangarajan, R., Galan, J.F., Whited, G., and Birge, R.R. (2007). Mechanism of spectral tuning in green-absorbing proteorhodopsin. Biochemistry *46*, 12679–12686.

Rayfield, G. (1989). Ultra high speed bacteriorhodopsin photodetectors. In Molecular Electronics, F.T. Hong, ed. (New York: Plenum), pp. 361–368.

Renner, T., and Hampp, N. (1992). Bacteriorhodopsin-films for dynamic time average interferometry. Opt. Commun. *96*, 142–149.

Richter, H.-T., Brown, L.S., Needleman, R., and Lanyi, J.K. (1996). A linkage of the pKas of asp-85 and glu-204 forms part of the reprotonation switch of bacteriorhodopsin. Biochemistry *35*, 4054–4062.

Roll, S., Müller-Nordhorn, J., Keil, T., Scholz, H., Eidt, D., Greiner, W., and Willich, S.N. (2008). Dacron® vs. PTFE as bypass materials in peripheral vascular surgery – systematic review and meta-analysis. BMC Surg. *8*, 1–8.

Rothschild, K.J., Cantore, W.A., and Marrero, H. (1983). Fourier transform infrared difference spectra of intermediates in rhodopsin bleaching. Science *219*, 1333–1335.

Ryazantsev, M.N., Altun, A., and Morokuma, K. (2012). Color tuning in rhodopsins: the origin of the spectral shift between the chloride-bound and anion-free forms of halorhodopsin. J. Am. Chem. Soc. *134*, 5520–5523.

Sakaguchi, H., Fujikado, T., Fang, X., Kanda, H., Osanai, M., Nakauchi, K., Ikuno, Y., Kamei, M., Yagi, T., Nishimura, S., *et al.* (2004). Transretinal electrical stimulation with a suprachoroidal multichannel electrode in rabbit eyes. Jpn. J. Opthalmol. *48*, 256–261.

Sanchez, C., Arribart, H., and Guille, M.M.G. (2005). Biomimetism and bioinspiration as tools for the design of innovative materials and systems. Nat. Mater. *4*, 277–288.

Sasaki, J., and Spudich, J.L. (1999). Proton circulation during the photocycle of sensory rhodopsin II. Biophys. J. *77*, 2145–2152.

Schick, G.A., Lawrence, A.F., and Birge, R.R. (1988). Biotechnology and molecular computing. Trends Biotechnol. *6*, 159–163.

Schobert, B., Brown, L.S., and Lanyi, J.K. (2003). Crystallographic structures of the M and N intermediates of bacteriorhodopsin: assembly of a hydrogen-bonded chain of water molecules between Asp-96 and the retinal Schiff base. J. Mol. Biol. *330*, 553–570.

Scholz, C. (2007). Perspectives on: Materials Aspects for Retinal Prostheses. J. Bioact. Compat. Pol. *22*, 539–568.

Seki, A., Kubo, I., Sasabe, H., and Tomioka, H. (1994). A new anion-sensitive biosensor using an ion-sensitive field effect transistor and a light-driven chloride pump, halorhodopsin. Appl. Biochem. Biotech. *48*, 205–211.

Sharma, A.K., Spudich, J.L., and Doolittle, W.F. (2006). Microbial rhodopsins: functional versatility and genetic mobility. Trends Microbiol. *14*, 463–469.

Shichida, Y., and Imai, H. (1998). Visual pigment: G-protein-coupled receptor for light signals. Cell Mol. Life Sci. *54*, 1299–1315.

Shintani, K., Shechtman, D.L., and Gurwood, A.S. (2009). Review and update: current treatment trends for patients with retinitis pigmentosa. Optometry *80*, 384–401.

Siebert, F., Maentele, W., and Gerwert, K. (1983). Fourier-transform infrared spectroscopy applied to rhodopsin. Eur. J. Biochem. *136*, 119–127.

Sineshchekov, O.A., Jung, K.-H., and Spudich, J. (2002). Two rhodopsins mediate phototaxis to low- and high-intensity light in Chlamydomonasreinhardtii. Proc. Natl. Acad. Sci. USA. *99*, 8689–8694.

Smith, L.E.H. (2004). Bone marrow-derived stem cells preserve cone vision in retinitis pigmentosa. J. Clin. Invest. *114*, 755–757.

Song, L., El-Sayed, M.A., and Lanyi, J.K. (1993). Protein catalysis of the retinal subpicosecond photoisomerization in the primary process of bacteriorhodopsin photosynthesis. Science *261*, 891–894.

Song, Q.W., Zhang, C., Blumer, R., Gross, R.B., Chen, Z., and Birge, R.R. (1993b). Chemically enhanced bacteriorhodopsin thin-film spatial light modulator. Opt. Lett. *18*, 1373–1375.

Song, Q.W., Zhang, C., Gross, R., and Birge, R.R. (1993c). Optical limiting by chemically enhanced bacteriorhodopsin films. Opt. Lett. *18*, 775–777.

Spudich, J. (2000). A Chloride Pump at Atomic Resolution. Science *288*, 1358–1359.

Spudich, J., Yang, C.-S., Jung, K.-H., and Spudich, E.N. (2000). Retinylidene proteins: structures and functions from archaea to humans. Annu. Rev. Cell Dev. Biol. *16*, 365–392.

Spudich, J.L., McCain, D.A., Nakanishi, K., Okabe, M., Shimizu, N., Rodman, H., Honig, B., and Bogomolni, R.A. (1986). Chromophore/protein interaction in bacterial sensory rhodopsin and bacteriorhodopsin. Biophys. J. *49*, 479–483.

Stewart, J.J.P. (1996). Application of localized molecular orbitals to the solution of semiempirical self-consistent field equations. Int. J. Quant. Chem. *58*, 133–146.

Stuart, J.A., and Birge, R.R. (1996). Characterization of the primary photochemical events in bacteriorhodopsin and rhodopsin. In Biomembranes, A.G. Lee, ed. (London: JAI Press), pp. 33–140.

Sugano, E., Isago, H., Wang, Z., Murayama, N., Tamai, M., and Tomita, H. (2011). Immune response to adeno-assoicated virus type 2 encoding channelrhodopsin-2 in a genetically blind rat model for gene therapy. Gene Ther. *18*, 266–274.

Sung, C.-H., and Chuang, J.-Z. (2010). The cell biology of vision. J. Cell Biol. *190*, 953–963.

Sweeney, A.E., Seal, S., and Vaidyanathan, P. (2003). The promises and perils of nanoscience and nanotechnology: exploring emerging social and ethical issues. B. Sci. Technol. Soc. *23*, 236–245.

Takei, H., Lewis, H., Chen, Z., and Nebenzahl, I. (1992). Implementing receptive fields with excitatory adn inhibitory optoelectrical responses of bacteriorhodopsin films. Appl. Optics *30*, 500–509.

Tallent, J.R., Stuart, J.A., Song, Q.W., Schmidt, E.J., Martin, C.H., and Birge, R.R. (1998). Photochemistry in dried polymer films incorporating the deionized blue membrane form of bacteriorhodopsin. Biophys. J. *75*, 1619–1634.

Tamerler, C., and Sarikaya, M. (2009). Molecular biomimetics: nanotechnology and bionanotechnology using genetically engineered peptides. Phil. T. Roy. Soc. A. *367*, 1705–1726.

Terakita, A. (2005). The opsins. Genome Biol. *6*, 213–222.

Theogarajan, L.S.K. (2007). Supramolecular Architectures for Neural Prostheses. In Department of Electrical Engineering and Computer Science (Massachusetts Institute of Technology), pp. 1–230.

Thoma, R., and Hampp, N. (1992). Real-time holographic correlation of two video signals by using bacteriorhodopsin films. Opt. Lett. *17*, 1158–1160.

Thoma, R., Hampp, N., Bräuchle, C., and Oesterhelt, D. (1991). Bacteriorhodopsin films as spatial light modulators for nonlinear optical filtering. Opt. Lett. *16*, 651–653.

Thomas, C.E., Ehrhardt, A., and Kay, M.A. (2003). Progress and problems with the use of viral vectors for gene therapy. Nat. Rev. Genet. *4*, 346–358.

Thomas, K., and Sayre, P. (2005). Research strategies for safety evaluation of nanomaterials, Part I: Evaluating the human health implications of exposure to nanoscale materials. Toxicol. Sci. *87*, 316–321.

Thomson, H., and Lotery, A. (2009). The promise of nanomedicine for ocular disease. Nanomed. Nanotechnol. Biol. Med. *4*, 599–604.

Tittor, J., Haupts, U., Haupts, C., Oesterhelt, D., Becker, A., and Bamberg, E. (1997). Chloride and proton transport in bacteriorhodopsin mutant D85T: different modes of ion translocation in a retinal protein. J. Mol. Biol. *271*, 405–416.

Tittor, J., and Oesterhelt, D. (1990). The quantum yield of bacteriorhodopsin. FEBS Lett. *263*, 269–273.

Tomsia, A.P., Launey, M.E., Lee, J.S., Mankani, M.H., Wegst, U.G.K., and Saiz, E. (2001). Nanotechnology Approaches for Better Dental Implants. Int J Oral Maxillofac Implants *16*, 25–49.

Tønnesen, J., Sørensen, A.T., Deisseroth, K., Lundberg, C., and Kokaia, M. (2009). Optogenetic control of epileptiform activity. Proc. Natl. Acad. Sci. U.S.A. *106*, 12162–12167.

Tu, R.S., and Tirrell, M. (2003). Bottom-up design of biomimetic assemblies. Adv. Drug Del. Rev. *56*, 1537–1563.

Uehara, K., Kawai, K., and Kouyama, T. (1993). Photoelectric response of oriented purple membrane electrodeposited onto poly(vinyl alcohol) film. Thin Solid Films *232*, 271–277.

Váró, G., and Keszthelyi, L. (1983). Photoelectric signals from dried oriented purple membranes of *Halobacterium Halobium*. Biophys. J. *43*, 47–51.

Verweij, J., Kamermans, M., and Spekreijse, H. (1996). Horizontal cells feed back to cones by shifting the cone calcium-current activation range. Vision Res. *36*, 3943–3953.

Weiland, J.D., Liu, W., and Humayun, M.S. (2005). Retinal Prosthesis. Annu. Rev. Biomed. Eng. *7*, 361–401.

Whitesides, G.M. (2003). The 'right' size in nanobiotechnology. Nat. Biotechnol. *21*, 1161–1165.

Whitesides, G.M. (2004). Nanoscience, Nanotechnology, and Chemistry. Small *1*, 172–179.

Wise, K.J., Gillespie, N.B., Stuart, J.A., Krebs, M.P., and Birge, R.R. (2002). Optimization of bacteriorhodopsin for bioelectronic devices. Trends Biotechnol. *20*, 387–394.

Xu, J., Stickrath, A.B., Bhattacharya, P., Nees, J., Varo, G., Hillebrecht, J.R., Ren, L., and Birge, R.R. (2003). Direct measurement of the photoelectric response time of bacteriorhodopsin via electro-optic sampling. Biophys. J. *85*, 1128–1134.

Yanai, D., Weiland, J.D., Mahadevappa, M., Greenberg, R.J., Fine, I., and Humayun, M.S. (2007). Visual performance using a retinal prothesis in three subjects with retinitis pigmentosa. Am. J. Opthalmol. *143*, 820–827.

Yang, L., Rongqin, H., and Chen, J. (2011). Non-viral gene delivery and therapeutics targeting to brain. Curr. Nanosci. *7*, 55–70.

Yao, B., Dalun, X., and Hou, X. (1997). Oriented bacteriorhodopsin film biomolecular devices and their photoelectric dynamics. Proc SPIE *2869*, 752–758.

Yu, B., and Meyyappan, M. (2005). Nanotechnology: Role in emerging nanoelectronics. Solid State Electron. *50*, 536–544.

Zhang, F., Aravanis, A.M., Adamantidis, A., de Lecea, L., and Deisseroth, K. (2007). Circuit-breakers: optical technologies for probing neural signals and systems. Nat. Rev. Neurosci. *8*, 577–581.

Zhang, F., Wang, L.-P., Boyden, E.S., and Deisseroth, K. (2006). Nat. Methods *3*, 785–792.

Zhang, L., Zeng, T., Cooper, K., and Claus, R.O. (2003). High-performance photovoltaic behavior of oriented purple membrane polymer composite films. Biophys. J. *84*, 2502–2507.

Zhao, H., Giver, L., Shao, Z., Affholter, J.A., and Arnold, F.H. (1998). Molecular evolution by staggered extention process (StEP) *in vitro* recombination. Nat. Biotechnol. *16*, 258–261.

Zimányi, L., Cao, Y., Chang, M., Ni, B., Needleman, R., and Lanyi, J.K. (1992). The two consecutive M substates in the photocycle of bacteriorhodopsin are affected specifically by the D85N and D96N replacements. Photochem. Photobiol. *56*, 1049–1055.

Zimányi, L., Váró, G., Chang, B., Needleman, R., and Lanyi, J. (1992). Pathways of proton release in the bacteriorhodopsin photocycle. Biochemistry *31*, 8535–8543.

Magnetosomes

Mathieu Bennet, Teresa Perez-Gonzalez, Dean Wood and Damien Faivre

10

Abstract

Magnetotactic bacteria are microorganisms that form chains of magnetic nanoparticles. This process represents one of the most advanced examples of biological self-assembly at the nano- and micrometre scale. In fact, the nanoparticle size and morphology, together with the arrangement are controlled at the genetic level. The resulting hierarchical structure bestowing its magnetic properties to the bacteria is of utter interest to the development of bio-inspired nanotechnological self-assemblies. In this chapter, we describe the characteristics of the bacterial magnetic assembly with reference to the latest model found in the scientific literature. The roles of the magnetic dipoles interactions and of bacterial membrane proteins to achieve a stable, optimized and effective magnetic assembly are assessed and the relevant bio-inspired self-assembly scientific works are reviewed.

Introduction

In 1975, Richard Blakemore observed in a sample collected near Woods Hole in Massachusetts a group of bacteria that swam consistently towards the same geographical direction, regardless of the instrument configurations (Blakemore, 1975). Placing a magnet in the vicinity of his sample altered their swimming direction. Following this discovery the characterization of magnetotactic bacteria (MTB) and their habitats (Blakemore, 1975; Denham *et al.*, 1980; Frankel *et al.*, 1979, 1981; Frankel and Blakemore, 1980; Kirschvink, 1980) was performed.

MTB are found in aquatic environment. They are prokaryotes exhibiting a diversity with respect to morphology, physiology and phylogeny (Faivre and Schueler, 2008). The common feature of MTB is their magnetic organelle referred to as magnetosome. Using an assembly of these magnetosomes, they are able to passively orientate along magnetic field lines. On the surface of the earth, with the exception of the equator, the magnetic field lines have a component pointing downwards (towards the centre of the planet). Therefore, the function inferred from magnetosomes is an enhance effectiveness in the quest for micro-oxygenated region usually found slightly below the sediment sublayers of aquatic environments. To date, all the studied MTB are motile by means of flagella and have a cell wall structure characteristic of Gram-negative bacteria (Bazylinski and Frankel, 2004).

Nowadays, researchers from a wide range of scientific background are interested in magnetotactic bacteria. This is mainly due to their unique magnetic assembly that is, among equally exciting applications (Corchero and Villaverde, 2009; Fan *et al.*, 2009;

Martel *et al.*, 2009), believed to be a biomarker for extra-terrestrial life (Weiss *et al.*, 2004). The most fascinating fact about MTB is the degree of complexity found in the chain assembly. From the high degree of control inferred from membrane proteins to the growth of ferromagnetic single-domain crystal and to the skeleton extending from pole to pole of the bacteria and on which the magnetosomes can anchor and align. The bacteria are excelling where humans are still stammering: the self-assembly of nano scale building blocks. Such a level of organization in bacteria was still beyond expectation just a decade ago before researchers discovered that actin-like filaments are not eukaryotic cell specific but can also be found in bacteria (Gitai, 2005, 2007; Wang *et al.*, 2010). The production of the bacterial magnetic assembly starts from the collection of sufficient iron ions from the surrounding environment to trigger the nucleation of magnetic nanoparticles and finishes when the fully grown magnetic nanocrystals align, typically along a proteinaceous filament. MTB can therefore inspire a multitude of bottom up approaches for the fabrication of structures by self-assembly.

This book chapter is divided into four sections. In the first section, the reader will be introduced to the magnetotactic bacteria and their different hierarchical level of organizations. Starting from a simplified model of the magnetosome chain in the bacteria, we have described the typical characteristics of the magnetosomes and their assembly *in vivo*, focusing particularly on their size, morphology and the atomic structure of magnetite crystals. In the second section, we will review the important laws that rule the interaction of dipolar magnetic particles. The third section is dedicated to the role of magnetosome membrane proteins in the control of crystal growth and their importance for the formation of a stable optimized magnetic chain. Finally, we will discuss the advances made in terms of biotechnological applications, paying particular attention to the state of the development of biomimetic assembly approaches inspired by magnetotactic bacteria.

Hierarchical structuring of magnetosomes chains

Chain

As will be described in the subsequent sections, different strains of magnetotactic bacteria grow crystals of varying size, shape and composition. This diversity is also true at a higher hierarchical level since a variety in the number, position and organization of magnetosome arrangement chains is also found. Most of the cultured strains exhibit a single chain. This is, for example, the case for the *Magnetospirillum gryphiswaldense* strain MSR-1, *Magnetospirillum magnetotacticum* strain MS-1, and the *Magnetospirillum magneticum* strain AMB-1 that all exhibit a linear or slightly bent chain of magnetosomes aligned along the cell axis. Other magnetotactic bacteria, such as the *Itaipu cocci* found in the marine sediments of Itaipu Lagoon (Brazil), contain two magnetosome chains (Lins and Farina, 2004). Whereas most of the observed magnetotactic bacteria exhibit a chain-like arrangement of their magnetosomes, there are rare exceptions to the rule such as the uncultured freshwater cocci *Bilophococcus magnetotacticus* that exhibits clustered arrangement of magnetosomes (Moench, 1988). On the other extreme in terms of complexity, up to 1000 bullet shaped crystals can be found within *Magnetobacterium bavaricum*. They form up to five rope-shaped twisted bundles of magnetosomes (Hanzlik *et al.*, 1996; Jogler *et al.*, 2009b, 2010; Spring *et al.*, 1993).

The assembly of the magnetosomes chain *in vivo* is highly controlled at the genetic and structural level (Faivre and Schueler, 2008). The specific role of proteins is presented in the next chapter. However, a general introduction to the overall organization of the magnetic assembly *in vivo* is given here. The reader may refer to Fig. 10.1 for the schematic representation of a magnetotactic bacterium showing the cellular organization of the magnetic chain and the formation of the chain. With reference to Fig. 10.1, a chronological description has been proposed as follows: Iron is taken up from the environment as Fe^{2+} or Fe^{3+} and converted into an intracellular ferrous species in the membrane where invagination compartments are being formed (Fig. 10.1) (Faivre *et al.*, 2007). In these compartments the iron ions are precipitated into magnetite. This step occurs in the close vicinity of the membrane of the bacteria where the vesicles are formed. Whereas the intracellular iron concentration and the medium concentration has to be kept relatively low (optimum at 20 μM, detrimental above 200 μM) for the bacteria to grow (Heyen and Schüler, 2003; Schüler and Baeuerlein, 1996), at least 30 mM iron must be present in the vesicle to trigger

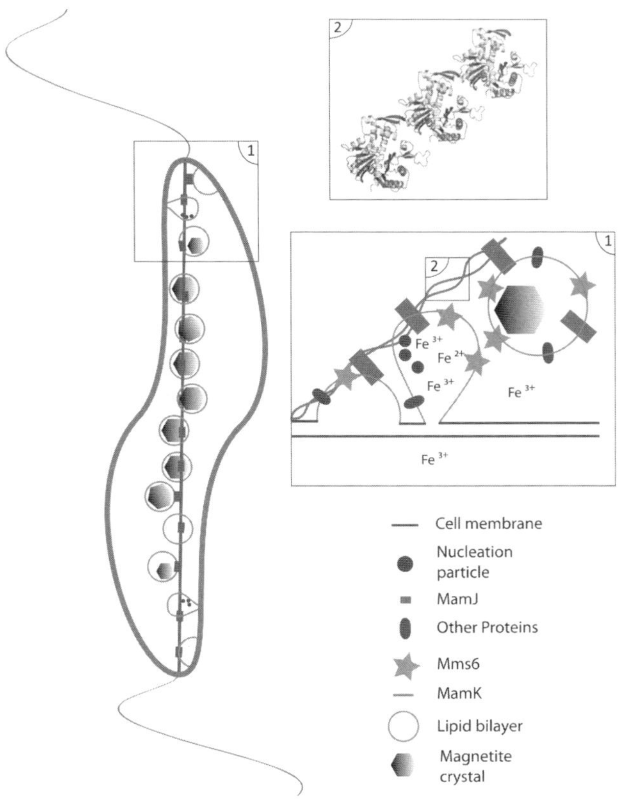

Figure 10.1 Schematic representation of the hierarchical organization of the magnetosome chain in magnetotactic bacteria (left) and the chronological formation of the magnetosomes (left to right in box 1). Box 2 is a modelled structure of MamK assembled into an actin-like filament (Sonkaria *et al.*, 2012). The proteins involved in the biological and chemical control of the formation of magnetosomes are currently being characterized, hence the extent to which their function is dedicated to the different processes is still unclear. A colour version of this figure is located in the plate section at the back of the book.

the nucleation of magnetite under basic conditions (Faivre *et al.*, 2004; Faivre and Zuddas, 2006, 2007). The fully grown crystals are observed at mid-cell. Biologically, this ensures the equal repartition of the magnetite crystals following cell division (Katzmann *et al.*, 2011). A magnetosome comprises the iron crystal bonded to a lipid bilayer membrane containing proteins. This membrane is referred to as magnetosome membrane. Whereas some of the membrane proteins are thought responsible for the crystals formation, others are important building blocks for the magnetosomes chain and serve as mechanical stabilizers. For example, Schüler *et al.* (Katzmann *et al.*, 2010; Scheffel and Schüler, 2007) have shown that the magnetosome membrane protein MamJ is essential for the presence of stable chains in the *Magnetospirillum gryphiswaldense* strains. The last and very interesting part of the magnetosome chain assembly is the actin-like filament (Komeili *et al.*, 2006). This filament is, at least partially, made of the magnetosome membrane protein MamK (Draper *et al.*, 2011). Until recently, actin-like protein cytoskeletons were thought to be eukaryotic cells specific. This assumption was based on the belief that prokaryotes were too simple organisms to have recourse to cellular functions based on complex protein networks (Gitai, 2005). Magnetosomes are anchored along the MamK filament found in *Magnetospirillum gryphiswaldense* (Katzmann *et al.*, 2010) most probably by interaction with the magnetosome membrane protein MamJ.

Size and morphology

Magnetosomes play an important role for the bacteria to find the optimal habitat in their natural environments. The production of magnetic minerals is thus supposed to have evolved, eventually resulting in particles with optimal magnetic properties. MTB can biomineralize either magnetite (Fe_3O_4) or greigite (Fe_3S_4) or both (Bazylinski and Frankel, 2004). These two minerals are ferrimagnetic at ambient temperature, hence useful for magnetotaxis.

Mature magnetosome crystals of magnetite and greigite lay usually between a range from 35 to about 120 nm in size when measured along their long axis (Bazylinski *et al.*, 1994; Devouard *et al.*, 1998; Faivre and Zuddas, 2006). Within this size range, there is a variation between the different species of MTB, with strain-specific properties. Magnetosomes from magnetospirilla typically range from 30 to 50 nm in length (Devouard *et al.*, 1998; Faivre and Schueler, 2008) and the *Desulfovibrio magneticus* magnetosomes length is about 40 nm (Posfai *et al.*, 2006). Magnetosomes from vibrios (MV-1) are slightly larger, with average dimensions varying from 40 to 60 nm (Devouard *et al.*, 1998; Meldrum *et al.*, 1993a). The largest magnetosomes from cultured strains occur in MC-1, with average sizes from 80 to 120 nm (Devouard *et al.*, 1998; Meldrum *et al.*, 1993a). The only greigite producing strain that has been isolated in pure culture can produce either magnetite or greigite magnetosomes depending on the amount of sulphide in the media. Its magnetite magnetosomes have tooth shape and are about 80 nm long. Its greigite magnetosomes are about 40 nm long (Lefèvre *et al.*, 2011a).

With the exception of large magnetosomes (250 nm) found in an uncultured coccoid MTB (Lins and Farina, 2004), all mature particles found in MTB are within the size range of single-magnetic domains (35–100 nm) (SD). These kinds of particles have a permanent magnetic moment at ambient temperature. They also have all the elementary magnetic dipoles aligned parallel and thus form a uniform magnetization that is maximal for a given volume. By producing SD particles, the bacterium has the maximum magnetization for the individual grains and by arranging the grains in chains, it also maximizes the magnetic moment of the cell (Bazylinski and Frankel, 2004).

Crystals inside magnetosomes have different morphologies or crystals habits. It is widely accepted that these morphologies are strain specific (Bazylinski *et al.*, 1994). Morphological variations have been also more frequently found between greigite crystals. In greigite-producing bacteria, samples collected from natural environments showed different particle morphologies even within a single cell (Posfai *et al.*, 1998). In addition, different growing conditions can affect the morphology of the crystals (Faivre *et al.*, 2008; Meldrum *et al.*, 1993a). Physical properties such as crystal-size distribution, aspect ratio, and morphology can significantly differ under different environmental parameters and by the rates of Fe uptake (Faivre *et al.*, 2008).

Mineral particles inside magnetosomes have high crystallographic perfection. Although defects such as screw dislocations are rare in magnetosome crystals, twinned and multiple-twinned crystals are frequently observed in several bacterial strains (ca. 40% of the crystals in some organisms) (Devouard *et al.*, 1998). The twinned crystals show the individuals related by rotations of 180° around the [1 1 1] direction, parallel to the chain direction, and sharing a common (1 1 1) contact plane (Devouard *et al.*, 1998). That does not affect the magnetic properties of the crystals because the easy axis of magnetization is [1 1 1], i.e. the direction of the twin axis (Mann *et al.*, 1989). Multiple twins have also been observed but are less common. Finally, elongation along a different axis than the axis of easy magnetization (the [1 1 1] axis) was also reported by several groups, mostly for tooth or bullet-shaped magnetosomes (Isambert *et al.*, 2007; Vali and Kirschvink, 1989).

Magnetite and greigite belong to the Fd3m space group and have face-centred spinel crystal structures (Palache, 1944). Magnetosomes can have different morphologies all depending on the crystal forms {1 0 0}, {1 1 0} and {1 1 1}, i.e. respectively the combinations of the (1 0 0) faces (cube), the (1 1 0) faces (dodecahedron), and the (1 1 1) faces (octahedron); and all the possible distortions and elongations (Devouard *et al.*, 1998).

The strain-specific morphologies are the following: In the case of the *Magnetospirillum* species, these are the well-known and studied *M. magnetotacticum* (AMB-1) and *M. gryphiswaldense* (MSR-1), the mature crystal produced has a cuboctahedral morphology, formed by {1 0 0} and {1 1 1} faces with equidimensional development of the six symmetry-related faces of the (1 0 0) form and of the eight symmetry-related faces of the (1 1 1) form (Fig. 10.2a and d). The marine spirillum MMS-1 also forms cuboctahedral but elongated crystals (Meldrum *et al.*, 1993a). Other species like the marine vibrios (MV-1 and MV-2) and the magnetotactic coccus MC-1 also have elongated morphologies (Fig. 10.2b and e). They form elongated pseudohexagonal prismatic magnetosomes that have been described as combinations of {1 0 0}, {1 1 1}, and {1 1 0} (Meldrum *et al.*, 1993a,b). *Desulfovibrio magneticus* (RS-1), the only delta-proteobacteria isolated in pure culture, has tooth-shape magnetosomes. RS-1 crystals appear to be conformed by two different parts: a base that has a triangular shape and an elongated, pointed second part. The straight edges of the triangles that conform the base are parallel to (111)-type planes. The second part is elongated along [100], and not in [111] in contrast to most types of bacterial magnetite (Posfai *et al.*, 2006) (Fig. 10.2c and f).

The only greigite-producing bacteria isolated in pure culture, BW-1, can produce either magnetite or greigite depending on the growing media. When sulphide is over 0.3 mM in the media, this strain produces greigite magnetosomes that have irregular outlines and lacked

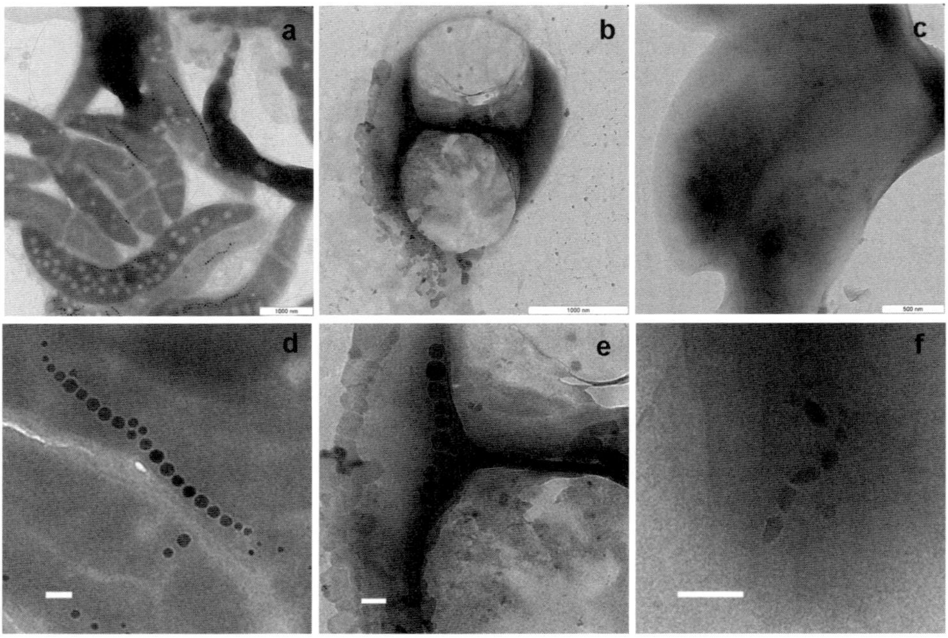

Figure 10.2 TEM images of some magnetotactic bacteria strains and their magnetosome chain. (a) and (d) *Magnetospirillum gryphiswaldense* and detail of its magnetosome chain. (b) and (e) MC-1 and detail of its magnetosome chain. (c) and (f) *Desulfovibrio magneticus* and detail of its magnetosome chain. Scale bars represent 100 nm.

a well-defined crystal habit. When the sulphide concentration goes below 0.3 mM this produces magnetite crystals that have bullet-shaped morphologies (Lefèvre *et al.*, 2011). In the case of the multicellular magnetic prokaryote (MMP), it contains a combination of greigite crystal morphologies, including pleiomorphic, pseudorectangular prismatic, tooth-shaped, and cubo-octahedral (Posfai *et al.*, 1998).

Individual greigite crystals, in contrast with magnetite crystals from magnetosomes, often exhibit irregular contrast when viewed with the electron microscope. Greigite crystals surfaces lack well-defined, distinct facets, and are often rounded and irregular (Posfai *et al.*, 1998). These defects have been interpreted as resulting from the mackinawite to greigite solid-state conversion process (Posfai *et al.*, 1998).

It was shown in the previous section that, for the purpose of magnetotaxis, the most efficient for a microorganism is to produce SD particles. The high crystallographical perfection and the anisotropic elongated morphologies in the easy magnetization axis also help to increase the particle magnetic moment.

Structure

Structural studies by electron diffraction have shown that the mineral present in magnetosomes was the inverse spinel of ferrous-diferric oxide, magnetite ($[Fe^{3+}]_{Td}[Fe^{2+}Fe^{3+}]_{Oh}O_4$, Fd3m space group, lattice parameter $a = 8.39$ Å) (Mann *et al.*, 1987; Matsuda and Egami, 1983). However, it is extremely difficult to differentiate magnetite from maghemite

(α-Fe_2O_3), a more oxidized and less magnetic iron oxide by conventional XRD and electron diffraction techniques. The lack of precision of electron diffraction with respect to lattice parameter determination makes the quantitative comparison between biogenic magnetite and abiotic magnetite difficult (Fischer *et al.*, 2011). Mossbauer spectroscopy was used to confirm the presence of magnetite inside magnetosomes, nevertheless this technique could only be performed on cultured strains, like the *Magnetospirillum* species and MV-1, as a large amount of sample is necessary (Faivre *et al.*, 2007; Frankel *et al.*, 1983). Lately, high resolution XRD measurements were performed at the m-spot beamline at the BESSY II synchrotron radiation facility [Helmholtz-Zentrum Berlin (HZB) by Fischer *et al.* (2011)]. They found significant differences between the biological and the synthetic materials and between isolated and non-treated biological materials. They concluded that cells are able to synthesize stoichiometric magnetite at ambient temperature, while stoichiometric inorganic nanomagnetite, not protected from oxidation, is more unstable.

Magnetite produced by magnetotactic bacteria is almost stoichiometrically pure. However the presence of cations other than iron in the magnetosomes has been reported. Trace amounts of titanium were reported in magnetite magnetosomes of an uncultured freshwater magnetotactic coccus collected from a waste water treatment pond (Towe and Moench, 1981). The presence of manganese has also been reported in magnetite magnetosomes from an uncultured MTB up to a 2.8 atom% of the total metal content (Fe + Mn) when the microorganisms are exposed to an environment enriched in manganese chloride (Keim *et al.*, 2009). The presence of 0.2–1.4% cobalt has been described for three different *Magnetospirilla* species in dedicated doping experiments (Staniland *et al.*, 2008). The mechanism of this addition of foreign cations is not clear yet. In particular, it is not known whether this happens via active or passive transport, during cell life or after cell death.

To date, the structure of the iron sulphide mineral has only been studied by electron diffraction in single cells. Other methods requiring larger volume of samples were not practicable due to the unavailability of pure culture. Nevertheless, a greigite-producing strain has recently been isolated in pure culture (Lefèvre *et al.*, 2011). This will probably lead to new studies and deeper knowledge of these magnetosomes. Apart from greigite, other non-magnetic minerals have been found in sulphide-containing magnetosomes. These minerals correspond to the different steps the microorganisms follow in their geochemical pathway to form greigite. This is now widely accepted to be originated in cubic FeS that transforms to mackinawite (tetragonal FeS) and finally turns into greigite (Fe_3S_4) as described by Posfai *et al.* (1998).

In greigite magnetosomes, significant amounts of copper have been found (up to 10%) (Posfai *et al.*, 1998). This has been also described for an uncultured multicellular prokaryote collected in a salt marsh (Bazylinski *et al.*, 1993a). Finally, both magnetite and greigite were reported to be present in the same cell, and recently, a pure isolated strain was shown to be able to synthesize magnetite or greigite depending on the sulphide concentration present in the media (Bazylinski *et al.*, 1993b; Lefèvre *et al.*, 2011).

Other trace elements, which include gold and silver and calcium and barium, were found in MTB but not associated to the magnetosomes (Gassman *et al.*, 1996; Keim *et al.*, 2005). In addition, iron/phosphorus-rich granules have been found in MTB, but they are most probably not related to magnetosomes (Byrne *et al.*, 2011).

Physical properties of dipolar magnetic nanoparticles

The magnetic properties of isolated bacterial magnetosomes and their behaviour are similar to that of other magnetic particles referred to as dipolar particles. Their behaviour in colloidal suspensions or as part of a polymer has been an ever expanding area of research (de Vicente *et al.*, 2011). This has been driven by the large number of applications these particles can be put to. Dipolar particles in bulk systems responding to external fields can dramatically change the behaviour of materials. This has led to suspensions and mixtures of dipolar particles finding applications in areas as diverse as audio amplification, medical applications (Lao and Ramanujan, 2004; Liu *et al.*, 2006), artificial muscles (Ramanujan and Lao, 2006) and tuning of material mechanical properties (Erb *et al.*, 2012).

Basic of magnetism

In order to understand the behaviour of these particles it is first necessary to review some basics of magnetism. If more information is required, the basic information on magnetism found here is expanded on in reference (Jackson, 1999).

The strength of a magnet can be stated in two ways, which are sometimes confusingly both called the magnetic field. The H-field (sometimes known as magnetic field strength or magnetic field intensity) is generally what is controlled in experiment as it is directly related to the current in a wire. The B-field (also known as flux density or magnetic induction) includes the contributions of any material placed in the magnetic field. This is the more fundamental quantity and is seen in some experimental studies. The relationship between the H-field and B-field is given by:

$$\vec{H} = \frac{\vec{B}}{\mu} - \vec{M}$$

where μ is a material-dependent property known as the *permeability* and $\vec{M}$ is known as *magnetization*. Magnetization is a measure of the density of magnetic dipoles in a material and is related to their alignment to an external magnetic field.

To add to the confusion there are also a number of units which may be used. As in many fields in physics, units were initially developed as a matter of convenience to simplify the mathematical calculations. The first measure of the B-field, in what are known as cgs (standing for 'centimetre, grams, seconds') units, was called the gauss (G). In 1960 an SI-derived unit (based on metre, kilograms and seconds) was introduced, known as the tesla (T). This has not gained complete adoption due to its large size for everyday usage (for example, 1 tesla is 10,000 gauss and the magnetic field of the earth at the equator is 3.1×10^{-5} T or 0.31 G) and magnet strengths can still be found in gauss. Likewise, the H-field has alternative units. In SI units it is measured in ampere-turn per metre (A/m) and in cgs units in oersteds (Oe).

The last piece of terminology to explain is the *susceptibility*. This dimensionless constant relates the response of a magnetic material in the magnetization to an applied external magnetic field. It is denoted by χ or χ_m and given by:

$$\vec{M} = \chi \vec{H}$$

When looking at the magnetic properties of particles, typically four distinct behaviours can be observed. These are inferred from different types of magnetic material that differ simply in terms of the moments of their constituent atoms.

Ferromagnetism occurs when the magnetic moments of the atoms are of equal strength and align with each other. This leads to a net dipole moment for the material. *Antiferromagnetism* is caused by equal neighbouring dipole moments being antiparallel and these materials will have no net magnetic moment. Ferrimagnetism occurs in materials whose atoms have unequal dipole moments. In this material, the moment of neighbouring atoms will be anti-parallel with one being stronger than the other leading to a net dipole moment. Magnetite, a common material for magnetic colloids and the most common constituent of the crystal of the magnetosomes, is *ferrimagnetic*. *Paramagnetic* materials have no long-range order in their magnetic moments but an external field will cause their moments to align, leading to an attraction to the field.

Finally, the last material property we will discuss here is *superparamagnetism*. This occurs when ferromagnetic or ferrimagnetic particles are small enough so that thermal fluctuations can flip their magnetic moments. The time between changes in the magnetic moment is called the *Néel relaxation time* and when this is small compared to the timescale of measurements, the magnetization will appear to be 0. Superparamagnetic materials will respond and align with an external magnetic field. The materials described above are summarized in Table 10.1.

Simple models of magnetic interactions

The full quantum mechanical picture of magnetism is complex and mathematically sophisticated. Calculations of bulk behaviour using this approach are extremely difficult and would require great effort. Luckily, the behaviour of these systems can be dealt with in a fully classical manner, greatly simplifying the mathematics and computation necessary to solve interesting systems. In fact, much of the behaviour of a system of dipolar particles can be naively explained by simple mechanics in uniform and non-uniform magnetic fields.

The movement of dipolar particles in an external field can be explained by the force felt on the particle, rotationally and linearly, and the energy of the dipole–dipole interaction. The forces experienced by a dipole vary if the field is uniform or non-uniform. Mathematically, the force experienced by a dipole in a magnetic field is given by:

$$\vec{F} = \nabla\left(\vec{m}.\vec{B}\right)$$

where $\vec{m}$ is the magnetic moment, ∇ is the gradient operator and $\vec{B}$ is the magnetic field. In a non-uniform field, $\vec{F}$ is non-zero and plays a significant role in the interactions of dipoles. Conversely, in a uniform field $\vec{F} = 0$ and the only forces felt are dipole–dipole interactions.

Table 10.1 Alignment of the dipole moments in different types of materials in the absence of external magnetic field

Ferromagnetic	↑↑↑↑↑↑↑↑↑↑↑↑ Equal aligned dipole moments. Net moment
Antiferromagnetic	↑↓↑↓↑↓↑↓↑↓↑↓ Equal antiparallel dipole moments. No net moment
Ferrimagnetic	↑↑↓↑↑↓↑↑↓↑↑↓ Unequal antiparallel moments. Net moment
Paramagnetic	←←→↓→↑↓↓→↑←↑ No long-range order. Aligns with external field

In both the uniform and non-uniform cases, a torque $\vec{\tau}$ will be felt by the dipoles. This is given by:

$$\vec{\tau} = \vec{m} \times \vec{B}$$

The cross-product of vectors here implies that the torque will be 0 when $\vec{m}$ and $\vec{B}$ point in the same direction and maximum when they are at right angles to each other. This torque explains the rotation of dipoles and their ability to align with a magnetic field. This is the property of magnetic materials which is exploited in compasses and by the magnetotactic bacteria.

In magnetotactic bacteria, the moments of magnetosomes are parallel to each other hence minimizing the magnetostatic energy (Frankel, 1984). Therefore the total magnetic moment m of the bacteria is the algebraic sum of the moments of the individual magnetosome and the torque inferred by an external magnetic field onto the bacteria is maximized. The fact that we can navigate with a compass and that the magnetotactic bacteria can passively align thanks to their chain of magnetosomes are visible examples that even a field as small as the earth's magnetic field can have a significant effect on magnetic material.

The one remaining part of this simplified understanding of the behaviour of dipoles is the dipole–dipole interaction. For simplicity, this discussion considers spherical particles containing one dipole. An easy way to understand this is by analogy with magnets. Dipoles can be thought of as a tiny bar magnets complete with north and south pole. Like bar magnets, like poles will repel and opposite poles will attract and this will guide the interactions between dipoles. The pure dipole–dipole interaction was heavily studied via simulations in the mid-1990s (Jund *et al.*, 1995; Levesque and Weis, 1994; Weis and Levesque, 1993). These seminal studies looked at systems of dipoles at cryogenic temperatures to remove any thermodynamic perturbations of the system and only consider pure dipolar interactions. Weis and Levesque (Levesque and Weis, 1994; Weis and Levesque, 1993) looked at large systems of dipoles at a low density in 3D. In these studies it was found that with no other effects involved, dipoles would choose to form chains with the dipoles lining up nose to tail. The same result can be observed at room temperature, as shown in Fig. 10.3.

The fact that dipoles formed chains was surprising, as the minimum energy conformation for a system of more than five dipoles would be a ring configuration (Jund *et al.*, 1995). To understand the reason for this lack of ring structures in 3D we must consider the thermodynamics of the system. By forming rings, the dipoles would reduce the entropy of the system. As we learn in high school physics, entropy of the universe always increases. This means there is a delicate balance between the minimization of energy and entropy effects which stop large chains of dipoles forming into rings.

In 2D, the equilibrium state is slightly different. This difference is significant enough to warrant consideration here and potentially more applicable as 2D systems are analogous to systems like thin films and could resemble the conditions found in magnetotactic bacteria. In 2D, dipoles will still line up 'nose to tail' but entropic effects are minimized in this case and the minimum energy conformation is accessible. This means that both chains and rings are seen (Butter *et al.*, 2003; Tripp *et al.*, 2003). The applicability of this to the magnetic particles found in magnetotactic bacteria has been directly observed. In 2002, Philipse and Maas extracted dipolar particles from magnetotactic bacteria (magnetosomes) and imaged them via transmission electron microscopy (TEM) (Philipse and Maas, 2002). They found the resulting particles tended to form rings and linear chains.

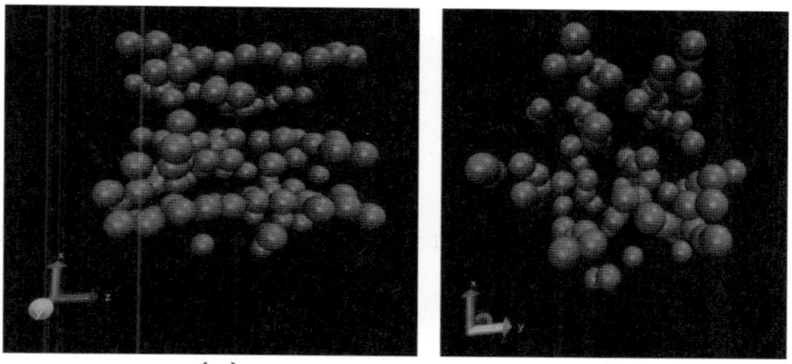

(a) (b)

Figure 10.3 Snapshot of a simulation of a toy system of dipoles showing their self-organization at room temperature in a weak magnetic field in three dimensions. The dipole moment of the particles has been increased to exaggerate the chaining in such a system. A colour version of this figure is located in the plate section at the back of the book.

In order to relate all of the above behaviour to the real world, it is necessary to add one small complication. All of the above has been considering simple spherical particles of one size. Systems such as this are known as monodisperse. In reality, systems may be more complicated, containing particles of different size and shape and are polydisperse. This can affect things in two ways: by changing the effective dipole–dipole interaction and making the particle multidomain.

Dipolar particles are quite sensitive to a change in shape. Any stretching of the particles in one direction can change the interaction energies such that a nose to tail configuration is equally likely to an antiparallel side by side structure (GilVillegas *et al.*, 1997; McGrother and Jackson, 1996) allowing clusters to develop. Any system which contains such 'stretched' particles will include clusters as well as chains and rings. Above a certain size, particles can no longer be thought of as containing a single dipole. Larger particles can contain independent domains, each containing a discrete dipole moment of its own. In the model of a dipole as a small bar magnet, a multidomain particle would be the equivalent sticking two bar magnets together with a shield between them to make one particle. The size at which this happens depends strongly on the material and can range from ~ 40 nm to ~ 200 nm (Butler and Banerjee, 1975; Das *et al.*, 2008; Sanchez *et al.*, 2002). Fig. 10.4 illustrates the formation of multidomain as the magnetic particles become larger. The transition from single domain to multidomain is accompanied by a decrease of the particle coercivity.

The coercivity quantifies the stability of the magnetic properties of a material or object. This is typically measured using a B–H analyser and relates the intensity of an applied magnetic field, B, required to reduce the magnetization, M, of a magnetically saturated material to zero. Typically, a superparamagnetic particle exhibits no coercivity, and the single domain particles have the highest coercivity which means that they are the most stable magnetically. For magnetite particles single domain limit is around 80 nm (Butler and Banerjee, 1975; Das et al., 2008; Sanchez et al., 2002), below this size and above around 25 nm they exhibit single domain characteristics and their coercivity is maximized. Whereas it is difficult to obtain stable, single domain, magnetic nanoparticles under natural conditions, magnetotactic bacteria excel in synthesizing and stabilizing those, hence making the best of their invaginated magnetic material. Multidomain particles are more complex and considerably less studied due to their unsuitability for most applications.

(a) (b) (c) (d)

Figure 10.4 Schematic of nanoparticles. Below a certain size, particles are superparamagnetic and do not have a dipole in the absence of an external magnetic field (a). Above a certain particle size (ca. 20 nm for magnetite), the dipole of particles is equivalent to a bar magnet and their magnetic characteristic referred to as single domain (b). When particles become larger (ca. larger than 100nm for magnetite), they exhibit an increasing number of magnetic domains and are referred to as multidomain (c, d). These can be thought of as a particles containing multiple bar magnets.

As opposed to other magnetic nanoparticle assembly processes, magnetotactic bacteria are growing magnetic crystals at the same time as they are assembling them. Therefore, the assembly partially occurs while the particles are still superparamagnetic. Furthermore, as opposed to synthetic processes, magnetotactic bacteria have to assemble the chain in a confined microenvironment that does not resemble the environment of the bulk material. These two specific properties of the assembly process found in magnetotactic bacteria means that forces induced by the Earths magnetic field alone are not sufficient to systematically result in the formation of a stable magnetic chain *in vivo*. This also holds true for harvested magnetosomes that form rings and folded chains (Philipse and Maas, 2002). In order to stabilize the magnetic chain, which is essential to ensure the biologically necessary magnetic properties, magnetotactic bacteria rely on biological stabilizers. Their functions are described in the following chapter.

Biology of chain formation

Magnetosome formation

The magnetosome is formed by a magnetic material nanocrystal surrounded by a lipid membrane. This membrane is responsible for the establishment of the necessary environment for the biomineralization of the magnetic material (Komeili, 2012). However, although this membrane has been observed to be present before crystal nucleation in some species of magnetotactic bacteria (MTB) (Blakemore, 1975; Gorby *et al.*, 1988) and even when the microorganism is lacking iron for several generations (Komeili *et al.*, 2004), it is not clearly distinguishable in all species of MTB. For example, *Desulfovibrio magneticus*, belonging to the Deltaproteobacteria, has the particularity of forming tooth-shape magnetosomes (Byrne *et al.*, 2011) in which this membrane could not been observed. This is possibly due to limitations in the resolution of the analytical technique used for the observation. Another tooth-shape crystals forming microorganism *Magnetobacterium bavaricum* was also supposed to be lacking this membrane (Hanzlik *et al.*, 2002). However it was recently suggested that the membrane could not be detected due to its size and tight bonding to the crystal

(Jogler *et al.*, 2010). Although not clearly present in the final stages of biomineralization, the presence of the membrane is probably necessary in the beginning of the crystal formation (Tanaka *et al.*, 1997).

This vesicle is composed of a lipid bilayer membrane that contains phospholipids, fatty acids, glycolipids and sulfolipids associated with specific proteins (Gorby *et al.*, 1988; Gruenberg *et al.*, 2004). The predominant phospholipids are phosphatidylserine, phosphatidylglycerol, phosphatidylethanolamine, ornithinamid lipid and an unidentified amino lipid (Gruenberg *et al.*, 2004). Phospholipids represent 58–65% of the total lipids of the magnetosome membrane (MM) of *M. magneticum*, 50% of them being phosphatidylethanolamine (Nakamura and Matsunaga, 1993). This fatty acids composition shows that the MM is similar to the cytoplasmic membrane (CM) but distinct from the outer membrane (OM) (Tanaka *et al.*, 2006). This contrasts with other inclusions in prokaryotes, which are generally surrounded by a single layer of protein (Bazylinski and Schübbe, 2007). The protein composition of the MM is strikingly different regarding to other prokaryotic subcellular compartments. Several biochemical and genomic studies confirm (Schübbe *et al.*, 2006, 2009; Tanaka *et al.*, 2006) that some magnetosome proteins are conserved among all MTB. Some of these proteins are unique to the MM and it is likely that they play key roles in magnetite crystal nucleation. These proteins have received different nomenclature, partly in relationship with their supposed function: Mam (magnetosome membrane), Mms (magnetic particle membrane specific), Mtx (magnetotaxis), or Mme (magnetosome membrane) (Schüler, 2008).

Most of these proteins are encoded in several clusters that are contained in a genomic region called the magnetosome island (MAI) (Schübbe *et al.*, 2003; Ullrich *et al.*, 2005). This is a hypervariable 130-kb genome fragment which is apparently conserved in all MTB and which may have been transferred horizontally (Fukuda *et al.*, 2006; Grünberg *et al.*, 2001; Jogler *et al.*, 2009a; Ullrich *et al.*, 2005). The most important proteins, and those that have a known function in the magnetosome formation, are: the ones contained in mam AB and mam GFDC clusters (Murat *et al.*; Scheffel *et al.*, 2008); Mms6, a small iron-binding protein, which has been isolated from the magnetite crystal surface of magnetosomes (Arakaki *et al.*, 2003); mamXY, a cluster containing proteins important in the first steps of the magnetosome formation (Tanaka *et al.*, 2010); and the tetratricopeptide repeat protein, MamA (identical to Mms24), that may play a role in the activation of magnetosome vesicles (Komeili *et al.*, 2004).

From all these gene clusters, only mam AB has been shown to be indispensable for magnetic particle production, as the deletion of this cluster causes the total loss of the magnetic particles in the strain *Magnetospirillum magneticum* (AMB-1). Mam AB contains 18 genes. The individual deletion of four of them (mamI, mamL, mamQ, and mamB) resulted in a loss of the magnetosome membrane from AMB-1 (Murat *et al.*, 2010). MamI and MamL are two small proteins (c. 17 amino acids, with two predicted transmembrane sections) only present in MTB and that do not have homology with any other known proteins. Mam L also has a 15 amino acid C-terminal tail that is predicted to form a helix on the cytoplasmic side of the inner membrane. This region rich in positively charged amino acids resembles peptides that are capable to interact and even go through membranes (Schmidt *et al.*, 2010). The possible interaction of these positively charged residues with the cytoplasmic side of the inner membrane creates an asymmetry that may trigger the bending of the membrane that will eventually form the magnetosome (Komeili, 2012).

MamQ and MamB the other two proteins codified by this operon formed part of cation diffusion facilitator (CDF) family of transporters. These proteins have coiled-coil repeat domains that may help to shape the magnetosome membrane (Murat *et al.*, 2010). However, although this explanation seems plausible, it is possible that these proteins could have another or a complementary role. They could be involved in the maintenance of the magnetosome membrane shape, once it is already formed, rather than in the formation of the vesicle itself (Komeili, 2012). It has been shown that these four proteins are essential, but not enough on their own, for the biogenesis of the magnetosome membrane. Murat *et al.* showed that when the genes coding for these protein are restored in a strain lacking the rest of the mamAB gene cluster no magnetosome membrane formation is observed (Murat *et al.*, 2010).

In the case of *Magnetospirillum gryphiswaldense* (MSR-1), Lohße *et al.* (2011) corroborated that only the mamAB operon contains genes which are essential for magnetosome formation. Besides, they were able to demonstrate that the mamAB operon is the only region of the MAI which is necessary and sufficient to restore the formation of nanoparticles (Lohße *et al.*, 2011), even in the absence of the mamGFDC, mms6, and mamXY clusters, what could not be revealed in previous studies. However, the nature of these nanoparticles has not been demonstrated yet.

In this cluster the MamA protein is also codified. This protein is not indispensable for biomineralization (Murat *et al.*, 2010). However it has an important role in magnetosome activation (Komeili *et al.*, 2004) and may act in bridging the interaction of some magnetosome proteins with the magnetosome membrane (Zeytuni *et al.*, 2011).

The mam GFDC cluster is another region conserved in the MAI of many alpha-proteobacteria MTB (AMB-1, MSR-1, MS-1, MC-1, and MV-1) but not in the delta-proteobacteria (RS-1) (Komeili, 2012; Nakazawa *et al.*, 2009). It codes for the small hydrophobic MamGFDC proteins, which represent nearly 35% of all proteins associated with the MM. Although its presence in large quantities and its ubiquity among MTB could suggest they have important functions in magnetosome formation they play a nonessential role in biomineralization (Scheffel *et al.*, 2008). In the case of MSR-1, Mam GFDC proteins are supposed to have partially overlapping roles, and control the growth of magnetite crystals by an unknown mechanism. The most plausible theory says that the MamGFDC proteins may regulate the crystal growth of magnetite crystals by changing physicochemical conditions inside the vesicles, such as, the charge distribution at the inner surfaces, the pH or the redox conditions (Scheffel *et al.*, 2008). They may also change the size of the organelle. In the case of AMB-1 (Murat *et al.*, 2010) the mutation of this operon does also provoke a reduction in the particle size.

Mms6 is a small acidic protein that is tightly associated with bacterial magnetite in *Magnetospirillum magneticum* AMB-1 (Arakaki *et al.*, 2003). It is an amphiphilic protein with a N-terminal hydrophobic region and C-terminal hydrophilic region containing multiplets of acidic amino acids. Its C-terminal region has been suggested as an iron-binding site (Arakaki *et al.*, 2003). This protein has shown its iron binding activity in allowing the generation of uniform magnetic crystals by co-precipitation of ferrous and ferric ions *in vitro* (Amemiya *et al.*, 2007; Arakaki *et al.*, 2010; Prozorov *et al.*, 2007). *In vivo* it has been demonstrated that its operon deletion in the case of AMB-1 causes smaller and elongated crystals (Tanaka *et al.*, 2011) while in MSR-1 58% of crystals within cells still had cubicle-shaped appearance (Lohße *et al.*, 2011).

Other proteins, such as mamE, mamO, mamM, and mamN, also have important roles in magnetosome formation. Knock-out mutants for these proteins in AMB-1 are completely non-magnetic and produce no electron-dense particles when analysed by electron microscopy(Murat *et al.*, 2010). Nevertheless, chains of empty magnetosomes membranes are still present (Murat *et al.*, 2010; Yang *et al.*, 2010). This phenotype may be due to a direct involvement of the proteins in the mineral production or to a wrong localization of the magnetosome proteins.

Another protein has been determined to have an important role in magnetosome formation at least in the early stages. This protein is MamY that is coded in the mam XY operon. In 2010 it was shown that magnetosomes having smaller crystals were enriched in one particular protein (Tanaka *et al.*, 2010). This protein was called Amb1018/MamY and the genes encoding for it are inside the magnetosome island. MamY has been showed to bind directly to liposomes, causing them to form long tubules (Tanaka *et al.*, 2010). It is also thought to be implicated in MM biogenesis in AMB-1 (Tanaka *et al.*, 2010). MamX has similarity to the serine like proteases MamE and MamS (Lohße *et al.*, 2011). The deletion of this operon in MSR-1 has a stronger effect than in AMB-1 (Lohße *et al.*, 2011). MamXY genes have a crucial function in magnetite biomineralization of MSR-1, loss of mamXY operon lead to poorly crystalline and elongated crystals (Murat *et al.*, 2010).

Chain formation

In order to serve most efficiently as a magnetic sensor, magnetosomes must be arranged in single or multiple linear chains in which the cellular magnetic dipole is the sum of the permanent magnetic dipole moments of the individual single-domain magnetosome particles (Frankel *et al.*, 2007). However, a row of magnetic dipoles will collapse easily to lower its magnetostatic energy without any mechanical stabilization (Kirschvink, 1982). Two complementary cryoelectron tomography studies have shown the presence of a network of filaments, 3–4 nm in diameter, which traverses cells of *M. gryphiswaldense* and *M. magneticum* (Komeili *et al.*, 2006; Scheffel *et al.*, 2006). These filaments are nearby the CM. Magnetosomes were closely arranged along this cytoskeletal structure, which has been tentatively called a magnetosome filament (MF) and is presumably at least partly formed by the MamK protein (Frankel and Bazylinski, 2006; Komeili *et al.*, 2006).

The assembly and maintenance of well-ordered chains *in vivo* are highly controlled at the genetic and structural levels. Several proteins are known to be responsible for magnetite magnetosome chain formation in *Magnetospirillum* species. The mamJ and mamK genes are located within the mamAB gene cluster in *Magnetospirillum* species and are cotranscribed (Schübbe *et al.*, 2006).

MamK has homology to the cytoskeletal actin-like MreB protein, that forms cytoskeletal structures in some non-magnetotactic bacteria and it is involved in a number of essential cellular processes in bacteria, such as cell shape determination, establishment of cell polarity, and chromosome segregation (Carballido-López, 2006; Figge *et al.*, 2004; Jones *et al.*, 2001; Van den Ent *et al.*, 2001). However, MamK proteins in magnetotactic bacteria are more similar to each other than they are to MreB homologues (Komeili *et al.*, 2006). Because of this intriguing similarity it was speculated that MamK might have a role in aligning and stabilizing the magnetosome chain (Schüler, 2004). In fact, the MF was absent in a mutant of *M. magneticum* from which the mamK gene was deleted (Komeili *et al.*, 2006). The magnetosome chains were less regular and dispersed throughout the cell, which led to the conclusion that the MF is involved in the stabilization and anchoring of the magnetosome chain within the cell (Komeili, 2007; Komeili *et al.*, 2006). MamK of *M. magneticum* formed straight filaments, structurally and functionally distinct from the known MreB and ParM filaments, rather than helical structures (Pradel *et al.*, 2006). The formation of MamK straight filaments in *M. gryphiswaldense* was independent of the presence of other magnetosome genes (Jogler and Schueler, 2009). MamK is presumed to have a more complex function than just providing a rigid scaffold for magnetosome alignment. It could also play a role in positioning and concatenating magnetosome chains (Katzmann *et al.*, 2010). This protein also has an important function in splitting magnetosome chains during cytokinesis and its presence is required for proper magnetosome position and segregation (Katzmann *et al.*, 2010; Klumpp and Faivre, 2012).

The MamJ protein is a strongly acidic protein with a repeating glutamate-rich section in its central domain (Scheffel *et al.*, 2006). The mamJ gene immediately precedes mamK within the mamAB operon, which is transcribed from a single promoter (Schübbe *et al.*, 2003; Schübbe *et al.*, 2006). MamJ is known for interacting with the MamK filament and directing the assembly and localization of the prokaryotic organelles along the chain in MSR-1 (Scheffel *et al.*, 2006). In this strain MamJ physically interacts with MamK. The deletion of MamJ causes magnetosomes not to assemble in linear chains but instead to arrange in three-dimensional clusters (Scheffel *et al.*, 2006). In the case of AMB-1 two proteins, MamJ and LimJ, perform a redundant role in promoting the dynamic behaviour of MamK filaments in wild-type cells. However, their deletion does not cause the collapse of the magnetosome chain. The absence of both MamJ and LimJ leads to static filaments, a disrupted magnetosome chain, and an anomalous build-up of cytoskeletal filaments between magnetosomes (Draper *et al.*, 2011).

Uses of magnetic chains

The magnetosomes and their assemblies indeed infer unequalled properties to the system which they are part of. Therefore they have potential use in a wide range of applications and they are highly inspiring for the development of biomimetic approaches and the fabrication of magnetic nano- and micro-objects (Faivre and Schueler, 2008; Fan *et al.*, 2009; Varadan *et al.*, 2008; Wang *et al.*, 2011).

Magnetotaxis

The most common strategy for bacteria motility is swimming. This is the means chosen by magnetotactic bacteria in search for beneficial environment. Bacterial swimming is provided by flagella, which are external organelles that serve as propellers. Bacteria have been reported to swim at speeds of up to 250 μm/s or ca. 150 times their body length per second (Martel *et al.*, 2006). Their motility had been established in the early years of microscopy. In 1880s, W. Engelmann and W. Pfeffer reported the observation of aerotaxis, an instance of chemotaxis whereby the stimulus is provided by oxygen levels. Since then, many other forms of stimuli have been characterized. They are named after their driving components. For example the motility driven by gravity is referred to as geotaxis, by light, phototaxis and by temperature, thermotaxis (Eisenbach, 2001). In their search for ideal environment, bacteria motility is characterized by a succession of swimming and tumble (Eisenbach, 2001) positively or negatively biased by a short life chemical memory provided by attractant or repellent, respectively. Therefore, the search for favoured-environment by swimming bacteria is a biased random walk in a 3dimensional spatial frame. In some cases, it is more efficient for bacteria to reduce this walk to one spatial dimension. This is the case when the environment exhibits a stratified-structure of attractant/repellent. The oxygen levels at the bottom of aqueous environment are an example of such chemical stratification. The oxic–anoxic interface, which is particularly prized by magnetotactic bacteria in view of colonization, is found at the water sediment interface or slightly below the sediment sublayer (Bazylinski, 1995; Stolz, 1993). With the exception of the equator, on the Earth's surface the geomagnetic field has a component pointing downwards. Thanks to their magnetosomes chain, the bacteria are passively orientated along the geomagnetic field, therefore reducing their walk towards suitably oxygenated region to a one-dimensional space.

Microrobots

The ability of magnetotactic bacteria to align along a magnetic field line is a very attractive property for development of microrobots. Such a development can be foreseen in two ways. On one hand, magnetotactic bacteria can inspire the biomimetic fabrication of microrobots (Dhar *et al.*, 2007; Dreyfus *et al.*, 2005; Kline *et al.*, 2005; Pawashe *et al.*, 2009b; Zhang *et al.*, 2009) and, on the other hand, researchers can use cultured bacteria and remote control them in order to perform tasks at the microscale (Martel *et al.*, 2006, 2009; Shechter and Martel, 2010).

Biomimetic approach

The development of microrobots (also referred as to nanorobots in the literature (Martel *et al.*, 2009)) has the potential to revolutionize the treatment of disease and become a cheaper, less painful and more flexible alternative to surgery (Sitti, 2009). However, such development remains at a stammering state and a lot of efforts are required for the scientific community to fabricate fully functional microrobots inspired by magnetotactic bacteria and effectively used them in medical applications, miniaturized manufacturing or other processes. In any case, the microrobots have to be able to perform at least one of the following tasks: actuation, sensing, signalling, information processing, intelligence, and swarm behaviour at the microscale (Mavroidis and Ferreira, 2009). The obvious technical challenges in the development of microrobots that are able to perform the aforementioned tasks are: how to fabricate them, how to power them and how to steer them (Sitti, 2009). Because magnetotactic bacteria can be steered by applying a controlled magnetic field and are propelled by means of flagella, they have inspired a number of scientists that have worked on the biomimetic development of microrobots. Owing to the multidisciplinary nature of such a development and the recent interest in the subject, it is difficult to track every advance reported. This is also because the relevant publications are disseminated in a large range of scientific journals from various scientific backgrounds. However, we have tried to present a fair account of the current development hereinafter.

On one hand, the shape of the bacteria has driven biomimetic fabrication. For example, M. Sitti's group have devised magnetic microrobots. These have been fabricated by laser micromachining of a magnetized piece of neodymium–iron-boron (Pawashe *et al.*, 2009a,b). The device is a rectangular prism 250 μm long. This can move at up to 32 times its body length in air. Another approach has consisted of attaching an artificial flagella (50 μm in length) to a magnetic material (4.5 μm × 4.5 μm) that induces the rotation of the helical artificial flagella under a rotating magnetic field (Zhang *et al.*, 2009), propelling the microrobot at up to 1.2 μm/s or 0.024 times its body length per second. However, both of the above-mentioned microrobots are much larger than magnetotactic bacteria. Furthermore, whereas the shape of the robot resemble that of the bacteria flagella, the unique magnetic properties of the chain of magnetosomes in magnetotactic bacteria which allows their orientation in weak magnetic field are not met. Hence, even when smaller magnetic artificial flagella are being fabricated (Ghosh and Fischer, 2009), a strong rotating magnetic field is required to propel the microrobots that can move at velocity of up to 14 μm/s or six times its body length. A schematic of this microrobot is shown in Fig. 10.5.

On another hand, researchers have also attempted to reproduce the magnetic properties of magnetotactic bacteria while using a different propulsion mechanism. In order to evaluate the relative merits of ferromagnetism and paramagnetism, Kline *et al.* (2005) have fabricated striped metallic nanorods made of gold, nickel and platinum. The nickel segments of each rod can be magnetized. Because of their size the nickel segments in the metallic rod are single domain. A unique characteristic of these nanorobots comes from the fact that the speed at which they move is not inferred by the magnetic field. Actually, the magnetic field only orientates the nanorods while their velocity is inferred by interfacial tension gradient that results from the catalysed decomposition of hydrogen peroxide into oxygen at the platinum end of the rod. In order to evaluate the capability of different type of magnetizations, Dhar *et al.* (2007) studied the movement of magnetotactic bacteria and of three nanorods that exhibit different magnetic properties on a magnetic garnet film. In a field corresponding to the saturation magnetization of the garnet film, the magnetic moment of a paramagnetic nanorod approximates that of the magnetotactic bacteria. Their results suggest that guidance of a magnetic microrobots is best achieved if they exhibit paramagnetic properties (This is because ferromagnetic navigators would be trapped by the domain walls where the magnetic energy overcomes the propulsion energy) (Dhar *et al.*, 2007).

With a different approach, Dreyfus *et al.* (2005) have fabricated, to the authors' opinion, the most advanced biomimetic magnetic microrobots to date. Their microrobots consist of a flexible magnetic filament made of superparamagnetic 1-μm-diameter colloids linked by several double-stranded DNAs. The flexibility required to induce motility of the microrobots is provided by the DNA strands properties (number, length) and by the particle diameter. By applying a time varying sinusoidal field on one directional axis and a constant field on another, the time-reversal invariance of the microrobot deformations is broken and a motion is induced. These microrobots were used to transport red blood cells.

The development of magnetic micro-objects has also been accompanied by the development of the appropriate control system. The integration of magnetic control on the

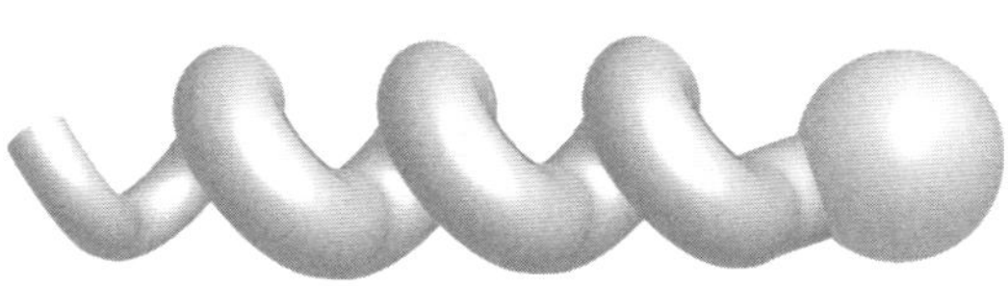

Figure 10.5 Schematic of a magnetic microrobot resembling that fabricated by Zhang *et al.* and that fabricated by Ghosh and Fischer. The head is typically magnetic and is used to infer a rotating motion to the flagella which in turns propel the micro-object (Dhar *et al.*, 2007; Zhang *et al.*, 2009).

micro-world has motivated the development of microelectromagnets (Lee *et al.*, 2001) and their use to assemble magnetosomes (Lee *et al.*, 2004). The parallel development of microfluidic devices have also led to a review on magnetism and microfluidics which describe the different approach useful for the fabrication of a magnetic controller on chip (Pamme, 2006).

In general, despite of their multiple inspiring properties, magnetosome assemblies have scarcely inspired the development of magnetically controlled microrobots. Furthermore, in the few instances where a biomimetic approach has been used, the fabricated microrobots did not retain the full panel of properties found with magnetotactic bacteria: size (micrometre range), velocity (30 times the body length) and ferromagnetism (passively orientated with magnetic field as weak as the earth's magnetic field). This is probably due to the difficulties encountered in the fabrication of such devices. In order to circumvent these difficulties, a few researchers have explored the capability of MTB as microrobots. Their work is presented below.

Use of bacteria as microrobots

As described in the previous section, the development of biomimetic microrobots that have equally interesting properties as the one exhibited by magnetotactic bacteria presents many challenges that have not been entirely addressed. In order to circumvent this shortcoming, researchers have remotely controlled magnetotactic bacteria using an applied magnetic field and use them as microrobots. A few papers have suggested that the magnetotactic bacteria are ideal candidates for the development of magnetically controlled microrobots although, in these publications, the bacteria have not been used to perform a task at the microscale (Bahaj *et al.*, 1998; Dhar *et al.*, 2007). However, the use of magnetotactic bacteria to perform microrobotic tasks have been demonstrated by a paper from Martel *et al.* (2006). In their experiment, they have used the *Magnetospirillum gryphiswaldense* bacteria to load and carry microbeads. When a solution of microbeads (3 μm in diameter) was mixed with a solution of cultivation medium containing magnetotactic bacteria, a very low percentage (around 1%) of the bacteria would passively bind to the microbeads. Using a homogeneous magnetic field of 3.5 G, the motion of swimming bacteria was controlled and imaged using a phase contrast microscope. The bacteria attached to the beads could carry the load at an average speed of 7.5 μm/s (when not attached to the bead, the average speed of a bacteria was 22.5 μm/s). From Stokes' law, the authors estimated the thrust of a single *Magnetospirillum gryphiswaldense* bacterium to be ca. 0.5 pN. Whereas this paper makes an account of the possible use of magnetotactic bacteria as microrobots, the authors pin-pointed some of the following steps to be accomplished. For example, they noticed that after 5 min of pushing the microbeads, a percentage of the few bacteria that were bound to microbeads did not exhibit casual motility any more possibly due to the flagella or the cell being attached to the bead. The use of other strains is expected to increase the maximum thrusts, for example up to 4 pN/bacterium for MC-1. Despite of their magnetic properties, this latter strain has solely be controlled using dissolved oxygen concentration (Shechter and Martel, 2010) and its intrinsic superiority over MSR-1 for their use as magnetically controlled microrobot has been extensively described (Martel *et al.*, 2009).

As for the biomimetic approach, the research devoted to the use of MTB as microrobot is extremely limited, partly due to the embryonic stage of the research in this area and in part because of the interdisciplinary nature such a project requires. Furthermore, the use of living microrobots leads to complications that are not encountered when using artificial microrobots.

Possible other applications that could result from the use of the magnetic properties of magnetosome assemblies are numerous. To name just a few, magnetotactic bacteria could be used as controlled microswitches, microvalves, micropistons, micromixers, micromotors and to propel autonomous microrobots (Martel *et al.*, 2006). These could be realized using MTB as a template for the fabrication of hollow micro-objects that would preserve the magnetic properties and shape of the bacteria while gaining the properties of hybrid material (Baumgartner *et al.*, 2012).

Other applications

The manipulation of magnetic nanoparticles has applications in spintronics (Black *et al.*, 2000), magnetic memory (Sun *et al.*, 2000), medicine (Ai *et al.*, 2005; Klem *et al.*, 2005; Pankhurst *et al.*, 2003; Tartaj *et al.*, 2003; Vartholomeos *et al.*, 2011), and biology (von Schonfedt *et al.*, 1999). In some instances, the fabrication of such nanoparticles has been inspired by the synthesis of the magnetosomes by magnetotactic bacteria. Researchers have involved the proteins that are believed to play a role in the *in vivo* synthesis of magnetosomes for the *in vitro* synthesis of nanomagnets. Furthermore, the assembly of magnetosomes in a chain and their function has inspired the researchers for the use of directed self-assembly and for the development of technologies that bear similar properties to the magnetotactic bacteria. An account of these works is given hereafter.

Biomimetic fabrication of nanomagnets

If a fully biomimetic approach for the fabrication of a magnetic chain by self-assembly is ever to be developed, this will most probably follow a bottom-up approach. Therefore, it is of primary importance to succeed in the biomimetic self-assembly of the building nanoblocks, and to ensure their fabrication with control over their characteristics using biomimetic approaches. This latter step, which is the first step in a fully biomimetic fabrication of magnetic assemblies, has given rise to important works.

Biomimetic synthesis of magnetic nanoparticles offers the possibility of controlling size, shape, crystal structure, orientation and organization using alternative synthesis routes that circumvent the complicated conditions required in the classical synthesis: high reaction temperature, costly reagents, and post processing requirements (Klem *et al.*, 2005). Matsunaga's group have investigated the role of mineral-associated proteins tightly bound to the magnetite crystals of AMB-1 in the initiation of nucleation and magnetic crystal growth (Arakaki *et al.*, 2003). After finding that the iron binding protein Mms6 provides nucleation sites for precipitation of iron oxide in the bacterial magnetic particles, they reported the use of Mms6 for the controlled formation of magnetite crystal (Amemiya *et al.*, 2007) and the use of Mms6 peptides for the growth of magnetite particles with controlled morphology and size following a synthesis route that does not require high temperature or organic solvents (Arakaki *et al.*, 2010). Whereas the analysis of the results obtained *in vitro* are not sufficient to determine the exact function of proteins *in vivo* (Lang *et al.*, 2007), the papers from Matsunaga's group show the use of proteins from magnetotactic bacteria for the biomimetic growth of magnetic nanoparticles (Galloway *et al.*, 2011; Tanaka *et al.*, 2011). Prozorov *et al.* (2007), who have also been working with

Mms6 to promote shape-specific magnetite growth *in vitro*, have reported the use of Mms6 for the bio-inspired synthesis of magnetic nanoparticles that have not been found in magnetotactic bacteria.

Bacterial magnetosomes and biotechnology

Owing to their unique magnetic and biochemical properties, magnetosomes provide numerous attractive possibilities. For example, the production and isolation of magnetosomes permit the development of various applications (Lang and Schueler, 2006). The ease of modification of the magnetosomes has been exploited by Matsugana *et al.* (2001) for the purification, isolation and detection of mRNA and DNA (Matsunaga *et al.*, 2001; Sode *et al.*, 1993; Yoza *et al.*, 2003a,b). Particularly, they observed the superiority of magnetosomes over synthetic magnetic nanoparticles for DNA recovery (Yoza *et al.*, 2003b). Other applications developed by Matsugana *et al.* (Matsunaga and Kamiya, 1987; Matsunaga *et al.*, 2001) arise from the functionalization of the surface of magnetosomes. Two types of approaches have been used. The conventional chemical approach to functionalization is achieved by cross-linking, conjugation of amine-modified oligonucleotides (Sode *et al.*, 1993), immobilization of myosin (Tanaka *et al.*, 1997) and biotinylation (Amemiya *et al.*, 2005; Ceyhan *et al.*, 2006). These functionalized magnetosomes can be used for the selective separation of biological/chemical targets, as analytical probe of chemical environment and for ultra-sensitive detection.

While chemical approaches offer a wide range of possible applications, in the lasts 10 years, researchers have reported superior functionalization using genetic approaches. This involves the construction of genetic fusions of magnetosome membrane anchor polypeptides with functional proteins and enzymes (Lang *et al.*, 2007). The superiority of the genetic approaches over the conventional chemical approaches arises from the better preservation of the protein activity of magnetosomes. This has permitted the display and screening of various enzymes, proteins and compounds (Matsunaga *et al.*, 2006; Matsunaga *et al.*, 2000; Yoshino *et al.*, 2005). For more details about these applications, the reader may refer to recently published extensive reviews about magnetosomes and their applications (Faivre and Schueler, 2008; Lang and Schueler, 2006; Lang *et al.*, 2007; Matsunaga *et al.*, 2007; Varadan *et al.*, 2008).

Genetic fusion on magnetosomes has also permitted the expression of fluorescent proteins by the bacterial proteins (Lang and Schüler, 2008). The fluorescence of the membrane of magnetosomes can be exploited in many screening and detection applications utilizing magnetosomes. Moreover, there are properties inferred by the localized expression of fluorophores in the bacteria that have been, to the author's opinion, seldom exploited. For example, the localization of the fluorophore on a specific protein of the magnetosome assembly, such as EGFP-MamK, could permit detailed imaging of this actin-like filament using super resolution optical microscopy (Izeddin *et al.*, 2011). Furthermore, the fluorescence characteristics (e.g. wavelength, fluorescence lifetime) of the fluorophore could be used to quantify the micro-environment of the proteins *in vivo* using techniques such as fluorescence lifetime imaging microscopy or wavelength ratiometry imaging (Iwai *et al.*, 2010; Olsen *et al.*, 2002) and the interaction of proteins forming the magnetosome assembly could be better understood using fluorescence resonance energy transfer. Because they would help better understanding the role of the individual blocks making up the magnetosome chains *in vivo*, such works would contribute greatly to the development of biomimetic approaches for the synthesis of magnetic nanoparticles and for the self-assembly of magnetic chains *in vitro* using a chemical approach and *in vivo* using a genetic approach.

Finally, the properties inferred by the magnetosome chain assembly on the magnetotactic bacteria permit the control of these latter using an externally applied magnetic field and their tracking using MRI (Martel *et al.*, 2009). They have therefore been proposed for medical nanorobotic applications. Other possible applications result from the use of magnetosomes as contrast agents for MR imaging. Their suitability for this task has been demonstrated by Herborn *et al.* (2003). In such application the synthetic ferrofluids conventionally used would be replaced by magnetosomes. They could also replace magnetic nanoparticles in hyperthermia treatment. Hyperthermia is a technique that requires magnetic nanoparticles that exhibit high power loss following the reorientation of their magnetic moment. The reorientation is typically induced by an externally applied alternating magnetic field and results in the cell necrosis of tumour cells. Hergt *et al.* (2005, 2006) have demonstrated the superior loss of power exhibited by magnetosomes over synthetic particles. In hyperthermia treatment, the use of a chain of magnetosomes rather than individual magnetic nanoparticles has also been reported to be beneficial (Alphandéry *et al.*, 2011a,b).

Assembly of magnetic nanoparticles

Nanoparticles self-assembly refers to the process by which nanoparticles spontaneously organize due to direct specific interaction or indirectly through their environment (Grzelczak *et al.*, 2010). As discussed in 'Physical properties of dipolar magnetic nanoparticles', the formation of a magnetic chain is associated with thermodynamic equilibrium. In this respect, the extent to which, under ambient conditions, the whole protein template is necessary for the formation of a chain of magnetosomes rather than a cluster is still unclear. Proteins play an important role in the formation of individual building blocks and in the stabilization of the magnetosome chain (Katzmann *et al.*, 2010; Scheffel *et al.*, 2006; Scheffel and Schüler, 2007). Therefore, there is a clear interest in the development of biomimetic method for the synthesis of magnetic nanoparticle chains (Klem *et al.*, 2005; Lang *et al.*, 2007; Tamerler and Sarikaya, 2008; Wang *et al.*, 2011) and important advances have been made in identifying proteins and characterizing their role in the assembly of magnetosome chains *in vivo* (Arakaki *et al.*, 2003; Ding *et al.*, 2010; Draper *et al.*, 2011; Katzmann *et al.*, 2010; Murat *et al.*, 2010; Quinlan *et al.*, 2011; Richter *et al.*, 2007; Scheffel *et al.*, 2006; Scheffel and Schüler, 2007; Tanaka *et al.*, 2011; Uebe *et al.*, 2011).

The self-assembly of magnetic nanoparticles have been subjected to bio-inspired approaches in a number of publications. Chen and Liu have reported a bio-inspired synthesis method and successfully fabricated magnetic chains of nanoparticles that exhibit ferromagnetic properties (Liu and Chen, 2008). Their assembly is obtained at high temperature and encapsulated in an amorphous carbon shell. Hartgerink *et al.* (2001) have worked on the development of peptide–amphiphile molecules that self-assemble into cylindrical fibres 6–8 nm in diameter. They have demonstrated the effectiveness of these fibres as template for the growth of magnetite crystals in a linear arrangement that resembles that of magnetotactic bacteria (Sone and Stupp, 2011). DNA, polymers, molecular linkers and sol–gel method have also been used to assist the formation of superparamagnetic and ferromagnetic nanoparticle chain structures (Kinsella and Ivanisevic, 2008; Korth *et al.*, 2006; Nakata *et al.*, 2008; Xiong *et al.*, 2007; Zhang and Wang, 2008; Zhou *et al.*, 2009). A singularly different approach was reported by Lee *et al.* who used a microelectrode to trap, position and orient magnetotactic bacteria. Subsequent removal of the cellular membrane by cell lysis effectively resulted in the deposition of a chain and a ring of magnetic nanoparticles onto a substrate (Lee *et al.*, 2004). Finally, researchers have encapsulated bacterial magnetic nanoparticles on peptide nanotubes (Banerjee *et al.*, 2005). This resulted in a linear chain assembly behaving like a magnetic nanowires that could not be fabricated by ordinary synthetic approaches.

Conclusions

Recently, bacteria have unravelled organization properties that were unsuspected in such primitive form of life. Among bacteria, our focus in this chapter was on magnetosomes, an organelle inferring magnetic properties to the cell. In order to effectively bias its search towards favoured colonizing environment the bacteria control the formation and growth of ferromagnetic crystals and organize the magnetosomes in a chain using bacterial proteins and magnetic interactions. In the last 10 years, numerous papers have reported advances in the understanding of the role and in the biomimetic fabrication of the building blocks of the magnetosome assembly found in magnetotactic bacteria. These have not been yet sufficient for the development of a fully biomimetic self-assembly fabrication of magnetic chains inspired by magnetotactic bacteria. Nonetheless, as shown in this chapter, there is a clear trend and interest towards such goal and many scientists are devoting their work to this field of research. Advances in the field will greatly benefit numerous microtechnological applications in microrobotics, medical sciences and material sciences.

References

Ai, H., Flask, C., Weinberg, B., Shuai, X., Pagel, M.D., Farrell, D., Duerk, J., and Gao, J.M. (2005). Magnetite-loaded polymeric micelles as ultrasensitive magnetic-resonance probes. Adv. Mater. *17*, 1949-+.

Alphandéry, E., Faure, S., Raison, L., Duguet, E., Howse, P.A., and Bazylinski, D.A. (2011a). Heat production by bacterial magnetosomes exposed to an oscillating magnetic field. J. Phys. Chem. C *115*, 18–22.

Alphandéry, E., Faure, S., Seksek, O., Guyot, F., and Chebbi, I. (2011b). Chains of magnetosomes extracted from amb-1 magnetotactic bacteria for application in alternative magnetic field cancer therapy. ACS Nano *5*, 6279–6296.

Amemiya, Y., Tanaka, T., Yoza, B., and Matsunaga, T. (2005). Novel detection system for biomolecules using nano-sized bacterial magnetic particles and magnetic force microscopy. J. Biotechnol. *120*, 308–314.

Amemiya, Y., Arakaki, A., Staniland, S.S., Tanaka, T., and Matsunaga, T. (2007). Controlled formation of magnetite crystal by partial oxidation of ferrous hydroxide in the presence of recombinant magnetotactic bacterial protein Mms6. Biomaterials *28*, 5381–5389.

Arakaki, A., Webb, J., and Matsunaga, T. (2003). A novel protein tightly bound to bacterial magnetic particles in *Magnetospirillum magneticum* strain AMB-1. J. Biol. Chem. *278*, 8745–8750.

Arakaki, A., Masuda, F., Amemiya, Y., Tanaka, T., and Matsunaga, T. (2010). Control of the morphology and size of magnetite particles with peptides mimicking the Mms6 protein from magnetotactic bacteria. J. Colloid Interface Sci. *343*, 65–70.

Bahaj, A.S., James, P.A.B., and Moeschler, F.D. (1998). Low magnetic-field separation system for metal-loaded magnetotactic bacteria. J. Magn. Magn. Mater. *177–181, Part 2*, 1453–145.

Banerjee, I.A., Shima, L.Y.M., Yoshino, T., Takeyama, H., Matsunaga, T., and Matsui, H. (2005). Magnetic nanotube fabrication by using bacterial magnetic nanocrystals. Advanced Materials *17*, 1128.

Baumgartner, J., Lesevic, P., Kumari, M., Halbmair, K., Bennet, M., Körnig, A., Widdrat, M., Andert, J., Wollgarten, M., Bertinetti, L., *et al.* (2012). From magnetotactic bacteria to hollow spirilla-shaped silica containing a magnetic chain. RSC Adv. *2*, 8007–8009.

Bazylinski, D.A. (1995). Structure and function of the bacterial magnetosome ASM News *61*, 337–343.

Bazylinski, D.A., and Frankel, R.B. (2004). Magnetosome formation in prokaryotes. Nat. Rev. Microbiol. *2*, 217–230.

Bazylinski, D.A., and Schübbe, S. (2007). Controlled Biomineralization by and Applications of Magnetotactic Bacteria, pp. 21–62.

Bazylinski, D.A., Garratt-Reed, A.J., Abedi, A., and Frankel, R.B. (1993a). Copper associated with iron sulfide magnetosomes in a magnetotactic barium. Archiv. Microbiol. *160*, 35–42.

Bazylinski, D.A., Heywood, B.R., Mann, S., and Frankel, R.B. (1993b). Fe3O4 and Fe3S4 in a bacterium [13]. Nature *366*, 218.

Bazylinski, D.A., Garratt-Reed, A.J., and Frankel, R.B. (1994). Electron microscopic studies of magnetosomes in magnetotactic bacteria. Microsc. Res. Tech. *27*, 389–401.

Black, C.T., Murray, C.B., Sandstrom, R.L., and Sun, S.H. (2000). Spin-dependent tunneling in self-assembled cobalt-nanocrystal superlattices. Science *290*, 1131–1134.

Blakemore, R.P. (1975). Magnetotactic bacteria. Science *190*, 377–379.

Butler, R.F., and Banerjee, S.K. (1975). Theoretical single-domain grain-size range in magnetite and titanomagnetite. J. Geophys. Res. *80*, 4049–4058.

Butter, K., Bomans, P.H.H., Frederik, P.M., Vroege, G.J., and Philipse, A.P. (2003). Direct observation of dipolar chains in iron ferrofluids by cryogenic electron microscopy. Nature Materials *2*, 88–91.

Byrne, M.E., Ball, D.A., Guerquin-Kern, J.L., Rouiller, I., Wu, T.D., Downing, K.H., Vali, H., and Komeili, A. (2011). Desulfovibrio magneticus RS-1 contains an iron- and phosphorus-rich organelle distinct from its bulletshaped magnetosomes. Proc. Natl. Acad. Sci. U.S.A. *107*, 12263–12268.

Carballido-López, R. (2006). The bacterial actin-like cytoskeleton. Microbiol. Mol. Biol. Rev.s *70*, 888–909.

Ceyhan, B., Alhorn, P., Lang, C., Schüler, D., and Niemeyer, C.M. (2006). Semisynthetic biogenic magnetosome nanoparticles for the detection of proteins and nucleic acids. Small 2, 1251–1255.

Corchero, J.L., and Villaverde, A. (2009). Biomedical applications of distally controlled magnetic nanoparticles. Trends Biotechnol. *27*, 468–476.

Das, R., Gupta, A., Kumar, D., Hoh, S., Pennycook, S.J., and Hebard, A.F. (2008). Dipolar interactions and their influence on the critical single domain grain size of Ni in layered Ni/Al(2)O(3) composites. J. Phys. Condensed Matter *20*.

de Vicente, J., Klingenberg, D.J., and Hidalgo-Alvarez, R. (2011). Magnetorheological fluids: a review. Soft Matter *7*, 3701–3710.

Denham, C.R., Blakemore, R.P., and Frankel, R.B. (1980). Bulk magnetic-properties of magnetotactic bacteria. IEEE Trans. Magn. *16*, 1006–1007.

Devouard, B., Posfai, M., Hua, X., Bazylinski, D.A., Frankel, R.B., and Buseck, P.R. (1998). Magnetite from magnetotactic bacteria: Size distributions and twinning. Am. Mineral. *83*, 1387–1398.

Dhar, P., Cao, Y., Kline, T., Pal, P., Swayne, C., Fischer, T.M., Miller, B., Mallouk, T.E., Sen, A., and Johansen, T.H. (2007). Autonomously moving local nanoprobes in heterogeneous magnetic fields. J. Phys. Chem. C *111*, 3607–3613.

Ding, Y., Li, J., Liu, J., Yang, J., Jiang, W., Tian, J., Li, Y., Pan, Y., and Li, J. (2010). Deletion of the ftsZ-Like Gene Results in the Production of Superparamagnetic Magnetite Magnetosomes in Magnetospirillum gryphiswaldense. J. Bacteriol. *192*, 1097–1105.

Draper, O., Byrne, M.E., Li, Z., Keyhani, S., Barrozo, J.C., Jensen, G., and Komeili, A. (2011). MamK, a bacterial actin, forms dynamic filaments in vivo that are regulated by the acidic proteins MamJ and LimJ. Mol. Microbiol. *82*, 342–354.

Dreyfus, R., Baudry, J., Roper, M.L., Fermigier, M., Stone, H.A., and Bibette, J. (2005). Microscopic artificial swimmers. Nature *437*, 862–865.

Eisenbach, M. (2001). Bacterial Chemotaxis. In eLS (John Wiley & Sons, Ltd).

Erb, R.M., Libanori, R., Rothfuchs, N., and Studart, A.R. (2012). Composites reinforced in three dimensions by using low magnetic fields. Science *335*, 199–204.

Faivre, D., and Schueler, D. (2008). Magnetotactic bacteria and magnetosomes. Chem. Rev. *108*, 4875–4898.

Faivre, D., and Zuddas, P. (2006). An integrated approach for determining the origin of magnetite nanoparticles. Earth Planet Sci. Lett. *243*, 53–60.

Faivre, D., and Zuddas, P. (2007). Mineralogical and Isotopic Properties of Biogenic NanocrystallineMagnetites. In Magnetoreception and Magnetosomes in Bacteria, D. Schüler, ed. (Heidelberg, Springer), pp. 175–196.

Faivre, D., Agrinier, P., Menguy, N., Zuddas, P., Pachana, K., Gloter, A., Laval, J.-Y., and Guyot, F. (2004). Mineralogical and isotopic properties of inorganic nanocrystalline magnetites. Geochim. Cosmochim. Acta *68*, 4395–4403.

Faivre, D., Böttger, L.H., Matzanke, B.F., and Schüler, D. (2007). intracellular magnetite biomineralization in bacteria proceeds by a distinct pathway involving membrane-bound ferritin and an iron(II) species. Angew. Chem. Int. Ed. *46*, 8495–8499.

Faivre, D., Menguy, N., Pósfai, M., and Schüler, D. (2008). Environmental parameters affect the physical properties of fast-growing magnetosomes. Am. Mineral. *93*, 463–469.

Fan, T.-X., Chow, S.-K., and Zhang, D. (2009). Biomorphic mineralization: From biology to materials. Prog. Mater. Sci. *54*, 542–659.

Figge, R.M., Divakaruni, A.V., and Gober, J.W. (2004). MreB, the cell shape-determining bacterial actin homologue, co-ordinates cell wall morphogenesis in *Caulobacter crescentus*. Mol. Microbiol. *51*, 1321–1332.

Fischer, A., Schmitz, M., Aichmayer, B., Fratzl, P., and Faivre, D. (2011). Structural purity of magnetite nanoparticles in magnetotactic bacteria. J. Roy. Soc. Interface *8*, 1011–1018.

Frankel, R., Williams, T., and Bazylinski, D. (2007). Magneto-Aerotaxis Magnetoreception and Magnetosomes in Bacteria. In, D. Schüler, ed. (Springer Berlin/Heidelberg), pp. 1–24.

Frankel, R.B. (1984). Magnetic Guidance of Organisms. Annu. Rev. Biophys. Bioeng. *13*, 85–103.

Frankel, R.B., and Bazylinski, D.A. (2006). How magnetotactic bacteria make magnetosomes queue up. Trends Microbiol. *14*, 329–331.

Frankel, R.B., and Blakemore, R.P. (1980). Navigational compass in magnetic bacteria. J. Magnet. Magnet. Mater. *15–18*, 1562–1564.

Frankel, R.B., Blakemore, R., and Wolfe, R.S. (1979). Magnetite in freshwater magnetotactic bacteria. Science *203*, 1355–1356.

Frankel, R.B., Blakemore, R.P., Dearaujo, F.F.T., Esquivel, D.M.S., and Danon, J. (1981). Magnetotactic bacteria at the geomagnetic equator. Science *212*, 1269–1270.

Frankel, R.B., Papaefthymiou, G.C., Blakemore, R.P., and O'Brien, W. (1983). Fe3O4 precipitation in magnetotactic bacteria. BBA – Mol. Cell Res. *763*, 147–159.

Fukuda, Y., Okamura, Y., Takeyama, H., and Matsunaga, T. (2006). Dynamic analysis of a genomic island in *Magnetospirillum* sp. strain AMB-1 reveals how magnetosome synthesis developed. FEBS Lett. *580*, 801–812.

Galloway, J.M., Arakaki, A., Masuda, F., Tanaka, T., Matsunaga, T., and Staniland, S.S. (2011). Magnetic bacterial protein Mms6 controls morphology, crystallinity and magnetism of cobalt-doped magnetite nanoparticles in vitro. J. Mater. Chem. *21*, 15244–15254.

Gassman, P.G., Tan, L., and Hoye, T.R. (1996). Intramolecular cationic cyclizations initiated by electrocyclic cleavage of cyclopropanes. Synthesis of trienic cyclopentane derivatives. Tetrahedron Letters *37*, 439–442.

Ghosh, A., and Fischer, P. (2009). Controlled propulsion of artificial magnetic nanostructured propellers. Nano Lett. *9*, 2243–2245.

GilVillegas, A., McGrother, S.C., and Jackson, G. (1997). Chain and ring structures in smectic phases of molecules with transverse dipoles. Chem. Phys. Lett. *269*, 441–447.

Gitai, Z. (2005). The new bacterial cell biology: Moving parts and subcellular architecture. Cell *120*, 577–586.

Gitai, Z. (2007). Diversification and specialization of the bacterial cytoskeleton. Curr. Opin. Cell Biol. *19*, 5–12.

Gorby, Y.A., Beveridge, T.J., and Blakemore, R.P. (1988). Characterization of the bacterial magnetosome membrane. J. Bacteriol. *170*, 834–841.

Gruenberg, K., Mueller, E.C., Otto, A., Reszka, R., Linder, D., Kube, M., Reinhardt, R., and Schueler, D. (2004). Biochemical and proteomic analysis of the magnetosome membrane in *Magnetospirillum gryphiswaldense*. Appl. Environ. Microbiol. *70*, 1040–1050.

Grünberg, K., Wawer, C., Tebo, B.M., and Schüler, D. (2001). A large gene cluster encoding several magnetosome proteins is conserved in different species of magnetotactic bacteria. Appl. Environ. Microbiol. *67*, 4573–4582.

Grzelczak, M., Vermant, J., Furst, E.M., and Liz-Marzan, L.M. (2010). Directed self-assembly of nanoparticles. ACS Nano *4*, 3591–3605.

Hanzlik, M., Winklhofer, M., and Petersen, N. (1996). Spatial arrangement of chains of magnetosomes in magnetotactic bacteria. Earth Planet. Sci. Lett. *145*, 125–134.

Hanzlik, M., Winklhofer, M., and Petersen, N. (2002). Pulsed-field-remanence measurements on individual magnetotactic bacteria. J. Magn. Magn. Mater. *248*, 258–267.

Hartgerink, J.D., Beniash, E., and Stupp, S.I. (2001). Self-assembly and mineralization of peptide-amphiphile nanofibers. Science *294*, 1684–1688.

Herborn, C.U., Papanikolaou, N., Reszka, R., Grünberg, K., Schüler, D., and Debatin, J.F. (2003). Magnetosomes as biological model for iron binding: relaxivity determination with MRI. Rofo Fortschr Geb Rontgenst Neuen Bildgeb Verfahr *175*, 830–834.

Hergt, R., Dutz, S., Müller, R., and Zeisberger, M. (2006). Magnetic particle hyperthermia:nanoparticle magnetism and materials development for cancer therapy. J. Phys: Condens. Matter *18*, S2919-S2934.

Hergt, R., Hiergeist, R., Zeisberger, M., Schüler, D., Heyen, U., Hilger, I., and Kaiser, W.A. (2005). Magnetic properties of bacterial magnetosomes as potential diagnostic and therapeutic tools. J. Magn. Magn. Mater. *293*, 80–86.

Heyen, U., and Schüler, D. (2003). Growth and magnetosome formation by microaerophilic *Magnetospirillum* strains in an oxygen-controlled fermentor. Appl. Microbiol. Biotechnol. *61*, 536–544.

Isambert, A., Menguy, N., Larquet, E., Guyot, F., and Valet, J.P. (2007). Transmission electron microscopy study of magnetites in a freshwater population of magnetotactic bacteria. Am. Mineral. *92*, 621–630.

Iwai, M., Yokono, M., Inada, N., and Minagawa, J. (2010). Live-cell imaging of photosystem II antenna dissociation during state transitions. Proc. Natl. Acad. Sci. U.S.A. *107*, 2337–2342.

Izeddin, I., Specht, C.G., Lelek, M., Darzacq, X., Triller, A., Zimmer, C., and Dahan, M. (2011). Super-resolution dynamic imaging of dendritic spines using a low-affinity photoconvertible actin probe. PLoS ONE *6*, e15611.

Jackson, J.D. (1999). Classical electrodynamics, 3rd edn (John Wiley & Sons).

Jogler, C., Kube, M., Schübbe, S., Ullrich, S., Teeling, H., Bazylinski, D.A., Reinhardt, R., and Schüler, D. (2009a). Comparative analysis of magnetosome gene clusters in magnetotactic bacteria provides further evidence for horizontal gene transfer. Environ. Microbiol. *11*, 1267–1277.

Jogler, C., Lin, W., Meyerdierks, A., Kube, M., Katzmann, E., Flies, C., Pan, Y., Amann, R., Reinhardt, R., and Schuler, D. (2009b). Towards cloning of the magnetotactic metagenome: identification of magnetosome island gene clusters in uncultivated magnetotactic bacteria from different aquatic sediments. Appl. Environ. Microbiol. *75*, 3972–3979.

Jogler, C., Niebler, M., Lin, W., Kube, M., Wanner, G., Kolinko, S., Stief, P., Beck, A.J., de Beer, D., Petersen, N., *et al.* (2010). Cultivation-independent characterization of '*Candidatus Magnetobacterium bavaricum*' via ultrastructural, geochemical, ecological and metagenomic methods. Environ. Microbiol. *12*, 2466–2478.

Jogler, C., and Schueler, D. (2009). Genomics, genetics, and cell biology of magnetosome formation. Annu. Rev. Microbiol. *63*, 501–521.

Jones, L.J., Carballido-López, R., and Errington, J. (2001). Control of cell shape in bacteria: helical, actin-like filaments in *Bacillus subtilis*. Cell *104*, 913922.

Jund, P., Kim, S.G., Tomanek, D., and Hetherington, J. (1995). Stability and fragmentation of complex structures in ferrofluids. Phys. Rev. Let. *74*, 3049–3052.

Katzmann, E., Mueller, F.D., Lang, C., Messerer, M., Winklhofer, M., Plitzko, J.M., and Schueler, D. (2011). Magnetosome chains are recruited to cellular division sites and split by asymmetric septation. Mol. Microbiol. *8*, 1316–1329.

Katzmann, E., Scheffel, A., Gruska, M., Plitzko, J.M., and Schüler, D. (2010). Loss of the actin-like protein MamK has pleiotropic effects on magnetosome formation and chain assembly in *Magnetospirillum gryphiswaldense*. Mol. Microbiol. *77*, 208–224.

Keim, C.N., Lins, U., and Farina, M. (2009). Manganese in biogenic magnetite crystals from magnetotactic bacteria. FEMS Microbiol. Lett. *292*, 250–253.

Keim, C.N., Solorzano, G., Farina, M., and Lins, U. (2005). Intracellular inclusions of uncultured magnetotactic bacteria. Int. Microbiol. *8*, 111–117.

Kinsella, J.M., and Ivanisevic, A. (2008). Magnetotransport of one-dimensional chains of CoFe2O4 nanoparticles ordered along DNA. J. Phys. Chem. C *112*, 3191–3193.

Kirschvink, J.L. (1980). South-seeking magnetic bacteria. J. Exp. Biol. *86*, 345–347.

Kirschvink, J.L. (1982). Paleomagnetic evidence for fossil biogenic magnetite in western crete. Earth Planet. Sci. Lett. *59*, 388–392.

Klem, M.T., Young, M., and Douglas, T. (2005). Biomimetic magnetic nanoparticles. Mater. Today *8*.

Kline, T.R., Paxton, W.F., Mallouk, T.E., and Sen, A. (2005). Catalytic nanomotors: Remote-controlled autonomous movement of striped metallic nanorods. Angewandte Chemie-International Edition *44*, 744–746.

Klumpp, S., and Faivre, D. (2012). Interplay of magnetic interactions and active movements in the formation of magnetosome chains. PLoS ONE *7*, e33562.

Komeili, A. (2007). Molecular mechanisms of magnetosome formation. Annu. Rev. Biochem. *76*, 351–366.

Komeili, A. (2012). Molecular mechanisms of compartmentalization and biomineralization in magnetotactic bacteria. FEMS Microbiol. Rev. *36*, 232–255.

Komeili, A., Li, Z., Newman, D.K., and Jensen, G.J. (2006). Magnetosomes are cell membrane imaginations organized by the actin-like protein MamK. Science *311*, 242–245.

Komeili, A., Vali, H., Beveridge, T.J., and Newman, D.K. (2004). Magnetosome vesicles are present before magnetite formation, and MamA is required for their activation. Proc. Natl. Acad. Sci. U.S.A. *101*, 3839–3844.

Korth, B.D., Keng, P., Shim, I. Bowles, S.E., Tang, C., Kowalewski, T., Nebesny, K.W., and Pyun, J. (2006). Polymer-coated ferromagnetic colloids from well-defined macromolecular surfactants and assembly into nanoparticle chains. J Am. Chem. Soc. *128*, 6562–6563.

Lang, C., and Schueler, D. (2006). Biogenic nanoparticles: production, characterization, and application of bacterial magnetosomes. J. Phys. Condensed Matter *18*, S2815-S2828.

Lang, C., Schueler, D., and Faivre, D. (2007). Synthesis of magnetite nanoparticles for bio- and nanotechnology: Genetic engineering and biomimetics of bacterial magnetosomes. Macromol. Biosci. *7*, 144–151.

Lang, C., and Schüler, D. (2008). Expression of green fluorescent protein fused to magnetosome proteins in microaerophilic magnetotactic bacteria. Appl Environ Microbiol *74* 4944–4953.

Lao, L.L., and Ramanujan, R.V. (2004). Magnetic and hydrogel composite materials for hyperthermia applications. Journal of Materials Science-Materials in Medicine *15*, 1061–1064.

Lee, C.S., Lee, H., and Westervelt, R.M. (2001). Microelectromagnets for the control of magnetic nanoparticles. Appl. Phys. Lett. *79*, 3308–3310.

Lee, H., Purdon, A.M., Chu, V., and Westervelt, R.M. (2004). Controlled assembly of magnetic nanoparticles from magnetotactic bacteria using microelectromagnets arrays. Nano Lett. *4*, 995–998.

Lefèvre, C.T., Menguy, N., Abreu, F., Lins, U., Pósfai, M., Prozorov, T., Pignol, D., Frankel, R.B., and Bazylinski, D.A. (2011). A cultured greigite-producing magnetotactic bacterium in a novel group of sulfate-reducing bacteria. Science *334*, 1720–1723.

Levesque, D., and Weis, J.J. (1994). Orientational and structural order in strongly interacting dipolar hard-spheres. Phys. Rev. E *49*, 5131–5140.

Lins, U., and Farina, M. (2004). Magnetosome chain arrangement and stability in magnetotactic cocci. Antonie Van Leeuwenhoek Int. J. Gen. Mol. Microbiol. *85*, 335–341.

Liu, T.-Y., Hu, S.-H., Liu, T.-Y., Liu, D.-M., and Chen, S.-Y. (2006). Magnetic-sensitive behaviour of intelligent ferrogels for controlled release of drug. Langmuir *22*, 5974–5978.

Liu, Y.R., and Chen, Q.W. (2008). Synthesis of magnetosome chain-like structures. Nanotechnology *19*, 475603.

Lohße, A., Ullrich, S., Katzmann, E., Borg, S., Wanner, G., Richter, M., Voigt, B., Schweder, T., and Schueler, D. (2011). Functional analysis of the magnetosome Island in magnetospirillum gryphiswaldense: The mamAB operon is sufficient for magnetite biomineralization. PLoS ONE *6*.

McGrother, S.C., and Jackson, G. (1996). Island of vapor-liquid coexistence in dipolar hard-core systems. Phys. Rev. Let. *76*, 4183–4186.

Mann, S., Sparks, N.H.C., and Blakemore, R.P. (1987). Ultrastructure and characterization of anisotropic magnetic inclusions in magnetotactic bacteria. Proc. Roy. Soc. Lond. B *231*, 469–476.

Mann, S., Sparks, N.H.C., Couling, S.B., Larcombe, M.C., and Frankel, R.B. (1989). Crystallochemical characterization of magnetic spinels prepared from aqueous solution. Journal of the Chemical Society, Faraday Transactions 1: Physical Chemistry in Condensed Phases *85*, 3033–3044.

Martel, S., Mohammadi, M., Felfoul, O., Lu, Z., and Pouponneau, P. (2009). Flagellated Magnetotactic Bacteria as Controlled MRI-trackable Propulsion and Steering Systems for Medical Nanorobots Operating in the Human Microvasculature. Int J Robot Res *28*, 571–582.

Martel, S., Tremblay, C.C., Ngakeng, S., and Langlois, G. (2006). Controlled manipulation and actuation of micro-objects with magnetotactic bacteria. Appl. Phys. Lett. *89*.

Matsuda, H., and Egami, T. (1983). Atomic structure of zero-magnetorestrictive amorphous Fe4.7Co70.3Si15B10 alloy. Materi. Sci. Eng. *60*, 87–93.

Matsunaga, T., and Kamiya, S. (1987). Use of magnetic particles isolated from magnetotactic bacteria for enzyme immobilization. Appl. Microbiol. Biotechnol. *26*, 328–332.

Matsunaga, T., Togo, H., Kikuchi, T., and Tanaka, T. (2000). Production of luciferase-magnetic particle complex by recombinant *Magnetospirillum* sp AMB-1. Biotechnol. Bioeng. *70*, 704–709.

Matsunaga, T., Nakayama, H., Okochi, M., and Takeyama, H. (2001). Fluorescent detection of cyanobacterial DNA using bacterial magnetic particles on a MAG-microarray. Biotechnol Bioeng *73*, 400–405.

Matsunaga, T., Takahashi, M., Yoshino, T., Kuhara, M., and Takeyama, H. (2006). Magnetic separation of CD14+ cells using antibody binding with protein A expressed on bacterial magnetic particles for generating dendritic cells. Biochem. Biophys. Res. Commun. *350*, 1019–1025.

Matsunaga, T., Suzuki, T., Tanaka, M., and Arakaki, A. (2007). Molecular analysis of magnetotactic bacteria and development of functional bacterial magnetic particles for nano-biotechnology. Trends Biotechnol *25*, 182–188.

Mavroidis, C., and Ferreira, A. (2009). Special Issue on Current State of the Art and Future Challenges in Nanorobotics. Int J Robot Res *28*, 419–420.

Meldrum, F.C., Mann, S., Heywood, B.R., Frankel, R.B., and Bazylinski, D.A. (1993a). Electron microscopy study of magnetosomes in a cultured coccoid magnetotactic bacterium. Proceedings of the Royal Society B: Biological Sciences *251*, 231–236.

Meldrum, F.C., Mann, S., Heywood, B.R., Frankel, R.B., and Bazylinski, D.A. (1993b). Electron microscopy study of magnetosomes in two cultured vibrioid magnetotactic bacteria. Proceedings of the Royal Society B: Biological Sciences *251*, 237–242.

Moench, T.T. (1988). Bilophoccus-magnetotacticus Gen-Nov-Sp-Nov, a motile, magnetic coccus. Antonie Van Leeuwenhoek Journal of Microbiology *54*, 483–496.

Murat, D., Quinlan, A., Vali, H., and Komeili, A. (2010). Comprehensive genetic dissection of the magnetosome gene island reveals the step-wise assembly of a prokaryotic organelle. Proc. Natl. Acad. Sci. U.S.A. *107*, 5593–5598.

Nakamura, N., and Matsunaga, T. (1993). Highly sensitive detection of allergen using bacterial magnetic particles. Analytica Chimica Acta *281*, 585–589.

Nakata, K., Hu, Y., Uzun, O., Bakr, O., and Stellacci, F. (2008). Chains of Superparamagnetic Nanoparticles. Adv *20* Mater 4294–4299.

Nakazawa, H., Arakaki, A., Narita-Yamada, S., Yashiro, I., Jinno, K., Aoki, N., Tsuruyama, A., Okamura, Y., Tanikawa, S., Fujita, N., *et al.* (2009). Whole genome sequence of Desulfovibrio magneticus strain RS-1 revealed common gene clusters in magnetotactic bacteria. Genome Research *19*, 1801–1808.

Olsen, K.N., Budde, B.B., Siegumfeldt, H., Rechinger, K.B., Jakobsen, M., and Ingmer, H. (2002). Noninvasive measurement of bacterial intracellular pH on a single-cell level with green fluorescent protein and fluorescence ratio imaging microscopy. Applied and Environmental *68* Microbiology 4145–4147.

Palache, C. (1944). Crystallography of calcite from Lake Superior copper mines. American Mineralogist *29*, 35–48.

Pamme, N. (2006). Magnetism and microfluidics. Lab on a Chip *6*, 24–38.

Pankhurst, Q.A., Connolly, J., Jones, S.K., and Dobson, J. (2003). Applications of magnetic nanoparticles in biomedicine. Journal of Physics D-Applied Physics *36*, R167-R181.

Pawashe, C., Floyd, S., and Sitti, M. (2009a). Modelling and Experimental Characterization of an Untethered Magnetic Micro-Robot. Int J Robot Res *28*, 1077–1094.

Pawashe, C., Floyd, S., and Sitti, M. (2009b). Multiple magnetic microrobot control using electrostatic anchoring. Appl. Phys. Lett. *94*.

Philipse, A.P., and Maas, D. (2002). Magnetic colloids from magnetotactic bacteria: Chain formation and colloidal stability. Langmuir *18*, 9977–9984.

Posfai, M., Buseck, P.R., Bazylinski, D.A., and Frankel, R.B. (1998). Iron sulfides from magnetotactic bacteria: Structure, composition, and phase transitions. Am. Mineral. *83*, 1469–1481.

Posfai, M., Moskowitz, B.M., Arato, B., Schueler, D., Flies, C., Bazylinski, D.A., and Frankel, R.B. (2006). Properties of intracellular magnetite crystals produced by *Desulfovibrio magneticus* strain RS-1. Earth Planet.y Sci. Lett. *249*, 444–455.

Pradel, N., Santini, C.L., Bernadac, A., Fukumori, Y., and Wu, L.F. (2006). Biogenesis of actin-like bacterial cytoskeletal filaments destined for positioning prokaryotic magnetic organelles. Proc. Natl. Acad. Sci. U.S.A. *103*, 17485–17489.

Prozorov, T., Mallapragada, S.K., Narasimhan, B., Wang, L., Palo, P., Nilsen-Hamilton, M., Williams, T.J., Bazlinski, D.A., Prozorov, R., and Canfield, D.E. (2007). Protein-mediated synthesis of uniform superparamagnetic magnetite nanocrystals. Adv. Funct. Mater. *17*, 951–957.

Quinlan, A., Murat, D., Vali, H., and Komeili, A. (2011). The HtrA/DegP family protease MamE is a bifunctional protein with roles in magnetosome protein localization and magnetite biomineralization. Mol. Microbiol. *80*, 1075–1087.

Ramanujan, R.V., and Lao, L.L. (2006). The mechanical behavior of smart magnet-hydrogel composites. Smart Materials & Structures *15,* 952–956.

Richter, M., Kube, M., Bazylinski, D.A., Lombardot, T., Glockner, F.O., Reinhardt, R., and Schüler, D. (2007). Comparative genome analysis of four magnetotactic bacteria reveals a complex set of group-specific genes implicated in magnetosome biomineralization and function. J. Bacteriol. *189,* 4899–4910.

Sanchez, R.D., Rivas, J., Vaqueiro, P., Lopez-Quintela, M.A., and Caeiro, D. (2002). Particle size effects on magnetic garnets prepared by a properties of yttrium iron sol–gel method. J. Magn. Magn. Mater. *247,* 92–98.

Scheffel, A., Gruska, M., Faivre, D., Linaroudis, A., Plitzko, J.M., and Schüler, D. (2006). An acidic protein aligns magnetosomes along a filamentous structure in magnetotactic bacteria. Nature *440,* 110–114.

Scheffel, A., Guerdes, A., Gruenberg, K., Wanner, G., and Schueler, D. (2008). The major magnetosome proteins MamGFDC are not essential for magnetite biomineralization in Magnetospirillum gryphiswaldense but regulate the size of magnetosome crystals. J. Bacteriol. *190,* 377–386.

Scheffel, A., and Schüler, D. (2007). The Acidic Repetitive Domain of the *Magnetospirillum gryphiswaldense* MamJ protein displays hypervariability but is not required for magnetosome chain assembly. J. Bacteriol. *189,* 6437–6446.

Schmidt, N., Mishra, A., Lai, G.H., and Wong, G.C.L. (2010). Arginine-rich cell-penetrating peptides. FEBS Lett.*584,* 1806–1813.

Schübbe, S., Kube, M., Scheffel, A., Wawer, C., Heyen, U., Meyerdierks, A., Madkour, M.H., Mayer, F., Reinhardt, R., and Schüler, D. (2003). Characterization of a spontaneous nonmagnetic mutant of Magnetospirillum gryphiswaldense reveals a large deletion comprising a putative magnetosome island. J. Bacteriol. *185,* 5779–5790.

Schübbe, S., Williams, T.J., Xie, G., Kiss, H.E., Brettin, T.S., Martinez, D., Ross, C.A., Schüler, D., Cox, B.L., Nealson, K.H., *et al.* (2009). Complete genome sequence of the chemolithoautotrophic marine *Magnetotactic coccus* strain MC-1. Appl. Environ. Microbiol. 75, 4835–4852.

Schübbe, S., Würdemann, C., Peplies, J., Heyen, U., Wawer, C., Gloeckner, F.O., and Schüler, D. (2006). Transcriptional organization and regulation of magnetosome operons in Magnetospirillum gryphiswaldense. Appl. Environ. Microbiol. *72,* 5757–5765.

Schüler, D. (2004). Molecular analysis of a subcellular compartment: The magnetosome membrane in Magnetospirillum gryphiswaldense. Archiv. Microbiol. *181,* 1–7.

Schüler, D. (2008). Genetics and cell biology of magnetosome formation in magnetotactic bacteria. FEMS Microbiol. Rev. *32,* 654–672.

Schüler, D., and Baeuerlein, E. (1996). Iron-limited growth and kinetics of iron uptake in Magnetospirilum gryphiswaldense. Arch Microbiol *166,* 301–307.

Shechter, E., and Martel, S. (2010). Principles of motion control of bacterial micro-robots using oxygen gradients. 2010 IEEE/ASME International Conference on Advanced Intelligent Mechatronics (AIM 2010).

Sitti, M. (2009). Miniature devices: Voyage of the microrobots. Nature *458,* 1121–1122.

Sode, K., Kudo, S., Sakaguchi, T., Nakamura, N., and Matsunaga, T. (1993). Application of bacterial magnetic particles for highly selective messenger-RNA recovery-system. Biotechnology Techniques *7,* 688–694.

Sone, E.D., and Stupp, S.I. (2011). Bioinspired Magnetite Mineralization of Peptide–Amphiphile Nanofibers. Chem Mater *23,* 2005–2007.

Sonkaria, S., Fuentes, G., Verma, C., Narang, R., Khare, V., Fischer, A., and Faivre, D. (2012). Insight into the assembly properties and functional organization of the magnetotatic bacterial actin-like homologue, MamK. PLoS ONE *7,* e34189.

Spring, S., Amann, R., Ludwig, W., Schleifer, K.H., Vangemerden, H., and Petersen, N. (1993). Dominating role of an unusual magnetotactic bacterium in the microaerobic zone of a fresh-water sediment. Appl. Environ. Microbiol. *59,* 2397–2403.

Staniland, S., Williams, W., Telling, N., Van Der Laan, G., Harrison, A., and Ward, B. (2008). Controlled cobalt doping of magnetosomes in vivo. Nature Nanotechnology *3,* 158–162.

Stolz, J.F. (1993). Magnetosomes. Journal of General Microbiology *139,* 1663–1670.

Sun, S.H., Murray, C.B., Weller, D., Folks, L., and Moser, A. (2000). Monodisperse FePt nanoparticles and ferromagnetic FePt nanocrystal superlattices. Science *287,* 1989–1992.

Tamerler, C., and Sarikaya, M. (2008). Molecular Biomimetics: Genetic Synthesis, Assembly, and Formation of Materials Using Peptides. MRS Bull 33, 504–512.

Tanaka, M., Arakaki, A., and Matsunaga, T. (2010). Identification and functional characterization of liposome tubulation protein from magnetotactic bacteria. Mol. Microbiol. *76*, 480–488.

Tanaka, M., Mazuyama, E., Arakaki, A., and Matsunaga, T. (2011). Mms6 protein regulates crystal morphology during nano-sized magnetite biomineralization in vivo. J. Biol. Chem. *286*, 6386–6392.

Tanaka, M., Okamura, Y., Arakaki, A., Tanaka, T., Takeyama, H., and Matsunaga, T. (2006). Origin of magnetosome membrane: proteomic analysis of magnetosome membrane and comparison with cytoplasmic membrane. Proteomics *6*, 5234–5247.

Tanaka, T., Yamasaki, H., Tsujimura, N., Nakamura, N., and Matsunaga, T. (1997). Magnetic control of bacterial magnetite–myosin conjugate movement on actin cables. Mater Sci & Eng C – Biomimetic Materials Sensors and Systems *5*, 121–124.

Tartaj, P., Morales, M.D., Veintemillas-Verdaguer, S., Gonzalez-Carreno, T., and Serna, C.J. (2003). The preparation of magnetic nanoparticles for applications in biomedicine. Journal of Physics D-Applied Physics *36*, R182-R197.

Towe, K.M., and Moench, T.T. (1981). Electron-optical characterization of bacterial magnetite. Earth Planet. Sci. Lett. *52*, 213–220.

Tripp, S.L., Dunin-Borkowski, R.E., and Wei, A. (2003). Flux closure in self-assembled cobalt nanoparticle rings. Angew. Chem. Int. Ed. *42*, 5591–5593.

Uebe, R., Junge, K., Henn, V., Poxleitner, G., Katzmann, E., Plitzko, J.M., Zarivach, R., Kasama, T., Wanner, G., Pósfai, M., *et al.* (2011). The cation diffusion facilitator proteins MamB and MamM of *Magnetospirillum gryphiswaldense* have distinct and complex functions, and are involved in magnetite biomineralization and magnetosome membrane assembly. Mol. Microbiol. *82*, 818–835.

Ullrich, S., Kube, M., Schübbe, S., Reinhardt, R., and Schüler, D. (2005). A hypervariable 130-kilobase genomic region of Magnetospirillum gryphiswaldense comprises a magnetosome island which undergoes frequent rearrangements during stationary growth. J. Bacteriol. *187*, 7176–7184.

Vali, H., and Kirschvink, J.L. (1989). Magnetofossil dissolution in a paleomagnetically unstable deep-sea sediment. Nature *339*, 203–206.

Van den Ent, F., Amos, L.A., and Löwe, J. (2001). Prokaryotic origin of the actin cytoskeleton. Nature *413*, 39–44.

Varadan, V.K., Chen, L., and Xie, J. (2008). Magnetosomes and their Biomedical Applications. In Nanomedicine (John Wiley & Sons, Ltd), pp. 175–213.

Vartholomeos, P., Fruchard, M., Ferreira, A., and Mavroidis, C. (2011). MRI-guided nanorobotic systems for therapeutic and diagnostic applications. Annu. Rev. Biomed. Eng. *13*, 157–184.

von Schonfedt, V., Krishnamurthy, H., Foppiani, L., and Schlatt, S. (1999). Magnetic cell sorting is a fast and effective method of enriching viable spermatogonia from djungarian hamster, mouse, and marmoset monkey testes. Biology of Reproduction *61*, 582–589.

Wang, H., Yu, Y.F., Sun, Y.B., and Chen, Q.W. (2011). Magnetic nanochains: a review. Nano *6*, 1–17.

Wang, S., Arellano-Santoyo, H., Combs, P.A., and Shaevitz, J.W. (2010). Actin-like cytoskeleton filaments contribute to cell mechanics in bacteria. Proc. Natl. Acad. Sci. U.S.A. *107*, 9182–9185.

Weis, J.J., and Levesque, D. (1993). Ferroelectric phases of dipolar hard-spheres. Phys. Rev. E *48*, 3728–3740.

Weiss, B.P., Kim, S.S., Kirschvink, J.L., Kopp, R.E., Sankaran, M., Kobayashi, A., and Komeili, A. (2004). Magnetic tests for magnetosome chains in Martian meteorite ALH84001. Proc. Natl. Acad. Sci. U.S.A. *101*, 8281–8284.

Xiong, Y., Chen, Q., Tao, N., Ye, J., Tang, Y., Feng, J., and Gu, X. (2007). The formation of legume-like structures of Co nanoparticles through a polymer-assisted magnetic-field-induced assembly. Nanotechnology *18*, 345301.

Yang, W., Li, R., Peng, T., Zhang, Y., Jiang, W., Li, Y., and Li, J. (2010). MamO and mamE genes are essential for magnetosome crystal biomineralization in *Magnetospirillum gryphiswaldense* MSR-1. Research in Microbiology *161*, 701–705.

Yoshino, T., Kato, F., Takeyama, H., Nakai, M., Yakabe, Y., and Matsunaga, T. (2005). Development of a novel method for screening of estrogenic compounds using nano-sized bacterial magnetic particles displaying estrogen receptor. Anal. Chim.Acta *532*, 105–111.

Yoza, B., Arakaki, A., Maruyama, K., Takeyama, H., and Matsunaga, T. (2003a). Fully automated DNA extraction from blood using magnetic particles modified with a hyperbranched polyamidoamine dendrimer. J. Biosci. Bioeng. *95*, 21–26.

Yoza, B., Arakaki, A., and Matsunaga, T. (2003b). DNA extraction using bacterial magnetic particles modified with hyperbranched polyamidoamine dendrimer. J. Biotechnol. *101*, 219–228.

Zhang, F., and Wang, C.-C. (2008). Fabrication of one-dimensional iron oxide/silica nanostructures with high magnetic sensitivity by dipole-directed self-assembly. J. Phys. Chem. C *112*, 15151–15156.

Zhang, L., Abbott, J.J., Dong, L., Kratochvil, B.E., Bell, D., and Nelson, B.J. (2009). Artificial bacterial flagella: fabrication and magnetic control. Appl. Phys. Lett. *94*, 064107.

Zhou, Z., Liu, G., and Han, D. (2009). Coating and structural locking of dipolar chains of cobalt nanoparticles. ACS Nano *3*, 165–172.

Liposome–Nanoparticle Assemblies

11

Matthew R. Preiss, Anju Gupta and Geoffrey D. Bothun

Abstract

Liposome–nanoparticle assemblies (LNAs) combine the demonstrated potential of clinically approved nanoparticles and liposomes to achieve multiple therapeutic and diagnostic objectives. Efficient and effective biomedical application requires assemblies to be stable, biocompatible, and bioavailable, while enhancing the properties of encapsulates. LNAs have been demonstrated to be effective for *in vivo* and *in vitro* providing targeting and stimuli-responsive delivery of therapeutic and imaging agents. The ability to design LNAs with nanoparticle encapsulation, bilayer-decoration, and surface coupling provides a variety of different structures and functions. While the potential of LNAs has been demonstrated, future investigation into the interaction between the lipid bilayer and nanoparticles is necessary to understand and develop LNAs for clinical applications. This section will discuss the current state of liposome-nanoparticle assembly design, characterization, and applications of liposome-nanoparticle assemblies.

Introduction

Only about 11% of new promising therapeutic compounds in clinical development are eventually approved. Nearly 70% of drug failures are attributed to poor pharmacokinetics, efficacy, toxicology, clinical safety, and formulation (Kola and Landis, 2004; Leeson and Davis, 2004). High drug attrition rates are the major cause of the recent decline in breakthrough drugs and the rise in costs of new drug therapies. Developments in nanotechnology have demonstrated potential for overcoming the issues related to drug pharmacokinetics and pharmacodynamics. Targeted and controlled delivery of therapeutic agents directly to targeted tissues can be achieved, improving efficacy, lowering the necessary dose, and reducing adverse effects. Nobel Laureate Paul Ehrlich's dream of a 'magic bullet' to fight disease may be realized through controlled and targeted nanoscale therapeutics (Koo *et al.*, 2005).

In 2004, the National Cancer Institute launched the Alliance for Nanotechnology in Cancer (Alliance). The Alliance's goal is development of nanotechnology-based cancer treatments and imaging. Specifically, the Alliance is emphasizing the development of drug delivery that targets tumour cells, the tumour microenvironment, and metastatic, recurrent, and drug-resistant cancers with nanotherapeutic delivery systems, theranostics, contrast agents, and complexes capable of providing multiple therapies (National Cancer Institute). The design of such multifunctional constructs is inherently complex as it requires combining different molecular, colloidal, and/or particulate agents. Furthermore, the construct must be stable, resistant to protein and immune system absorption, and capable of targeting.

Novel liposome–nanoparticle assemblies (LNAs) provide a biologically inspired route for designing multifunctional targeted therapeutics and imaging. The LNA structure is inspired by the early development of magnetoliposomes (liposomes with magnetic nanoparticles encapsulated in the aqueous core). Recent literature has referred to LNAs as 'liposome–nanoparticle hybrids' (Al-Jamal and Kostarelos, 2007). LNAs are liposome structures in which nanoparticles (NPs) are encapsulated in the aqueous core, embedded in the lipid bilayer, or coupled to the bilayer surface. Liposomes are a well-established vehicle for the administration of therapeutic and diagnostic agents (Bangham and Horne, 1964; Bangham *et al.*, 1965; Gregoriadis, 1973; Papahadjopoulos and Ohki, 1969). As a biocompatible carrier, liposomes provide a stable means for the transportation and protection of hydrophilic and/or hydrophobic molecules. Nanoparticles are nanoscale moieties that have been demonstrated to be effective transportation vehicles, contrast agents, and agents responsive to external stimuli (such as electromagnetic fields and light). LNAs combine the advantageous properties of liposomes with functional nanoparticles to create a multifunctional therapeutic and diagnostic construct (Zhang *et al.*, 2008).

LNAs have several advantages when utilized for drug delivery, hyperthermia, imaging, and diagnostic applications. LNAs are able to delivery hydrophobic and/or hydrophilic molecules and NPs (Zhang *et al.*, 2009). The liposome can be modified to protect encapsulated agents from biomolecule absorption and functionalized for targeting. LNAs can also be used to concentrate encapsulates, increasing the efficiency of delivery. Also, the strategies for processing, stabilizing, and targeting liposomes are well established (Immordino *et al.*, 2006). NPs can be magnetically guided for *in vivo* targeting and provide a mechanism for stimuli-responsive triggering. Surface-bound NPs also enhance the colloidal stability of LNAs and bilayer-embedded NPs can reduce spontaneous leakage (Chen *et al.*, 2010; Paasonen *et al.*, 2007; Yu *et al.*, 2007; Zhang and Granick, 2006). LNAs harness the intrinsic advantages of a liposomal carrier, enhancing stability, bioavailability, and biocompatibility, and adds the imaging and/or responsive functionality of a NP (Zhang *et al.*, 2008).

Drug delivering liposomes and nanoparticles have both been approved separately for clinical use by the US Food and Drug Administration (FDA) (see Tables 11.1 and 11.2, respectively). These approved therapies represent the first generation of development for nano-scale therapeutics and diagnostics. Combining liposomes and nanoparticles, to form multimodal LNAs, is the natural evolution for these technologies. This section will focus on a review of LNA design and structure, characterization techniques, and biomedical applications, such as controlled drug release, imaging, and hyperthermia, expanding on our group's review on stimuli-responsive LNAs (Preiss and Bothun, 2011). Recent reviews focusing on the therapeutic and diagnostic applications of liposomes are provided in references (Goyal *et al.*, 2005; Immordino *et al.*, 2006; Kshirsagar *et al.*, 2005; Maurer *et al.*, 2001; Mulder *et al.*, 2006; Puri *et al.*, 2009; Samad *et al.*, 2007; Torchilin, 2005) and NPs are provided in references (Corchero and Villaverde, 2009; Emerich and Thanos, 2006; Fukumori and Ichikawa, 2006; Groneberg *et al.*, 2006; Jin and Ye, 2007; Laurent *et al.*, 2008; McCarthy and Weissleder, 2008; Michalet *et al.*, 2005; Polyak and Friedman, 2009; Rotomskis *et al.*, 2006; Wang *et al.*, 2008).

Liposome

Since the early work by Bangham, Papahadjopoulos, and Gregoriadis in the 1960s and 1970s, liposomes have become one of the most highly investigated nano-structures. Today,

Table 11.1 FDA-approved liposomal formulations

Drug	Trade name	Manufacturer	Nano component	Treatment	Route of administration	FDA approval
Amphotericin B	AmBisome	Astellas	Liposome	Fungal and protozoal infections	Injectable	1997
Amphotericin B	Amphotec	Aldopharma USA	Lipid colloidal dispersion	Fungal and protozoal infections	Injectable	1996
Daunorubicin	DaunoXome	Galen	Liposome	Advanced HIV Kaposi's sarcoma; ovarian cancer	Injectable	1996
Cytarabine	DepoCyt	Pacira Pharms	Liposome	Malignant lymphomatous meningitis	Injectable	1999
Morphine	DepoDur	EKR Theraputics	Liposome	Postsurgical analgesia	Epidural	2004
Doxorubicin	Doxil	Janssen	PEGylated liposomes	Metastatic ovarian cancer; AIDS-related Kaposi's sarcoma	Injectable	1995
Verteporfin	Visudyne	QLT	Liposome	Age-related macular degeneration, pathological myopia, ocular histoplasmosis	Injectable	2000
Propofol	Diprivan	APP Pharms	Liposome	Anaesthetic	Injectable	1989
Bupivacaine	Exparel	Pacira Pharmaceuticals	Liposome	Analgesic	Injectable	2011

Table 11.2 FDA-approved therapeutic nanoparticles

Drug	Trade name	Manufacturer	Nano component	Treatment	Route of administration	FDA approval
Superparamagnetic iron oxide	Feridex	AMAG Pharms	Nanoparticle	Liver tumour imaging	Injectable	1996
Superparamagnetic iron oxide	Gastromark	AMAG Pharms	Nanoparticle	Gastrointestinal imaging	Oral	1996
Superparamagnetic iron oxide	Feraheme	AMAG Pharms	Nanoparticle	Iron-deficiency anaemia	i.v.	2009
Gold	Verigene	Nanosphere	Nanoparticle	Nucleic acid detection	Medical device	2007
Silver	Acticoat	Smith & Nephew	Nanoparticle	Antimicrobial dressing	Medical device	1996
Silver	Allevyn Ag	Smith & Nephew	Nanoparticle	Antimicrobial dressing	Medical device	2007

liposomes are used as model biological membranes and for therapeutic and diagnostic agent delivery (Bangham and Horne, 1964; Bangham *et al.*, 1965; Gregoriadis, 1973; Papahadjopoulos and Ohki, 1969). Liposomes are reliable systemic drug delivery systems because they are non-toxic, biocompatible, capable of prolonging bioavailability of encapsulated agents by reducing or preventing drug degradation and enhancing solubility and stability (Al-Jamal and Kostarelos, 2007).

Liposomes, as depicted in Fig. 11.1A, are composed of self-assembled spherical vesicles consisting of one or multiple lipid bilayers surrounding an internal aqueous core. Bilayer thickness (l_b) is ~5 nm thick (l_b), composed of a hydrophobic acyl lipid tail region (~3 nm) and a hydrophilic headgroup region. Liposomes can be prepared with zwitterionic, anionic, or cationic lipids, and the net liposome surface charge can be adjusted by mixing different ratios of these components. From a morphological aspect, liposomes are distinguished according to their diameter, small (<100 nm), large (100–1000 nm, or giant (>1000 nm), and number of bilayers, single (unilamellar) or multiple (multilamellar) (Sivashankar, 2011). Fig. 11.1B is a cryo-transmission electron microscope (cryo-TEM) image of liposomes depicting the structures that can be formed. For drug delivery and diagnostics, liposomes are attractive because of their ability to encapsulate both hydrophilic (in the aqueous core or bound to the liposome surface) and hydrophobic (in the lipid bilayer) molecules. Liposomes also open the therapeutic window, reducing adverse effects, by altering the pharmacokinetic and pharmacodynamic characteristics of the encapsulated agent (Al-Jamal and Kostarelos, 2007).

FDA-approved liposomal amphotericin B formulations (Abelcet®, AmBisome®, and Amphotec®) are good examples of the effectiveness of liposomes for drug delivery.

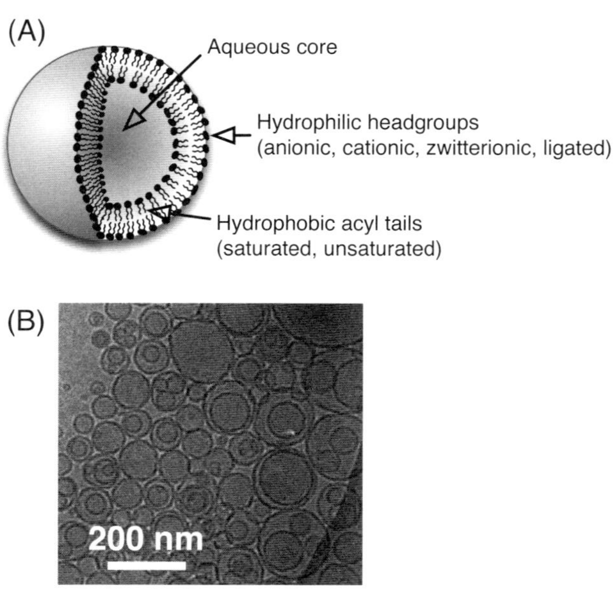

Figure 11.1 (A) Liposome schematic depicting the aqueous core, hydrophilic headgroup, and hydrophobic tail regions and (B) a cryogenic transmission electron micrograph of dipalmitoylphosphatidylcholine liposomes (DPPC, 10 mM) prepared in phosphate-buffered saline. A colour version of this figure is located in the plate section at the back of the book.

Amphotericin B is considered the 'gold standard' for systemic treatment of fungal infections. However, amphotericin B is hydrophobic and nephrotoxic, limiting its stability and administered dosage. Encapsulation of amphotericin B in the lipid bilayer reduced the concentration of amphotericin B in the kidneys, providing similar efficacy as conventional amphotericin B while significantly reducing adverse side-effects (Gibaldi *et al.*, 2007; Moen *et al.*, 2009).

Release of encapsulated molecules from liposomes is controlled by the permeability through the lipid bilayer, which can be achieved by transbilayer diffusion or transient pore formation triggered by bilayer disruption or phase separation. Phase separation can be induced by 'melting' the liposomal bilayers – i.e. heating to a temperature greater than the characteristic main phase transition or melting temperature of the lipids (T_m). Below T_m the lipids are in the solid or gel phase in which the lipids are rigid and highly organized. Above T_m the lipids are disordered in a liquid crystalline or fluid phase. Permeability is high at the interface between gel and fluid phases. Phase separation and bilayer permeability can be manipulated by adjusting the lipid bilayer composition. A simple example illustrating this principle can be made with dipalmitoylphosphatidylcholine (DPPC, $T_m = 42°C$) and dimyristoylphosphatidylcholine (DMPC, $T_m = 23°C$). At a DPPC/DMPC molar ratio of 74:26 the melting temperature occurs at physiological temperature (37°C). Furthermore, cholesterol is commonly incorporated into the bilayer to reduce membrane fluidity above the melting temperature. Membrane fluidity has been shown to be affected by pH, ion concentration, and the presence of molecules (such as nanoparticles) absorbed into the bilayer (Al-Jamal and Kostarelos, 2007; Bothun, 2008; Chen *et al.*, 2010).

A major limitation to liposomal drug delivery is the short half-life (Zhang *et al.*, 2008). Within minutes, the reticuloendothelial system (RES) will eliminate the liposomes from the blood, limiting the drug's efficacy and ability to accumulate at target sites (Moghimi and Szebeni, 2003). Proteins, called opsonins, recognize and target foreign agents (such as untargeted liposomes) for elimination by the mononuclear phagocyte system (MPS) or by hepatocyte uptake. Other proteins are capable of lysing liposomes directly by compromising the stability of the lipid bilayer (Ishida *et al.*, 2002; Maurer *et al.*, 2001; Yan *et al.*, 2005). Liposome residence time is dependent on liposome size, surface charge, lipid packing, bilayer composition, and surface modifiers (Maurer *et al.*, 2001; Samad *et al.*, 2007). Attaching polyethylene glycol (PEG) to the liposome, forming 'stealth liposomes', can increase half-life to 2–24 h and increase liposome stability (Medina *et al.*, 2004; Moghimi and Szebeni, 2003; Zhang *et al.*, 2008). The first FDA-approved liposomal drug formulation (and FDA-approved 'nanodrug') was Doxil® in 1995. Doxil® is doxorubicin, the most commonly used anthracycline anticancer drug, encapsulated within a PEGylated liposome. The elimination half-life for Doxil® is 55 h and an area under the plasma concentration–time curve of 900 Mg h/ml, compared with 0.2 h and 4 Mg h/ml for free doxorubicin (Barenholz, 2012; Chang and Yeh, 2012).

Drug delivery from liposomes can also be accomplished by cellular uptake, which can occur by adsorption, endocytosis, fusion, and/or lipid transfer (Pagano and Weinstein, 1978; Samad *et al.*, 2007; Torchilin, 2005). Adsorption is the association of liposome bilayer with cell bilayer without destroying the liposome bilayer or being internalized by the cell. Adsorption can be specific (assisted by targeting ligands such as antibodies) or nonspecific (controlled by intermolecular and surface forces). Endocytosis involves the uptake of liposomes into the cell by encapsulation within endosomes. Release of drugs to the cytoplasm can occur by membrane destabilization of the encapsulating endosome or by delivery to lysosomes. Lysosomes have an acidic pH and contain lysing enzymes. Drug release is accomplished when lysosome enzymes hydrolyse the lipid bilayer releasing the drug. Lysosome drug release is only effective when the encapsulated drugs are not susceptible to lysosome enzymes and pH. Fusion involves the adsorption and incorporation of the liposome bilayer with the cell membrane, releasing the payload into the cytoplasm. Finally, lipid transfer involves the exchange of lipids between the liposome bilayer and the cell membrane without enveloping the liposome (Samad *et al.*, 2007; Torchilin, 2005).

Nanoparticles

Nanoparticles are nanoscale moieties having magnetic and optical properties for use in therapeutic and imaging applications. The high surface area-to-volume ratio, stability, functionalization, and size (1–100 nm, on the order of biological macromolecules) of nanoparticles make them particularly attractive for biomedical applications. Nanoparticles have shown to be particularly effective as a contrast agent, a heat source, and as a targeting agent. Clinical application of nanoparticles can be hindered by poor colloidal stability, hydrophobicity, protein absorption, immune system uptake, and cytotoxicity. LNAs provide a carrier to take advantage of the properties of nanoparticles for controlled release, targeted therapies, hyperthermia, diagnostics, and imaging applications (Al-Jamal and Kostarelos, 2007; Huang *et al.*, 2011). A number of different inorganic nanoparticles have been used in LNAs, such as quantum dots (Al-Jamal *et al.*, 2008b; Bothun *et al.*, 2009; Gopalakrishnan *et al.*, 2006), fullerenes (fullerenosomes) (Babincova *et al.*, 2003, 2004; Chen and Bothun, 2009; Doi *et al.*, 2008; Hwang and Mauzerall, 1993; Ikeda and Kikuchi, 2008; Ikeda *et al.*, 2005, 2009; Jeng *et al.*, 2005; Niu and Manzerall, 1996), silver (Bothun, 2008; Park *et al.*, 2005), superparamagnetic iron oxide (SPIO) (Bothun and Priess, 2011; Chen *et al.*, 2010), and gold (Park *et al.*, 2006; Von White II *et al.*, 2012). This section will discuss several nanoparticles that have been utilized in LNA applications.

Despite their applications in drug and gene delivery and cosmetics, cytotoxicity remains a major concern. Understanding the interactions between nanoparticles and cell membranes is crucial to NP biomedical applications and provides insight into their toxicity. NPs can be designed to bind on the cell surface, adsorb within the membrane, and translocate across the cell membrane. NPs can be exploited for novel applications by controlling the interaction between the NP and bilayer. A common way to achieve this interaction is by modifying the surface of the NP, specifically by adding positive or negative charges onto NP surface (Li, 2006; Legrand, 2008). Binding interaction between superparamagnetic iron oxide particles and stem cells are being used in cell selection process (Pavon, 2008). NPs used in drug delivery applications can be modified to avoid drug degradation by increasing the circulation period which in turn results in cell uptake efficiency (Jin and Ye, 2007).

Quantum dots

Quantum dots (QDs), 2–10 nm fluorescent semiconductor nanocrystals, have been demonstrated as effective imaging and diagnostics agents. QDs can provide a highly sensitive contrast agent capable of exhibiting fluorescence that is 10–20 times greater than conventional imaging agents, such as organic dyes and fluorescent proteins. QDs are also 100 times more stable against photobleaching than organic dyes (Chan, 1998). The optical properties of QDs can be tuned by adjusting their size and composition. Commonly used quantum dots for biomedical applications include cadmium selenide (CdSe), cadmium telluride (CdTe), indium phosphide (InP), and indium arsenide (InAs) (Bharali and Mousa, 2010). Clinical application of quantum dots is limited due to their inherent hydrophobicity and potential cytotoxicity. Conjugation of quantum dots with liposomes have shown to be effective to overcome these limitations (Al-Jamal and Kostarelos, 2007; Bothun *et al.*, 2009; Dudu *et al.*, 2008; Smith *et al.*, 2006; Walling *et al.*, 2009; Weng *et al.*, 2008).

Gold nanoparticles

Imaging and photothermal effects of gold NPs stem from their enhanced surface plasmon resonance (SPR), where visible or near-infrared light is absorbed causing oscillation of surface electrons (Huang *et al.*, 2010). SPR absorbance and the wavelength range are dependent upon nanoparticle size, core/shell configuration (e.g. silica core/gold shell (Oldenburg *et al.*, 1999)), and geometry. Shifts in these properties are indicative of the degree of NP aggregation and/or molecular adsorption on the NP surface (Li and Gu, 2010). For photothermal therapy, absorbed light energy is converted into local heat that thermally diffuses into the surrounding medium. Varying NP size and core/shell configuration provides a means of tuning the frequency window for photothermal therapy. It is generally accepted that gold NP-mediated phototherapy is attributed to heat or resulting bubble nucleation depending on the light intensity and mode of exposure (Li and Gu, 2010). However, recent work by Krpetic *et al.* (2010) at low light energies suggests that photochemical effects – the formation of free radicals during NP irradiation – may play an important role. In addition to photothermal heating, electromagnetic fields operating at RF can be used to heat gold NPs. For example, Gannon *et al.* (2008) examined the effect of NP concentration and RF field strength on the heating rates of 5 nm Au NPs in water. A rate of ~74°C/min was measured using an 800 W RF field at a NP concentration of 67 μM.

Superparamagnetic iron oxide nanoparticles

Superparamagnetic iron oxide (SPIO) NPs are 4–20 nm nanoparticles typically composed of magnetite (Fe_3O_4) or maghemite (γ-Fe_2O_3). SPIO NPs demonstrate physical and magnetic properties, such as low toxicity and paramagnetism, making them advantageous for *in vitro* and *in vivo* applications. Due to the nanoscale crystal size of iron oxide, a single magnetic domain forms making the particle superparamagnetic. The atomic magnetic dipoles of paramagnetic materials are randomly oriented due to Brownian fluctuation in the absence of a magnetic field. Presence of a magnetic field causes the crystals to align in the direction of the field. After removal of the magnetic field, Brownian fluctuation will cause the random orientation, leaving no magnetic reminisce (Thorek *et al.*, 2006). The superparamagnetic characteristics of SPIO NPs allow them to be used as contrast agents for magnetic resonance imaging (MRI), targeted therapeutic agents capable of being directed under a static magnetic field, and a heat source from when exposed to alternating current electromagnetic fields (AC EMF) (Brezovich, 1988; Teja and Koh, 2009). SPIO NPs also have low toxicity because the iron oxide is broken down naturally by the liver and spleen (Laurent *et al.*, 2008; Mornet *et al.*, 2004; Pankhurst *et al.*, 2003; Rivera Gil *et al.*, 2010). The characteristics of SPIO NPs allow for the development of multifunctional LNAs capable of simultaneous targeting, imaging, hyperthermia and/or drug delivery.

Formation, structure, and design strategies

The functionality of a LNA is determined by the liposome composition, liposome and NP surface modifiers, NPs employed, intermolecular and surface interactions, and colloidal stability. LNA design strategies include the encapsulation of individual or multiple NPs within the aqueous core of the liposome, embedding hydrophobic NPs in the lipid bilayer, and binding or conjugating NPs to the liposome surface (Fig. 11.2). Table 11.3 contains a list of Au, iron oxide, and γ-iron oxide LNAs reported in the literature since 2008. LNAs can be used to protect NPs and encapsulated agents from the adsorption of exogenous molecules, enhancing bioavailability and reducing the need for complex surface chemistries. Concentration of NPs and therapeutic agents within the liposome can increase intracellular delivery, providing greater contrast for imaging, more efficient drug delivery, and enhanced heating capability for hyperthermia applications. Functionality can also be added by modifying the LNA bilayer with functional lipids or surface coatings for improved stability and providing targeting capability.

LNA functionality extends beyond that of a traditional liposome. Liposome delivery requires not just creating a stable system capable of retaining cargo during both storage

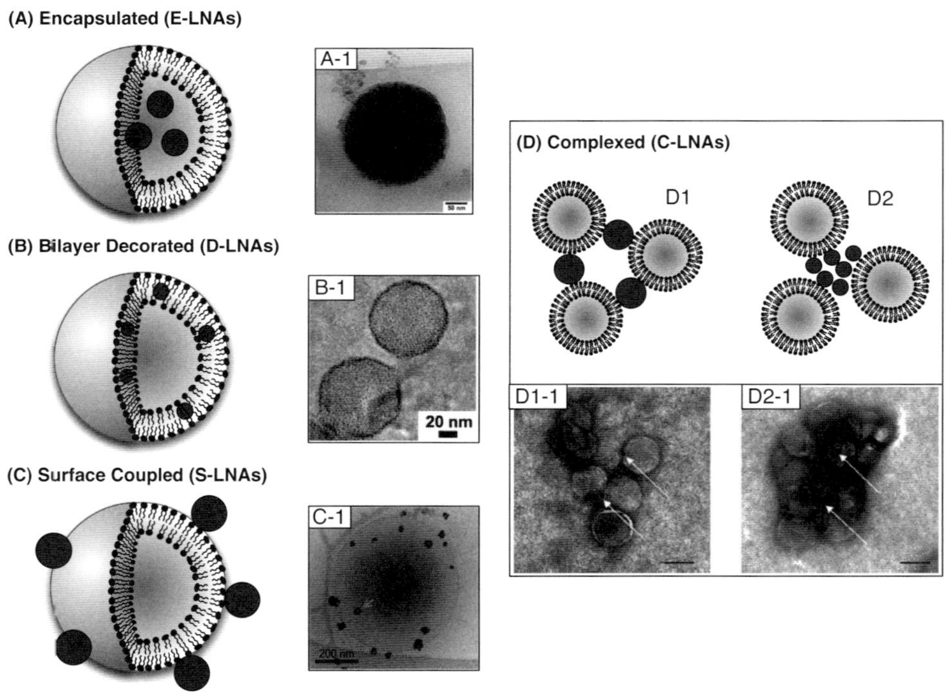

Figure 11.2 Schematic and representative transmission electron micrographs of LNAs formed by encapsulating hydrophilic nanoparticles (E-LNAs; A, A-1; Wijaya and Hamad-Schifferli, 2007), bilayer decorating hydrophobic nanoparticles (D-LNAs; B, B-1; Rasch *et al.*, 2010b) or surface coupling hydrophilic nanoparticles (S-LNAs; C C-1; Wu *et al.*, 2008). Surface coupling (C) can also be used to create controlled aggregates or complexes (D1, D2; D1-1, D2-1; Volodkin *et al.*, 2009). Structures and proportions are not to scale. Reprinted from Rasch *et al.* (2010b), Volodkin *et al.* (2009), Wijaya and Hamad-Schifferli, (2007) and Wu *et al.* (2008) with permission. A colour version of this figure is located in the plate section at the back of the book.

and circulation, but also the ability to release encapsulates at a target site. Efficient release can be achieved by using environmental responsive liposomes that melt near physiological temperature or through chemical mechanisms, such as pH-sensitive lipids. Controlled and triggered-release from LNAs can be achieved by taking advantage of these liposomal responsive properties and NP RF and photothermal heating capability. Multifunctional LNAs capable of targeting, imaging, hyperthermia, and/or controlled release can be constructed by combining the advantageous properties of the nanoparticles and lipids used.

Encapsulated liposome–nanoparticle assembly

Encapsulated liposome–nanoparticle assemblies (E-LNAs) are formed by encapsulating NPs within the aqueous core of liposomes (Fig. 11.2A). The first investigation of LNAs was inspired by the use of liposomes as a carrier for hydrophilic drugs. E-LNAs, by encapsulating NPs in the liposome core, force NPs to cluster together at a high density. High density nanoparticle loading is advantageous to hyperthermia and drug delivery because heating and drug release can be localized preventing damage to adjacent tissues. Also, high density loading provides a strong contrast agent for biomedical imaging (Wijaya and Hamad-Schifferli, 2007).

Magnetoliposomes (MLs), liposomes encapsulating superparamagnetic NPs, are one of the simplest and first developed LNA configurations (De Cuyper and Joniau, 1988; Shinkai *et al.*, 1996). They can be prepared by encapsulating preformed NPs in solution or by forming NPs within the liposome core, as first shown by Papahadjopoulos in 1983 (Hong *et al.*, 1983). E-LNAs can be prepared by thin film hydration (TFH), double emulsion (DE) (Zheng *et al.*, 1994), or reverse phase evaporation (REV) (Szoka and Papahadjopoulos, 1978). Extrusion or sonication of post-formation liposomes can be employed to control the size of E-LNAs. Supported lipid bilayers (SLBs), NPs coated with a lipid bilayer, are formed when $d_{core} = d_{NP}$. E-LNA formation requires the use of colloidal stable nanoparticles with a diameter (d) that is smaller than the inner diameter of the aqueous liposome core, $d_{core} > d_{NP}$ (Fig. 11.3A). The maximum theoretical number of encapsulated NPs is $n \approx 0.74\ (V_{core}/V_{NP}$; V represents the volume of the core or NP), due to the close packing of spheres and $d_{core} >> d_{NP}$. Wijaya and Hamad-Schifferli (2007) demonstrated that it is possible to approach this limit, demonstrating high-density encapsulation of Fe_3O_4 NPs ($d_{NP} = 12.5$ nm) within DPPC liposomes (Fig. 11.2A, A-1). With this design the available core volume for co-encapsulating aqueous drug molecules decreases within increasing NP concentration. However, the ability for embedding hydrophobic molecules within the bilayer is unaffected by NP concentration.

The osmotic pressure differential across the lipid bilayer and the attractive or repulsive forces between the bilayer and the NPs determine the structure of E-LNAs. The elasticity of the bilayer determines how the LNA will deform in response to these forces. Attractive forces can include van der Waals, hydrophobic, and electrostatic interactions; and repulsive forces can include electrostatic, depletion, hydration, and steric interactions. The physical stability of a liposome–NP system can be determined by the Deryaguin–Landau–Verwey–Overbreek (DLVO) theory. The DLVO theory balances the opposing forces to provide a total energy of interaction between the particles. Liposome-NP systems are characterized by three types of interactions that take place, repulsion between liposome–liposome and NP–NP and attractive forces between liposome and NP. Electrostatic repulsion becomes significant when nanoparticles and liposomes approach each other and their double layers

Table 11.3 Review of LNAs formed with gold (Au) or iron oxide nanoparticles

Lipids (ratio)[a]	Charge	NP diameter (nm)	NP surface coating	Lipid:NP[a]	LNA	Reference
Gold						
DOPC DOPC:DOPC+ (90:10) DOPC:DOPP (9:1)	Zwitterionic Cationic Anionic	NR	Ascorbic acid	NR	S-LNA	Sau *et al.* (2009)
DOPC:DOTAP (8:2)	Cationic	80	Citrate	NR[f]	E-LNA	Anderson *et al.* (2010)
DPPC:Chol (55:40)	Zwitterionic	1.4	n.r.	500:1–2000:1 (DPPE-AuNP:Liposome)	S-LNA	Chithrani *et al.* (2010)
DPPC:DPTAP:Chol (6:3:1 w/w)	Cationic	20	n.r.		C-LNA	Volodkin *et al.* (2009)
DPPC:DSPC (9:1)	Zwitterionic	2.5	hexanethiol	17.2:1 (w/w)	D-LNA	Paasonen *et al.* (2010b)
DPPC:DSPC (9:1)	Zwitterionic	4	mercaptosuccinic acid	10:1 (w/w)	E-LNA	Paasonen *et al.* (2010b)
DPPC:DSPC (9:1)	Zwitterionic	1.4	DPPE-Nanogold™	NR	E-LNA, D-LNA	Paasonen *et al.* (2007a)
Egg PC	Cationic	10	Chitosan	NR	S-LNA	Pornpattananangkul *et al.* (2011)
Egg PC	Zwitterionic	2	Dodecanethiol	100:1–1500:1	D-LNA	Rasch *et al.* (2010a)
Egg PC:DOTAP (9:1 w/w)	Cationic	4	Mercaptopropionic acid	$\geq 3.6 \times 10^{-3}$:1 (mol/mol)	S-LNA	Pornpattananangkul *et al.* (2010)
EYPC EYPC:DDAB (9:1) EYPC:PEG-DSPE (95:5)	Zwitterionic Cationic Zwitterionic	13	Citrate	10:1, 1:1 (mol/mol)	S-LNA	Kojima *et al.* (2008b)
Iron oxide (Fe_3O_4)						
[maleimide]PEG-DSPE:FAM-DOPE (10:1 w/w)		10–14	Heptanoic acid, acetic acid	1:1.8 (w/w)	E-LNA	Larsen *et al.* (2008)

DMPC:Chol:XL[e] (47.5:47.5:5) DPPC:DMPC:XL[e] (9.5:85.5:5)	Zwitterionic	10	Catechol	≥8.3:1 (mol/w)	C-LNA	Mart *et al.* (2009)
DMPC:DMTAP:Chol:DMPE-PEG (35:50:10:5)	Cationic	NR	NR	NR	E-LNA	Dandamudi *et al.* (2009)
DPPC:Chol:PEG-DMPE:Fol-PEG-DSPE (80:20:4.2:0.5)	Zwitterionic	10	Lauric acid	NR	E-LNA	Pradhan *et al.* (2010)
PC	Zwitterionic	12.5	NR		E-LNA	Sabate *et al.* (2008)
PC:PE (2:1)	Zwitterionic	10	NR	NR	E-LNA	Kikumori *et al.* (2008)
Iron oxide (γ-Fe_2O_3)						
DOPC:DPTAP:Chol:DPPE:PEG-DMPE:Fol-PEG-DSPE (47.5:18.9:28.5:1:4:0.1 w/w)	Cationic	16	Oleic acid	NR	E-LNA	Bothun *et al.* (2011)
DOPE:Chol	Catonic	20	Tartaric acid	0.75–3 mg/ml	E:LNAs	Zheng *et al.* (2009)
DOTAP:DOPE	Cationic	10	(3-Aminopropyl)triethoxysilane	NR	E-LNA	Yang *et al.* (2008)
DPPC	Zwitterionic	5	Oleic acid	1000:1–10000:1	D-LNA	Chen *et al.* (2010)
DPPC:Chol (5:1, 15:3 w/w) DPPC:DSPC:Chol (10:5:3 w/w)	Zwitterionic	43	Dextran	NR	E-LNA	Tai *et al.* (2009)
DPPC:Chol (67:33)	Zwitterionic	10	Glutamic acid	NR	E-LNA	Zhu *et al.* (2009)
DPPC:DPTAP	Cationic	16, 30	Oleic acid		S-LNA	Chen and Bothun (2011)

[a]Molar ratios provided unless noted otherwise. NR, not reported.
[b]Cationic peptide lipid (Murakami *et al.*, 1984).
[c]Contained within a cyclodextran cavity and embedded via fullerene exchange method (Ikeda *et al.*, 2005).
[d]Derasome (ceramic-coated liposome).
[e]Cross-linking molecule (adhesive lipid).
[f]Estimated at 4 liposomes per NP.

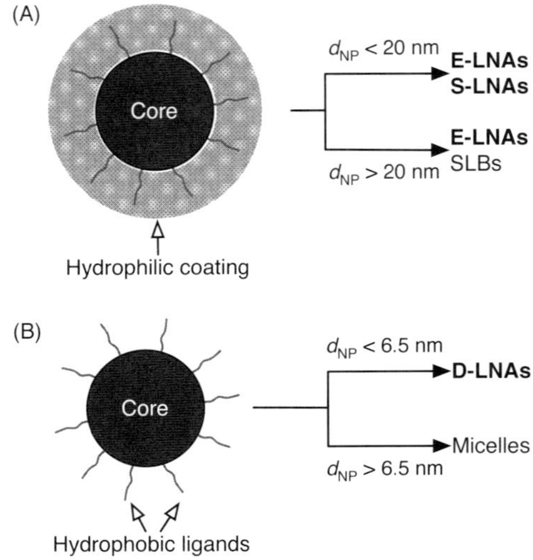

Figure 11.3 Dependence of LNA formation on nanoparticle size and surface chemistry. (A) Hydrophilic nanoparticles can be encapsulated to form E-LNAs or, if strong adhesive forces exist between the nanoparticle and lipid bilayer, surface coupled to form S-LNAs. For larger nanoparticles, strong adhesive forces can also lead to the formation of supported lipid bilayers (SLBs) on the nanoparticle surface. (B) Hydrophobic nanoparticles can embed within the liposomal bilayer to form D-LNAs or be coated with a lipid monolayer to form micelle-like structures.

begin to interfere. Electrostatic energy curve represents the energy required to overcome the repulsion. The maximum energy corresponds to the situation when the surfaces are touching each other and is zero outside the double layer (Leckband, 2001).

The adhering and non-adhering characteristics of nanoparticles can lead to changes in bilayer curvature, which can impact liposome size, shape, and phase homogeneity (Lipowsky and Dobereiner, 1998). Generally, this will occur when encapsulates are different from molecules present outside (e.g. sugars or proteins) liposomes.

LNAs are generally formed with small non-adhering NPs because NP adhesion to bilayers can significantly alter LNA structure and morphology. The exception to this is LNAs formed by coating a single large NP with an adsorbed or supported lipid bilayer. For non-adhering encapsulated particles, the bilayer can curve towards the larger particles. As an example of a non-adhesive system, Pradhan *et al.* (2007) compared the encapsulation efficiency of 10 nm $MnFe_2O_4$ NPs coated with lauric acid composed of egg PC–cholesterol (at molar ratios of 1:0, 2:1, 3:2, 1:1, and 1:2) and formed by TFH and DE. In general, TFH resulted in higher encapsulation efficiency with smaller ML diameter compared with DE owing to stripping of lauric acid during the DE process. In both cases, the observation that an egg PC–cholesterol ratio of 2:1 yielded the best encapsulation efficiency (70% via TFH) was attributed to cholesterol inducing a single liquid ordered bilayer phase.

In contrast, for small adhering encapsulated particles (attractive) where $d_{core} >> d_{NP}$ and $d_{NP} < 2l_b$ the bilayer can curve away from the particles. For large adhering particles, where $d_{NP} >> 2l_b$, the bilayer can curve around or engulf the particles. For example, Sabate *et al.* (2008) examined the effect of Fe_3O_4 NP concentration coated with tetramethylammonium hydroxide (58 nm hydrodynamic d_{NP}) on the encapsulation efficiency of extruded soybean

PC MLs. The encapsulation efficiency decreased from 96.6% at 1.22 g Fe_3O_4/mol PC to 18.5% at 119.95 g Fe_3O_4/mol PC. This was attributed to electrostatic interactions (attraction) between the cationic NPs and the PC bilayers. The size of the MLs increased from 140 to 197 nm, consistent with lower curvature due to NP adhesion at the inner bilayer surface.

Electrolytes can also effect the curvature of lipid bilayers (Lipowsky and Dobereiner, 1998). Gomes *et al.* (2009) prepared polyelectrolyte-coated MLs by encapsulating 8 nm anionic γ-Fe_2O_3 NPs within egg PC liposomes and then coating with alternating poly(allylamine hydrochloride) and poly(sodium 4-styrenesulfonate) layers. The final coating determined the surface charge (anionic PSS or cationic PAH). The size ranged from 200 to 400 nm and two or more polyelectrolyte coatings sufficiently protected the lipid bilayer from detergent-induced disruption.

Bilayer-decorated liposome-nanoparticle assembly

Bilayer decorated liposome-nanoparticle assemblies (D-LNA) are liposomes with hydrophobic nanoparticles embedded in the lipid bilayer (Figs. 11.2B and 11.3B). Similar to the ability of cells to accommodate membrane proteins, liposomes can distort to accommodate hydrophobic NPs that exceed the thickness of hydrophobic acyl region of the bilayer (~3 nm) (Al-Jamal *et al.*, 2008b; Bothun, 2008; Chen *et al.*, 2010; Jang *et al.*, 2003). Embedded NPs can affect lipid packing, lipid phase behaviour, transbilayer permeability, and LNA structure and morphology (Binder *et al.*, 2007; Bothun, 2008; Bothun *et al.*, 2009; Chen *et al.*, 2010; Chen and Bothun, 2009; Jeng *et al.*, 2005; Park *et al.*, 2005, 2006; Rasch *et al.*, 2010). Cryo-TEM can be used to observe the structure and morphology of these nano-scale systems in solution. Atomic force microscopy has also been used to observe phase-separated domains and monitor membrane remodelling and alteration due to the presence and distribution of nanoparticles within the bilayer (Kirat, 2010). The diameter of embedded nanoparticles (core and surface coating) is similar to the thickness of the lipid bilayer (~5 nm). Theoretically, the diameter of an embedded nanoparticle must be less than 6.5 nm in order for the lipid bilayer to maintain its structure. Hydrophobic nanoparticles with diameters greater than 6.5 nm form micelles because they are more energetically favourable due to the high local curvature strain on the bilayer, as described in Fig. 11.4 (Ginzburg and Balijepalli, 2007; Wi *et al.*, 2008).

The characteristics of LNAs are directly affected by embedded nanoparticles. Embedded NPs interact with the lipid acyl tails changing the fluidity of the membrane bilayer. Physical obstruction of the movement of lipid tails alters lipid bilayer ordering. The melting temperature of LNAs with embedded silver (Bothun, 2008), gold (Mady *et al.*, 2011), and SPIO NPs (Chen *et al.*, 2010) have shown to reduce the transition temperature of the bilayer with increased loading by fluorescence anisotropy and differential scanning calorimetry. Recent FTIR measurements of embedded gold NPs have validated the interaction of NPs and the acyl tail groups. Changes in the frequency of CH_2 stretching indicated conformational change in acyl tails of bilayers with embedded NPs. Also, dynamic light scattering of citrate-stabilized gold NPs embedded in DPPC liposomes demonstrated an increase in the negative charge. Surface charge can hinder aggregation of LNAs affecting the overall stability of LNAs (Mady *et al.*, 2011). Therefore, changes to the membrane alter LNA properties for delivery of therapeutic and diagnostic agents. The interactions between NPs and liposomes still remains misunderstood. Further investigation into the effect of nanoparticles is necessary to develop more efficient multimodal LNAs

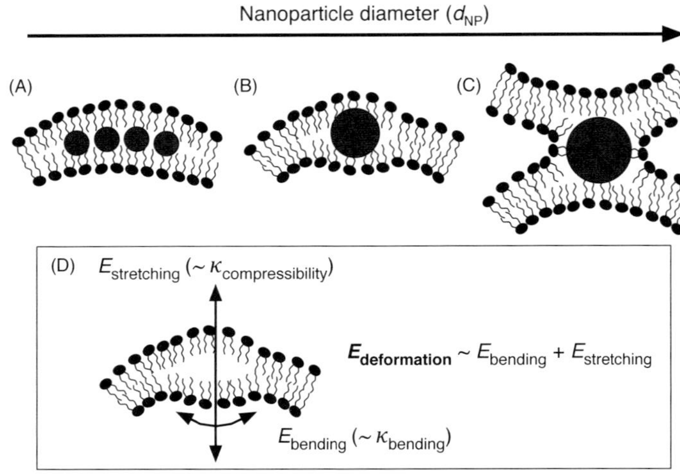

Figure 11.4 Changes in bilayer decoration mechanism of D-LNAs with increasing nanoparticle diameter (d_{NP}, particle core + surface coating). (A) Small nanoparticles (defined herein as d_{NP} < 2 nm) can cluster together to minimize bilayer bending energy (Rasch *et al.*, 2010b). (B) Larger nanoparticles (defined herein as d_{NP} = 2–6.5 nm) can create 'pockets' within the bilayer or (C) bridge adjoining liposomes (Al-Jamal *et al.*, 2008b; Chen *et al.*, 2010). (D) D-LNA assembly is governed in part by the bilayer deformation energy, which can be evaluated based on the energy penalty associated with lipid stretching (related to the compressibility modulus) and monolayer curvature (related to bending modulus).

Clustering of embedded NPs has been observed by Rasch *et al.* (2010) in LNAs with dodecanethiol-coated Au (d_{NP} = 1.6–1.8 nm) (Fig. 11.4A). They showed that high NP loading with uniform distribution can be achieved in PC liposomes via thin film hydration (with sonication and extrusion). Janus particles can be prepared with embedded NPs clustered in approximately one half of the liposomes via detergent loading followed by dialysis. Clustering occurs as the liposomes minimize the energy penalty for bilayer deformation – i.e. for a given concentration of embedded NPs the periodic bilayer bending energy needed to accommodate individual particles is greater than that needed to accommodate nanoparticle clusters. Park *et al.* (2006) and Chen *et al.* (2010) have observed a similar clustering phenomenon with stearylamine-coated 3–4 nm Au and oleic acid-coated 5 nm γ-Fe_2O_3 NPs in DPPC liposomes (Fig. 11.2B-1), respectively. This suggests that NP clustering is not restricted to d_{NP} < 2 nm. Increasing the loading of nanoparticles in the bilayer causes turbidity to change, which can be detected by higher shifts in absorption bands in UV-vis. The shifts are attributed to stronger interactions on dipole moments on aggregate nanoparticles (Marchenko, 2010).

In addition to clustering, embedded NPs with d_{NP} = 2–6.5 nm can reside in bilayer 'pockets' within individual (Fig. 11.4B) or neighbouring (Fig. 11.4C) LNAs. These cases arise when the lipid–NP ratio is high (~1000:1 or greater). This has been observed by Al Jamal *et al.* (2008a) for 4 nm hydrophobic CdSe/ZnS core/shell quantum dots in DOPC bilayers. Embedded SPIO NPs clusters have also led to merging of adjacent liposomes. Chen *et al.* hypothesized that optimal nanoparticle embedding for controlled release is a 'trade-off between structural changes and aggregation, which affects the liposome surface area, bilayer stabilization, bilayer permeability, and the localized heating in a EMF (Chen *et al.*, 2010).

Surface-coupled liposome-nanoparticle assembly and complexation

Surface-coupled magnetoliposomes (S-LNAs) are formed when hydrophilic NPs are absorbed onto or coupled to the outer surface of the lipid bilayer (Fig. 11.2C). This is achieved through attractive surface interactions, notably long-range electrostatic attraction. An advantage of S-LNAs is the ease in which they can be prepared – adding NPs to pre-existing liposome dispersions. Similar to bilayer embedment, decorated bilayers also provide direct heating to the bilayer in the presence of external stimuli. The design constraint for forming S-LNAs is dependent on bilayer NP adhesion and curvature. Recent investigations have shown that NPs with $d_{NP} > \sim 20\,nm$ lead to the formation of SLBs due to liposome adsorption and rupture, followed by the bilayer curving around the particle (Fig. 11.3A) (Chen and Bothun, 2011). The critical NP diameter under which S-LNAs can be formed is $d_{NP} < 2(k_b/w)^{1/2}$, where k_b is the bilayer bending elasticity, which is dependent on lipid composition and phase state, and w is the adhesion energy (Roiter *et al.*, 2008).

The Granick group has shown that stable S-LNA dispersions can be formed using zwitterionic liposomes with decorated cationic or anionic NPs ($< 20\,nm$) with a NP surface coverage above ~25% (Yu *et al.*, 2007; Zhang and Granick, 2006). This was achieved by electrostatic attraction. Lower surface coverage led to aggregation, which demonstrates the need to balance the lipid:NP ratio. It was shown with isothermal titration calorimetry that upon binding the nanoparticles could restructure the lipid bilayer, inducing gel phases in fluid liposomes and fluid phases in gel liposomes (Wang *et al.*, 2008). This observation shows that, even without external stimuli, bound NPs can induce changes in lipid phase behaviour and, presumably, permeability.

NP adhesion to the outer bilayer can affect the morphology and structure of S-LNAs similar to E-LNAs. Cationic nanoparticle adhesion to the outer surface of GUVs has been shown to cause pearling. The head group area of zwitterionic lipids was increased due to the use of charged particles. Attraction of the head group and electrostatic repulsion to the cationic nanoparticles caused a mismatch of the outer and inner curvature of liposomes. These interactions resulted in the pearling structure of the liposome (Yu and Granick, 2009).

Sau *et al.* (2009) have also used electrostatic binding to prepare S-LNAs with Au NPs. High NP surface coverage was achieved by using anionic Au NPs with physisorbed ascorbic acid and cationic liposomes (9:1 DOPC to ethyl-DOPC; $T_m = -20°C$). This high surface coverage was accompanied by NP aggregation due to the high local concentration and (likely) to charge screening via cationic lipids between bound particles (Kojima *et al.*, 2008a). Binding was also achieved on zwitterionic and anionic liposomes with decreasing coverage (and NP aggregation), respectively. Pornpattananangkul *et al.* (2010) have taken this one step further and have shown that pH can be used to control carboxyl-modified (anionic) Au NP binding to cationic liposomes and, in turn, liposome stability. Above the pK_a of the carboxyl groups the bound NPs stabilize the S-LNAs and prevent aggregation and fusion, while below the pK_a the NPs detach and liposome fusion resumes.

Lastly, LNAs can be formed by complexation (C-LNAs) if the liposomes surround NP aggregates (Fig. 11.2D, D2) or the NPs bind to multiple liposomes and act as 'bridges' (Fig. 11.2D, D1). Voldokin *et al.* (2009) have shown that either structure can be formed from the same anionic Au NP-cationic liposome by manipulating electrostatic interactions using via salt concentration. High NaCl concentration (75 mM) enhanced NP aggregation (Fig. 11.2D, D2-1) and low salt concentrations inhibited it (Fig. 11.2D, D1-1).

In addition to non-specific physical interactions (electrostatic), cross-linking can be used to create C-LNAs. Mart *et al.* (2009) used Fe_3O_4 NPs coated with histidine groups to bind to and complex zwitterionic/cholesterol liposomes containing Cu(iminodiacetate)-functionalized lipid. The objective was to demonstrate a potential method using histidine-Cu(IDA) binding to form C-LNAs, thereby concentrating a therapeutic and an imaging agents at a target site. The resulting aggregates ranged from 20–100 Mm in diameter. The C-LNAs demonstrated EMF heating and controlled release of model drugs.

Controlled release

Controlled release of encapsulated payloads from LNAs can be induced by heating of nanoparticles raising temperature of the lipid bilayer. *In vivo* heating of magnetic nanoparticles has been demonstrated with external stimuli such as alternating current electromagnetic fields, microwaves, light irradiation, and lasers (Brazel, 2009). As discussed earlier, the bilayer permeability rises with temperature allowing release of encapsulates. Bothun and Preiss (2009) have demonstrated local bilayer heating due of Fe_3O_4 nanoparticles heated by RF causing phase transition. However, there was negligible difference between the bulk and local bilayer temperatures. Therefore, controlled release is likely due to both thermally induced phase transition and mechanical rupture of the bilayer caused by NP heating. LNA controlled release with gold and iron oxide nanoparticle will be discussed herein.

Gold nanoparticles and photothermal effects

Heating of Au NPs is caused by the SPR properties that convert absorbed light into heat. SPR heating makes Au NPs efficient heat sources for LNA controlled release (Sassaroli *et al.*, 2009). Utilizing the photothermal heating of Au NPs, Paasonen *et al.* (2007) demonstrated the ability to control the release of calcein (622.6 MW) from Au LNAs composed of DPPC/DSPC at 9:1 (T_m = 44.9°C) with E-, D-, and S-LNAs (Fig. 11.2A–C). Leakage was examined with and without UV light at a wavelength of 250 nm over 30 min at 37°C. Without UV exposure, spontaneous calcein release was observed for E-LNAs with encapsulated mercaptosuccinic acid-coated NPs and D-LNAs with embedded hexanethiol-coated NPs. This was attributed to NP–lipid interactions at the bilayer/water interface and within the acyl tail region, respectively, which reduced bilayer integrity. With UV exposure, direct contact between NPs and the liposomes via bilayer-embedment led to the greatest release (~90% at 30 min). Intuitively, direct contact would improve the local heat transfer from the NPs to the liposomal bilayers relative to encapsulation. This led to a gel-fluid phase transition where calcein release was presumably enhanced by diffusion at the interface between coexisting gel and fluid domains.

Volodkin *et al.* (2009) demonstrated the release of 5(6)-carboxyfluorescein (CF; 376.3 MW) from C-LNAs formed by the complexation of 128 nm cationic liposomes (DPPC/DPTAP/chol, T_m ~ 40–45°C) and 20 nm anionic Au NPs. Low NaCl concentration (7.5 mM) yielded LNAs with NP-mediated liposome bridges (Fig. 11.2D, D1-1) and high NaCl concentration (75 mM) yielded LNAs with liposome-coated NP aggregates (Fig. 11.2D, D-1). CF release from the C-LNAs was observed within 5 s after near-IR irradiation (Fig. 11.5).

Anderson *et al.* (2010) utilized the principle of plasmonic nanobubble (PNB) formation to control the release of 104 and 240 kDa proteins from cationic LNAs (~1 Mm) containing

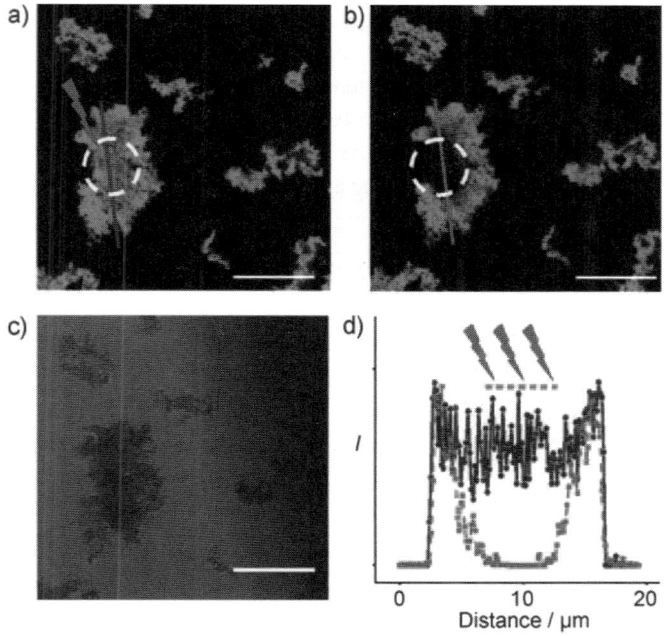

Figure 11.5 Remote dye release (5(6)-carboxyfluorescein) triggered by near IR laser illumination from C-LNAs prepared with gel-phase cationic liposomes and 20 nm hydrophilic gold nanoparticles. The scale bars represent 10 mm. These results show that near complete release can be achieved from the complexes after 5 s of illumination. Reprinted from Volodkin *et al.* (2009) with permission. A colour version of this figure is located in the plate section at the back of the book.

encapsulated 80 nm anionic Au NPs (Fig. 11.6). Irradiation was achieved using a single pump laser at 532 nm over 0.5 ns. Local vapour bubble formation led to mechanical disruption, as opposed to thermal, of the LNA bilayer and rapid protein release. The advantage of PNB formation is heating is isolated within the LNA triggering the immediate release of all encapsulated cargo.

Wu *et al.* (2008) used hollow gold nanoshells (HGNs) encapsulated within or decorating the surface of DPPC liposomes to trigger CF release by near-IR pulses (800 nm) via microbubble formation and collapse (Fig. 11.2C, C-1). LNA release was dependent on the proximity of the HGNs to the liposomes (decorated or tethered HGNs yielded the greatest response) and the laser power. Their results strongly suggest that release was attributed to transient disruption or poration of the lipid bilayer via transient bubble cavitation.

Iron oxide nanoparticles and alternating magnetic fields

AC EMF operating at RF heating is due to magnetic losses being converted to heat, typically at low frequencies between 100 and 400 kHz. The magnetic losses for NPs < ~30 nm are due to Néel relaxation, arising from rapidly alternating magnetic dipole moments, and Brownian relaxation, arising from nanoparticle rotation and viscous losses (friction). RF heating is advantageous because it is non-invasive, easily penetrates the body, and is physiologically acceptable for up to 1 h if the product *Hf*, where *H* is the field amplitude (current ′ number of coils per length) and *f* is the frequency, is below 4.85×10^5 kA/m/s (Brezovich, 1988).

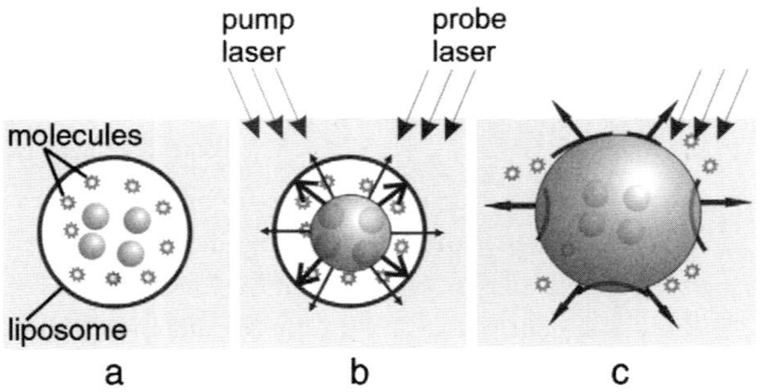

Figure 11.6 Proposed 'plasmonic nanobubble' release mechanism of encapsulated molecules from E-LNAs prepared with gold nanoparticles. Pulsed laser irradiation of the nanoparticles nucleates vapour bubbles that expand and mechanically disrupt the liposomal bilayer and trigger permeabilization. Reprinted from Anderson *et al.* (2010) with permission. A colour version of this figure is located in the plate section at the back of the book.

NP heating capability is based on the inherent specific absorbance rate (SAR, W/g) of the nanoparticles:

$$SAR = \frac{c_p}{m_{NP}} \frac{\Delta T}{\Delta t}$$

where c_p is the average heat capacity of the sample, m_{NP} is the NP mass and $\Delta T/\Delta t$ is the initial heating rate of the sample. SAR values up to approximately 700 W/g can be obtained depending on the NP size, composition, and surface coating (Mornet *et al.*, 2004).

Tai *et al.* (2009) examined CF release from thermosensitive zwitterionic liposomes containing encapsulated dextran-coated 43 nm γ-Fe_2O_3 NPs (Resovist™) using a high-frequency generator (6.4 kW, 750–1150 kHz) operating for 5–25 min. CF release from DPPC:Chol (5:1) liposomes without encapsulated NPs was initiated between 35°C and 37°C. In contrast, the LNA analogues exhibited initial release at 34°C and 32°C with 7 and 14 mg Fe/ml, respectively. This initial release temperature was further tuned by increasing the cholesterol content (DPPC:Chol at 15:3) and incorporating a higher melting lipid (DSPC, $T_{\mathrm{m}} = 55$°C). Release was attributed to NP heating. Using a rat model, they demonstrated that release could also be achieved *in vivo*.

Chen *et al.* (2010) have examined the release of CF from LNAs formed with DPPC and bilayer-embedded oleic acid-coated 5 nm γ-Fe_2O_3 NPs at lipid:NP ratios of 10000:1, 5000:1, and 1000:1 as a function of RF energy (1 kW; 50–250 A, 281 kHz). Experiments were conducted at non-invasive RF energies up to 40 min and the greatest release rate was observed at 5000:1, indicating an optimal NP loading for triggering release. This optimum reflected a balance between NP loading and LNA structure – high loading is needed for triggering bilayer release, but can lead to NP aggregation and can compromise LNA structure and stability. A unique observation was the fact that increasing NP loading reduced or eliminated spontaneous leakage by increasing bilayer stability. CF release was attributed to bilayer disruption via local heating and/or LNA rupture, which produced transient voids or pores. Although, these findings demonstrated that higher nanoparticle embedding does not strictly equate to greater controlled release during EMF heating.

Targeted therapy

Optimal drug delivery and biomedical imaging involves the distribution of drug and/or imaging agent to the diseased tissue while minimizing adverse side-effects to healthy tissues. Adverse side-effects limit the drug dosage that may be used during treatment, potentially requiring the dosage to be reduced, delayed, and/or discontinued. Targeting LNAs limit adverse side-effects to healthy tissues and enhance drug delivery and uptake by localizing drug delivery to specific target sites (Malam *et al.*, 2009; Minko *et al.*, 2006).

Effective therapeutic and diagnostic treatment requires sufficient bioavailability of LNAs to diseased cells. A major limitation of LNAs, and liposomal drug delivery in general, is that the residence time is not long enough to enable the non-modified liposome assemblies to accumulate at tumour sites (Zhang *et al.*, 2008). LNA residence time is directly related to size and surface properties (discussed herein) (Maurer *et al.*, 2001; Samad *et al.*, 2007).

Passive targeting

Passive targeting of lipid–nanoparticle assemblies is facilitated by the enhanced permeability and retention (EPR) effect. Tumour growth requires a sufficient supply of oxygen and nutrients. Therefore, during angiogenesis, tumours will produce an increased amount of vascular endothelial growth factors routing new blood vessels and/or existing blood vessels directly to growing tumours. These tumour-associated neovasculatures are typically disorganized and highly permeable allowing leakage of larger entities, such as lipid–nanoparticle assemblies, into the tumour. The entities are retained by the tumour because the defective lymphatic drainage prevents larger entities from returning to the circulatory system. LNAs accumulate in the tumour providing an effective targeting mechanism that can be utilized for drug delivery, imaging, and/or hyperthermia (Cho *et al.*, 2008; Corchero and Villaverde, 2009; van Vlerken and Amiji, 2006).

In order for LNAs to accumulate within a tumour by the EPR effect, the residence time of LNAs in the blood must be sufficient to permeate into the tumour. Longer circulating assemblies provide a greater opportunity to reach the tumour vascular system and enter the tumour for drug delivery and/or imaging. Nanoparticle elimination is primarily based upon the reticuloendothelial system (RES) uptake. LNAs (as mentioned earlier) can increase their blood residence times by reducing clearance and absorption through PEGylation.

Passive targeting also exploits the unique environment created by tumours. As stated above, rapidly proliferating tumour cells require sufficient oxygen and nutrients. However, the supply of oxygen and nutrients is typically insufficient to maintain the rapid cell growth, forcing tumour cells to breakdown sugars through glycolysis (Cho *et al.*, 2008; Corchero and Villaverde, 2009; Minko *et al.*, 2006; van Vlerken and Amiji, 2006). Glycolysis, under anaerobic conditions, leads to lactic acid build-up, creating an acidic environment (Pelicano *et al.*, 2006). LNAs comprise pH-responsive lipids facilitating drug release in the acidic tumour environment (Cho *et al.*, 2008).

Passive targeting can also be utilized for antibacterial applications. Pornpattananangkul *et al.* (2011) attached chitosan-coated gold to vancomycin encapsulated egg PC liposomes. The attached gold nanoparticles prevented fusing and spontaneous leakage increasing stability and shelf-life. Antibiotic release was triggered by the presence of bacterial toxins that caused pore formation. After 24 h of exposure, 100% of vancomycin was released from the liposome. Vancomycin is used as a 'last resort' for MRSA infections (Crossley, 2010). Such regulated LNA delivery systems reduce side effects and dosage by being capable of releasing encapsulates at the delivery site and in response to the severity of the infection.

Overall LNA size between 50–150 nm is optimal in order to avoid renal elimination (nanoparticles smaller than 6 nm) and prevent uptake by the reticuloendothelial system (nanoparticles larger than 200 nm) (Gullotti and Yeo, 2009; Veiseh *et al.*, 2010). LNA residence time is also dependent on surface charge. Cationic LNAs demonstrate a much higher affinity to bind to the vasculature of tumours than to normal healthy tissues. Tumour vessel area targeted by cationic LNAs was approximately 5 times greater than the vessels of normal healthy tissues (Campbell *et al.*, 2002). Cationic LNA uptake has been shown to be almost10 times higher than similar anionic and neutral LNAs (Ito *et al.*, 2005). The preferential binding can be exploited for targeted hyperthermia, drug delivery, and/or imaging.

Anionic and neutral entities and bodies larger than 200 nm, upon entering the blood stream, tend to absorb antibodies, called opsonins, activating the reticuloendothelial system and inducing mononuclear phagocyte system (MPS) or by hepatocyte uptake and elimination (Ishida *et al.*, 2002; Maurer *et al.*, 2001; Yan *et al.*, 2005). Incorporation of polyethylene glycol (PEG), a neutral hydrophilic polyether, onto the surface of LNAs is very effective for preventing opsonization and increasing the residence time. PEG has proven to be one of the most effective surface modifications and is probably the most used for producing long-circulating entities. For example, Doxil and Myocet in Table 11.1 are PEG-liposome encapsulated formulations of doxorubicin. Myocet and Doxin have an elimination half-life of 2.5 h and 55 h, respectively, whereas free doxorubicin has elimination half-life of 0.2 h (Danhier *et al.*, 2010). Other surface modifiers and ligands, such as polysaccharides, dysopsonins, dextran, etc., have been incorporated into the bilayer to increase liposome stability, residence time, and enabling targeting capability (Zhang *et al.*, 2008).

The final passive targeting method is topical delivery. Topical delivery is localized drug administration directly to a tumour by injection or surgery. This approach bypasses drugs administration through the circulatory system. By avoiding the circulatory system, adverse side-effects can be limited to the administration site (Minko *et al.*, 2006).

Active targeting

Active targeting involves incorporating a targeting ligand, most often antibodies, antibody fragments, vitamins, glycolipids, or peptides (Jørgensen and Nielsen, 2009), onto the surface of LNAs. In order to be effective, the targeting ligand must specifically bind to a receptor that is exclusive to tumours or is overexpressed by tumours compared to normal healthy tissue. Active targeting can be split into two categories: cancer cell targeting and tumoral endothelium targeting. Cancer cell targeting promotes tumour internalization of LNAs and intracellular delivery of therapeutic agents. The most common receptors for cancer cell targeting are transferrin, folate receptor, cell surface glycoproteins, and the epidermal growth factor. Tumoral endothelium targeting restricts the blood supply available to tumour cells by attacking angiogenic blood vessels. The most common entities subject to tumoral endothelium targeting are vascular endothelial growth factors, vascular endothelial growth factors receptors (VEGFR-1 and VEGFR-1), $\alpha_v\beta_3$ integrin, vascular cell adhesion molecule-1, and matrix metalloproteinases (Danhier *et al.*, 2010).

Active targeting can also be accomplished with magnetic drug targeting (MDT). MDT utilizes a static magnetic field to concentration of LNAs at selected delivery sites. LNAs with encapsulated or embedded magnetic or superparamagnetic NPs can be forced to aggregate at sites with applied magnetic fields. MDT allows LNAs to be localized at an identified disease site for hyperthermia and drug delivery, minimizing the effect to adjacent tissues.

Ligand targeting

Active targeting of tumours by LNAs has been accomplished by targeting folate receptor (FR). Tumour cells express FRs almost 100 times normal cells. Also, the FR expression rate is proportional to the how advanced the cancer is. Therefore, FR targeting can be an effective method for identifying and treating tumours (Lu and Low, 2002). FR targeting has been demonstrated both *in vitro* and *in vivo*. Folate targeted liposomes have been used with nanoparticles for tumour imaging, drug delivery, and hyperthermia (Kamaly *et al.*, 2009; Pradhan *et al.*, 2010).

FR targeted MLs have demonstrated selective drug delivery *in vitro*. PEGylated MLs incorporating FR targeting-lipids were constructed with co-encapsulated doxorubicin and maghemite. Human cervical cancer cell (HeLa), which overexpress FR, bound to the cell surface and uptake from HeLa cells was observed, whereas the control showed no surface binding or uptake. FR targeting was combined with AC EMF at RF nanoparticle heating to induce doxorubicin release. Doxorubicin release under EMF treatment for 2 h was three times greater than without EMF (Bothun *et al.*, 2011).

Pradhan *et al.* (2010) developed multifunctional thermosensitive PEGylated magnetoliposomes with a folic acid ligand capable of binding to FR. The magnetoliposomes were encapsulated with doxorubicin and iron oxide nanoparticles (Fig. 11.7). Folate receptor mediated uptake of magnetoliposomes into tumour cells was demonstrated. *In vitro* hyperthermia and controlled release of encapsulated doxorubicin was achieved by subjecting cell lines to EMF.

Liposomes with incorporated antibodies or antibody fragments are referred to as immunoliposomes. A number of studies have utilized immunoliposomes and nanoparticles for biomedical applications. Antibody-conjugated LNAs has been employed to target cells that overexpress epidermal growth factor receptor 2 (Her-2). Overexpression of Her-2 frequently occurs in cervical, colon, breast, prostate, brain, bladder, and lung cancers because it allows cells to grow more rapidly. Current cancer treatments tend to lack specificity administered causing adjacent tissues to be damaged. Her-2 targeted treatment may encounter similar adverse effects to adjacent tissues because Her-2 is a naturally occurring protein. Kullberg *et al.* developed a two-component drug delivery system to ensure that treatment would be concentrated to cells that overexpress Her-2. The two-component system requires more than one LNA to bind to a tumour cell in order to release the encapsulated drugs. The two-component system could also have two or more sets of LNAs that encapsulate different substances that are only effective when they are both present. The specificity of the two-component system is equal to the difference of tumour and healthy tissue Her-2 expression squared. Therefore, the two-component system could potentially deliver 3600 times more drugs to glioblastomas (capable of a 60-fold overexpression) than normal cells. The two-fold system could hypothetically be expanded to three or more components increasing the specificity exponentially (Kullberg *et al.*, 2005, 2009). Her-2 antibodies have also been utilized to construct targeted magnetoliposomes for hyperthermia treatment. The magnetoliposomes demonstrated localized hyperthermia to Her-2 overexpressing tumours under EMF (Kikumori *et al.*, 2008).

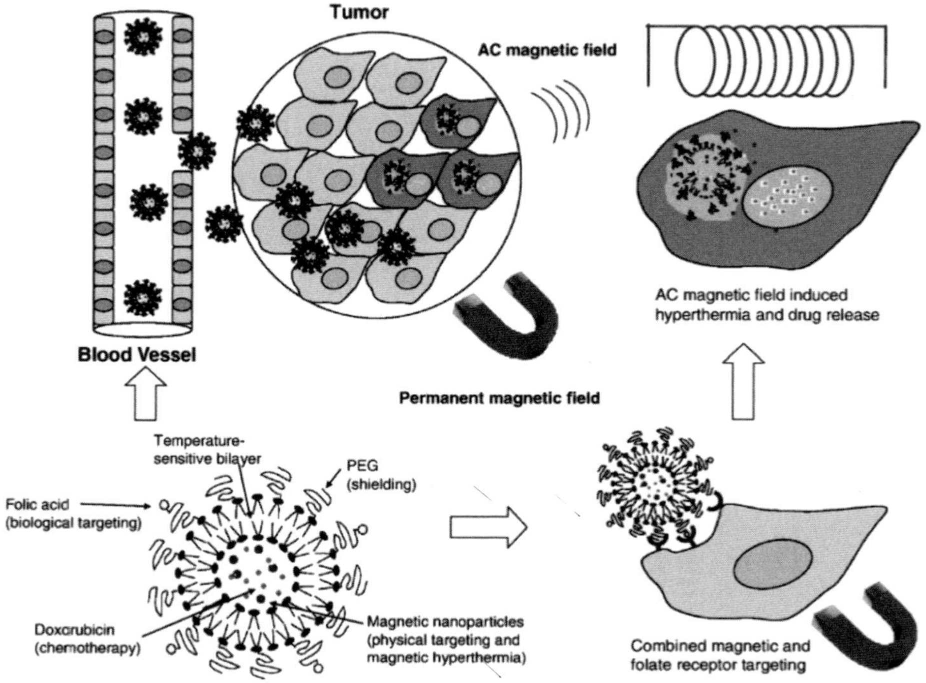

Figure 11.7 A multifunctional E-LNA for cancer thermo-chemotherapy from Pradhan *et al.* (2010). In this example the LNA is a temperature-sensitive magnetoliposome containing co-encapsulated iron oxide nanoparticles and doxorubicin. Passive targeting can be achieved through the enhanced permeation and retention (EPR) effect of tumour vasculature, and active targeting can be achieved via folate receptor and by applying a permanent magnetic field. The application of an AC electromagnetic field can be used to release the drug and achieve hyperthermia treatment. Reprinted from Pradhan *et al.* (2010) with permission. A colour version of this figure is located in the plate section at the back of the book.

Cationic MLs incorporating transferrin receptor targeting antibodies (anti-transferrin receptor single-chain antibody fragment or TfRscFv) have been developed by Yang *et al.* (2008). SPIO NPs were encapsulated into DOTAP-DOPE liposomes with TfRscFv incorporated onto the bilayer surface. *In vitro* breast and pancreatic uptake of SPIO liposomes with TfRscFv antibodies was at least ~2.5 times and 11 times greater than non-targeting SPIO liposomes and free SPIO NPs, respectively.

Magnetic drug targeting

Accumulation with MDT is dependent on the properties of the encapsulated magnetic nanoparticles, tumour depth, blood flow, vasculature, drug binding, and concentration. MDT is also highly dependent on the external magnetic field. Penetration depth of magnets has been shown to be on the millimetre scale making MDT only effective for superficial tissues. Although neodymium iron boron magnets have demonstrated magnetic field depths of up to 15 cm on SPIOs (Neuberger *et al.*, 2005). Interest in MDT has declined in recent years due to lack of effectiveness and issues such as embolization (Sun *et al.*, 2008).

The feasibility of magnetic targeting of with magnetoliposomes *in vitro* and *in vivo* with external magnet has been demonstrated (Fortin-Ripoche *et al.*, 2006). MLs (E-LNAs) encapsulating 7.5 nm maghemite have been used as effective MRI contrast agents *in vivo*. MDT delivery of MLs had a MRI negative enhancement of −52% compared with −7% for without MDT. Enhancement was maximized after 3 h and after 24 h maghemite only remained in the targeted tumour (Fortin-Ripoche *et al.*, 2006). Rivière *et al.* (2007) also demonstrated that real-time *in vivo* imaging and drug delivery of LNAs with a neodymium magnet to the brain vasculature of mice was achievable.

Magnetic targeting has also been applied to gene delivery. Transfection of pDNA was shown to be greater under the influence of a static magnetic field. After 10 min of magnetic induction, was 2.8 and 1.9 times greater for MLs (E-LNAs) with 3.0 mg/ml and 1.5 mg/ml magnetite concentration, respectively. Also, transfection at 0.75 mg/ml magnetite concentration was 2.6-fold greater with 30 min of magnetic induction than without (Zheng *et al.*, 2009).

Dandamudi *et al.* (2009) utilized MDT to deliver vinblastine-loaded MLs. After application of an external magnet for 2 h, PEGylated cationic liposomes encapsulating magnetite showed enhanced delivery of vinblastine (a chemotherapeutic drug). Tumour uptake with the magnet was ~16% as compared to ~5% with no magnet. MDT of vinblastine showed a significant showed an antitumour effect versus the control, reducing tumour nodules and hindering the growth of the tumour vasculature.

Some studies incorporated MDT with other targeting techniques. Pradhan *et al.* examined the effect of MDT on human epidermoid carcinoma (KB) and human cervical carcinoma (HeLa) cell uptake of doxorubicin encapsulated liposomes with and without folate receptor targeting (Fig. 11.7). Uptake was 2-fold higher for KB and 4-fold higher for HeLa with MDT than without. Also, the combination of MDT and folate receptor targeting uptake was 8-fold greater in KB and 42-fold greater in HeLa than without either active targeting techniques (Pradhan *et al.*, 2010). These results suggest that further research into MDT as for supplementary targeting may be beneficial.

Diagnostics and imaging

Magnetic resonance imaging is a noninvasive imaging technique that can be enhanced by the use of LNAs with superparamagnetic nanoparticles, such as iron oxide. These particles are capable of being manipulated by magnetic fields allowing them to be used as a contrast agent form magnetic resonance imaging (MRI). MRI is based upon the nuclear magnetic resonance of hydrogen protons of water in the body. A strong magnetic field (B_0) is applied to the body causing these hydrogen protons align with the magnetic field. The body is then exposed to a radiofrequency (RF) pulse, transverse to B_0, perturbing proton alignment with B_0. Relaxation, or the realignment of protons with B_0 after the RF pulse, releases energy absorbed during the RF pulse. Two separate relaxations are monitored; longitudinal relaxation, or T_1-recovery, and transverse relaxation, or T_2-decay. T_1-recovery (also referred to as 'spin-lattice' relaxation) is the release of energy to adjacent tissue as hydrogen realigns with B_0. T_2-decay (also referred to as 'spin-spin' relaxation) is the energy released from the exchange of energy between separate protons as nuclei goes from spinning in phase with each other (due to the RF pulse) to out of phase. 2D Fourier transform is applied to the relaxation data to generate an image. Differences in density and physical properties of tissues are related to the relaxation time and therefore can be distinguished in MRI images. Magnetic nanoparticles and LNAs can shorten the relaxation times of surrounding tissue and therefore be used as a contrast agent. A compound's effect on the relaxation of surrounding tissues, and therein its effectiveness on contrast agent, is called relaxivity. Relaxivity is defined as $R_1 = 1/T_1$ and $R_2 = 1/T_2$ (Mornet *et al.*, 2004; Sun *et al.*, 2008). Enhancement of relaxivity by LNA contrast agents is based on the concentration and inherent relaxivity of nanoparticles delivered (Fattahi *et al.*, 2011).

MRI has also provided to be a very useful tool for observing and understanding interstitial drug delivery. Ponce *et al.* (2007) measured doxorubicin concentration by measuring the change in T_1 relaxation of the co-encapsulated contrast agent. Spatial and temporal distribution profiles of doxorubicin and manganese from temperature-sensitive liposomes during hyperthermia is crucial to understanding tumour–drug interaction. Similar real time observations will provide pharmacodynamic and pharmacokinetic information to optimize drug efficacy from LNAs.

QDs have also been utilized for to measure and observe the cellular delivery. Bothun *et al.* (2009) demonstrated the ability to deliver different types of QDs. Hydrophobic and hydrophilic QDs were co-encapsulated in the bilayer and aqueous core of a cationic liposome. Uptake by human hepatocellular carcinoma cells demonstrated multi-fluorescent imaging. Subsequent work by Al-Jamal *et al.* (2011) demonstrated the release characteristics of D-LNAs containing QDs (for imaging) and encapsulated doxorubicin from gel (DSPC-QD) and fluid phase (EPC-QD) LNA bilayers. Their results showed that fluid phase D-LNAs were able to achieve 80% release over 6 h and that the cytotoxicity of doxorubicin was reduced through D-LNA encapsulation. The QD D-LNAs without doxorubicin exhibited minimal cytotoxicity (Fig. 11.8) (Al-Jamal and Kostarelos, 2011; Tian *et al.*, 2011).

Hyperthermia

Hyperthermia therapy to treat cancer dates back to Ancient Egyptian medical texts from about 3000 BC (Bolomey *et al.*, 1995). Hyperthermia involves heating the malignant tissues to temperatures between ~40 and 45°C. During hyperthermia, heat denatures intracellular proteins inducing death by necrosis or apoptosis. Hyperthermia has also demonstrated the ability to make tumour cells more vulnerable to therapies. Therefore, when hyperthermia is used, it is frequently used in conjunction with chemotherapy and radiotherapy. While the side-effects of hyperthermia are far less severe than those of chemotherapy and radiotherapy, there is a tendency to damage adjacent tissues. Use of hyperthermia has been tempered in recent years because of difficulty in applying heat to deeper tumours and delivering targeted heating (Hsu and Su, 2008; Ito *et al.*, 2004; Pradhan *et al.*, 2010). Development of new therapies may reenergize the field as an effective treatment for malignant tumours. Local hyperthermia has been demonstrated with NPs capable of photothermal and RF heating treatment of malignant tissues (Hildebrandt *et al.*, 2002). Targeting LNAs potentially can provide a local heat source for both hyperthermia and drug release without adversely effecting adjacent tissue (Hsu and Su, 2008; Huang *et al.*, 2011; Ito *et al.*, 2004; Smith *et al.*, 2006).

Heat transfer within tissues via NP heating can be described by a modified Pennes' bioheat transfer model (Pennes, 1948)

$$\rho_t c_{p,t} \frac{\delta T}{\delta t} = \nabla \cdot (k_t \nabla T) + \rho_b c_{p,b} \omega_b (T - T_b) + Q_m + Q_{NP}$$

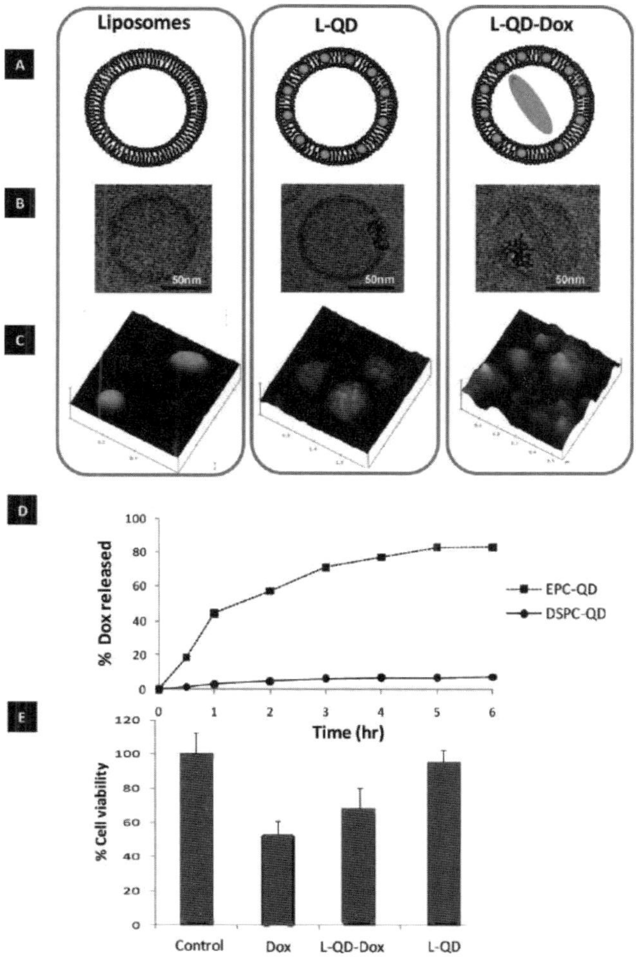

Figure 11.8 Combined imaging and doxorubicin release from D-LNAs prepared with hydrophobic quantum dots (green). QD embedment and D-LNA structure was verified by cryogenic transmission electron microscopy and atomic force microscopy. The results show that bilayer phase behaviour is a key parameter in doxorubicin release; fluid-phase D-LNAs achieved sustained release while gel-phase D-LNAs did not. Reprinted from Al-Jamal and Kostarelos (2011) with permission. A colour version of this figure is located in the plate section at the back of the book.

where ρ_t is the tissue density and $c_{p,t}$ is the tissue heat capacity. The first term on the right-hand side (RHS) of the equation describes the conductive heat transfer (k_t is the tissue thermal conductivity) and the second term describes the convective heat transfer (ρ_b is the blood density, $c_{p,b}$ is the blood heat capacity, ω_b is the blood perfusion rate and T_b is the blood temperature). Q_m is the rate of heat generated metabolically and Q_{NP} is the rate of heat generated from the power dissipation by the NPs, which accounts for the concentration of NPs (e.g. represented as *SAR*). This equation represents the case where the temperature profile in a tissue mass (macro-scale) can be determined as a function of NP concentration and applied field strength (e.g. laser or RF). It has been shown theoretically that sufficient iron oxide NP heating can be achieved (>42°C) at low blood perfusion rates to achieve tissue-level hyperthermia (Rast and Harrison, 2010). For cellular-level heating (nano- or micro-scale), convective heat transfer due to blood transfusion and Q_m are eliminated from the equation, yielding the expression reported by Keblinski *et al.* (2006) for RF NP heating and Xu *et al.* (2007) for Fe_3O_4 NP hyperthermia *in vitro*.

In vivo and *in vitro* biomedical applications

Cellular uptake and drug delivery

Chithrani *et al.* (2010) prepared anionic Au decorated liposomes with 105 nm d_h by incorporating Au-conjugated DPPE (DPPE-Nanogold; 1.4 nm Au particles) into preformed DPPC/cholesterol liposomes. Incorporation of DPPE-Nanogold was confirmed by TEM and EDS. Liposome uptake by HeLa cells in vitro was independent of the DPPE-Nanogold concentration (2000:1, 1000:1, and 500:1 Au NPs per liposome). This key discovery suggests that the presence of DPPE-Nanogold does not influence cell uptake and that high nanoparticle loadings can be achieved in such LNAs without compromising internalization. Examining the intracellular fate revealed that the Au-liposomes were present in lysosomes and accumulated near the nuclear membrane after incubating for 45 min.

Paasonen *et al.* (2010) examined uptake and light-induced calcein release of D-LNAs prepared hexanethiol-coated Au NPs (extension of Paasonen *et al.*, 2007). The LNAs were composed of DSPC:DPPC (9:1 molar ratio) and were multilamellar with sizes ranging from 250 to 370 nm. *In vitro* studies with human retinal pigment epithelial cell line (ARPE-19) showed that the D-LNAs were internalized by endocytosis and localized in endosomes. Exposure to UV light at 400 mW/cm^2 for 300+ s led to calcein release from D-LNAs, but not from liposomes that did not contain embedded nanoparticles.

Drug delivery and hyperthermia

Pradhan *et al.* (2010) recently prepared folate (Fol) ligated magnetoliposomes (E-LNAs) via thin film hydration that contained co-encapsulated 60 nm iron oxide NPs and doxorubicin in PBS (Fig. 11.8). The lipid components included DPPC, Chol, PEG_{2000}-DSPE, and Fol-PEG_{2000}-DSPE and the liposomes exhibited a melting temperature near 41°C. The liposomes were 361 nm in diameter and polydispersed (0.289), with a nanoparticle encapsulation efficiency of 24% (low, presumably attributed to the use of the thin film hydration method). The liposomes exhibited temperature-dependent doxorubicin release (>2-fold increase from 37°C to 43°C). Folate-receptor targeted uptake was demonstrated in HeLa and KB cell cultures, and reductions in cell viability were attributed to a synergistic effect of doxorubicin and hyperthermia treatment achieved in the presence of an EMF. While not mentioned specifically, we speculate that liposomal heating by the nanoparticles may have aided doxorubicin release.

Babincova *et al.* (2005) prepared DPPC/PEG_{2000}-DSPE LNAs via REV that contained both dextran-coated Fe_3O_4 and C_{60} fullerene (referred to as magnetofullerenosomes) to achieve magnetic targeting and photodynamic therapy (PDT), respectively. Bis(di-isobutyloctadecylsiloxy)-2,3-naphthalocyanato silicon (isoBO-SiNc), a photosensitizing agent, was also encapsulated. LNA performance for treating B16 pigmented melanoma was examined *in vivo* using a female mouse model. Without magnetic targeting, C_{60}-mediated

PDT followed by isoBO-SiNc-mediated PDT led to a significant reduction in tumour growth over 19 days relative to no treatment or isoBO-SiNc-mediated PDT alone. When magnetic targeting was applied (0.32 T magnet placed on the tumour surface for 24 h) there was negligible tumour growth over the same duration.

Conclusion

Liposomes and NPs are both well-established therapeutic and diagnostic agents. As both have been approved for clinical use, the next stage of development is to combine these two systems. LNAs combine the therapeutic advantages of these two nanotechnology systems creating a unique opportunity for achieving multi-functional therapeutic objectives. The liposomes can act to concentrate small NPs and shield them from the immune system. In turn, the NPs can be used to initiate and control drug release when exposed to external stimuli. Recent work demonstrates that there is a range of options for the design of LNAs to provide desired structures and functions. LNAs have been demonstrated to enhance the qualities of encapsulated payloads, by providing a system capable of targeting, responding to external stimuli, and concentrating encapsulates. Also, LNAs are capable of both therapeutic and diagnostic functions.

Nanotechnology is a burgeoning new field, providing solutions to problems that were once considered unsolvable. The multi-functional quality of LNAs makes them a very exciting development in the field of nanotheranostics. However, the design and use of LNAs is still in its infancy. Further investigation of the interactions between nanoparticles and the lipid bilayer is necessary to fully understand the formation, structure, and stability of LNAs. For biomedical applications, LNA bioavailability and toxicity need to be studied to grasp the full potential as multimodal nanotechnology therapeutics and diagnostics.

References

Al-Jamal, W.T., and Kostarelos, K. (2007). Liposome–nanoparticle hybrids for multimodal diagnostic and therapeutic applications. Nanomedicine *2*, 85–98.

Al-Jamal, W.T., and Kostarelos, K. (2011). Liposomes: from a clinically established drug delivery system to a nanoparticle platform for theranostic nanomedicine. Accounts Chem. Res. *44*, 1094–1104.

Al-Jamal, W.T., Al-Jamal, K.T., Bomans, P.H., Frederik, P.M., and Kostarelos, K. (2008a). Functionalized–quantum-dot-liposome hybrids as multimodal nanoparticles for cancer. *4* Small 1406–1415.

Al-Jamal, W.T., Al-Jamal, K.T., Tian, B., Lacerda, L., Bomans, P.H., Frederik, P.M., and Kostarelos, K. (2008b). Lipid-quantum dot bilayer vesicles enhance tumour cell uptake and retention in vitro and in vivo. ACS Nano *2*, 408–418.

Anderson, L.J.E., Hansen, E., Lukianova-Hleb, E.Y., Hafner, J.H., and Lapotko, D.O. (2010). Optically guided controlled release from liposomes with tunable plasmonic nanobubbles. Journal of Controlled Release *144*, 151–158.

Babincova, M., Sourivong, P., Leszczynska, D., and Babinec, P. (2003). Fullerenosomes: design of a novel nanomaterial for laser controlled topical drug release. Physica Medica *19*, 213–216.

Babincova, M., Sourivong, P., Leszczynska, D., and Babinec, P. (2004). Photodynamic therapy of pigmented melanoma B16 using sterically stabilized fullerenosomes. Laser Phys Lett *1*, 476–478.

Babinec, P., Babincova, M., Sourivong, P., and Leszczynska, D. (2005). Efficient treatment of pigmented B16 melanoma using photosensitized long-circulating magnetofullerenosomes. J Magn Magn Mater *293*, 394–397.

Bangham, A.D., and Horne, R.W. (1964). Negative staining of phospholipids and their structural modification by surface active agents as observed in the electron microscope. J. Mol. Biol. *8*, 660–668.

Bangham, A.D., Standish, M.M., and Watkins, J.C. (1965). Diffusion of univalent ions across the lamellae of swollen phospholipids. J. Mol. Biol. *13*, 238–252.
Barenholz, Y. (2012). Doxil® – The first FDA-approved nano-drug: Lessons learned. J. Control. Rel. *160*, 117–134.
Bharali, D.J., and Mousa, S.A. (2010). Emerging nanomedicines for early cancer detection and improved treatment: current perspective and future promise. Pharmacol. Ther. *128*, 324–335.
Binder, W.H., Sachsenhofer, R., Farnik, D., and Blaas, D. (2007). Guiding the location of nanoparticles into vesicular structures: a morphological study. Phys. Chem. Chem Phys. *9*, 6435–6441.
Bolomey, J.C., Seegenschmiedt, M.H., Fessenden, P., and Vernon, C.C. (1995). Thermoradiotherapy and Thermochemotherapy (Berlin, Springer).
Bothun, G.D. (2008). Hydrophobic silver nanoparticles trapped in lipid bilayers: Size distribution, bilayer phase behavior, and optical properties. J. Nanobiotechnol. *6*, 13.
Bothun, G.D., Lelis, A., Chen, Y., Scully, K., Anderson, L.E., and Stoner, M.A. (2011). Multicomponent folate-targeted magnetoliposomes: design, characterization, and cellular uptake. Nanomed. Nanotechnol. Biol. Med. *7*, 797–805.
Bothun, G.D., and Priess, M.R. (2011). Bilayer heating in magnetite nanoparticle–liposome dispersions via fluorescence anisotropy. J Coll Inter Sci *in press*.
Bothun, G.D., Rabideau, A.E., and Stoner, M.A. (2009). Hepatoma cell uptake of cationic multifluorescent quantum dot liposomes. J Phys Chem B *113*, 7725–7728.
Brazel, C.S. (2009). Magnetothermally-responsive nanomaterials: combining magnetic nanostructures and thermally-sensitive polymers for triggered drug release. Pharm Res *26*, 644–656.
Brezovich, I.A. (1988). Low frequency hyperthermia. Med Phys Monograph *16*, 82–111.
Campbell, R.B., Fukumura, D., Brown, E.B., Mazzola, L.M., Izumi, Y., Jain, R.K., Torchilin, V.P., and Munn, L.L. (2002). Cationic charge determines the distribution of liposomes between the vascular and extravascular compartments of tumours. Cancer Res *62*, 6831–6836.
Chan, W.C. (1998). Quantum dot bioconjugates for ultrasensitive nonisotopic detection. Science *281*, 2016–2018.
Chang, H.I., and Yeh, M.K. (2012). Clinical development of liposome-based drugs: formulation, characterization, and therapeutic efficacy. Int. J. Nanomed. *7*, 49–60.
Chen, Y., and Bothun, G.D. (2011). Cationic gel-phase liposomes with 'decorated' anionic SPIO nanoparticles: morphology, colloidal, and bilayer properties. Langmuir *27*, 8645–8652.
Chen, Y.J., Bose, A., and Bothun, G.D. (2010). Controlled release from bilayer-decorated magnetoliposomes via electromagnetic heating. ACS Nano. *4*, 3215–3221.
Chen, Y.J., and Bothun, G.D. (2009). Lipid-assisted formation and dispersion of aqueous and bilayer-embedded nano-C_{60}. Langmuir *25*, 4875–4879.
Chithrani, D.B., Dunne, M., Stewart, J., Allen, C., and Jaffray, D.A. (2010). Cellular uptake and transport of gold nanoparticles incorporated in a liposomal carrier'. Nanomed. Nanotechnol. *6*, 161–169.
Cho, K., Wang, X., Nie, S., Chen, Z.G., and Shin, D.M. (2008). Therapeutic nanoparticles for drug delivery in cancer. Clin. Cancer Res. *14*, 1310–1316.
Corchero, J.L., and Villaverde, A. (2009). Biomedical applications of distally controlled magnetic nanoparticles. Trends Biotechnol. *27*, 468–476.
Crossley, K.B. (2010). Staphylococci in Human Disease, 2nd edn (Chichester, Wiley-Blackwell).
Dandamudi, S., Patil, V., Fowle, W., Khaw, B.-A., and Campbell, R.B. (2009). External magnet improves antitumor effect of vinblastine and the suppression of metastasis. Cancer Science *100*, 1537–1543.
Danhier, F., Feron, O., and Préat, V. (2010). To exploit the tumor microenvironment: Passive and active tumor targeting of nanocarriers for anti-cancer drug delivery. J. Control. Rel. *148*, 135–146.
De Cuyper, M., and Joniau, M. (1988). Magnetoliposomes: Formation and structural characterization. Eur. Biophys. J. *15*, 311–319.
Doi, Y., Ikeda, A., Akiyama, M., Nagano, M., Shigematsu, T., Ogawa, T., Takeya, T., and Nagasaki, T. (2008). Intracellular uptake and photodynamic activity of water-soluble [60]- and [70]fullerenes incorporated in liposomes. Chem. Eur. J. *14*, 8892–8897.
Dudu, V., Ramcharan, M., Gilchrist, M.L., Holland, E.C., and Vazquez, M. (2008). Liposome delivery of quantum dots to the cytosol of live cells. J. Nanosci. Nanotechnol. *8*, 2293–2300.
Emerich, D.F., and Thanos, C.G. (2006). The pinpoint promise of nanoparticle-based drug delivery and molecular diagnosis. Biomol. Eng. *23*, 171–184.

Fattahi, H., Laurent, S., Liu, F., Arsalani, N., Elst, L.V., and Muller, R.N. (2011). Magnetoliposomes as multimodal contrast agents for molecular imaging and cancer nanotheragnostics. Nanomedicine *6*, 529–544.

Fortin-Ripoche, J.P., Martina, M.S., Gazeau, F., Menager, C., Wilhelm, C., Bacri, J.C., Lesieur, S., and Clement, O. (2006). magnetic targeting of magnetoliposomes to solid tumors with mr imaging monitoring in mice: feasibility. Radiology *239*, 415–424.

Fukumori, Y., and Ichikawa, H. (2006). Nanoparticles for cancer therapy and diagnosis. Adv. Powder Technol. *17*, 1–28.

Gannon, C.J., Patra, C.R., Bhattacharya, R., Mukherjee, P., and Curley, S.A. (2008). Intracellular gold nanoparticles enhance non-invasive radiofrequency thermal destruction of human gastrointestinal cancer cells. J Nanobiotechnology *6*, 2.

Gibaldi, M., Lee, M., Desai, A., and American Society of Health-System Pharmacists. (2007). Gibaldi's drug delivery systems in pharmaceutical care (Bethesda, MD, American Society of Health-System Pharmacists).

Ginzburg, V.V., and Balijepalli, S. (2007). Modelling the thermodynamics of the interaction of nanoparticles with cell membranes. Nano Lett. *7*, 3716–3722.

Gomes, J.F.P.S., Rank, A., Kronenberger, A., Fritz, J., Winterhalter, M., and Ramaye, Y. (2009). Polyelectrolyte-coated unilamellar nanometer-sized magnetic liposomes. Langmuir *25*, 6793–6799.

Gopalakrishnan, G., Danelon, C., Izewska, P., Prummer, M., Bolinger, P.-Y., Geissbühler, I., Demurtas, D., Dubochet, J., and Vogel, H. (2006). Multifunctional lipid/quantum dot hybrid nanocontainers for controlled targeting of live cells. Angew. Chem. Int. Ed. *45*, 5478–5483.

Goyal, P., Goyal, K., Vijaya Kumar, S.G., Singh, A., Katare, O.P., and Mishra, D.N. (2005). Liposomal drug delivery systems – clinical applications. Acta Pharm. *55*, 1–25.

Gregoriadis, G. (1973). Drug entrapment in liposomes. FEBS Lett *36*, 292–296.

Groneberg, D.A., Giersig, M., Welte, T., and Pison, U. (2006). Nanoparticle-based diagnosis and therapy. Curr. Drug Targets *7*, 643–648.

Gullotti, E., and Yeo, Y. (2009). Extracellularly activated nanocarriers: a new paradigm of tumour targeted drug delivery. Mol.Pharmaceut. *6*, 1041–1051.

Hildebrandt, B., Wust, P., Ahlers, O., Dieing, A., Sreenivasa, G., Kerner, T., Felix, R., and Riess, H. (2002). The cellular and molecular basis of hyperthermia. Crit. Rev. Oncol. Hematol. *43*, 33–56.

Hong, K., Friend, D.S., Glabe, C.G., and Papahadjopoulos, D. (1983). Liposomes containing colloidal gold are a useful probe of liposome–cell interactions. Biochim. Biophys. Acta *732*, 320–323.

Hsu, M.H., and Su, Y.C. (2008). Iron-oxide embedded solid lipid nanoparticles for magnetically controlled heating and drug delivery. Biomed. Microdevices *10*, 785–793.

Huang, H.-C., Barua, S., Sharma, G., Dey, S.K., and Rege, K. (2011). Inorganic nanoparticles for cancer imaging and therapy. J. Control. Rele. *155*, 344–357.

Huang, X., El-Sayed, I.H., and El-Sayed, M.A. (2010). Applications of gold nanorods for cancer imaging and photothermal therapy. Methods Mol. Biol. *624*, 343–357.

Hwang, K.C., and Mauzerall, D. (1993). Photoinduced electron-transport across a lipid bilayer mediated by C70. Nature *361*, 138–140.

Ikeda, A., and Kikuchi, J.-I. (2008). Fullerene C70-containing liposome, method for producing the same, and use of the same (National University Corporation Nara Institute of Science and Technology, Japan.), pp. 47pp.

Ikeda, A., Nagano, M., Akiyama, M., Matsumoto, M., Ito, S., Mukai, M., Hashizume, M., Kikuchi, J.-i., Katagiri, K., Ogawa, T., *et al.* (2009). Photodynamic activity of C70 caged within surface-cross-linked liposomes. Chem. Asian J. *4*, 199–205.

Ikeda, A., Sato, T., Kitamura, K., Nishiguchi, K., Sasaki, Y., Kikuchi, J., Ogawa, T., Yogo, K., and Takeya, T. (2005). Efficient photocleavage of DNA utilizing water-soluble lipid membrane-incorporated [60]fullerenes prepared using a [60]fullerene exchange method. Org Biomol Chem *3*, 2907–2909.

Immordino, M.L., Dosio, F., and Cattel, L. (2006). Stealth liposomes: review of the basic science, rationale, and clinical applications, existing and potential. Int J Nanomed *1*, 297–315.

Ishida, T., Harashima, H., and Kiwada, H. (2002). Liposome clearance. Biosci Rep *22*, 197–224.

Ito, A., Kuga, Y., Honda, H., Kikkawa, H., Horiuchi, A., Watanabe, Y., and Kobayashi, T. (2004). Magnetite nanoparticle-loaded anti-HER2 immunoliposomes for combination of antibody therapy with hyperthermia. Cancer Lett. *212*, 167–175.

Ito, A., Shinkai, M., Honda, H., and Kobayashi, T. (2005). Medical application of functionalized magnetic nanoparticles. J. Biosci. Bioeng. *100*, 1–11.

Jang, H., Pell, L.E., Korgel, B.A., and English, D.S. (2003). Photoluminescence quenching of silicon nanoparticles in phospholipid vesicle bilayers. J. Photochem. Photobiol. A *158*, 111–117.

Jeng, U.-S., Hsu, C.-H., Lin, T.-L., Wu, C.-M., Chen, H.-L., Tai, L.-A., and Hwang, K.-C. (2005). Dispersion of fullerenes in phospholipid bilayers and the subsequent changes in the host bilayers. Physica B *357*, 193–198.

Jin, S., and Ye, K. (2007). Nanoparticle-mediated drug delivery and gene therapy. Biotechnol. Prog. *23*, 32–41.

Jørgensen, L., and Nielsen, H.M. (2009). Delivery Technologies for Biopharmaceuticals: Peptides, Proteins, Nucleic Acids and Vaccines (Chichester, Wiley).

Kamaly, N., Kalber, T., Thanou, M., Bell, J.D., and Miller, A.D. (2009). Folate receptor targeted bimodal liposomes for tumour magnetic resonance imaging. Bioconjug. Chem. *20*, 648–655.

Keblinski, P., Cahill, D.G., Bodapati, A., Sullivan, C.R., and Taton, T.A. (2006). Limits of localized heating by electromagnetically excited nanoparticles. J. Appl. Phys. *100*, doi.org/10.1063/1.2335783.

Kikumori, T., Kobayashi, T., Sawaki, M., and Imai, T. (2008). Anti-cancer effect of hyperthermia on breast cancer by magnetite nanoparticle-loaded anti-HER2 immunoliposomes. Breast Cancer Research and Treatment *113*, 435–441.

Kirat, K.E., Morandat, S., and Dufrene, Y.F. (2010). Nanoscale analysis of supported lipid bilayers using atomic force microscopy. Biochim. Biophys. Acta *1798*, 750–765.

Kojima, C., Hirano, Y., Yuba, E., Harada, A., and Kono, K. (2008a). Preparation and characterization of complexes of liposomes with gold nanoparticles. Coll. Surf. B *66*, 246–252.

Kojima, C., Hirano, Y., Yuba, E., Harada, A., and Kono, K. (2008b). Preparation and characterization of complexes of liposomes with gold nanoparticles. Coll. Surf. B *66*, 246–252.

Kola, I., and Landis, J. (2004). Opinion: Can the pharmaceutical industry reduce attrition rates? Nature Rev. Drug Disc. *3*, 711–716.

Koo, O.M., Rubinstein, I., and Onyuksel, H. (2005). Role of nanotechnology in targeted drug delivery and imaging: a concise review. Nanomed. Nanotechnol. Biol. Med. *1*, 193–212.

Krpetic, Z., Nativo, P., See, V., Prior, I.A., Brust, M., and Volk, M. (2010). Inflicting controlled nonthermal damage to subcellular structures by laser-activated gold nanoparticles. Nano Lett. *10*, 4549–4554.

Kshirsagar, N.A., Pandya, S.K., Kirodian, G.B., and Sanath, S. (2005). Liposomal drug delivery system from laboratory to clinic. J. Postgrad. Med. *51* (Suppl. 1), S5–15.

Kullberg, M., Mann, K., and Owens, J.L. (2005). Improved drug delivery to cancer cells: a method using magnetoliposomes that target epidermal growth factor receptors. Med. Hypotheses *64*, 468–470.

Kullberg, M., Mann, K., and Owens, J.L. (2009). A two-component drug delivery system using Her-2-targeting thermosensitive liposomes. J. Drug Target. *17*, 98–107.

Larsen, B.A., Haag, M.A., Serkova, N.J., Shroyer, K.R., and Stoldt, C.R. (2008). Controlled aggregation of superparamagnetic iron oxide nanoparticles for the development of molecular magnetic resonance imaging probes. Nanotechnology *19*, doi:10.1088/0957-4484/19/26/265102.

Laurent, S., Forge, D., Port, M., Roch, A., Robic, C., Vander Elst, L., and Muller, R.N. (2008). Magnetic iron oxide nanoparticles: synthesis, stabilization, vectorization, physicochemical characterizations, and biological applications. Chem. Rev. *108*, 2064–2110.

Leckband, D., Israelachvili, J. (2001). Intermolecular forces in biology. Q. Rev. Biophys. *34*, 105–267.

Legrand, S., Catheline, A., Kind, L., Constable, E.C., Housecroft, C.E., Landmann, L.. Banse, P., Pielesa, U., and Wirth-Heller A. (2008). Controlling silica nanoparticle properties for biomedical applications through surface modification. New J. Chem. *32*, 588–593.

Leeson, P.D., and Davis, A.M. (2004). Time-related differences in the physical property profiles of oral drugs. J. Med. Chem. *47*, 6338–6348.

Li, J.L., and Gu, M. (2010). Gold-nanoparticle-enhanced cancer photothermal therapy. IEEE J. Sel. Top. Quant. *16*, 989–996.

Li, N., Kommireddy, D.S., Lvov, Y., Liebenberg, W., Tiedt, L.R., and De Villiers, M.M. (2006). Nanoparticle multilayers: surface modification of photosensitive drug microparticles for increased stability and in vitro bioavailability. J. Nanosci. Nanotechnol. *6*, 3252–3260.

Lipowsky, R., and Dobereiner, H.G. (1998). Vesicles in contact with nanoparticles and colloids. Europhys. Lett. *43*, 219–225.

Lu, Y., and Low, P.S. (2002). Folate-mediated delivery of macromolecular anticancer therapeutic agents. Adv. Drug Deliv. Rev. *54*, 675–693.

McCarthy, J., and Weissleder, R. (2008). Multifunctional magnetic nanoparticles for targeted imaging and therapy. Adv. Drug Deliv. Rev. *60*, 1241–1251.
Mady, M.M., Fathy, M.M., Youssef, T., and Khalil, W.M. (2011). Biophysical characterization of gold nanoparticles-loaded liposomes. Physica Medica. *28*, 288–295.
Malam, Y., Loizidou, M., and Seifalian, A.M. (2009). Liposomes and nanoparticles: nanosized vehicles for drug delivery in cancer. Trends Pharmacol. Sci. *30*, 592–599.
Marchenko, I., Yashchenok,A., German,S., Inozemtseva,O., Gorin,D., Bukreeva,T., Mohwald, H., Skirtach,A. (2010). Polyelecrolytes: influence of evaporative self-assembly of particle and assembly of multilayres with polymers, nanoparticle, and carnon nanotubes. Polymers *2*, 690–708.
Mart, R.J., Liem, K.P., and Webb, S.J. (2009). Creating functional vesicle assemblies from vesicles and nanoparticles. Pharm. Res.-Dord. *26*, 1701–1710.
Maurer, N., Fenske, D.B., and Cullis, P.R. (2001). Developments in liposomal drug delivery systems. Exp. Opin. Bio.l Ther. *1*, 923–947.
Medina, O.P., Zhu, Y., and Kairemo, K. (2004). Targeted liposomal drug delivery in cancer. Curr. Pharm. Des. *10*, 2981–2989.
Michalet, X., Pinaud, F.F., Bentolila, L.A., Tsay, J.M., Doose, S., Li, J.J., Sundaresan, G., Wu, A.M., Gambhir, S.S., and Weiss, S. (2005). Quantum dots for live cells, in vivo imaging, and diagnostics. Science *307*, 538–544.
Minko, T., Pakunlu, R.I., Wang, Y., Khandare, J.J., and Saad, M. (2006). New generation of liposomal drugs for cancer. Anticancer Agents Med. Chem. *6*, 537–552.
Moen, M.D., Lyseng-Williamson, K.A., and Scott, L.J. (2009). Liposomal amphotericin B: a review of its use as empirical therapy in febrile neutropenia and in the treatment of invasive fungal infections. Drugs *69*, 361–392.
Moghimi, S.M., and Szebeni, J. (2003). Stealth liposomes and long circulating nanoparticles: critical issues in pharmacokinetics, opsonization and protein-binding properties. Prog. Lipid Res. *42*, 463–478.
Mornet, S., Vasseur, S., Grasset, F., and Duguet, E. (2004). Magnetic nanoparticle design for medical diagnosis and therapy. J. Mater. Chem. *14*, 2161–2175.
Mulder, W.J., Strijkers, G.J., van Tilborg, G.A., Griffioen, A.W., and Nicolay, K. (2006). Lipid-based nanoparticles for contrast-enhanced MRI and molecular imaging. NMR Biomed. *19*, 142–164.
Murakami, Y., Nakano, A., Yoshimatsu, A., Uchitomi, K., and Matsuda, Y. (1984). Characterization of Molecular Aggregates of Peptide Amphiphiles and Kinetics of Dynamic Processes Performed by Single-Walled Vesicles. J. Am. Chem. Soc. *106*, 3613–3623.
Neuberger, T., Schöpf, B., Hofmann, H., Hofmann, M., and von Rechenberg, B. (2005). Superparamagnetic nanoparticles for biomedical applications: possibilities and limitations of a new drug delivery system. J. Magn. Magn. Mater. *293*, 483–496.
Niu, S.F., and Manzerall, D. (1996). Fast and efficient charge transport across a lipid bilayer is electronically mediated by C-70 fullerene aggregates. J. Am. Chem. Soc. *118*, 5791–5795.
Oldenburg, S.J., Jackson, J.B., Westcott, S.L., and Halas, N.J. (1999). Infrared extinction properties of gold nanoshells. Appl. Phys. Lett. *75*, 2897–2899.
Paasonen, L., Laaksonen, T., Johans, C., Yliperttula, M., Kontturi, K., and Urth, A. (2007). Gold nanoparticles enable selective light-induced contents release from liposomes. J. Control. Rel. *122*, 86–93.
Paasonen, L., Sipila, T., Subrizi, A., Laurinmaki, P., Butcher, S.J., Rappolt, M., Yaghmur, A., Urtti, A., and Yliperttula, M. (2010). Gold-embedded photosensitive liposomes for drug delivery: Triggering mechanism and intracellular release. J. Control. Rel. *147*, 136–143.
Pagano, R.E., and Weinstein, J.N. (1978). Interactions of liposomes with mammalian-cells. Annu. Rev. Biophys. Biol. *7*, 435–468.
Pankhurst, Q.A., Connolly, J., Jones, S.K., and Dobson, J. (2003). Applications of magnetic nanoparticles in biomedicine. J. Phys. D *36*, 167–181.
Papahadjopoulos, D., and Ohki, S. (1969). Stability of asymmetric phospholipid membranes. Science *164*, 1075–1077.
Park, S.-H., Oh, S.-G., Mun, J.-Y., and Han, S.-S. (2005). Effects of silver nanoparticles on the fluidity of bilayer in phospholipid liposome. Coll. Surf. B *44*, 117–122.
Park, S.-H., Oh, S.-G., Mun, J.-Y., and Han, S.-S. (2006). Loading of gold nanoparticles inside the DPPC bilayers of liposome and their effects on membrane fluidities. Coll. Surf. B *48*, 112–118.

Pavon, L.F., Gamarra, L.F., Marti, L.C., Amaro, E., Moreira, C.A., Camargo-Mathias, M.I., and Okamoto, O.K. (2008). Ultrastructural characterization of CD133+ stem cells bound to superparamagnetic nanoparticles: possible biotechnological applications. J. Microsc. *231*, 374–383.

Pelicano, H., Martin, D.S., Xu, R.H., and Huang, P. (2006). Glycolysis inhibition for anticancer treatment. Oncogene *25*, 4633–4646.

Pennes, H.H. (1948). Analysis of tissue and arterial temperature in the resting human forearm. J. Appl. Physiol. *1*, 93–122.

Polyak, B., and Friedman, G. (2009). Magnetic targeting for site-specific drug delivery: applications and clinical potential. Exp. Opin. Drug Deliv. *6*, 53–70.

Ponce, A.M., Viglianti, B.L., Yu, D., Yarmolenko, P.S., Michelich, C.R., Woo, J., Bally, M.B., and Dewhirst, M.W. (2007). magnetic resonance imaging of temperature-sensitive liposome release: drug dose painting and antitumor effects. J. Natl. Cancer Inst. *99*, 53–63.

Pornpattananangkul, D., Olson, S., Aryal, S., Sartor, M., Huang, C.M., Vecchio, K., and Zhang, L. (2010). Stimuli-responsive liposome fusion mediated by gold nanoparticles. ACS Nano *4*, 1935–1942.

Pornpattananangkul, D., Zhang, L., Olson, S., Aryal, S., Obonyo, M., Vecchio, K., Huang, C.-M., and Zhang, L. (2011). Bacterial Toxin-Triggered Drug Release from Gold Nanoparticle-Stabilized Liposomes for the Treatment of Bacterial Infection. J. Am. Chem. Soc. *133*, 4132–4139.

Pradhan, P., Giri, J., Banerjee, R., Bellage, J., and Bahadur, D. (2007). Preparation and characterization of manganese ferrite-based magnetic liposomes for hyperthermia treatment of cancer. J. Magn. Magn. Mater. *311*, 208–215.

Pradhan, P., Giri, J., Rieken, F., Koch, C., Mykhaylyk, O., Döblinger, M., Banerjee, R., Bahadur, D., and Plank, C. (2010). Targeted temperature sensitive magnetic liposomes for thermo-chemotherapy. J. Control. Rel. *142*, 108–121.

Preiss, M.R., and Bothun, G.D. (2011). Stimuli-responsive liposome-nanoparticle assemblies. Exp. Opin. Drug Deliv. *8*, 1025–1040.

Puri, A., Loomis, K., Smith, B., Lee, J.H., Yavlovich, A., Heldman, E., and Blumenthal, R. (2009). Lipid-based nanoparticles as pharmaceutical drug carriers: from concepts to clinic. Crit. Rev. Ther. Drug Carrier Syst. *26*, 523–580.

Rasch, M.R., Rossinyol, E., Hueso, J.L., Goodfellow, B.W., Arbiol, J., and Korgel, B.A. (2010). Hydrophobic gold nanoparticle self-assembly with phosphatidylcholine lipid: membrane-loaded and Janus vesicles. Nano Letters *10*, 3733–3739.

Rast, L., and Harrison, J.G. (2010). Computational modelling of electromagnetically induced heating of magnetic manoparticle materials for hyperthermic cancer treatment. PIERS Online *6*, 690–694.

Rivera Gil, P., Hühn, D., del Mercato, L.L., Sasse, D., and Parak, W.J. (2010). Nanopharmacy: Inorganic nanoscale devices as vectors and active compounds. Pharmacol. Res. *62*, 115–125.

Riviere, C., Martina, M.S., Tomita, Y., Wilhelm, C., Tran Dinh, A., Menager, C., Pinard, E., Lesieur, S., Gazeau, F., and Seylaz, J. (2007). Magnetic targeting of nanometric magnetic fluid loaded liposomes to specific brain intravascular areas: a dynamic imaging study in mice. Radiology *244*, 439–448.

Roiter, Y., Ornatska, M., Rammohan, A.R., Balakrishnan, J., Heine, D.R., and Minko, S. (2008). Interaction of Nanoparticles with Lipid Membrane. Nano Lett. *8*, 941–944.

Rotomskis, R., Streckyte, G., and Karabanovas, V. (2006). [Nanoparticles in diagnostics and therapy: towards nanomedicine]. Medicina (Kaunas) *42*, 542–558.

Sabate, R., Barnadas-Rodriguez, R., Callejas-Fernandez, J., Hidalgo-Alvarez, R., and Estelrich, J. (2008). Preparation and characterization of extruded magnetoliposomes. Int. J. Pharm. *347*, 156–162.

Samad, A., Sultana, Y., and Aqil, M. (2007). Liposomal drug delivery systems: an update review. Curr. Drug Deliv. *4*, 297–305.

Sassaroli, E., Li, K.C.P., and O'Neill, B.E. (2009). Numerical investigation of heating of a gold nanoparticle and the surrounding microenvironment by nanosecond laser pulses for nanomedicine applications. Phys. Med. Biol. *54*, 5541–5560.

Sau, T.P., Urban, A.S., Dondapati, S.K., Fedoruk, M., Horton, M.R., Rogach, A.L., Stefani, F.D., Radler, J.O., and Feldmann, J. (2009). Controlling loading and optical properties of gold nanoparticles on liposome membranes. Coll. Surf. A *342*, 92–96.

Shinkai, M., Yanase, M., Honda, H., Wakabayashi, T., Yoshida, J., and Kobayashi, T. (1996). Intracellular hyperthermia for cancer using magnetite cationic liposomes: In vitro study. Jpn J. Cancer Res. *87*, 1179–1183.

Sivashankar, M., and Katyayani, T. (2011). Liposomes-the future of formulations. Int. J. Res. Pharm. Chem. *1,* 259–267.

Smith, A.M., Dave, S., Nie, S., True, L., and Gao, X. (2006). Multicolor quantum dots for molecular diagnostics of cancer. Expert Review of Molecular Diagnostics *6,* 231–244.

Sun, C., Lee, J.S., and Zhang, M. (2008). Magnetic nanoparticles in MR imaging and drug delivery. Adv. Drug Deliv. Rev. *60,* 1252–1265.

Szoka, F., Jr., and Papahadjopoulos, D. (1978). Procedure for preparation of liposomes with large internal aqueous space and high capture by reverse-phase evaporation. Proc. Natl. Acad. Sci. U.S.A. *75,* 4194–4198.

Tai, L.A., Tsai, P.J., Wang, Y.C., Wang, Y.J., Lo, L.W., and Yang, C.S. (2009). Thermosensitive liposomes entrapping iron oxide nanoparticles for controllable drug release. Nanotechnology *20* [Epub].

Teja, A.S., and Koh, P.-Y. (2009). Synthesis, properties, and applications of magnetic iron oxide nanoparticles. Prog. Crystal Growth Charact. Mater. *55,* 22–45.

Thorek, D.L.J., Chen, A.K , Czupryna, J., and Tsourkas, A. (2006). Superparamagnetic iron oxide nanoparticle probes for molecular imaging. Ann. Biomed. Eng. *34,* 23–38.

Tian, B., Al-Jamal, W.T., Al-Jamal, K.T., and Kostarelos, K. (2011). Doxorubicin-loaded lipid-quantum dot hybrids: Surface topography and release properties. Int. J. Pharm. *416,* 443–447.

Torchilin, V.P. (2005). Recent advances with liposomes as pharmaceutical carriers. Nat. Rev. Drug Discov. *4,* 145–160.

van Vlerken, L.E., and Amiji, M.M. (2006). Multi-functional polymeric nanoparticles for tumour-targeted drug delivery. Exp. Opin. Drug Deliv. *3,* 205–216.

Veiseh, O., Gunn, J.W., and Zhang, M. (2010). Design and fabrication of magnetic nanoparticles for targeted drug delivery and imaging. Adv. Drug Deliv. Rev. *62,* 284–304.

Volodkin, D.V., Skirtach, A.G., and Mohwald, H. (2009). Near-IR remote release from assemblies of liposomes and nanoparticles. Angew. Chem. Int. Ed. *48,* 1807–1809.

Von White II, G., Chen, Y., Roder-Hanna, J., Bothun, G.D., and Kitchens, C.L. (2012). Structural and thermal analysis of lipid vesicles encapsulating hydrophobic gold nanoparticles. ACS Nano *6,* 4678–4685.

Walling, M.A., Novak, J.A., and Shepard, J.R.E. (2009). Quantum dots for live cell and in vivo imaging. Int. J. Mol. Sci. *10,* 441–491.

Wang, B., Zhang, L.F., Bae, S.C., and Granick, S. (2008). Nanoparticle-induced surface reconstruction of phospholipid membranes. Proc. Natl. Acad. Sci. U.S.A. *105,* 18171–18175.

Weng, K.C., Noble, C.O , Papahadjopoulos-Sternberg, B., Chen, F.F., Drummond, D.C., Kirpotin, D.B., Wang, D., Hom, Y.K., Hann, B., and Park, J.W. (2008). Targeted tumour cell internalization and imaging of multifunctional quantum dot-conjugated immunoliposomes in vitro and in vivo. Nano Lett. *8,* 2851–2857.

Wi, H.S., Lee, K., and Pak, H.K. (2008). Interfacial energy consideration in the organization of a quantum dot-lipid mixed system. J. Phys. Cond. Mat. *20,* 1–6.

Wijaya, A., and Hamad-Schifferli, K. (2007). High-density encapsulation of Fe_3O_4 nanoparticles in lipid vesicles. Langmuir *23,* 9546–9550.

Wu, G.H., Milkhailovsky, A., Khant, H.A., Fu, C., Chiu, W., and Zasadzinski, J.A. (2008). Remotely triggered liposome release by near-infrared light absorption via hollow gold nanoshells. J. Am. Chem. Soc. *130,* 8175–8177.

Xu, R.Z., Zhang, Y., Ma, M., Xia, J.G., Liu, J.W., Guo, Q.Z., and Gu, N. (2007). Measurement of specific absorption rate and thermal simulation for arterial embolization hyperthermia in the Maghemite-Gelled model. Ieee T Magn *43,* 1078–1085.

Yan, X., Scherphof, G.L., and Kamps, J.A.A.M. (2005). Liposome Opsonization. Journal of Liposome Research *15,* 109–139.

Yang, C., Rait, A., Pirollo, K.F., Dagata, J.A., Farkas, N., and Chang, E.H. (2008). Nanoimmunoliposome delivery of superparamagnetic iron oxide markedly enhances targeting and uptake in human cancer cells in vitro and in vivo. Nanomed. Nanotechnol. Biol. Med. *4,* 318–329.

Yu, Y., and Granick, S. (2009). Pearling of lipid vesicles induced by nanoparticles. J. Am. Chem. Soc. *131,* 14158–14159.

Yu, Y., Anthony, S.M., Zhang, L., Bae, S.C., and Granick, S. (2007). Cationic nanoparticles stabilize zwitterionic liposomes better than anionic ones. J. Phys. Chem. C *111,* 8233–8236.

Zhang, L.F., and Granick, S. (2006). How to stabilize phospholipid liposomes (using nanoparticles). Nano Lett. *6,* 694–698.

Zhang, L., Gu, F.X., Chan, J.M., Wang, A.Z., Langer, R.S., and Farokhzad, O.C. (2008). Nanoparticles in medicine: therapeutic applications and developments. Clin. Pharmacol. Ther. *83*, 761–769.

Zhang, S.L., Li, J., Lykotrafitis, G., Bao, G., and Suresh, S. (2009). Size-dependent endocytosis of nanoparticles. Adv. Mater. *21*, 419–424.

Zheng, S., Zheng, Y., Beissinger, R.L., and Fresco, R. (1994). Microencapsulation of hemoglobin in liposomes using a double emulsion, film dehydration rehydration approach. BBA-Biomembr. *1196*, 123–130.

Zheng, X., Lu, J., Deng, L., Xiong, Y., and Chen, J. (2009). Preparation and characterization of magnetic cationic liposome in gene delivery. Int J Pharm *366*, 211–217.

Zhu, L., Huo, Z.L., Wang, L.L., Tong, X., Xiao, Y., and Ni, K.Y. (2009). Targeted delivery of methotrexate to skeletal muscular tissue by thermosensitive magnetoliposomes. Int. J. Pharmaceut. *370*, 136–143.

Index

S

T

V

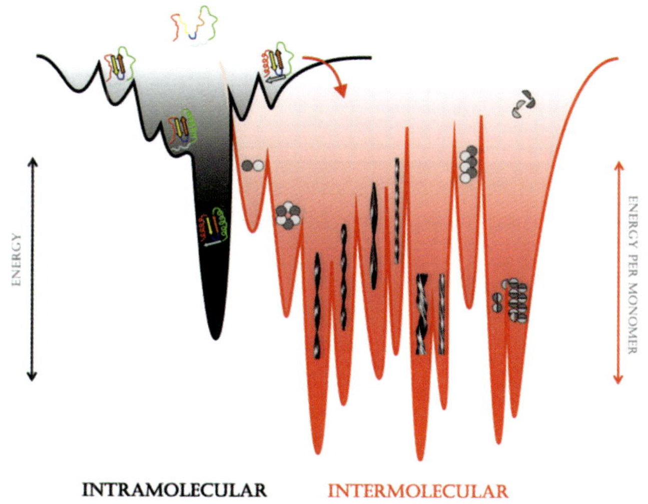

Figure 4.3 A schematic energy landscape of protein folding and aggregation/amyloid fibril formation. Reprinted from Eichner and Radford (2011), with permission from Elsevier.

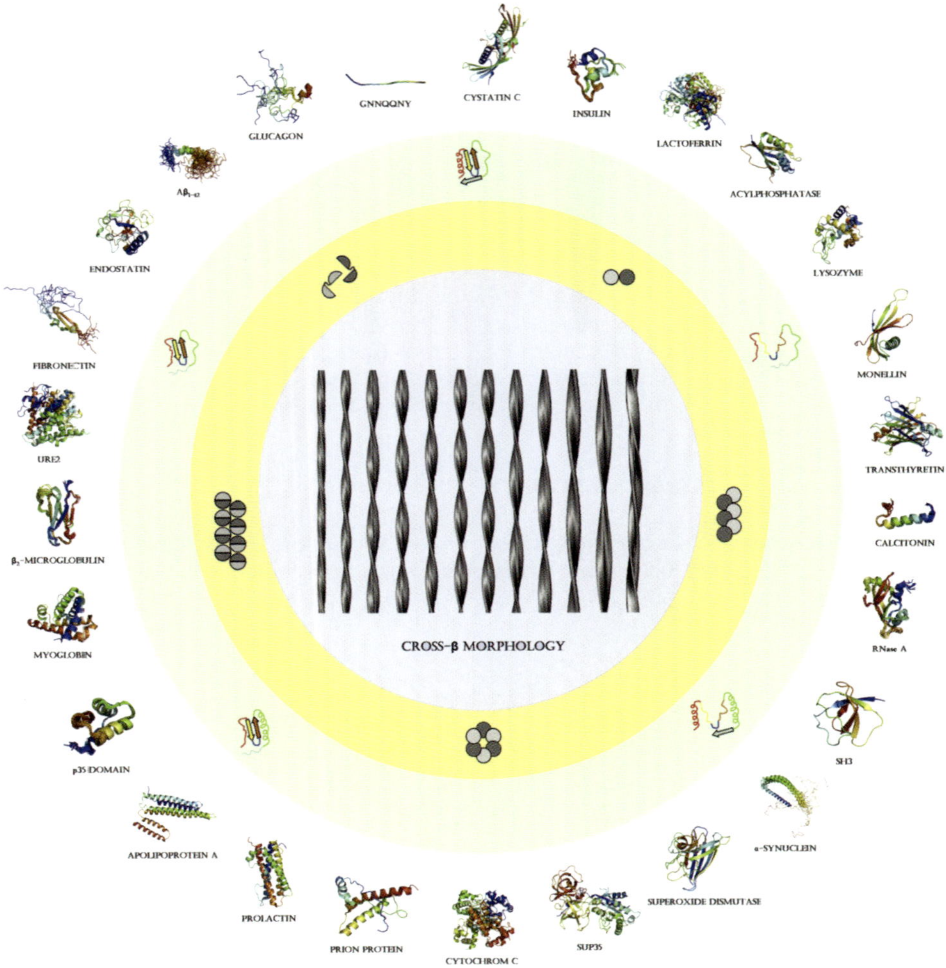

Figure 4.4 The diversity of amyloid precursor protein tertiary structures (reprinted from Eichner and Radford, 2011, with permission from Elsevier.) derived from X-ray crystallography and solution NMR are shown in rainbow colours. The centre (purple area) shows 12 different fibrillar polymorphs of Aβ_{1-40} determined by cryo-EM (reprinted from Meinhardt and Fändrich (2009), with permission from Elsevier). The green and yellow areas depict the monomers and oligomers that may be accessible by a polypeptide chain en route to the amyloid fibril fold.

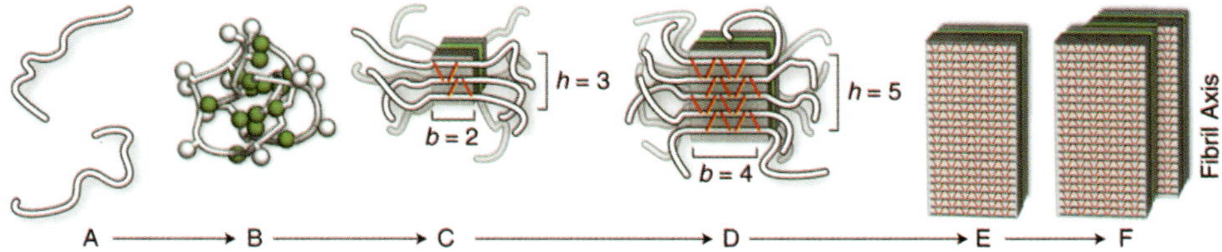

Figure 4.6 A model process of amyloid fibril formation via a nucleated growth mechanism. Reprinted from Schmit *et al.* (2011), with permission from Elsevier. '(State A) Isolated peptide monomers in solution. (State B) Oligomeric assembly of a few peptide chains. (State C) Nucleus of β-sheet structure. The peptide backbone runs perpendicular to the fibre axis. (State D) Postcritical nucleus structure showing more β-structure. (State E) A protofilament is a single long thread of β-structure consisting of a β-sandwich and two face-to-face β-sheet planes. (State F) The full fibril, a bundle of protofilaments shown here to contain p = 2 protofilament threads.'

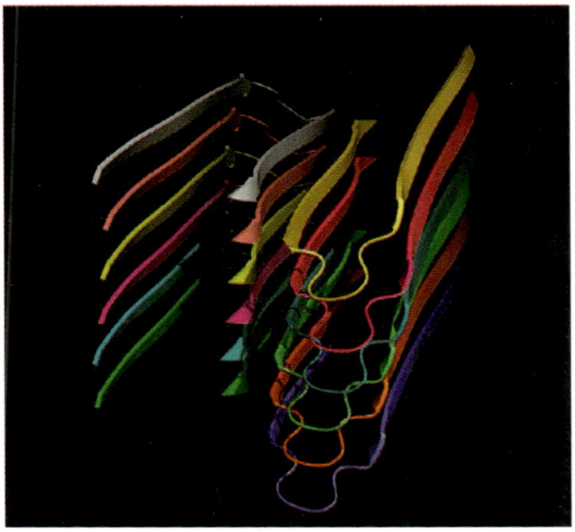

Figure 4.7 Structural model of mature $A\beta_{1-40}$ amyloid fibrils. Reprinted from Bertini *et al.* (2011), with permission from the American Chemical Society.

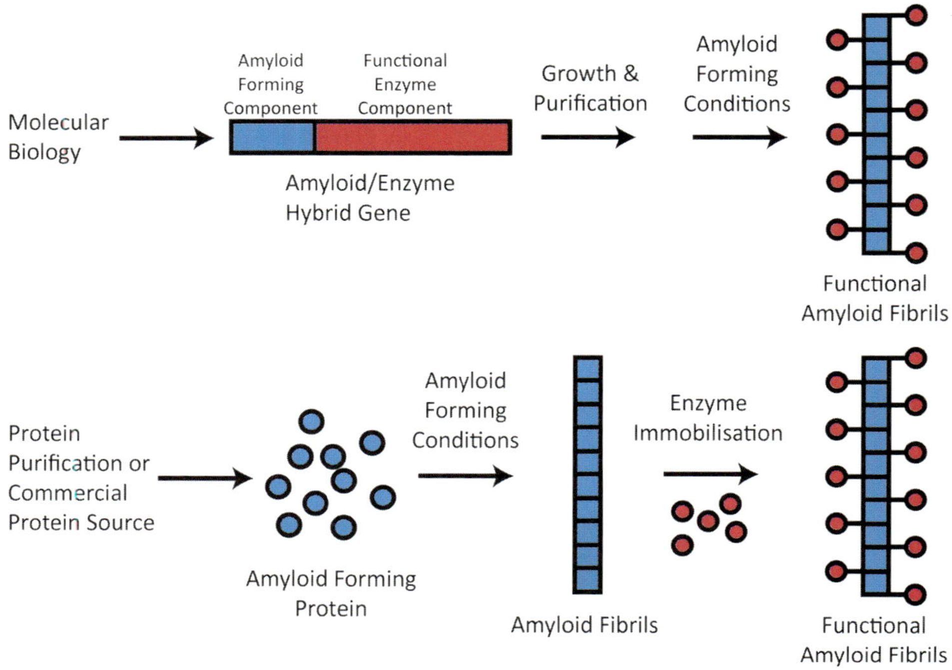

Figure 4.13 Formation of functional amyloid fibrils by (top) gene fusion of an amyloid forming peptide/protein sequence with a functional enzyme/protein sequence or (bottom) enzyme immobilization post amyloid formation.

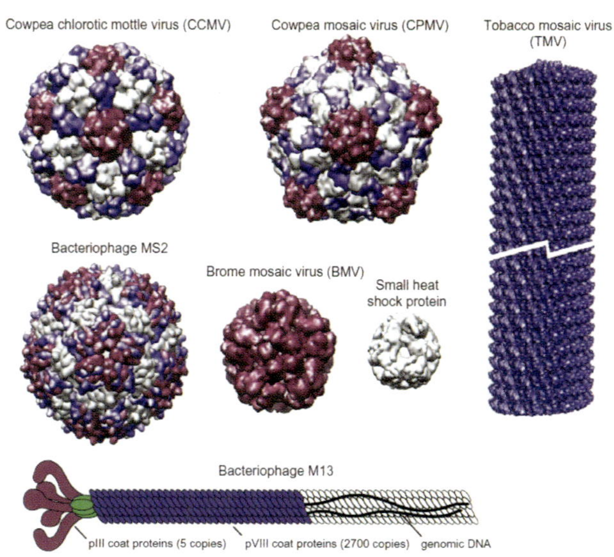

Figure 6.1 Structure of several viral capsids used for materials applications. (Reproduced with permission from Dedeo *et al.*, 2011.)

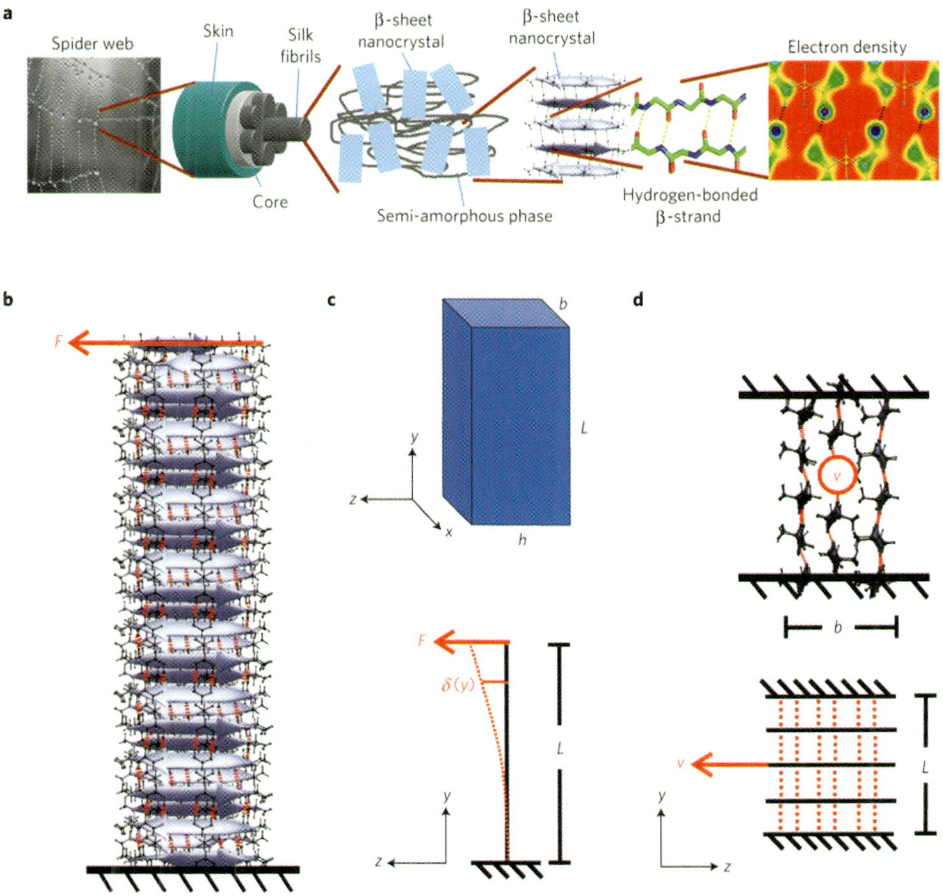

Figure 6.2 Hierarchical structure of spider silk, simulation set-up and theoretical considerations. (a) Schematic that highlights the key features of the hierarchical structure of spider silk from nano to macro, including the electron density at the Angstrom scale, hydrogen bonded β-strands, β-sheet nanocrystals, a hetero-nanocomposite of stiff nanocrystals embedded in a softer semi-amorphous phase and silk fibrils, which assembles into macroscopic silk fibres. (b) The atomistic structure of the silk β-sheet nanocrystal obtained from the Protein Data Bank (identification code 2slk) and replicated to build β-sheet nanocrystals of different sizes. In the first set of simulations, the β-sheet nanocrystal is subject to loading conditions similar to a cantilever beam with a constant tip loading, used to identify the bending rigidity and other structural properties. This loading mimics the characteristic lateral loading relevant to silk mechanics. c) Schematic representation of the β-sheet nanocrystal and definition of coordinates used here (upper part, where parameters b and h describe geometric parameters related to the number of sheets and the length of strands in the nanocrystal, and L is the size of the nanocrystal in the y direction). The lower part shows the geometry of the bending study and defines the displacement variable. d) Set-up for pull-out simulations, where to characterize fracture resistance of the nanocrystals, the central strand of the middle sheet is pulled out with constant velocity, while the top and bottom strands are restrained. (Reproduced with permission from Keten *et al.*, 2010.)

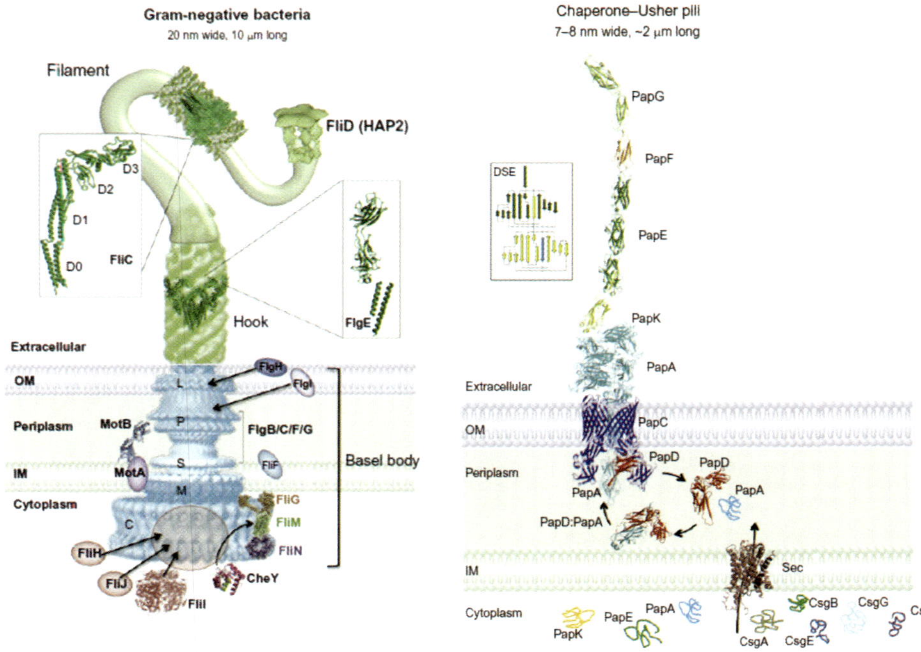

Figure 6.3 Bacterial flagella (left). The bacterial flagellum is a large macromolecular complex composed of about 30 different proteins and consisting of three major substructures: the filament, the hook, and the basal body. The flagellar basal body is formed by oligomeric ring complexes joined together by a rod and represented here by the cryo-electron microscopy 3-D structure (light blue). The filament itself consists of a single flagellin protein. The flagellin subunit of S. typhimurium FliC has an overall shape of the Greek letter 'Γ' and consists of four linearly connected domains, D0 to D3. Chaperone–Usher pili (right). Schematic representation of the biosynthetic pathway of chaperone–usher (CU) pili, here exemplified by P pili. P pili are composed of two major subassemblies: a thick and rigid rod made up of subunits of PapA (light blue) arranged in a right-handed helical cylinder; and a thin, flexible tip fibre in the distal end of the rod. In P pili, the tip fibre is composed of 5–10 PapE subunits (dark green) flanked by two minor subunits, PapF (orange) and PapK (yellow). The adhesin subunit PapG is responsible for interaction with the host tissues and present at the tip of the fibre. (Reproduced with permission from Van Gerven *et al.*, 2011.)

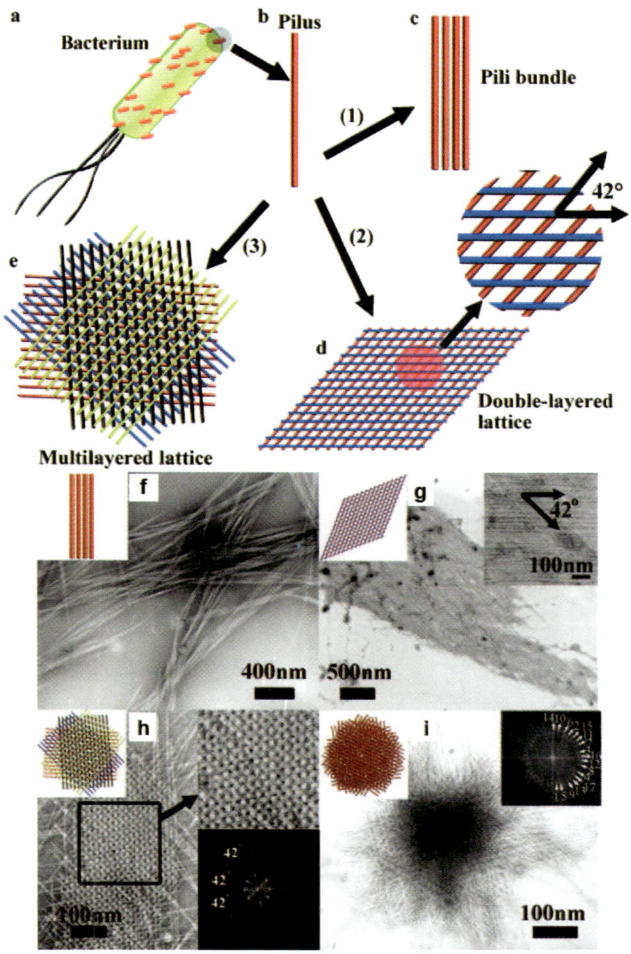

Figure 6.7 Self-assembly of pili into 1-D bundles, 2-D lattices, and 3-D lattices. (a) Illustration of a bacterial cell with many pili protruding from its surface. (b) An individual pilus. (c) Pili bundles. (d) Double-layer pili lattices. The two neighbouring layers of pili have a fixed angle of 428°. (e) Four-layer pili lattices. A red layer of parallel pili was formed, with subsequent assembly of a blue layer of pili on top of the red layer with a twist angle of 428°. After that, a black and a yellow layer are deposited in turn on top of the blue layer and black layer, with the neighbouring layer twisted at the same angle of 428°, resulting in the formation of a four-layer pili lattice. (f) TEM images of pili bundles. (g) A double-layer lattice with a fixed twist angle of 428° between two layers. (h) A four-layer pili lattice. FFT (inset, lower right) of selected area confirms the highly ordered structure and the twist angle between neighbouring layers. (i) A 17-layer pili lattice. FFT image (inset, upper right) confirms there are 17 layers of pili in the lattice with a fixed twist angle of 428° between adjacent layers. The spot corresponding to each layer is marked in the 2D-FFT image. The inset in the upper left corner of each TEM image illustrates the corresponding pili assemblies. (Reproduced with permission from Cao *et al.*, 2011.)

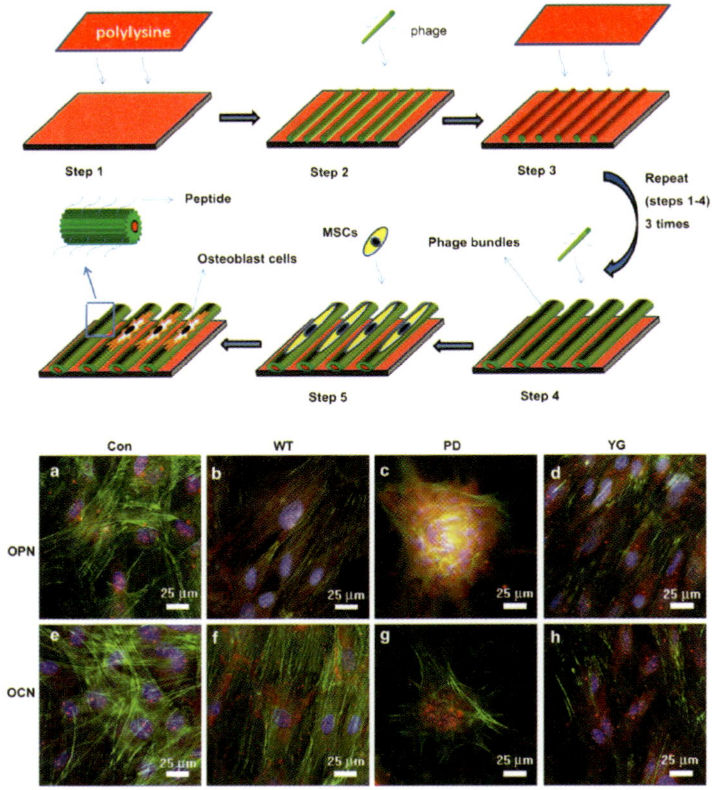

Figure 6.8 Schematic illustration (top) of the formation of the phage film by the layer-by-layer method and the MSCs growth on the phage film. After several layers were alternately deposited, the phage nanofibres assembled into parallel bundles separated by grooves. The BMSCs were oriented and elongated along the phage bundles. Fluorescent images (a–h) show the effect of phages with different surface chemistries on the differentiation of MSCs into osteoblasts. The phage film (b–d, f–h) show strong osteocalcin and osteopontin expression compared to those in the control groups (i.e. cells on glass slides coated with polylysine) (a, e). The phage film derived from phage displaying the OGP peptide (PDPLEPRREVCE) can stimulate MSCs aggregation to form a calcified nodule-like substance. Cell nuclei are stained by DAPI (blue) and F-actin is stained by FITC-labelled phalloidin (green). Osteopontin and osteocalcin are stained by rhodamine-labelled antibody (red). (Reproduced with permission from Zhu *et al.*, 2011.)

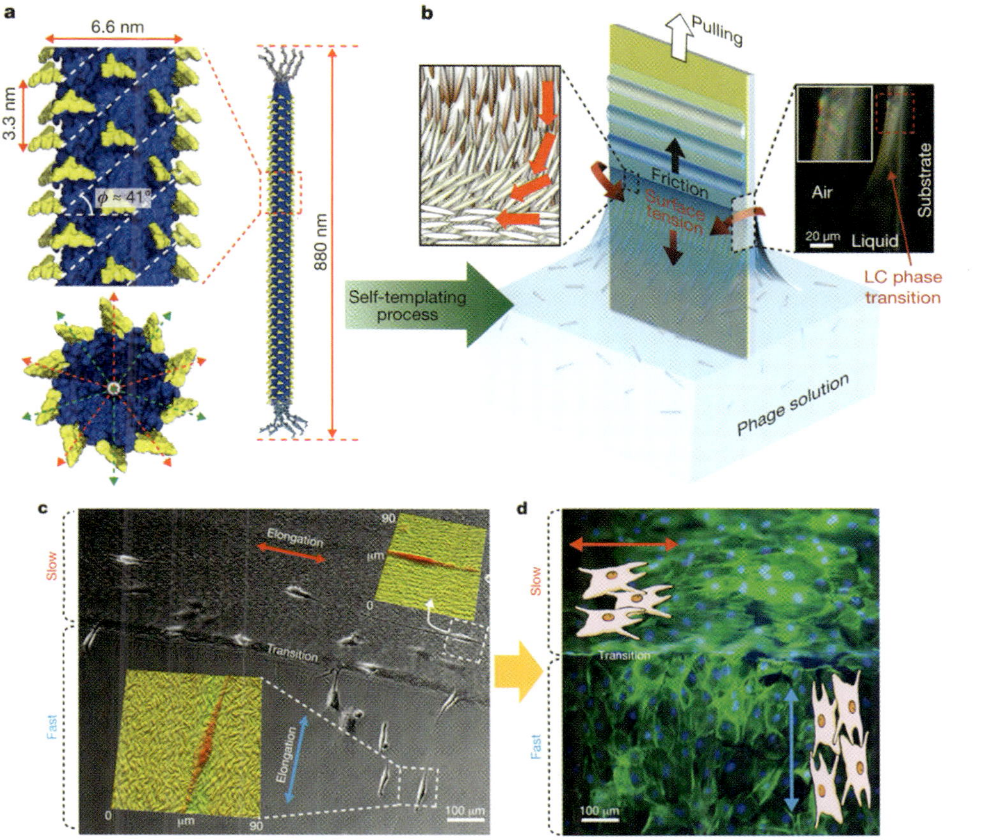

Figure 6.9 (a) Schematic illustration of the phage structure covered by about 2700 α-helical major coat (pVIII) protein subunits having five-fold helical symmetry and a two-fold screw rotation axis (red and green arrows). The phage has a helical angle ( ) of 41 with periodic subunit spacing of 3.3 nm. (b) Schematic illustration of the helical self-templating assembly of phage particles controlled by competing interfacial forces at the meniscus where liquid crystal phase transitions occur. The polarized optical microscopy image shows iridescent colours originating from liquid crystal phase formation at the air–liquid–solid interface. (c) Growth of biomimetic soft and hard tissue on the self-templated structures. Phase-contrast optical microscopy image of preosteoblast cells (MC3T3-E1) cultured on RGD phage films with SHN (top, slow pulling) and nematically ordered (bottom, fast pulling) structures. Insets: AFM images showing the directional guidance of cellular growth in response to the orientation of the underlying film microstructure. (d) Composite fluorescence image showing that the pre-osteoblast cells (MC-3T3 E1) grew to confluence (day 3) in the desired directions on the RGD phage films. This image was merged from two fluorescence images taken at two different focal planes. (Reproduced with permission from Chung *et al.*, 2011.)

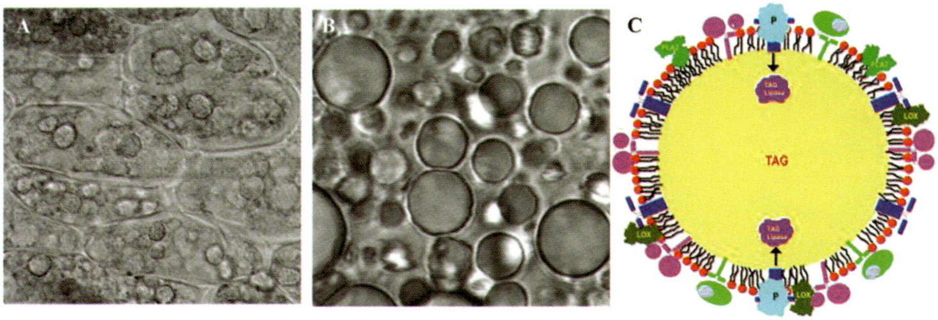

Figure. 8.1 (A) Distribution of oil bodies in intact cells of sunflower seedling (2-day-old) cotyledon cross section (400×). (B) Isolated oil bodies from 2-day-old sunflower seedling cotyledons (1000×). (C) Representation of the detailed structure of the OB from seedling cotyledon, associated with various structural and functional proteins (modified from Bhatla *et al.*, 2009), namely oleosin, caleosin, steroleosin, protease, phospholipase, TAG lipase and lipoxygenase.

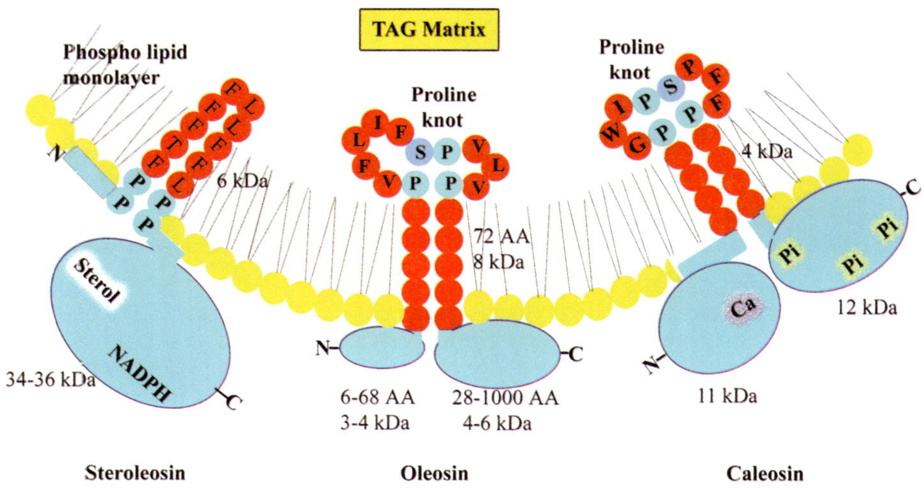

Figure 8.2 Detailed structure of steroleosin (A), oleosin (B) and caleosin (C) anchored to the phospholipid monolayer of oil body with their conserved hydrophobic stretch and their hydrophilic N and C terminals facing the cytosol. The amino acid F represents phenylalanine, L (leucine), T (threonine), P (proline), S (serine), V (valine), I (isoleucine), W (tryptophan), and G (glycine). The phosphorylation sites in the hydrophilic C-terminus of caleosin are depicted as 'Pi' and the calcium binding EF hand as 'Ca'.

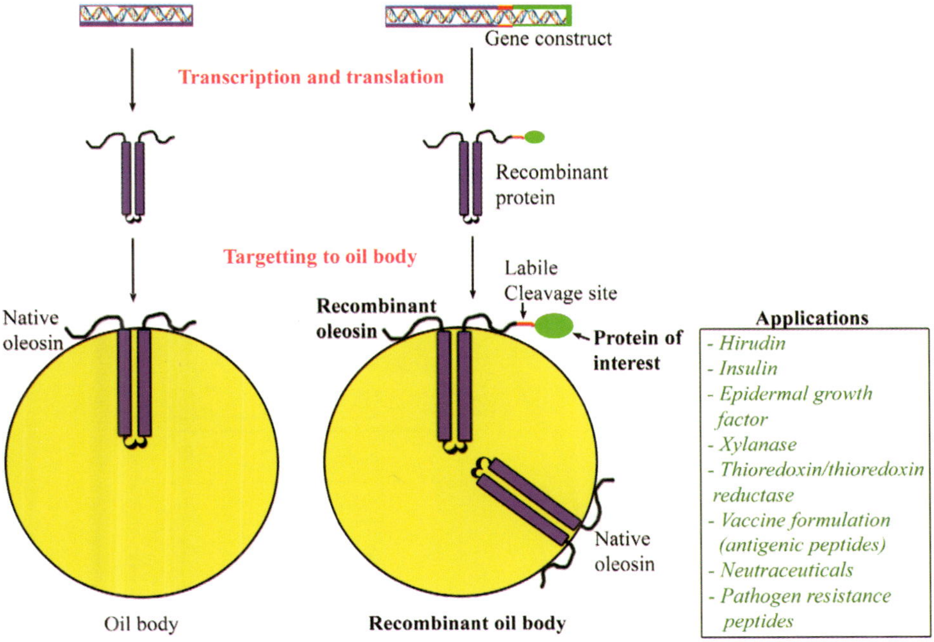

Figure. 8.3 Synthesis of OB with recombinant protein and their biotechnological applications.

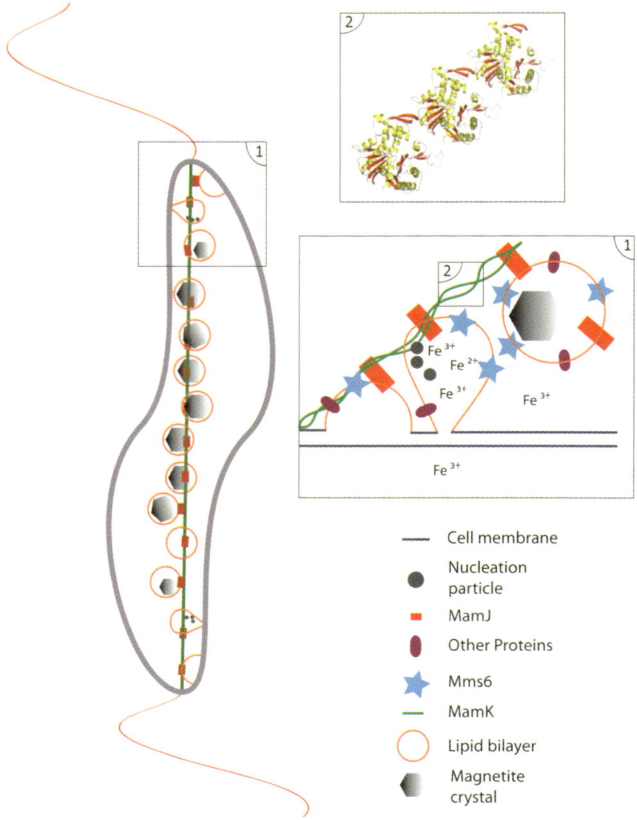

Figure 10.1 Schematic representation of the hierarchical organization of the magnetosome chain in magnetotactic bacteria (left) and the chronological formation of the magnetosomes (left to right in box (1)). Box 2 is a modelled structure of MamK assembled into an actin-like filament (Sonkaria *et al.*, 2012). The proteins involved in the biological and chemical control of the formation of magnetosomes are currently being characterized, hence the extent to which their function is dedicated to the different processes is still unclear.

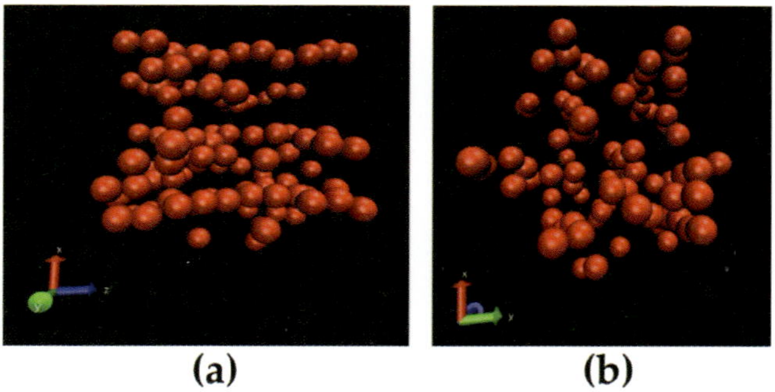

Figure 10.3 Snapshot of a simulation of a toy system of dipoles showing their self-organization at room temperature in a weak magnetic field in three dimensions. The dipole moment of the particles has been increased to exaggerate the chaining in such a system.

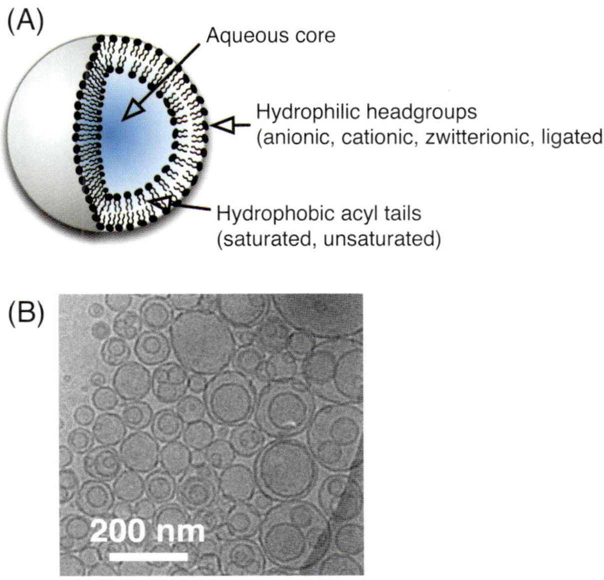

Figure 11.1 (A) Liposome schematic depicting the aqueous core, hydrophilic headgroup, and hydrophobic tail regions and (B) a cryogenic transmission electron micrograph of dipalmitoylphosphatidylcholine liposomes (DPPC, 10 mM) prepared in phosphate-buffered saline.

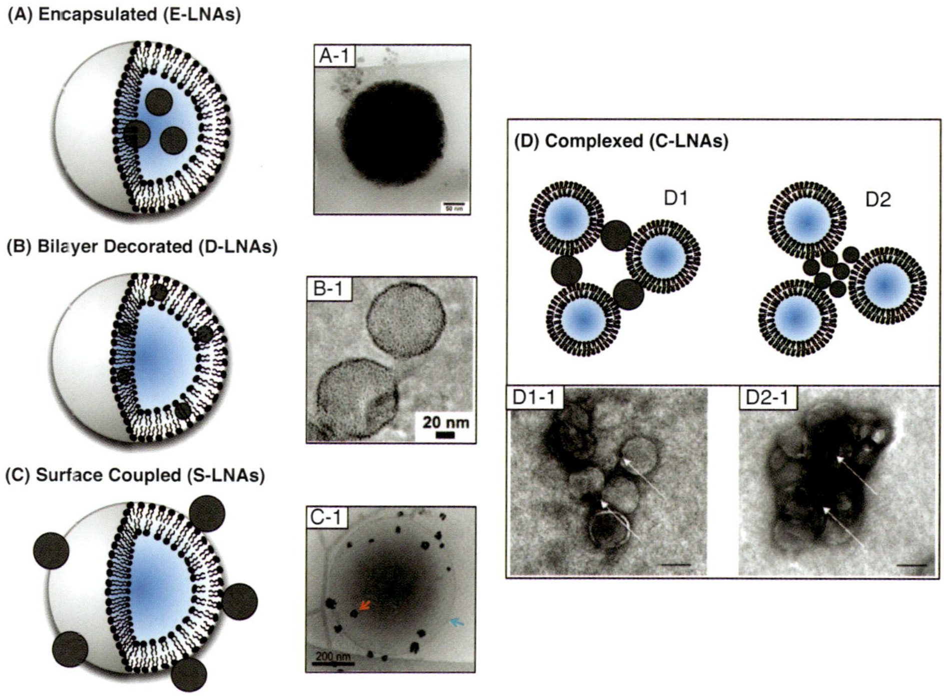

Figure 11.2 Dependence of LNA formation on nanoparticle size and surface chemistry. (A) Hydrophilic nanoparticles can be encapsulated to form E-LNAs or, if strong adhesive forces exist between the nanoparticle and lipid bilayer, surface coupled to form S-LNAs. For larger nanoparticles, strong adhesive forces can also lead to the formation of supported lipid bilayers (SLBs) on the nanoparticle surface. (B) Hydrophobic nanoparticles can embed within the liposomal bilayer to form D-LNAs or be coated with a lipid monolayer to form micelle-like structures.

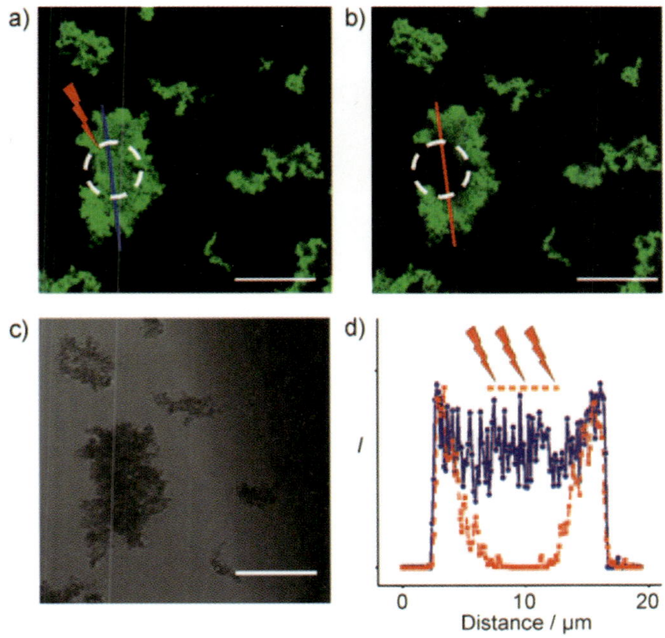

Figure 11.5 Remote dye release (5(6)-carboxyfluorescein) triggered by near IR laser illumination from C-LNAs prepared with gel-phase cationic liposomes and 20 nm hydrophilic gold nanoparticles. The scale bars represent 10 mm. These results show that near complete release can be achieved from the complexes after 5 s of illumination. Reprinted from Volodkin *et al.* (2009) with permission.

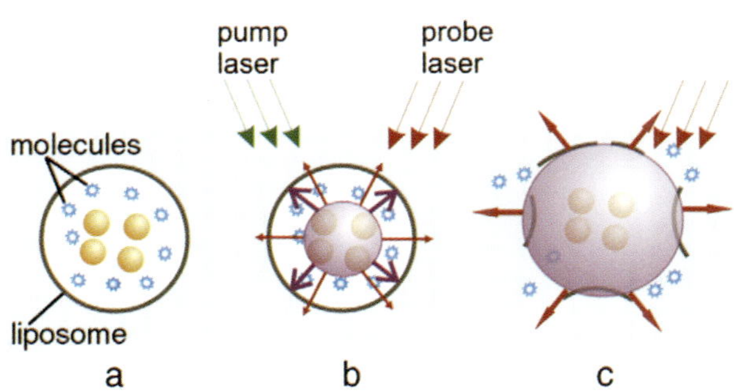

Figure 11.6 Proposed 'plasmonic nanobubble' release mechanism of encapsulated molecules from E-LNAs prepared with gold nanoparticles. Pulsed laser irradiation of the nanoparticles nucleates vapour bubbles that expand and mechanically disrupt the liposomal bilayer and trigger permeabilization. Reprinted from Anderson *et al.* (2010) with permission.

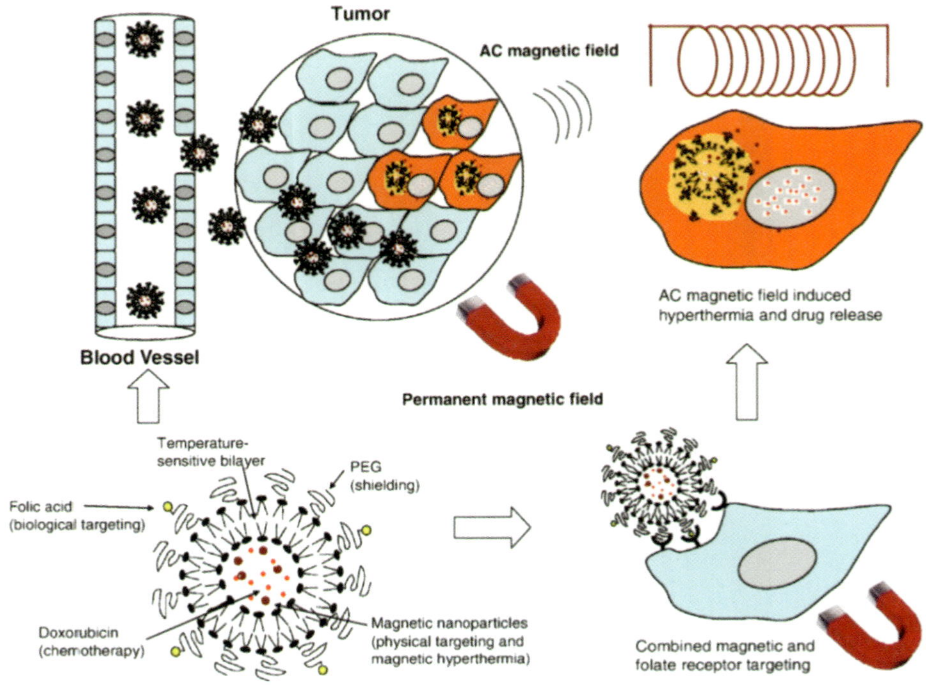

Figure 11.7 A multifunctional E-LNA for cancer thermo-chemotherapy from Pradhan *et al.* (2010). In this example the LNA is a temperature-sensitive magnetoliposome containing co-encapsulated iron oxide nanoparticles and doxorubicin. Passive targeting can be achieved through the enhanced permeation and retention (EPR) effect of tumour vasculature, and active targeting can be achieved via folate receptor and by applying a permanent magnetic field. The application of an AC electromagnetic field can be used to release the drug and achieve hyperthermia treatment. Reprinted from Pradhan *et al.* (2010) with permission.

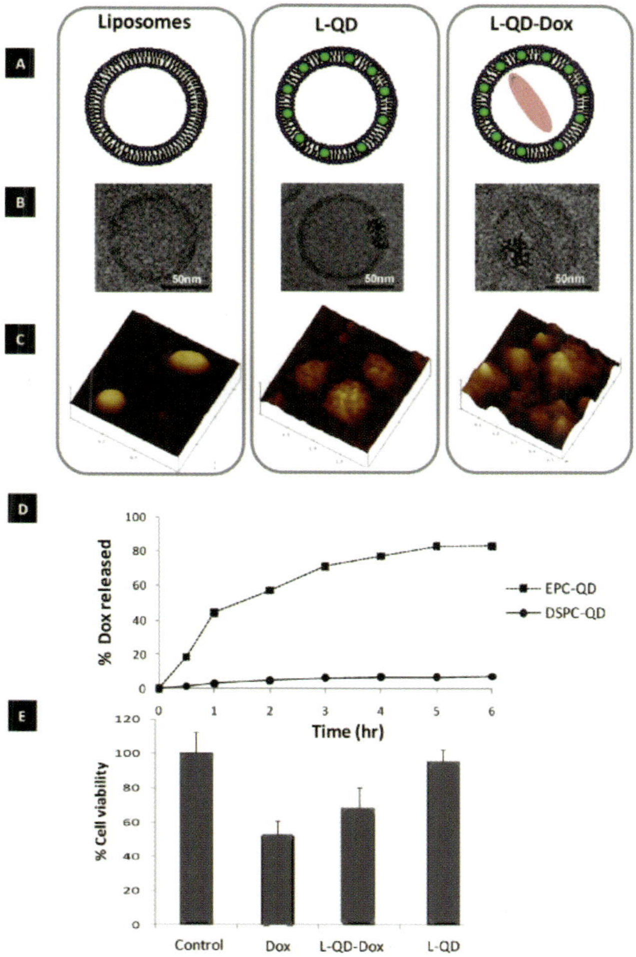

Figure 11.8 Combined imaging and doxorubicin release from D-LNAs prepared with hydrophobic quantum dots (green). QD embedment and D-LNA structure was verified by cryogenic transmission electron microscopy and atomic force microscopy. The results show that bilayer phase behaviour is a key parameter in doxorubicin release; fluid-phase D-LNAs achieved sustained release while gel-phase D-LNAs did not. Reprinted from Al-Jamal and Kostarelos (2011) with permission.